Two-dimensional Crossing-Variable Cubic Nonlinear Systems

Albert C. J. Luo

Two-dimensional Crossing-Variable Cubic Nonlinear Systems

Springer

Albert C. J. Luo
Department of Mechanical and Mechatronics Engineering
Southern Illinois University Edwardsville
Edwardsville, IL, USA

ISBN 978-3-031-62809-2 ISBN 978-3-031-62810-8 (eBook)
https://doi.org/10.1007/978-3-031-62810-8

This Springer imprint is published by the registered company Springer Nature Switzerland AG
The registered company address is: Gewerbestrasse 11, 6330 Cham, Switzerland

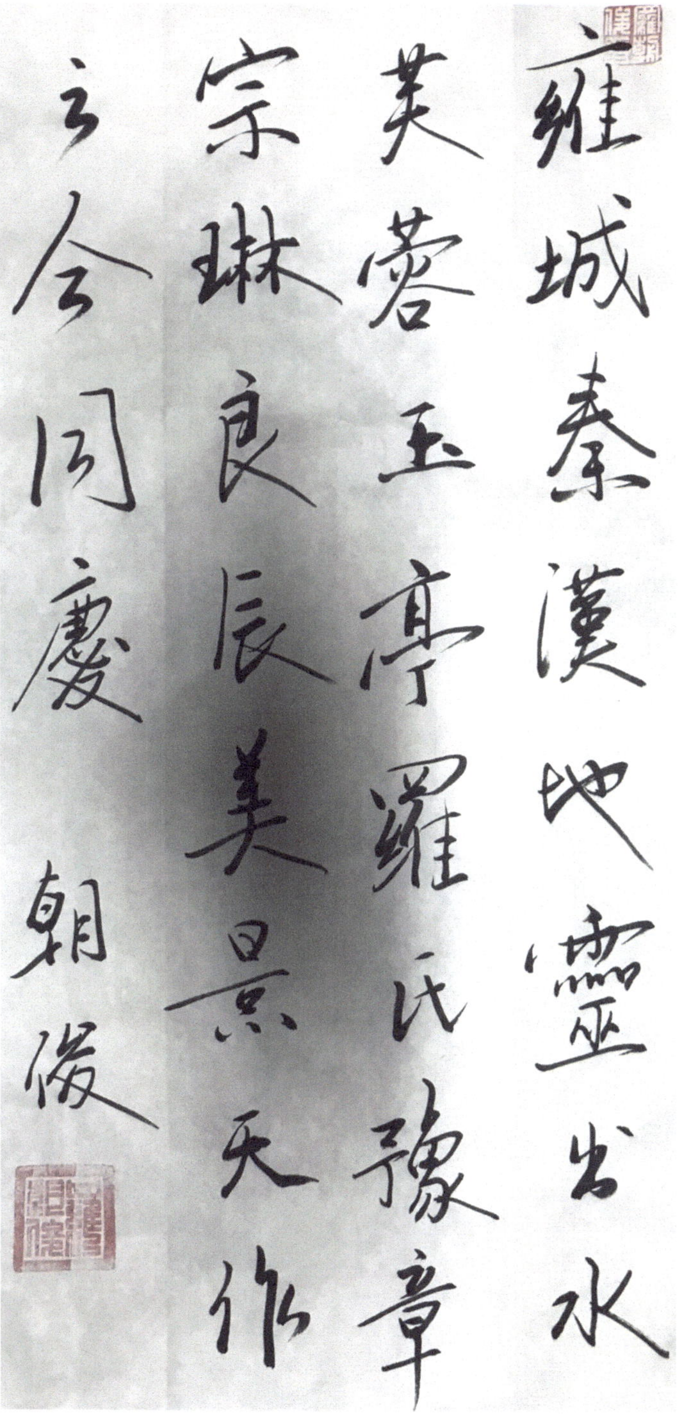

雍城秦汉地灵，出水芙蓉玉亭，罗氏豫章宗琳，良辰美景，天作之合同庆。

—朝俊【恭贺宗琳、秦佳喜结连理】

Preface

This book is about nonlinear dynamics of dynamical systems possessing two crossing-variable univariate cubic vector fields. The dynamical systems with a constant vector field and a variable-crossing univariate cubic vector field will be presented first, and the 1-dimensional flow bifurcations are illustrated for a better understanding of 1-dimensional flows. After that, dynamical systems with crossing-linear and cubic vector fields are discussed, and the corresponding bifurcation and global dynamics are presented. The saddle-center bifurcations are presented from parabola-saddle bifurcations. Furthermore, dynamical systems with crossing-quadratic and cubic vector fields are discussed, and the switching and appearing bifurcations for saddles and centers are discussed through the first integral manifolds, and the homoclinic networks with centers are first presented. Double-inflection saddle bifurcations are for appearing of networks of saddles and centers, and the homoclinic networks with centers are constructed. The saddle-center networks with limit cycles are developed from the first integral manifolds. Finally, dynamical systems with two crossing-cubic vector fields are analyzed, and the corresponding bifurcations and global dynamics are discussed. The appearing bifurcations of the third-order centers and saddles are for a network of counter-clockwise and clockwise centers with positive and negative saddles. The two third-order centers and saddles are for appearing bifurcations of a network of centers and saddles.

In Chap. 1, nonlinear dynamics of dynamical systems with constant and variable-crossing univariate cubic vector fields are discussed. The dynamical systems with a constant vector field with a variable-crossing univariate cubic vector field is presented. The 1-dimensional flows are presented. The third-order up-parabola flow is an appearing bifurcation from an up-parabola flow to up-parabola, down-parabola, up-parabola flows. The third-order down-parabola flow is an appearing bifurcation from a down-parabola flow to down-parabola, up-parabola, down-parabola flows. The third-order parabola flows are the switching bifurcations of parabola and inflection flows.

In Chap. 2, a cubic dynamical system possessing variable-crossing-univariate linear and cubic vector fields is discussed, and the corresponding dynamical behaviors are studied. The parabola-saddles are the appearing and switching of saddle and center. The third-order centers are the appearing bifurcation from a center to center, saddle, and center. The third-order saddles are the appearing bifurcation from a saddle to saddle, center, and saddle. The third-order saddles and centers are the switching bifurcations of saddle and center with parabola-saddles. The series of saddle and centers are presented through the first integral manifold.

In Chap. 3, a cubic dynamical system possessing variable-crossing-univariate quadratic and cubic vector fields is discussed, and the corresponding dynamical behaviors are discussed. As in constant and cubic variable-crossing systems, the third-order parabola flows exist for the appearing bifurcation from a parabola flow to a series of three parabola flows, and the inflection flows are the appearing and switching bifurcations of up-parabola and down parabola flows. In addition to parabola-saddles and third-order saddle and centers, the third-order parabola-saddle and double inflection-saddles exists. The (3,2)-third-order parabola-saddle are for the appearing bifurcations of a 2×3-network of saddle and center. The (2,2)-double inflection-saddles are the appearing bifurcations of a 2×2 network of saddle and center. The (3,2)-third-order parabola-saddles are also for the switching bifurcations of a double inflection-saddle with saddle (center). The networks of saddles and centers with homoclinic orbits are presented through the first integral manifolds.

In Chap. 4, a cubic dynamical system possessing two crossing-variable cubic vector fields is discussed. The appearing and switching bifurcations for centers and saddles are studied, and the limit cycles and homoclinic networks are presented through the first integral manifolds. The corresponding equilibriums include simple centers and saddles; parabola-saddle; third-order saddles and centers; (2,2)-double-diagonal inflection saddles; (3,2)-parabola-saddle; and (3,3)-double-third-order saddles and centers. The (3,3)-double third-order saddle and centers are the appearing bifurcations of a 3×3 network of saddles and centers with homoclinic orbits. The (3,3)-double-third-order saddles and centers are also the switching bifurcation for the networks of center (saddle), parabola-saddles, and inflection saddles. The (3,3)-double-third-order saddles and centers are also the switching bifurcations of parabola-saddle with saddles and centers. The networks of saddle and center for such nonlinear systems with two cubic variable-crossing vector fields are presented.

Finally, the author hopes the materials presented herein can provide a better understanding of variable-crossing cubic nonlinear systems in science and engineering.

Edwardsville, IL, USA Albert C. J. Luo

Contents

Chapter 1
Constant and Crossing-Cubic Vector Fields

In this chapter, nonlinear dynamics of dynamical systems with two variable-crossing univariate cubic vector fields is discussed. The dynamical systems with a constant vector field and a variable-crossing univariate cubic vector field is presented first, and the one-dimensional flows are presented. The third-order up-parabola flow is an appearing bifurcation from an up-parabola flow to up-parabola, down-parabola, and up-parabola flows. The third-order down-parabola flow is an appearing bifurcation from a down-parabola flow to down-parabola, up-parabola, and down-parabola flows. The third-order parabola flows are the switching bifurcations of parabola and inflection flows.

Consider a two-dimensional, nonlinear dynamical system as

$$\begin{aligned} \dot{x}_{j_1} &= f_{j_1}(\mathbf{x}, \mathbf{p}) = A_{j_1 j_2 3} + A_{j_1 j_2 2} x_{j_2} + A_{j_1 j_2 1} x_{j_2}^2 + A_{j_1 j_2 0} x_{j_2}^3, \\ \dot{x}_{j_2} &= f_{j_2}(\mathbf{x}, \mathbf{p}) = A_{j_2 j_1 3} + A_{j_2 j_1 2} x_{j_1} + A_{j_2 j_1 1} x_{j_1}^2 + A_{j_2 j_1 0} x_{j_1}^3. \end{aligned} \tag{1.1}$$

For $A_{j_2 j_1 2} = A_{j_2 j_1 1} = A_{j_2 j_1 0} = 0\,(j_1, j_2 \in \{1,2\}, j_1 \neq j_2)$, the two-dimensional dynamical system has constant and crossing-cubic vector fields as

$$\begin{aligned} \dot{x}_{j_2} &= A_{j_2 j_1 3}, \\ \dot{x}_{j_1} &= a_{j_1 j_2 0}(x_{j_2} - b_{j_1 j_2 1})(x_{j_2}^2 + B_{j_1 j_2} x_{j_2} + C_{j_1 j_2}) \end{aligned} \tag{1.2}$$

with

$$\Delta_{j_1 j_2} = (B_{j_1 j_2})^2 - 4C_{j_1 j_2} \ \text{ for } j_1, j_2 \in \{1,2\}, j_1 \neq j_2. \tag{1.3}$$

For this case, the properties of such a dynamical system are the same as single crossing-variable cubic system with a constant vector field. The corresponding theorem is stated as in Luo (2024).

A. C. J. Luo, *Two-dimensional Crossing-Variable Cubic Nonlinear Systems*,
https://doi.org/10.1007/978-3-031-62810-8_1

The parabola flows will not be demonstrated herein, which were discussed by Luo (2024). Herein, only the appearing bifurcations for the up-parabola and down-parabola flows are presented.

For $\Delta_{j_1j_2} < 0,$ consider a dynamical system as

$$\begin{aligned} \dot{x}_{j_2} &= A_{j_2j_1 3}, \\ \dot{x}_{j_1} &= a_{j_1j_2 0}(x_{j_2} - a_{j_1j_2 1})\left[(x_{j_2} - a_{j_1j_2})^2 + b_{j_1j_2}\right]. \end{aligned} \tag{1.4}$$

The corresponding first integral manifold is

$$\begin{aligned} A_{j_2j_1 3}(x_{j_1} - x_{j_1 0}) &= a_{j_1j_2 0}\{\frac{1}{4}\left[(x_{j_2} - a_{j_1j_2 1})^4 - (x_{j_2 0} - a_{j_1j_2 1})^4\right] \\ &\quad + \frac{2}{3}(a_{j_1j_2 1} - a_{j_1j_2})\left[(x_{j_2} - a_{j_1j_2 1})^3 - (x_{j_2 0} - a_{j_1j_2 1})^3\right] \\ &\quad + \frac{1}{2}\left[(a_{j_1j_2 1} - a_{j_1j_2})^2 + b_{j_1j_2}\right]\left[(x_{j_2} - a_{j_1j_2 1})^2 - (x_{j_2 0} - a_{j_1j_2 1})^2\right]\}. \end{aligned} \tag{1.5}$$

The equilibrium at $x^*_{j_2} = a_{j_1j_2 1}$ has the following properties:

(i) For $a_{j_1j_2 0} > 0$ and $A_{j_2j_1 3} > 0,$

$$(a_{j_2j_1 1}, \dot{x}_{j_2}) = \underbrace{(\text{UP}, \text{pF})}_{\text{up-parabola flow }(+)}. \tag{1.6}$$

(ii) For $a_{j_1j_2 0} < 0$ and $A_{j_2j_1 3} > 0,$

$$(a_{j_2j_1 1}, \dot{x}_{j_2}) = \underbrace{(\text{DP}, \text{pF})}_{\text{down-parabola flow }(+)}. \tag{1.7}$$

(iii) For $a_{j_1j_2 0} > 0$ and $A_{j_2j_1 3} < 0,$

$$(a_{j_2j_1 1}, \dot{x}_{j_2}) = \underbrace{(\text{DP}, \text{nF})}_{\text{down-parabola flow }(-)}. \tag{1.8}$$

(iv) For $a_{j_1j_2 0} < 0$ and $A_{j_2j_1 3} < 0,$

$$(a_{j_2j_1 1}, \dot{x}_{j_2}) = \underbrace{(\text{UP}, \text{nF})}_{\text{up-parabola flow }(-)}. \tag{1.9}$$

For $\Delta_{j_1j_2}=0$ and $a_{j_1j_2s_1}\neq a_{j_1j_2s_2}$ (s_1, $s_2\in\{1,2\}$ and $s_1\neq s_2$), consider a dynamical system as

$$\begin{aligned}\dot{x}_{j_2}&=A_{j_2j_13},\\ \dot{x}_{j_1}&=a_{j_1j_20}(x_{j_2}-a_{j_1j_2s_1})(x_{j_2}-a_{j_1j_2s_2})^2.\end{aligned}\tag{1.10}$$

The corresponding first integral manifold is

$$\begin{aligned}A_{j_2j_13}(x_{j_1}-x_{j_10})&=a_{j_1j_20}\Big\{\frac{1}{4}\big[(x_{j_2}-a_{j_2j_1s_1})^4-(x_{j_20}-a_{j_1j_2s_1})^4\big]\\&+\frac{2}{3}(a_{j_1j_2s_1}-a_{j_1j_2s_2})\big[(x_{j_2}-a_{j_1j_2s_1})^3-(x_{j_20}-a_{j_1j_2s_1})^3\big]\\&+\frac{1}{2}(a_{j_1j_2s_1}-a_{j_1j_2s_2})^2\big[(x_{j_2}-a_{j_1j_2s_1})^2-(x_{j_20}-a_{j_1j_2s_1})^2\big]\Big\}.\end{aligned}\tag{1.11}$$

The equilibrium at $x_{j_2}^*=a_{j_2j_1s_1}$ has the following properties:

(i) For $a_{j_1j_20}(a_{j_1j_2s_1}-a_{j_1j_2s_2})^2>0$ and $A_{j_2j_13}>0$,

$$(a_{j_1j_2s_1},\dot{x}_{j_2})=\underbrace{(\mathrm{UP},\mathrm{pF})}_{\text{up-parabola flow }(+)}.\tag{1.12}$$

(ii) For $a_{j_1j_20}(a_{j_1j_2s_1}-a_{j_1j_2s_2})^2<0$ and $A_{j_2j_13}>0$,

$$(a_{j_1j_2s_1},\dot{x}_{j_2})=\underbrace{(\mathrm{DP},\mathrm{pF})}_{\text{down-parabola flow }(+)}.\tag{1.13}$$

(iii) For $a_{j_1j_20}(a_{j_1j_2s_1}-a_{j_1j_2s_2})^2>0$ and $A_{j_2j_13}<0$,

$$(a_{j_1j_2s_1},\dot{x}_{j_2})=\underbrace{(\mathrm{DP},\mathrm{nF})}_{\text{down-parabola flow }(-)}.\tag{1.14}$$

(iv) For $a_{j_1j_20}(a_{j_1j_2s_1}-a_{j_1j_2s_2})^2<0$ and $A_{j_2j_13}<0$,

$$(a_{j_1j_2s_1},\dot{x}_{j_2})=\underbrace{(\mathrm{UP},\mathrm{nF})}_{\text{up-parabola flow }(-)}.\tag{1.15}$$

The equilibrium at $x_{j_2}^*=a_{j_2j_1s_2}$ has the following properties:

(i) For $a_{j_1j_20}(a_{j_1j_2s_2} - a_{j_1j_2s_1}) > 0$ and $A_{j_2j_13} > 0$,

$$(a_{j_1j_2s_2}, \dot{x}_{j_2}) = \underbrace{(\mathrm{II}, \mathrm{pF})}_{\text{increasing-inflection flow } (+)}. \tag{1.16}$$

(ii) For $a_{j_1j_20}(a_{j_1j_2s_2} - a_{j_1j_2s_1}) < 0$ and $A_{j_2j_13} > 0$,

$$(a_{j_1j_2s_2}, \dot{x}_{j_2}) = \underbrace{(\mathrm{DI}, \mathrm{pF})}_{\text{decreasing-inflection flow } (+)}. \tag{1.17}$$

(iii) For $a_{j_1j_20}(a_{j_1j_2s_2} - a_{j_1j_2s_1}) > 0$ and $A_{j_2j_13} < 0$,

$$(a_{j_1j_2s_2}, \dot{x}_{j_2}) = \underbrace{(\mathrm{DI}, \mathrm{nF})}_{\text{decreasing-inflection flow } (-)}. \tag{1.18}$$

(iv) For $a_{j_1j_20}(a_{j_1j_2s_2} - a_{j_1j_2s_1}) < 0$ and $A_{j_2j_13} < 0$,

$$(a_{j_1j_2s_2}, \dot{x}_{j_2}) = \underbrace{(\mathrm{II}, \mathrm{nF})}_{\text{Increasing-inflection flow } (-)}. \tag{1.19}$$

The inflection flows are the appearing bifurcations for the up-parabola and down-parabola flows.

(i) For $a_{j_1j_20}(a_{j_1j_2s_2} - a_{j_1j_2s_1}) > 0$ and $A_{j_2j_13} > 0$,

$$(a_{j_1j_2s_2}, \dot{x}_{j_2}) = \underbrace{(\mathrm{II}, \mathrm{pF})}_{\text{appearing bifurcation}} \rightleftharpoons \left\{ \begin{array}{c} \underbrace{(\mathrm{DP}, \mathrm{pF})}_{\text{down-parabola flow } (+)} \\ \underbrace{(\mathrm{UP}, \mathrm{pF})}_{\text{up-parabola flow } (+)} \end{array} \right\}. \tag{1.20}$$

(ii) For $a_{j_1j_20}(a_{j_1j_2s_2} - a_{j_1j_2s_1}) < 0$ and $A_{j_2j_13} > 0$,

$$(a_{j_1j_2s_2}, \dot{x}_{j_2}) = \underbrace{(\mathrm{DI}, \mathrm{pF})}_{\text{appearing bifurcation}} \rightleftharpoons \left\{ \begin{array}{c} \underbrace{(\mathrm{UP}, \mathrm{pF})}_{\text{up-parabola flow } (+)} \\ \underbrace{(\mathrm{DP}, \mathrm{pF})}_{\text{down-parabola flow } (+)} \end{array} \right\}. \tag{1.21}$$

(iii) For $a_{j_1j_20}(a_{j_1j_2s_2} - a_{j_1j_2s_1}) > 0$ and $A_{j_2j_13} < 0$,

$$(a_{j_1j_2s_2}, \dot{x}_{j_2}) = \underbrace{(\text{DI}, \text{nF})}_{\text{appearing bifurcation}} \rightleftharpoons \left\{ \begin{array}{c} \underbrace{(\text{UP}, \text{nF})}_{\text{up-parabola flow } (-)} \\ \underbrace{(\text{DP}, \text{nF})}_{\text{down-parabola flow } (-)} \end{array} \right\}. \tag{1.22}$$

(iv) For $a_{j_1j_20}(a_{j_1j_2s_2} - a_{j_1j_2s_1}) < 0$ and $A_{j_2j_12} < 0$,

$$(a_{j_1j_2s_2}, \dot{x}_{j_2}) = \underbrace{(\text{II}, \text{nF})}_{\text{appearing bifurcation}} \rightleftharpoons \left\{ \begin{array}{c} \underbrace{(\text{DP}, \text{nF})}_{\text{down-parabola flow } (-)} \\ \underbrace{(\text{UP}, \text{nF})}_{\text{up-parabola flow } (-)} \end{array} \right\}. \tag{1.23}$$

The inflection flows are also the switching bifurcations for up-parabola and down-parabola flows.

For $\Delta_{j_1j_2} = 0$ and $a_{j_1j_2s_1} = a_{j_1j_2s_2} = a_{j_1j_21}$ (s_1, $s_2 \in \{1, 2\}$), consider a dynamical system as

$$\begin{aligned} \dot{x}_{j_2} &= A_{j_2j_13}, \\ \dot{x}_{j_1} &= a_{j_1j_20}(x_{j_2} - a_{j_2j_11})^3. \end{aligned} \tag{1.24}$$

The corresponding first integral manifold is

$$A_{j_2j_13}(x_{j_1} - x_{j_10}) = \frac{1}{4} a_{j_1j_20} \left[(x_{j_2} - a_{j_2j_1s_1})^4 - (x_{j_20} - a_{j_2j_1s_1})^4 \right]. \tag{1.25}$$

The equilibrium at $x_{j_2}^* = a_{j_1j_21}$ has the following properties:

(i) For $a_{j_1j_20} > 0$ and $A_{j_2j_13} > 0$,

$$(a_{j_1j_21}, \dot{x}_{j_2}) = \underbrace{(3^{\text{rd}}\text{UP}, \text{pF})}_{\text{third-order up-parabola flow } (+)}. \tag{1.26}$$

(ii) For $a_{j_1j_20} < 0$ and $A_{j_2j_13} > 0$,

$$(a_{j_1j_21}, \dot{x}_{j_2}) = \underbrace{(3^{\text{rd}}\text{DP}, \text{pF})}_{\text{third-order down-parabola flow } (+)}. \tag{1.27}$$

(iii) For $a_{j_1 j_2 0} > 0$ and $A_{j_2 j_1 3} < 0$,

$$(a_{j_1 j_2 1}, \dot{x}_{j_2}) = \underbrace{(3^{\text{rd}}\text{DP}, \text{nF})}_{\text{third-order down-parabola flow } (-)}. \tag{1.28}$$

(iv) For $a_{j_1 j_2 0} < 0$ and $A_{j_2 j_1 3} < 0$,

$$(a_{j_1 j_2 1}, \dot{x}_{j_2}) = \underbrace{(3^{\text{rd}}\text{UP}, \text{nF})}_{\text{third-order up-parabola flow } (-)}. \tag{1.29}$$

For $\Delta_{j_1 j_2} > 0$, consider a dynamical system as

$$\begin{aligned} \dot{x}_{j_2} &= A_{j_2 j_1 3}, \\ \dot{x}_{j_1} &= a_{j_1 j_2 0}(x_{j_2} - a_{j_1 j_2 1})(x_{j_2} - a_{j_1 j_2 2})(x_{j_2} - a_{j_1 j_2 3}). \end{aligned} \tag{1.30}$$

The corresponding first integral manifold is

$$\begin{aligned} A_{j_2 j_1 3}(x_{j_1} - x_{j_1 0}) = a_{j_1 j_2 0}\Big\{&\frac{1}{4}\left[(x_{j_2} - a_{j_2 j_1 s_1})^4 - (x_{j_2 0} - a_{j_2 j_1 s_1})^4\right] \\ &+ \frac{1}{3}(2a_{j_2 j_1 s_1} - a_{j_2 j_1 s_2} - a_{j_2 j_1 s_3})\left[(x_{j_2} - a_{j_2 j_1 s_1})^3 - (x_{j_2 0} - a_{j_2 j_1 s_1})^3\right] \\ &+ \frac{1}{2}(a_{j_2 j_1 s_1} - a_{j_2 j_1 s_2})(a_{j_2 j_1 s_1} - a_{j_2 j_1 s_3}) \\ &\times\left[(x_{j_2} - a_{j_2 j_1 s_1})^3 - (x_{j_2 0} - a_{j_2 j_1 s_1})^3\right]\Big\}. \end{aligned} \tag{1.31}$$

The equilibrium at $x_{j_2}^* = a_{j_2 j_1 s_1}$ has the following properties:

(i) For $a_{j_1 j_2 0}\prod_{s_2 = 1, s_2 \neq s_1}^{3}(a_{j_1 j_2 s_1} - a_{j_1 j_2 s_2}) > 0$ and $A_{j_2 j_1 3} > 0$,

$$(a_{j_1 j_2 s_1}, \dot{x}_{j_2}) = \underbrace{(\text{UP}, \text{pF})}_{\text{up-parabola flow } (+)}. \tag{1.32}$$

(ii) For $a_{j_1 j_2 0}\prod_{s_2 = 1, s_2 \neq s_1}^{3}(a_{j_1 j_2 s_1} - a_{j_1 j_2 s_2}) < 0$ and $A_{j_2 j_1 3} > 0$,

$$(a_{j_1 j_2 s_1}, \dot{x}_{j_2}) = \underbrace{(\text{DP}, \text{pF})}_{\text{down-parabola flow } (+)}. \tag{1.33}$$

(iii) For $a_{j_1j_20}\prod_{s_2=1,s_2\neq s_1}^{3}(a_{j_1j_2s_1}-a_{j_1j_2s_2})>0$ and $A_{j_2j_13}<0$,

$$(a_{j_1j_2s_1},\dot{x}_{j_2})=\underbrace{(\text{DP},\text{nF})}_{\text{down-parabola flow }(-)}. \tag{1.34}$$

(iv) For $a_{j_1j_20}\prod_{s_2=1,s_2\neq s_1}^{3}(a_{j_1j_2s_1}-a_{j_1j_2s_2})<0$ and $A_{j_2j_13}<0$,

$$(a_{j_1j_2s_1},\dot{x}_{j_2})=\underbrace{(\text{UP},\text{nF})}_{\text{up-parabola flow }(-)}. \tag{1.35}$$

The third-order parabola flows are the appearing bifurcations for the up-parabola and down-parabola flows from a parabola flow.

(i) For $a_{j_1j_20}>0$ and $A_{j_2j_13}>0$,

$$\underbrace{(\text{UP},\text{pF})}_{\text{up-parabola flow }(+)}\rightleftharpoons\underbrace{(3^{\text{rd}}\text{UP},\text{pF})}_{\text{appearing bifurcation}}\rightleftharpoons\left\{\begin{array}{c}\underbrace{(\text{UP},\text{pF})}_{\text{up-parabola flow }(+)}\\ \underbrace{(\text{DP},\text{pF})}_{\text{down-parabola flow }(+)}\\ \underbrace{(\text{UP},\text{pF})}_{\text{up-parabola flow }(+)}\end{array}\right\}. \tag{1.36}$$

(ii) For $a_{j_1j_20}<0$ and $A_{j_2j_13}>0$,

$$\underbrace{(\text{DP},\text{pF})}_{\text{down-parabola flow }(+)}\rightleftharpoons\underbrace{(3^{\text{rd}}\text{DP},\text{pF})}_{\text{appearing bifurcation}}\rightleftharpoons\left\{\begin{array}{c}\underbrace{(\text{DP},\text{pF})}_{\text{down-parabola flow }(+)}\\ \underbrace{(\text{UP},\text{pF})}_{\text{up-parabola flow }(+)}\\ \underbrace{(\text{DP},\text{pF})}_{\text{down-parabola flow }(+)}\end{array}\right\}. \tag{1.37}$$

(iii) For $a_{j_1j_20}>0$ and $A_{j_2j_13}<0$,

$$\underbrace{(\text{DP},\text{nF})}_{\text{down-parabola flow }(-)}\rightleftharpoons\underbrace{(3^{\text{rd}}\text{DP},\text{nF})}_{\text{appearing bifurcation}}\rightleftharpoons\left\{\begin{array}{c}\underbrace{(\text{DP},\text{nF})}_{\text{down-parabola flow }(-)}\\ \underbrace{(\text{UP},\text{nF})}_{\text{up-parabola flow }(-)}\\ \underbrace{(\text{DP},\text{nF})}_{\text{down-parabola flow }(-)}\end{array}\right\}. \tag{1.38}$$

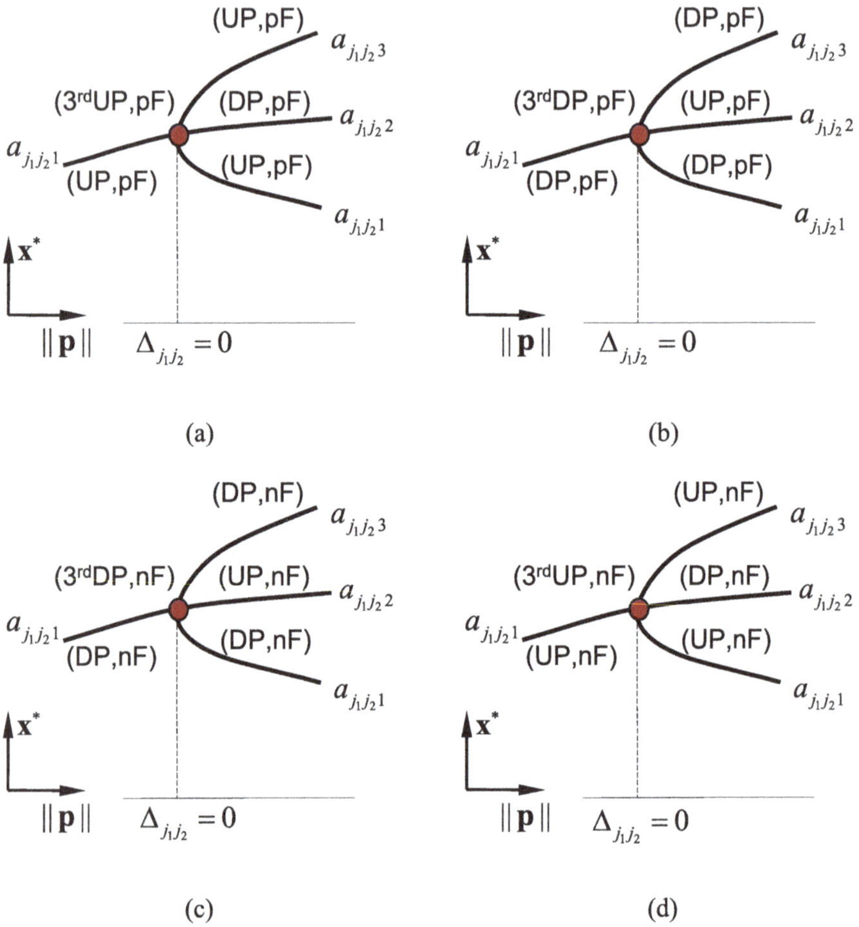

Fig. 1.1 The appearing bifurcations of the inflection flows for $x^*_{j_2} = a_{j_1j_2 1}, a_{j_1j_2 2}, a_{j_1j_2 3}$ and $\dot{x}_{j_2} = A_{j_2j_1 3}$: (**a**) (3rdUP,pF)-flow appearing bifurcation ($a_{j_1j_2 0} > 0, A_{j_2j_1 3} > 0$), (**b**) (3rdDP,pF)-flow appearing bifurcation ($a_{j_1j_2 0} < 0, A_{j_2j_1 3} > 0$), (**c**) (3rdDP,nF)-flow appearing bifurcation ($a_{j_1j_2 0} > 0, A_{j_2j_1 3} < 0$), and (**d**) (3rdUP,nF)-flow appearing bifurcation ($a_{j_1j_2 0} < 0, A_{j_2j_1 3} < 0$). DP: down parabola, UP: up-parabola, pF: positive flow, nF: negative flow, (3rdUP,pF): positive third-order up-parabola flow, (3rdDP,pF): positive third-order down-parabola flow, (3rdDP,nF): negative third-order down-parabola flow, (3rdUP,nF): negative third-order up-parabola flow, (UP,pF): positive upper-parabola flow, (DP,pF): positive down-parabola flow, (UP,nF): negative up-parabola flow, (DP,nF): negative down-parabola flow

(iv) For $a_{j_1 j_2 0} < 0$ and $A_{j_2 j_1 3} < 0$,

$$\underbrace{(\text{UP}, \text{nF})}_{\text{up-parabola flow } (-)} \rightleftharpoons \underbrace{(3^{\text{rd}}\text{UP}, \text{nF})}_{\text{appearing bifurcation}} \rightleftharpoons \left\{ \begin{array}{c} \underbrace{(\text{UP}, \text{nF})}_{\text{up-parabola flow } (-)} \\ \underbrace{(\text{DP}, \text{nF})}_{\text{down-parabola flow } (-)} \\ \underbrace{(\text{UP}, \text{nF})}_{\text{up-parabola flow } (-)} \end{array} \right\}. \tag{1.39}$$

The third-order parabola flow appearing bifurcation for three up-parabola and down-parabola flows from a parabola are sketched in Fig. 1.1. In Fig. 1.1a, the (3rdUP,pF)-up-parabola flow appearing bifurcation for positive up-parabola, down-parabola, and up-parabola flows are presented for ($a_{j_1 j_2 0} > 0,\ A_{j_2 j_1 3} > 0$). In Fig. 1.1b, the (3rdDP,pF)-down-parabola flow appearing bifurcation for the positive down-parabola, up-parabola, and down-parabola flows are presented for ($a_{j_1 j_2 0} < 0, A_{j_2 j_1 3} > 0$). In Fig. 1.1c, the (3rdDP,nF)-down-parabola flow appearing bifurcation for the negative down-parabola, up-parabola, and down-parabola flows are presented for ($a_{j_1 j_2 0} > 0, A_{j_2 j_1 3} < 0$). In Fig. 1.1d, the (3rdUP,nF)-up-parabola flow appearing bifurcation for the negative up-parabola, down parabola, and up-parabola flows are presented for ($a_{j_1 j_2 0} < 0, A_{j_2 j_1 3} < 0$).

Reference

Luo, ACJ (2024), *Two-dimensional Single-variable Cubic Nonlinear Systems, Vol. II: A Crossing-variable Cubic Vector Field*, Springer.

Chapter 2
Crossing-Linear and Crossing-Cubic Vector Fields

In this chapter, a cubic dynamical system possessing crossing-linear and crossing-cubic vector fields is considered, and the corresponding dynamical behaviors are discussed. The parabola-saddles are for the appearing and switching bifurcations of saddle and center. The third-order centers are the appearing bifurcations from a center to center, saddle, and center. The third-order saddles are the appearing bifurcation from a saddle to saddle, center, and saddle. The third-order saddles and centers are the switching bifurcations of saddle and center with parabola-saddles. The series of saddle and center are presented through the first integral manifold.

2.1 Crossing-Linear and Crossing-Cubic Systems

The global dynamics of a linear and cubic variable-crossing univariate dynamical system is discussed, and the singularity and appearing and switching bifurcations for saddles and centers are presented through the following theorem:

Theorem 2.1 *Consider a two-dimensional, dynamical system with variable-crossing univariate cubic vector fields as*

$$\begin{aligned}\dot{x}_{j_1} &= f_{j_1}(\mathbf{x},\mathbf{p}) = A_{j_1j_23} + A_{j_1j_22}x_{j_2} + A_{j_1j_21}x_{j_2}^2 + A_{j_1j_20}x_{j_2}^3,\\ \dot{x}_{j_2} &= f_{j_2}(\mathbf{x},\mathbf{p}) = A_{j_2j_13} + A_{j_2j_12}x_{j_1} + A_{j_2j_11}x_{j_1}^2 + A_{j_2j_10}x_{j_1}^3.\end{aligned} \tag{2.1}$$

For $A_{j_2j_10} = A_{j_2j_11} = 0, A_{j_2j_12} \neq 0,$ *the two-dimensional dynamical system possesses crossing-linear and crossing-cubic vector fields as*

$$\begin{aligned}\dot{x}_{j_2} &= a_{j_2j_10}(x_{j_1} - a_{j_2j_11}),\\ \dot{x}_{j_1} &= a_{j_1j_20}(x_{j_2} - b_{j_1j_21})(x_{j_2}^2 + B_{j_1j_2}x_{j_2} + C_{j_1j_2}).\end{aligned} \tag{2.2}$$

A. C. J. Luo, *Two-dimensional Crossing-Variable Cubic Nonlinear Systems*,
https://doi.org/10.1007/978-3-031-62810-8_2

with

$$\Delta_{j_1j_2} = (B_{j_1j_2})^2 - 4C_{j_1j_2} \ \text{ for } j_1, j_2 \in \{1,2\}, j_1 \neq j_2. \tag{2.3}$$

(i) If

$$A_{j_2j_1 2} \neq 0, A_{j_2j_1 3} \neq 0 \text{ and } \Delta_{j_1j_2} < 0, \tag{2.4}$$

the corresponding dynamical system is

$$\begin{aligned} \dot{x}_{j_2} &= a_{j_2j_1 0}(x_{j_1} - a_{j_2j_1 1}), \\ \dot{x}_{j_1} &= a_{j_1j_2 0}(x_{j_2} - a_{j_1j_2 1})\left[(x_{j_2} - a_{j_1j_2})^2 + b_{j_1j_2}\right] \end{aligned} \tag{2.5}$$

where

$$\begin{aligned} &a_{j_2j_1 1} = -\frac{1}{A_{j_2j_1 2}} A_{j_2j_1 3}, a_{j_2j_1 0} = A_{j_2j_1 2}; \\ &a_{j_1j_2 1} = b_{j_1j_2 1}; a_{j_1j_2} = -\frac{1}{2} B_{j_1j_2 1}, a_{j_1j_2 0} = A_{j_1j_2 0}; \\ &b_{j_1j_2} = \frac{1}{4}(-\Delta_{j_1j_2}). \end{aligned} \tag{2.6}$$

The first integral manifold is

$$\begin{aligned} &\frac{1}{2} a_{j_2j_1 0}\left[(x_{j_1} - a_{j_2j_1 1})^2 - (x_{j_1 0} - a_{j_2j_1 1})^2\right] \\ &= a_{j_1j_2 0}\Big\{\frac{1}{4}\left[(x_{j_2} - a_{j_1j_2 1})^4 - (x_{j_2 0} - a_{j_1j_2 1})^4\right] \\ &+ \frac{2}{3}(a_{j_1j_2 1} - a_{j_1j_2})\left[(x_{j_2} - a_{j_1j_2 1})^3 - (x_{j_2 0} - a_{j_1j_2 1})^3\right] \\ &+ \frac{1}{2}\left[(a_{j_1j_2 1} - a_{j_1j_2})^2 + b_{j_1j_2}\right]\left[(x_{j_2} - a_{j_1j_2 1})^2 - (x_{j_2 0} - a_{j_1j_2 1})^2\right]\Big\}. \end{aligned} \tag{2.7}$$

The equilibrium of $(x_{j_2}^*, x_{j_1}^*) = (a_{j_1j_2 1}, a_{j_2j_1 1})$ *has the following properties:*

- *For* $a_{j_1j_2 0} > 0$ *and* $a_{j_2j_1 0} > 0$,

$$(a_{j_1j_2 1}, a_{j_2j_1 1}) = \underbrace{(\mathrm{UP}_+, \mathrm{UP}_+)}_{\text{positive saddle}}. \tag{2.8}$$

  The equilibrium of $(x_{j_2}^*, x_{j_1}^*) = (a_{j_1j_2 1}, a_{j_2j_1 1})$ *is a* $(\mathrm{UP}_+,\mathrm{UP}_+)$*-positive saddle.*

- *For* $a_{j_1j_2 0} < 0$ *and* $a_{j_2j_1 0} > 0$,

$$(a_{j_1j_2 1}, a_{j_2j_1 1}) = \underbrace{(\mathrm{DP}_+, \mathrm{DP}_-)}_{\text{counter-clockwise center}}. \tag{2.9}$$

  The equilibrium of $(x_{j_2}^*, x_{j_1}^*) = (a_{j_1j_2 1}, a_{j_2j_1 1})$ *is a* $(\mathrm{DP}_+,\mathrm{DP}_-)$*-counter-clockwise center.*

- *For $a_{j_1j_20} > 0$ and $a_{j_2j_10} < 0$,*

$$(a_{j_1j_21}, a_{j_2j_11}) = \underbrace{(\mathrm{DP}_{-}, \mathrm{UP}_{+})}_{\text{clockwise center}} . \tag{2.10}$$

The equilibrium of $(x^*_{j_2}, x^*_{j_1}) = (a_{j_1j_21}, a_{j_2j_11})$ is a (DP$_-$,DP$_+$)*-clockwise center.*

- *For $a_{j_1j_20} < 0$ and $a_{j_2j_10} < 0$,*

$$(a_{j_1j_21}, a_{j_2j_11}) = \underbrace{(\mathrm{UP}_{-}, \mathrm{UP}_{-})}_{\text{negative saddle}} . \tag{2.11}$$

The equilibrium of $(x^*_{j_2}, x^*_{j_1}) = (a_{j_1j_21}, a_{j_2j_11})$ is a (UP$_-$,UP$_-$)*-negative saddle.*

(ii) If

$$A_{j_2j_11} \neq 0, A_{j_2j_12} \neq 0 \text{ and } \Delta_{j_1j_2} = 0, a_{j_1j_2s_1} \neq a_{j_1j_2s_2}, \tag{2.12}$$

the two-dimensional dynamical system becomes

$$\begin{aligned} \dot{x}_{j_1} &= a_{j_1j_20}(x_{j_2} - a_{j_1j_2s_1})(x_{j_2} - a_{j_1j_2s_2})^2, \\ \dot{x}_{j_2} &= a_{j_2j_10}(x_{j_1} - a_{j_2j_11}). \end{aligned} \tag{2.13}$$

For $a_{j_1j_2s_1} \neq a_{j_1j_2s_2}$, the first integral manifold is

$$\begin{aligned} &\frac{1}{2}a_{j_2j_10}\left[(x_{j_1} - a_{j_2j_11})^2 - (x_{j_10} - a_{j_2j_11})^2\right] \\ &= a_{j_1j_20}\Big\{\frac{1}{4}\left[(x_{j_2} - a_{j_2j_1s_1})^4 - (x_{j_20} - a_{j_1j_2s_1})^4\right] \\ &+ \frac{2}{3}(a_{j_1j_2s_1} - a_{j_1j_2s_2})\left[(x_{j_2} - a_{j_1j_2s_1})^3 - (x_{j_20} - a_{j_1j_2s_1})^3\right] \\ &+ \frac{1}{2}(a_{j_1j_2s_1} - a_{j_1j_2s_2})^2\left[(x_{j_2} - a_{j_1j_2s_1})^2 - (x_{j_20} - a_{j_1j_2s_1})^2\right]\Big\}. \end{aligned} \tag{2.14}$$

(ii_1) The equilibrium of $(x^*_{j_2}, x^*_{j_1}) = (a_{j_1j_2s_1}, a_{j_2j_11})$ has the following properties:

- *For $a_{j_1j_20}(a_{j_1j_2s_1} - a_{j_1j_2s_2})^2 > 0$ and $a_{j_2j_10} > 0$,*

$$(a_{j_1j_2s_1}, a_{j_2j_11}) = \underbrace{(\mathrm{UP}_{+}, \mathrm{UP}_{+})}_{\text{positive saddle}} . \tag{2.15}$$

The equilibrium of $(x^*_{j_2}, x^*_{j_1}) = (a_{j_1j_2s_1}, a_{j_2j_11})$ is a (UP$_+$,UP$_+$)*-positive saddle.*

- *For* $a_{j_1j_20}(a_{j_1j_2s_1}-a_{j_1j_2s_2})^2<0$ *and* $a_{j_2j_10}>0$,

$$(a_{j_1j_2s_1},a_{j_2j_11})=\underbrace{(\mathrm{DP}_+,\mathrm{DP}_-)}_{\text{counter-clockwise center}}. \tag{2.16}$$

The equilibrium of $(x^*_{j_2},x^*_{j_1})=(a_{j_1j_2s_1},a_{j_2j_11})$ *is a* (DP$_+$,DP$_-$)-counter-*clockwise center.*

- *For* $a_{j_1j_20}(a_{j_1j_2s_1}-a_{j_1j_2s_2})^2>0$ *and* $a_{j_2j_10}<0$,

$$(a_{j_1j_2s_1},a_{j_2j_11})=\underbrace{(\mathrm{DP}_-,\mathrm{UP}_+)}_{\text{clockwise center}}. \tag{2.17}$$

The equilibrium of $(x^*_{j_2},x^*_{j_1})=(a_{j_1j_2s_1},a_{j_2j_11})$ *is a* (DP$_-$,DP$_+$)-*clockwise center.*

- *For* $a_{j_1j_20}(a_{j_1j_2s_1}-a_{j_1j_2s_2})^2<0$ *and* $a_{j_2j_10}<0$,

$$(a_{j_1j_2s_1},a_{j_2j_11})=\underbrace{(\mathrm{UP}_-,\mathrm{UP}_-)}_{\text{negative saddle}}. \tag{2.18}$$

The equilibrium of $(x^*_{j_2},x^*_{j_1})=(a_{j_1j_2s_1},a_{j_2j_11})$ *is a* (UP$_-$,UP$_-$)-*negative saddle.*

(ii$_2$) The equilibrium of $(x^*_{j_2},x^*_{j_1})=(a_{j_1j_2s_2},a_{j_2j_11})$ *has the following properties:*

- *For* $a_{j_1j_20}(a_{j_1j_2s_2}-a_{j_1j_2s_1})>0$ *and* $a_{j_2j_10}>0$,

$$(a_{j_1j_2s_2},a_{j_2j_11})=\underbrace{(\mathrm{US},\mathrm{UP})}_{\text{up-parabola upper-saddle}}. \tag{2.19}$$

The equilibrium of $(x^*_{j_1},x^*_{j_2})=(a_{j_1j_2s_2},a_{j_2j_11})$ *is a* (US,UP)-*up-parabola upper-saddle.*

- *For* $a_{j_1j_20}(a_{j_1j_2s_2}-a_{j_1j_2s_1})<0$ *and* $a_{j_2j_10}>0$,

$$(a_{j_1j_2s_2},a_{j_2j_11})=\underbrace{(\mathrm{LS},\mathrm{DP})}_{\text{down-parabola lower-saddle}}. \tag{2.20}$$

The equilibrium of $(x^*_{j_1},x^*_{j_2})=(a_{j_1j_2s_2},a_{j_2j_11})$ *is an* (LS,DP)-*down-parabola lower-saddle.*

- *For* $a_{j_1j_20}(a_{j_1j_2s_2}-a_{j_1j_2s_1})>0$ *and* $a_{j_2j_10}<0$,

$$(a_{j_1j_2s_2},a_{j_2j_11})=\underbrace{(\mathrm{US},\mathrm{DP})}_{\text{down-parabola upper-saddle}}. \tag{2.21}$$

The equilibrium of $(x^*_{j_1},x^*_{j_2})=(a_{j_1j_2s_2},a_{j_2j_11})$ *is a* (US,DP)-*down-parabola upper-saddle.*

- *For* $a_{j_1j_20}(a_{j_1j_2s_2}-a_{j_1j_2s_1})<0$ *and* $a_{j_2j_10}<0$,

$$(a_{j_1j_2s_2},a_{j_2j_11}) = \underbrace{(\text{LS},\text{UP})}_{\text{up-parabola lower-saddle}}. \tag{2.22}$$

The equilibrium of $(x_{j_1}^*,x_{j_2}^*)=(a_{j_1j_2s_2},a_{j_2j_11})$ *is an* (LS,UP)-*up-parabola lower-saddle.*

(iii) If

$$A_{j_2j_12}\neq 0 \text{ and } \Delta_{j_1j_2}=0, a_{j_1j_2s_1}=a_{j_1j_2s_2}\equiv a_{j_1j_21}, \tag{2.23}$$

the two-dimensional dynamical system becomes

$$\begin{aligned}\dot{x}_{j_1}&=a_{j_1j_20}(x_{j_2}-a_{j_1j_21})^3,\\ \dot{x}_{j_2}&=a_{j_2j_10}(x_{j_1}-a_{j_2j_11}).\end{aligned} \tag{2.24}$$

The first integral manifold is

$$\begin{aligned}&\frac{1}{2}a_{j_2j_10}\left[(x_{j_1}-a_{j_2j_11})^2-(x_{j_10}-a_{j_2j_11})^2\right]\\ &=\frac{1}{4}a_{j_1j_20}\left[(x_{j_2}-a_{j_1j_21})^4-(x_{j_20}-a_{j_1j_21})^4\right].\end{aligned} \tag{2.25}$$

The equilibrium of $(x_{j_2}^*,x_{j_1}^*)=(a_{j_1j_21},a_{j_2j_11})$ *has the following properties:*

- *For* $a_{j_1j_20}>0$ *and* $a_{j_2j_10}>0$,

$$(a_{j_1j_21},a_{j_2j_11}) = \underbrace{(3^{\text{rd}}\text{UP}_+,\text{UP}_+)}_{\text{third-order positive saddle}}. \tag{2.26}$$

The equilibrium of $(x_{j_2}^*,x_{j_1}^*)=(a_{j_1j_21},a_{j_2j_11})$ *is a* (3^{rd}UP_+,UP_+)-*third-order positive saddle.*

- *For* $a_{j_1j_20}<0$ *and* $a_{j_2j_10}>0$,

$$(a_{j_1j_21},a_{j_2j_11}) = \underbrace{(3^{\text{rd}}\text{DP}_+,\text{DP}_-)}_{\text{third-order CCW center}}. \tag{2.27}$$

The equilibrium of $(x_{j_2}^*,x_{j_1}^*)=(a_{j_1j_21},a_{j_2j_11})$ *is a* (3^{rd}DP_+,DP_-)-*third-order counter-clockwise center.*

- *For* $a_{j_1j_20}>0$ *and* $a_{j_2j_10}<0$,

$$(a_{j_1j_21},a_{j_2j_11}) = \underbrace{(3^{\text{rd}}\text{DP}_-,\text{UP}_+)}_{\text{third-order CW center}}. \tag{2.28}$$

The equilibrium of $(x_{j_2}^*,x_{j_1}^*)=(a_{j_1j_21},a_{j_2j_11})$ *is a* (3^{rd}DP_-,DP_+)-*third-order clockwise center.*

- *For $a_{j_1j_2 0} < 0$ and $a_{j_2j_1 0} < 0$,*

$$(a_{j_1j_2 1}, a_{j_2j_1 1}) = \underbrace{(3^{\text{rd}}\text{UP}_-, \text{UP}_-)}_{\text{third-order negative saddle}}. \tag{2.29}$$

The equilibrium of $(x^*_{j_2}, x^*_{j_1}) = (a_{j_1j_2 1}, a_{j_2j_1 1})$ is a $(3^{\text{rd}}\text{UP}_-,\text{UP}_-)$-third-order negative saddle.

(iv) If

$$A_{j_2j_1 2} \neq 0 \text{ and } \Delta_{j_1j_2} > 0, \tag{2.30}$$

the corresponding dynamical system becomes

$$\begin{aligned} \dot{x}_{j_2} &= a_{j_2j_1 0}(x_{j_1} - a_{j_2j_1 1}), \\ \dot{x}_{j_1} &= a_{j_1j_2 0}(x_{j_2} - a_{j_1j_2 1})(x_{j_2} - a_{j_1j_2 2})(x_{j_2} - a_{j_1j_2 3}) \end{aligned} \tag{2.31}$$

where

$$\begin{aligned} \{a_{j_1j_2 1}, a_{j_1j_2 2}, a_{j_1j_2 2}\} &= \text{sort}\{b_{j_1j_2 1}, b_{j_1j_2 2}, b_{j_1j_2 3}\}, a_{j_1j_2 1} < a_{j_1j_2 2} < a_{j_1j_2 3}; \\ b_{j_1j_2 2}, b_{j_1j_2 3} &= -\frac{1}{2}(B_{j_1j_2} \pm \sqrt{\Delta_{j_1j_2}}), a_{j_1j_2 0} = A_{j_1j_2 0}. \end{aligned} \tag{2.32}$$

The first integral manifold is

$$\begin{aligned} &\frac{1}{2} a_{j_2j_1 0}\left[(x_{j_1} - a_{j_2j_1 1})^2 - (x_{j_1 0} - a_{j_2j_1 1})^2\right] \\ &= a_{j_1j_2 0}\Big\{\frac{1}{4}\left[(x_{j_2} - a_{j_1j_2 s_1})^4 - (x_{j_2 0} - a_{j_1j_2 s_1})^4\right] \\ &+ \frac{1}{3}\sum\nolimits_{s_2=1, s_2 \neq s_1}^{3} (a_{j_1j_2 s_1} - a_{j_1j_2 s_2})\left[(x_{j_2} - a_{j_1j_2 s_1})^3 - (x_{j_2 0} - a_{j_1j_2 s_1})^3\right] \\ &+ \frac{1}{2}\prod\nolimits_{s_2=1, s_2 \neq s_1}^{3} (a_{j_1j_2 s_1} - a_{j_1j_2 s_2})\left[(x_{j_2} - a_{j_1j_2 s_1})^2 - (x_{j_2 0} - a_{j_1j_2 s_1})^2\right]\Big\}. \end{aligned} \tag{2.33}$$

The three equilibriums are

$$\begin{aligned} x^*_{j_2} &= a_{j_1j_2 s_1} \ (s_1 = 1, 2, 3) \text{ for the } x_{j_1}\text{-direction}, \\ x^*_{j_1} &= a_{j_2j_1 1} \text{ for the } x_{j_2}\text{-direction}. \end{aligned} \tag{2.34}$$

The equilibrium of $(x^*_{j_2}, x^*_{j_1}) = (a_{j_1j_2 s_1}, a_{j_2j_1 1})$ $(s_1,\ s_2 \in \{1, 2, 3\},\ s_1 \neq s_2)$ has the following properties:

- *For $a_{j_1j_2 0}\prod\nolimits_{s_2=1, s_2 \neq s_1}^{3}(a_{j_1j_2 s_1} - a_{j_1j_2 s_2}) > 0$ and $a_{j_2j_1 0} > 0$,*

$$(a_{j_1j_2 s_1}, a_{j_2j_1 1}) = \underbrace{(\text{UP}_+, \text{UP}_+)}_{\text{positive saddle}}. \tag{2.35}$$

The equilibrium of $(x^*_{j_2}, x^*_{j_1}) = (a_{j_1j_2 s_1}, a_{j_2j_1 1})$ is a $(\text{UP}_+,\text{UP}_+)$-positive saddle.

- *For* $a_{j_1j_20}\prod_{s_2=1,s_2\neq s_1}^{3}(a_{j_1j_2s_1}-a_{j_1j_2s_2})<0$ *and*$a_{j_2j_10}>0$,

$$(a_{j_1j_2s_1},a_{j_2j_11},)=\underbrace{(\mathrm{DP}_+,\mathrm{DP}_-)}_{\text{CCW center}}. \tag{2.36}$$

The equilibrium of $(x_{j_2}^*,x_{j_1}^*)=(a_{j_1j_2s_1},a_{j_2j_11})$ *is a* (DP$_+$,DP$_-$)*-counter-clockwise center.*

- *For* $a_{j_1j_20}\prod_{s_2=1,s_2\neq s_1}^{3}(a_{j_1j_2s_1}-a_{j_1j_2s_2})>0$ *and* $a_{j_2j_10}<0$,

$$(a_{j_1j_2s_1},a_{j_2j_11})=\underbrace{(\mathrm{DP}_-,\mathrm{DP}_+)}_{\text{CW center}}. \tag{2.37}$$

The equilibrium of $(x_{j_2}^*,x_{j_1}^*)=(a_{j_1j_2s_1},a_{j_2j_11})$ *is a* (DP$_-$,DP$_+$)*-clockwise center.*

- *For* $a_{j_1j_20}\prod_{s_2=1,s_2\neq s_1}^{3}(a_{j_1j_2s_1}-a_{j_1j_2s_2})<0$ *and* $a_{j_2j_10}<0$,

$$(a_{j_1j_2s_1},a_{j_2j_11})=\underbrace{(\mathrm{UP}_-,\mathrm{UP}_-)}_{\text{negative saddle}}. \tag{2.38}$$

The equilibrium of $(x_{j_2}^*,x_{j_1}^*)=(a_{j_1j_2s_1},a_{j_2j_11})$ *is a* (UP$_-$,UP$_-$)*-negative saddle.*

(v) Under

$$A_{j_2j_12}\neq 0 \text{ and } \Delta_{j_1j_2}=0, \tag{2.39}$$

the two-dimensional dynamical system is

$$\begin{aligned}\dot{x}_{j_1}&=a_{j_1j_20}(x_{j_2}-a_{j_1j_2s_1})(x_{j_2}-a_{j_1j_2s_2})^2,\\ \dot{x}_{j_2}&=a_{j_2j_10}(x_{j_1}-a_{j_2j_11}).\end{aligned} \tag{2.40}$$

The equilibrium of $(x_{j_2}^*,x_{j_1}^*)=(a_{j_1j_2s_2},a_{j_2j_11})$ *has the following appearing bifurcations:*

- *For* $a_{j_1j_20}(a_{j_1j_2s_2}-a_{j_1j_2s_1})>0$ *and* $a_{j_2j_10}>0$,

$$\underbrace{(a_{j_1j_2s_2},a_{j_2j_11})}_{\substack{\text{(US, UP)-parabola-saddle}\\ \text{appearing bifurcation}}}\rightleftharpoons\left\{\begin{array}{c}\underbrace{(a_{j_1j_2(s_2)_1},a_{j_2j_11})}_{(\mathrm{DP}_+,\mathrm{DP}_-)\text{-CCW center}}\\ \underbrace{(a_{j_1j_2(s_2)_2},a_{j_2j_11})}_{(\mathrm{UP}_+,\mathrm{UP}_+)\text{-positive saddle}}\end{array}\right\}. \tag{2.41}$$

A (US,UP)*-parabola-saddle of* $(x_{j_2}^*,x_{j_1}^*)=(a_{j_1j_2s_2},a_{j_2j_11})$ *is a* (US, UP)*-appearing bifurcation for a* (DP$_+$,DP$_-$)*-counter-clockwise center and a* (UP$_+$,UP$_+$)*-positive-saddle.*

- *For* $a_{j_1j_20}(a_{j_1j_2s_2} - a_{j_1j_2s_1}) < 0$ *and* $a_{j_2j_10} > 0$,

$$\underbrace{(a_{j_1j_2s_2}, a_{j_2j_1 1})}_{\substack{\text{(LS, DP)-parabola-saddle} \\ \text{appearing bifurcation}}} \rightleftharpoons \left\{ \begin{array}{c} \underbrace{(a_{j_1j_2(s_2)_1}, a_{j_2j_1 1})}_{(\text{UP}_+,\text{UP}_+)\text{-positive saddle}} \\ \underbrace{(a_{j_1j_2(s_2)_2}, a_{j_2j_1 1})}_{(\text{DP}_+,\text{DP}_-)\text{-CCW center}} \end{array} \right\}. \tag{2.42}$$

An (LS,DP)*-parabola-saddle of* $(x_{j_2}^*, x_{j_1}^*) = (a_{j_1j_2s_2}, a_{j_2j_1 1})$ *is an* (LS, DP)*-appearing bifurcation for a* (UP$_+$,UP$_+$)*-positive saddle and a* (DP$_+$,DP$_-$)*-counter-clockwise center.*

- *For* $a_{j_1j_20}(a_{j_1j_2s_2} - a_{j_1j_2s_1}) > 0$ *and* $a_{j_2j_10} < 0$,

$$\underbrace{(a_{j_1j_2s_2}, a_{j_2j_1 1})}_{\substack{\text{(US, DP)-parabola-saddle} \\ \text{appearing bifurcation}}} \rightleftharpoons \left\{ \begin{array}{c} \underbrace{(a_{j_1j_2(s_2)_1}, a_{j_2j_1 1})}_{(\text{UP}_-,\text{UP}_-)\text{-negative saddle}} \\ \underbrace{(a_{j_1j_2(s_2)_2}, a_{j_2j_1 1})}_{(\text{DP}_-,\text{DP}_+)\text{-CW center}} \end{array} \right\}. \tag{2.43}$$

A (US,DP)*-saddle-parabola of* $(x_{j_2}^*, x_{j_1}^*) = (a_{j_1j_2s_2}, a_{j_2j_1 1})$ *is a* (US, DP)*-appearing bifurcation for a* (UP$_-$,UP$_-$)*-negative saddle and a* (DP$_-$,DP$_+$)*-clockwise center.*

- *For* $a_{j_1j_20}(a_{j_1j_2s_2} - a_{j_1j_2s_1}) < 0$ *and* $a_{j_2j_10} < 0$,

$$\underbrace{(a_{j_1j_2s_2}, a_{j_2j_1 1})}_{\substack{\text{(LS, UP)-parabola-saddle} \\ \text{appearing bifurcation}}} \rightleftharpoons \left\{ \begin{array}{c} \underbrace{(a_{j_1j_2(s_2)_1}, a_{j_2j_1 1})}_{(\text{DP}_-,\text{DP}_+)\text{-CW center}} \\ \underbrace{(a_{j_1j_2(s_2)_2}, a_{j_2j_1 1})}_{(\text{UP}_-,\text{UP}_-)\text{-negative saddle}} \end{array} \right\}. \tag{2.44}$$

An (LS,UP)*-parabola-saddle of* $(x_{j_2}^*, x_{j_1}^*) = (a_{j_1j_2s_2}, a_{j_2j_1 1})$ *is an* (LS, DP)*-appearing bifurcation for a* (DP$_-$,DP$_+$)*-clockwise center and a* (UP$_-$,UP$_-$)*-negative saddle.*

(vi) Consider

$$\begin{aligned} \dot{x}_{j_1} &= a_{j_1j_20}(x_{j_2} - a_{j_1j_2s_1})(x_{j_2} - a_{j_1j_2(s_2)_1})(x_{j_2} - a_{j_1j_2(s_2)_2}), \\ \dot{x}_{j_2} &= a_{j_2j_10}(x_{j_2} - a_{j_2j_1 1}), \end{aligned} \tag{2.45}$$

with

$$a_{j_1j_2(s_2)_1}^- \rightleftharpoons a_{j_1j_2(s_2)_2}^+ \text{ and } a_{j_1j_2(s_2)_2}^- \rightleftharpoons a_{j_1j_2(s_2)_1}^+. \tag{2.46}$$

Under

$$A_{j_2j_11} \neq 0, \text{ and } \Delta_{j_1j_2} = (a_{j_1j_2(s_2)_1} - a_{j_1j_2(s_2)_2})^2 = 0, \tag{2.47}$$

The equilibrium of $(x^*_{j_2}, x^*_{j_1}) = (a_{j_1j_2s_2}, a_{j_1j_21})$ *has the following switching bifurcations:*

- *For* $a_{j_1j_20}(a_{j_1j_2s_2} - a_{j_1j_2s_1}) > 0$ *and* $a_{j_2j_10} > 0$,

$$\left\{\begin{array}{c} \underbrace{(a^-_{j_1j_2(s_2)_1}, a^-_{j_2j_11})}_{(\text{UP}_+,\text{DP}_-)\text{-CCW center}} \\ \underbrace{(a^-_{j_1j_2(s_2)_2}, a^-_{j_2j_11})}_{(\text{UP}_+,\text{UP}_+)\text{-positive saddle}} \end{array}\right\} \rightleftharpoons \underbrace{(a_{j_1j_2s_2}, a_{j_2j_11})}_{\substack{(\text{US, UP})\text{-parabola-saddle} \\ \text{switching bifurcation}}} \rightleftharpoons \left\{\begin{array}{c} \underbrace{(a^+_{j_1j_2(s_2)_1}, a^+_{j_2j_11})}_{(\text{DP}_+,\text{DP}_-)\text{-CCW center}} \\ \underbrace{(a^+_{j_1j_2(s_2)_2}, a^+_{j_2j_11})}_{(\text{UP}_+,\text{UP}_+)\text{-positive saddle}} \end{array}\right\}. \tag{2.48}$$

A (US,UP)-*parabola-saddle of* $(x^*_{j_2}, x^*_{j_1}) = (a_{j_1j_2s_2}, a_{j_1j_21})$ *is a* (US, UP)-*switching bifurcation for a* (DP$_+$,DP$_-$)-*counter-clockwise-center and a* (UP$_+$,UP$_+$)-*positive-saddle.*

- *For* $a_{j_1j_20}(a_{j_1j_2s_2} - a_{j_1j_2s_1}) < 0$ *and* $a_{j_2j_10} > 0$,

$$\left\{\begin{array}{c} \underbrace{(a^-_{j_1j_2(s_2)_1}, a^-_{j_2j_11})}_{(\text{UP}_+,\text{UP}_+)\text{-positive saddle}} \\ \underbrace{(a^-_{j_1j_2(s_2)_2}, a^-_{j_2j_11})}_{(\text{UP}_+,\text{DP}_-)\text{-CCW center}} \end{array}\right\} \rightleftharpoons \underbrace{(a_{j_1j_2s_2}, a_{j_2j_11})}_{\substack{(\text{LS, DP})\text{-parabola-saddle} \\ \text{switching bifurcation}}} \rightleftharpoons \left\{\begin{array}{c} \underbrace{(a^+_{j_1j_2(s_2)_1}, a^+_{j_2j_11})}_{(\text{UP}_+,\text{UP}_+)\text{-positive saddle}} \\ \underbrace{(a^+_{j_1j_2(s_2)_2}, a^+_{j_2j_11})}_{(\text{DP}_+,\text{DP}_-)\text{-CCW center}} \end{array}\right\}. \tag{2.49}$$

An (LS,DP)-*parabola-saddle of* $(x^*_{j_2}, x^*_{j_1}) = (a_{j_1j_2s_2}, a_{j_1j_21})$ *is an* (LS, DP)-*switching bifurcation for a* (UP$_+$,UP$_+$)-*positive saddle and a* (DP$_-$,DP$_+$)-*counter-clockwise center.*

- *For* $a_{j_1j_20}(a_{j_1j_2s_2} - a_{j_1j_2s_1}) > 0$ *and* $a_{j_2j_10} < 0$,

$$\left\{\begin{array}{c} \underbrace{(a^-_{j_1j_2(s_2)_1}, a^-_{j_2j_11})}_{(\text{UP}_-,\text{UP}_-)\text{-negative saddle}} \\ \underbrace{(a^-_{j_1j_2(s_2)_2}, a^-_{j_2j_11})}_{(\text{DP}_-,\text{DP}_+)\text{-CW center}} \end{array}\right\} \rightleftharpoons \underbrace{(a_{j_1j_2s_2}, a_{j_2j_11})}_{\substack{(\text{US, DP})\text{-parabola-saddle} \\ \text{switching bifurcation}}} \rightleftharpoons \left\{\begin{array}{c} \underbrace{(a^+_{j_1j_2(s_2)_1}, a^+_{j_2j_11})}_{(\text{UP}_-,\text{UP}_-)\text{-negative saddle}} \\ \underbrace{(a^+_{j_1j_2(s_2)_2}, a^+_{j_2j_11})}_{(\text{DP}_-,\text{DP}_+)\text{-CW center}} \end{array}\right\}. \tag{2.50}$$

A (US,DP)-*parabola-saddle of* $(x_{j_2}^*, x_{j_1}^*) = (a_{j_1 j_2 s_2}, a_{j_1 j_2 1})$ *is* a (US, DP)-*switching bifurcation for* a (UP$_-$,UP$_-$)-*negative saddle and a* (DP$_-$,DP$_+$)-*clockwise center.*

- *For* $a_{j_1 j_2 0}(a_{j_1 j_2 s_2} - a_{j_1 j_2 s_1}) < 0$ *and* $a_{j_2 j_1 0} < 0$,

$$\left\{\begin{array}{c}\underbrace{(a^-_{j_1 j_2 (s_2)_1}, a^-_{j_2 j_1 1})}_{(\mathrm{DP}_-,\mathrm{DP}_+)\text{-CW center}} \\ \underbrace{(a^-_{j_1 j_2 (s_2)_2}, a^-_{j_2 j_1 1})}_{(\mathrm{UP}_-,\mathrm{UP}_-)\text{-negative saddle}}\end{array}\right\} \rightleftharpoons \underbrace{(a_{j_1 j_2 s_2}, a_{j_2 j_1 1})}_{\substack{(\mathrm{LS},\mathrm{UP})\text{-parabola-saddle} \\ \text{switching bifurcation}}} \rightleftharpoons \left\{\begin{array}{c}\underbrace{(a^+_{j_1 j_2 (s_2)_1}, a^+_{j_2 j_1 1})}_{(\mathrm{DP}_-,\mathrm{DP}_+)\text{-CW center}} \\ \underbrace{(a^-_{j_1 j_2 (s_2)_2}, a^+_{j_2 j_1 1})}_{(\mathrm{UP}_-,\mathrm{UP}_-)\text{-negative saddle}}\end{array}\right\}. \tag{2.51}$$

An (LS,UP)-*parabola-saddle of* $(x_{j_2}^*, x_{j_1}^*) = (a_{j_1 j_2 s_2}, a_{j_1 j_2 1})$ *is an* (LS, UP)-*switching bifurcation for a* (DP$_-$,DP$_+$)-*clockwise center and a* (UP$_-$,UP$_-$)-*negative saddle.*

(vii) Under

$$A_{j_2 j_1 2} \neq 0 \text{ and } \Delta_{j_1 j_2} = 0, a_{j_1 j_2 s_1} = a_{j_1 j_2 s_2} \equiv a_{j_1 j_2 1}, \tag{2.52}$$

the two-dimensional dynamical system is

$$\begin{aligned} \dot{x}_{j_1} &= a_{j_1 j_2 0}(x_{j_2} - a_{j_1 j_2 1})^3, \\ \dot{x}_{j_2} &= a_{j_2 j_1 0}(x_{j_1} - a_{j_2 j_1 1}). \end{aligned} \tag{2.53}$$

The equilibrium of $(x_{j_2}^*, x_{j_1}^*) = (a_{j_1 j_2 1}, a_{j_2 j_1 1})$ *has the following appearing bifurcations:*

- *For* $a_{j_1 j_2 0} > 0$ *and* $a_{j_2 j_1 0} > 0$,

$$\underbrace{(a_{j_1 j_2 1}, a_{j_2 j_1 1})}_{(\mathrm{UP}_+,\mathrm{UP}_+)\text{-positive saddle}} \rightleftharpoons \underbrace{(a_{j_1 j_2 1}, a_{j_2 j_1 1})}_{\substack{(3^{\mathrm{rd}}\mathrm{UP}_+,\mathrm{UP}_+)\text{-positive saddle} \\ \text{appearing bifurcation}}} \rightleftharpoons \left\{\begin{array}{c}\underbrace{(a_{j_1 j_2 1}, a_{j_2 j_1 1})}_{(\mathrm{UP}_+,\mathrm{UP}_+)\text{-positive saddle}} \\ \underbrace{(a_{j_1 j_2 2}, a_{j_2 j_1 1})}_{(\mathrm{DP}_+,\mathrm{DP}_-)\text{-CCW center}} \\ \underbrace{(a_{j_1 j_2 3}, a_{j_2 j_1 1})}_{(\mathrm{UP}_+,\mathrm{UP}_+)\text{-positive saddle}}\end{array}\right\}. \tag{2.54}$$

A (3^{rd}UP$_+$,UP$_+$)-positive *saddle of* $(x_{j_2}^*, x_{j_1}^*) = (a_{j_1 j_2 1}, a_{j_2 j_1 1})$ *is a* (3^{rd}UP$_+$,UP$_+$)-*appearing bifurcation from a* (UP$_+$,UP$_+$)-*positive-saddle to a* (UP$_+$,UP$_+$)-*positive-saddle, a* (DP$_+$,DP$_-$)-*counter-clockwise-center, and a* (UP$_+$,UP$_+$)-*positive-saddle.*

- *For* $a_{j_1j_2 0} < 0$ *and* $a_{j_2j_1 0} > 0$,

$$\underbrace{(a_{j_1j_2 1}, a_{j_2j_1 1})}_{(\mathrm{DP}_+,\mathrm{DP}_-)\text{-CCW center}} \rightleftharpoons \underbrace{(a_{j_1j_2 1}, a_{j_2j_1 1})}_{\substack{(3^{\mathrm{rd}}\mathrm{DP}_+,\mathrm{DP}_-)\text{-CCW center}\\ \text{appearing bifurcation}}} \rightleftharpoons \left\{ \begin{array}{c} \underbrace{(a_{j_1j_2 1}, a_{j_2j_1 1})}_{(\mathrm{DP}_+,\mathrm{DP}_-)\text{-CCW center}} \\ \underbrace{(a_{j_1j_2 2}, a_{j_2j_1 1})}_{(\mathrm{UP}_+,\mathrm{UP}_+)\text{-positive saddle}} \\ \underbrace{(a_{j_1j_2 3}, a_{j_2j_1 1})}_{(\mathrm{DP}_+,\mathrm{DP}_-)\text{-CCW center}} \end{array} \right\}. \tag{2.55}$$

A $(3^{\mathrm{rd}}\mathrm{DP}_+,\mathrm{UP}_-)$-counter-clockwise center *of* $(x^*_{j_2}, x^*_{j_1}) = (a_{j_1j_2 1}, a_{j_2j_1 1})$ *is a* $(3^{\mathrm{rd}}\mathrm{DP}_+,\mathrm{UP}_-)$*-appearing bifurcation from a* $(\mathrm{DP}_+,\mathrm{UP}_-)$*-counter-clockwise center to a* $(\mathrm{DP}_+,\mathrm{UP}_-)$*-counter-clockwise center, a* $(\mathrm{UP}_+,\mathrm{UP}_+)$*-positive saddle, and a* $(\mathrm{DP}_+,\mathrm{UP}_-)$*-counter-clockwise center*

- *For* $a_{j_1j_2 0} > 0$ *and* $a_{j_2j_1 0} < 0$,

$$\underbrace{(a_{j_1j_2 1}, a_{j_2j_1 1})}_{(\mathrm{DP}_-,\mathrm{DP}_+)\text{-CW center}} \rightleftharpoons \underbrace{(a_{j_1j_2 1}, a_{j_2j_1 1})}_{\substack{(3^{\mathrm{rd}}\mathrm{DP}_-,\mathrm{DP}_+)\text{-CW center}\\ \text{appearing bifurcation}}} \rightleftharpoons \left\{ \begin{array}{c} \underbrace{(a_{j_1j_2 1}, a_{j_2j_1 1})}_{(\mathrm{DP}_-,\mathrm{DP}_+)\text{-CW center}} \\ \underbrace{(a_{j_1j_2 2}, a_{j_2j_1 1})}_{(\mathrm{UP}_-,\mathrm{UP}_-)\text{-negative saddle}} \\ \underbrace{(a_{j_1j_2 3}, a_{j_2j_1 1})}_{(\mathrm{DP}_-,\mathrm{DP}_+)\text{-CW center}} \end{array} \right\}. \tag{2.56}$$

A $(3^{\mathrm{rd}}\mathrm{DP}_-,\mathrm{UP}_+)$-clockwise center *of* $(x^*_{j_2}, x^*_{j_1}) = (a_{j_1j_2 1}, a_{j_2j_1 1})$ *is a* $(3^{\mathrm{rd}}\mathrm{DP}_-,\mathrm{UP}_+)$*-appearing bifurcation from a* $(\mathrm{DP}_-,\mathrm{UP}_+)$*-clockwise center to a* $(\mathrm{DP}_-,\mathrm{UP}_+)$*-clockwise center, a* $(\mathrm{UP}_-,\mathrm{UP}_-)$*-negative saddle, and a* $(\mathrm{DP}_-,\mathrm{UP}_+)$*-clockwise center.*

- *For* $a_{j_1j_2 0} < 0$ *and* $a_{j_2j_1 0} < 0$,

$$\underbrace{(a_{j_1j_2 1}, a_{j_2j_1 1})}_{(\mathrm{UP}_-,\mathrm{UP}_-)\text{-negative saddle}} \rightleftharpoons \underbrace{(a_{j_1j_2 1}, a_{j_2j_1 1})}_{\substack{(3^{\mathrm{rd}}\mathrm{UP}_-,\mathrm{UP}_-)\text{-negative saddle}\\ \text{appearing bifurcation}}} \rightleftharpoons \left\{ \begin{array}{c} \underbrace{(a_{j_1j_2 1}, a_{j_2j_1 1})}_{(\mathrm{UP}_-,\mathrm{UP}_-)\text{-negative saddle}} \\ \underbrace{(a_{j_1j_2 2}, a_{j_2j_1 1})}_{(\mathrm{DP}_-,\mathrm{DP}_+)\text{-CW center}} \\ \underbrace{(a_{j_1j_2 3}, a_{j_2j_1 1})}_{(\mathrm{UP}_-,\mathrm{UP}_-)\text{-negative saddle}} \end{array} \right\}. \tag{2.57}$$

A $(3^{\mathrm{rd}}\mathrm{UP}_-,\mathrm{UP}_-)$-negative *saddle of* $(x^*_{j_2}, x^*_{j_1}) = (a_{j_1j_2 1}, a_{j_2j_1 1})$ *is a* $(3^{\mathrm{rd}}\mathrm{UP}_-,\mathrm{UP}_-)$*-appearing bifurcation from a* $(\mathrm{UP}_-,\mathrm{UP}_-)$*-negative-saddle to a* $(\mathrm{UP}_-,\mathrm{UP}_-)$*-negative-saddle, a* $(\mathrm{DP}_-,\mathrm{DP}_+)$*-clockwise-center, and a* $(\mathrm{UP}_-,\mathrm{UP}_-)$*-negative saddle.*

2.2 Proof of Theorem 2.1

Consider

$$\begin{aligned}\dot{x}_{j_1} &= f_{j_1}(\mathbf{x},\mathbf{p}) = A_{j_1j_2 3} + A_{j_1j_2 2}x_{j_2} + A_{j_1j_2 1}x_{j_2}^2 + A_{j_1j_2 0}x_{j_2}^3,\\ \dot{x}_{j_2} &= f_{j_2}(\mathbf{x},\mathbf{p}) = A_{j_2j_1 3} + A_{j_2j_1 2}x_{j_1} + A_{j_2j_1 1}x_{j_1}^2 + A_{j_2j_1 0}x_{j_1}^3.\end{aligned}$$

For $A_{j_2j_1 0} = A_{j_2j_1 1} = 0, A_{j_2j_1 2} \neq 0,$ the two-dimensional dynamical system possesses crossing-linear and cubic vector fields as

$$\begin{aligned}\dot{x}_{j_2} &= a_{j_2j_1 0}(x_{j_1} - a_{j_2j_1 1}),\\ \dot{x}_{j_1} &= a_{j_1j_2 0}(x_{j_2} - b_{j_1j_2 1})(x_{j_2}^2 + B_{j_1j_2}x_{j_2} + C_{j_1j_2})\end{aligned}$$

with

$$\Delta_{j_1j_2} = (B_{j_1j_2})^2 - 4C_{j_1j_2} \text{ for } j_1, j_2 \in \{1,2\}, j_1 \neq j_2.$$

(i) If

$$A_{j_2j_1 2} \neq 0, A_{j_2j_1 3} \neq 0 \text{ and } \Delta_{j_1j_2} < 0,$$

a dynamical system is

$$\begin{aligned}\dot{x}_{j_2} &= a_{j_2j_1 0}(x_{j_1} - a_{j_2j_1 1}),\\ \dot{x}_{j_1} &= a_{j_1j_2 0}(x_{j_2} - a_{j_1j_2 1})\left[(x_{j_2} - a_{j_1j_2})^2 + b_{j_1j_2}\right]\end{aligned}$$

where

$$\begin{aligned}&a_{j_2j_1 1} = -\frac{1}{A_{j_2j_1 2}}A_{j_2j_1 3}, a_{j_2j_1 0} = A_{j_2j_1 2};\\ &a_{j_1j_2 1} = b_{j_1j_2 1}; a_{j_1j_2} = -\frac{1}{2}B_{j_1j_2 1}, a_{j_1j_2 0} = A_{j_1j_2 0};\\ &b_{j_1j_2} = \frac{1}{4}(-\Delta_{j_1j_2}).\end{aligned}$$

In phase space,

$$\frac{dx_{j_2}}{dx_{j_1}} = \frac{a_{j_2j_1 0}(x_{j_1} - a_{j_2j_1 1})}{a_{j_1j_2 0}(x_{j_2} - a_{j_1j_2 1})\left[(x_{j_2} - a_{j_1j_2})^2 + b_{j_1j_2}\right]},$$

and

$$\begin{aligned}a_{j_2j_10}(x_{j_1}-a_{j_2j_11})dx_{j_1}&=a_{j_1j_20}(x_{j_2}-a_{j_1j_21})\left[(x_{j_2}-a_{j_1j_2})^2+b_{j_1j_2}\right]dx_{j_2}\\&=a_{j_1j_20}\{(x_{j_2}-a_{j_1j_21})^3+2(a_{j_1j_21}-a_{j_1j_2})(x_{j_2}-a_{j_1j_21})^2\\&\quad+\left[(a_{j_1j_21}-a_{j_1j_2})^2+b_{j_1j_2}\right](x_{j_2}-a_{j_1j_21})\}dx_{j_2}.\end{aligned}$$

With an initial condition, (x_{j_10}, x_{j_20}) the first integral manifold is

$$\begin{aligned}&\frac{1}{2}a_{j_2j_10}\left[(x_{j_1}-a_{j_2j_11})^2-(x_{j_{10}}-a_{j_2j_11})^2\right]\\&=a_{j_1j_20}\{\frac{1}{4}\left[(x_{j_2}-a_{j_1j_21})^4-(x_{j_20}-a_{j_1j_21})^4\right]\\&+\frac{2}{3}(a_{j_1j_21}-a_{j_1j_2})\left[(x_{j_2}-a_{j_1j_21})^3-(x_{j_20}-a_{j_1j_21})^3\right]\\&+\frac{1}{2}\left[(a_{j_1j_21}-a_{j_1j_2})^2+b_{j_1j_2}\right]\left[(x_{j_2}-a_{j_1j_21})^2-(x_{j_20}-a_{j_1j_21})^2\right]\}.\end{aligned}$$

Consider two cases (I) $x^*_{j_1}=a_{j_2j_11}$ and (II) $x^*_{j_2}=a_{j_1j_21}$.

(I) In phase space, for $x^*_{j_1}=a_{j_2j_11}$ and $\overline{x}_{j_2}\neq a_{j_1j_21}$,

$$\left.\frac{dx_{j_2}}{dx_{j_1}}\right|_{x^*_{j_1}=a_{j_2j_11}}=\left.\frac{a_{j_2j_10}(x_{j_1}-a_{j_2j_11})}{a_{j_1j_20}(\overline{x}_{j_2}-a_{j_1j_21})\left[(\overline{x}_{j_2}-a_{j_1j_2})^2+b_{j_1j_2}\right]}\right|_{x^*_{j_1}=a_{j_2j_11}}=0.$$

If

$$\left.\frac{d^2x_{j_2}}{dx^2_{j_1}}\right|_{x^*_{j_1}=a_{j_2j_11}}=\frac{a_{j_2j_10}}{a_{j_1j_20}(\overline{x}_{j_2}-a_{j_1j_21})\left[(\overline{x}_{j_2}-a_{j_1j_2})^2+b_{j_1j_2}\right]}>0,$$

the flow at $x^*_{j_1}=a_{j_2j_11}$ is an up-parabola flow in the x_{j_2}-direction. If

$$\left.\frac{d^2x_{j_2}}{dx^2_{j_1}}\right|_{x^*_{j_1}=a_{j_2j_11}}=\frac{a_{j_2j_10}}{a_{j_1j_20}(\overline{x}_{j_2}-a_{j_1j_21})\left[(\overline{x}_{j_2}-a_{j_1j_2})^2+b_{j_1j_2}\right]}<0,$$

the flow at $x^*_{j_1}=a_{j_2j_11}$ is a down-parabola flow in the x_{j_2}-direction. However,

$$\dot{x}_{j_1}=a_{j_1j_20}(\overline{x}_{j_2}-a_{j_1j_21})\left[(\overline{x}_{j_2}-a_{j_1j_2})^2+b_{j_1j_2}\right].$$

For $a_{j_1j_20}(\overline{x}_{j_2}-a_{j_1j_21})>0$ and $a_{j_1j_20}(\overline{x}_{j_2}-a_{j_1j_21})<0$, the flows at $x^*_{j_1}=a_{j_2j_11}$ are positive and negative in the x_{j_1}-direction, respectively.

Thus, the simple equilibrium of $x^*_{j_1}=a_{j_2j_11}$ with $\overline{x}_{j_2}\neq a_{j_1j_21}$ in the x_{j_2}-direction has the following properties:

- For $a_{j_1 j_2 0}(\overline{x}_{j_2} - a_{j_1 j_2 1}) > 0$ and $a_{j_2 j_1 0} > 0$,

$$(\dot{x}_{j_1}, a_{j_2 j_1 1}) = \underbrace{(\text{pF}, \text{UP})}_{\text{up-parabola flow } (+)}.$$

- For $a_{j_1 j_2 0}(\overline{x}_{j_2} - a_{j_1 j_2 1}) < 0$ and $a_{j_2 j_1 0} > 0$,

$$(\dot{x}_{j_1}, a_{j_2 j_1 1}) = \underbrace{(\text{nF}, \text{DP})}_{\text{down-parabola flow } (-)}.$$

- For $a_{j_1 j_2 0}(\overline{x}_{j_2} - a_{j_1 j_2 1}) > 0$ and $a_{j_2 j_1 0} < 0$,

$$(\dot{x}_{j_1}, a_{j_2 j_1 1}) = \underbrace{(\text{pF}, \text{DP})}_{\text{down-parabola flow } (+)}.$$

- For $a_{j_1 j_2 0}(\overline{x}_{j_2} - a_{j_1 j_2 1}) < 0$ and $a_{j_2 j_1 0} < 0$,

$$(\dot{x}_{j_1}, a_{j_2 j_1 1}) = \underbrace{(\text{nF}, \text{UP})}_{\text{up-parabola flow } (-)}.$$

(II) In phase space, for $x_{j_2}^* = a_{j_1 j_2 1}$ and $\overline{x}_{j_1} \neq a_{j_2 j_1 1}$,

$$\frac{dx_{j_1}}{dx_{j_2}}\bigg|_{x_{j_2}^* = a_{j_1 j_2 1}} = \frac{a_{j_1 j_2 0}(x_{j_2} - a_{j_1 j_2 1})\left[(x_{j_2} - a_{j_1 j_2})^2 + b_{j_1 j_2}\right]}{a_{j_2 j_1 0}(\overline{x}_{j_1} - a_{j_2 j_1 1})}\bigg|_{x_{j_2}^* = a_{j_1 j_2 1}} = 0.$$

If

$$\frac{d^2 x_{j_1}}{dx_{j_2}^2}\bigg|_{x_{j_2}^* = a_{j_1 j_2 1}} = \frac{a_{j_1 j_2 0}\left[(a_{j_1 j_2 1} - a_{j_1 j_2})^2 + b_{j_1 j_2}\right]}{a_{j_2 j_1 0}(\overline{x}_{j_1} - a_{j_2 j_1 1})} > 0,$$

the flow at $x_{j_2}^* = a_{j_1 j_2 1}$ is an up-parabola flow in the x_{j_1}-direction. If

$$\frac{d^2 x_{j_1}}{dx_{j_2}^2}\bigg|_{x_{j_2}^* = a_{j_1 j_2 1}} = \frac{a_{j_1 j_2 0}\left[(a_{j_1 j_2 1} - a_{j_1 j_2})^2 + b_{j_1 j_2}\right]}{a_{j_2 j_1 0}(\overline{x}_{j_1} - a_{j_2 j_1 1})} < 0,$$

the flow at $x_{j_2}^* = a_{j_1 j_2 1}$ is a down-parabola flow in the x_{j_1}-direction. However,

$$\dot{x}_{j_2} = a_{j_2 j_1 0}(\overline{x}_{j_1} - a_{j_2 j_1 1}).$$

For $a_{j_2j_10}(\overline{x}_{j_1} - a_{j_2j_11}) > 0$ and $a_{j_2j_10}(\overline{x}_{j_1} - a_{j_2j_11}) < 0$, the flows at $x^*_{j_2} = a_{j_1j_21}$ are positive and negative in the x_{j_2}-direction, respectively.

Thus, the simple equilibrium of $x^*_{j_2} = a_{j_1j_21}$ with $\overline{x}_{j_1} \neq a_{j_2j_11}$ in the x_{j_2}-direction has the following properties:

- For $a_{j_1j_20} > 0$ and $a_{j_2j_10}(\overline{x}_{j_1} - a_{j_2j_11}) > 0$,

$$(a_{j_1j_21}, \dot{x}_{j_2}) = \underbrace{(\mathrm{UP}, \mathrm{pF})}_{\text{up-parabola flow } (+)}.$$

- For $a_{j_1j_20} < 0$ and $a_{j_2j_10}(\overline{x}_{j_1} - a_{j_2j_11}) > 0$,

$$(a_{j_1j_21}, \dot{x}_{j_2}) = \underbrace{(\mathrm{DP}, \mathrm{pF})}_{\text{down-parabola flow } (+)}.$$

- For $a_{j_1j_20} > 0$ and $a_{j_2j_10}(\overline{x}_{j_1} - a_{j_2j_11}) < 0$,

$$(a_{j_1j_21}, \dot{x}_{j_2}) = \underbrace{(\mathrm{DP}, \mathrm{nF})}_{\text{down-parabola flow } (-)}.$$

- For $a_{j_1j_20} < 0$ and $a_{j_2j_10}(\overline{x}_{j_1} - a_{j_2j_11}) < 0$,

$$(a_{j_1j_21}, \dot{x}_{j_2}) = \underbrace{(\mathrm{UP}, \mathrm{nF})}_{\text{up-parabola flow } (-)}.$$

Therefore, from cases (I) and (II), the equilibrium of $(x^*_{j_2}, x^*_{j_1}) = (a_{j_1j_21}, a_{j_2j_11})$ has the following properties as in Eqs. (2.8)–(2.11):

- For $a_{j_1j_20} > 0$ and $a_{j_2j_10} > 0$,

$$(a_{j_1j_21}, a_{j_2j_11}) = \underbrace{(\mathrm{UP}_+, \mathrm{UP}_+)}_{\text{positive saddle}}.$$

  The equilibrium of $(x^*_{j_2}, x^*_{j_1}) = (a_{j_1j_21}, a_{j_2j_11})$ is a $(\mathrm{UP}_+,\mathrm{UP}_+)$-positive saddle.
- For $a_{j_1j_20} < 0$ and $a_{j_2j_10} > 0$,

$$(a_{j_1j_21}, a_{j_2j_11}) = \underbrace{(\mathrm{DP}_+, \mathrm{DP}_-)}_{\text{counter-clockwise center}}.$$

  The equilibrium of $(x^*_{j_2}, x^*_{j_1}) = (a_{j_1j_21}, a_{j_2j_11})$ is a $(\mathrm{DP}_+,\mathrm{DP}_-)$-counter-clockwise center.

- For $a_{j_1 j_2 0} > 0$ and $a_{j_2 j_1 0} < 0$,

$$(a_{j_1 j_2 1}, a_{j_2 j_1 1}) = \underbrace{(\mathrm{DP}_-, \mathrm{UP}_+)}_{\text{clockwise center}}.$$

The equilibrium of $(x_{j_2}^*, x_{j_1}^*) = (a_{j_1 j_2 1}, a_{j_2 j_1 1})$ is a $(\mathrm{DP}_-,\mathrm{DP}_+)$-clockwise center.

- For $a_{j_1 j_2 0} < 0$ and $a_{j_2 j_1 0} < 0$,

$$(a_{j_1 j_2 1}, a_{j_2 j_1 1}) = \underbrace{(\mathrm{UP}_-, \mathrm{UP}_-)}_{\text{negative saddle}}.$$

The equilibrium of $(x_{j_2}^*, x_{j_1}^*) = (a_{j_1 j_2 1}, a_{j_2 j_1 1})$ is a $(\mathrm{UP}_-,\mathrm{UP}_-)$-negative saddle.

(ii) If

$$A_{j_2 j_1 1} \neq 0, A_{j_2 j_1 2} \neq 0 \text{ and } \Delta_{j_1 j_2} = 0, a_{j_1 j_2 s_1} \neq a_{j_1 j_2 s_2},$$

the two-dimensional dynamical system becomes

$$\begin{aligned}
\dot{x}_{j_1} &= a_{j_1 j_2 0}(x_{j_2} - a_{j_1 j_2 s_1})(x_{j_2} - a_{j_1 j_2 s_2})^2, \\
\dot{x}_{j_2} &= a_{j_2 j_1 0}(x_{j_1} - a_{j_2 j_1 1}).
\end{aligned}$$

In phase space

$$\frac{dx_{j_2}}{dx_{j_1}} = \frac{a_{j_2 j_1 0}(x_{j_1} - a_{j_2 j_1 1})}{a_{j_1 j_2 0}(x_{j_2} - a_{j_1 j_2 s_1})(x_{j_2} - a_{j_1 j_2 s_2})^2},$$

and

$$\begin{aligned}
a_{j_2 j_1 0}(x_{j_1} - a_{j_2 j_1 1})dx_{j_1} &= a_{j_1 j_2 0}(x_{j_2} - a_{j_1 j_2 s_1})(x_{j_2} - a_{j_1 j_2 s_2})^2 dx_{j_2} \\
&= a_{j_1 j_2 0}[(x_{j_2} - a_{j_1 j_2 s_1})^3 + 2(a_{j_1 j_2 s_1} - a_{j_1 j_2 s_2})(x_{j_2} - a_{j_1 j_2 s_1})^2 \\
&\quad + (a_{j_1 j_2 s_1} - a_{j_1 j_2 s_2})^2 (x_{j_2} - a_{j_1 j_2 s_1})] dx_{j_2}.
\end{aligned}$$

With an initial condition $(x_{j_1 0}, x_{j_2 0})$, the first integral manifold is

$$\begin{aligned}
&\frac{1}{2} a_{j_2 j_1 0}\left[(x_{j_1} - a_{j_2 j_1 1})^2 - (x_{j_1 0} - a_{j_2 j_1 1})^2\right] \\
&= a_{j_1 j_2 0}\Big\{\frac{1}{4}\left[(x_{j_2} - a_{j_2 j_1 s_1})^4 - (x_{j_2 0} - a_{j_1 j_2 s_1})^4\right] \\
&+ \frac{2}{3}(a_{j_1 j_2 s_1} - a_{j_1 j_2 s_2})\left[(x_{j_2} - a_{j_1 j_2 s_1})^3 - (x_{j_2 0} - a_{j_1 j_2 s_1})^3\right] \\
&+ \frac{1}{2}(a_{j_1 j_2 s_1} - a_{j_1 j_2 s_2})^2\left[(x_{j_2} - a_{j_1 j_2 s_1})^2 - (x_{j_2 0} - a_{j_1 j_2 s_1})^2\right]\Big\}.
\end{aligned}$$

(ii$_1$) Consider two cases (I) $x^*_{j_1} = a_{j_2 j_1 1}$ and (II) $x^*_{j_2} = a_{j_1 j_2 s_1}$.

(I) For $x^*_{j_1} = a_{j_2 j_1 1}$ but $\overline{x}_{j_2} \neq a_{j_1 j_2 s_1}, a_{j_1 j_2 s_2}$, in phase space,

$$\frac{dx_{j_2}}{dx_{j_1}}\Big|_{x^*_{j_1}=a_{j_2 j_1 1}} = \frac{a_{j_2 j_1 0}(x_{j_1} - a_{j_2 j_1 1})}{a_{j_1 j_2 0}(\overline{x}_{j_2} - a_{j_1 j_2 s_1})(\overline{x}_{j_2} - a_{j_1 j_2 s_2})^2}\Big|_{x^*_{j_1}=a_{j_2 j_1 1}} = 0.$$

If

$$\frac{d^2 x_{j_2}}{dx^2_{j_1}}\Big|_{x^*_{j_1}=a_{j_2 j_1 1}} = \frac{a_{j_2 j_1 0}}{a_{j_1 j_2 0}(\overline{x}_{j_2} - a_{j_1 j_2 s_1})(\overline{x}_{j_2} - a_{j_1 j_2 s_2})^2} > 0,$$

the flow at $x^*_{j_1} = a_{j_2 j_1 1}$ is an up-parabola flow in the x_{j_2}-direction. If

$$\frac{d^2 x_{j_2}}{dx^2_{j_1}}\Big|_{x^*_{j_1}=a_{j_2 j_1 1}} = \frac{a_{j_2 j_1 0}}{a_{j_1 j_2 0}(\overline{x}_{j_2} - a_{j_1 j_2 s_1})(\overline{x}_{j_2} - a_{j_1 j_2 s_2})^2} < 0,$$

the flow at $x^*_{j_1} = a_{j_2 j_1 1}$ is a down-parabola flow in the x_{j_2}-direction. However,

$$\dot{x}_{j_1} = a_{j_1 j_2 0}(\overline{x}_{j_2} - a_{j_1 j_2 s_1})(\overline{x}_{j_2} - a_{j_1 j_2 s_2})^2.$$

For $\dot{x}_{j_1} > 0$ and $\dot{x}_{j_1} < 0$, the flows at $x^*_{j_1} = a_{j_2 j_1 1}$ are positive and negative in the x_{j_1}-direction, respectively. Thus, at $x^*_{j_1} = a_{j_2 j_1 1}$ with $\overline{x}_{j_2} \neq a_{j_1 j_2 s_1}, a_{j_1 j_2 s_2}$, the following properties exist:

- For $a_{j_1 j_2 0}(\overline{x}_{j_2} - a_{j_1 j_2 s_1})(\overline{x}_{j_2} - a_{j_1 j_2 s_2})^2 > 0$ and $a_{j_2 j_1 0} > 0$,

$$(\dot{x}_{j_1}, a_{j_2 j_1 1}) = \underbrace{(\text{pF}, \text{UP})}_{\text{up-parabola flow } (+)}.$$

- For $a_{j_1 j_2 0}(\overline{x}_{j_2} - a_{j_1 j_2 s_1})(\overline{x}_{j_2} - a_{j_1 j_2 s_2})^2 < 0$ and $a_{j_2 j_1 0} > 0$,

$$(\dot{x}_{j_1}, a_{j_2 j_1 1}) = \underbrace{(\text{nF}, \text{DP})}_{\text{down-parabola flow } (-)}.$$

- For $a_{j_1 j_2 0}(\overline{x}_{j_2} - a_{j_1 j_2 s_1})(\overline{x}_{j_2} - a_{j_1 j_2 s_2})^2 > 0$ and $a_{j_2 j_1 0} < 0$,

$$(\dot{x}_{j_1}, a_{j_2 j_1 1}) = \underbrace{(\text{pF}, \text{DP})}_{\text{down-parabola flow } (+)}.$$

- For $a_{j_1 j_2 0}(\overline{x}_{j_2} - a_{j_1 j_2 s_1})(\overline{x}_{j_2} - a_{j_1 j_2 s_2})^2 < 0$ and $a_{j_2 j_1 0} < 0$,

$$(\dot{x}_{j_1}, a_{j_2 j_1 1}) = \underbrace{(\text{nF}, \text{UP})}_{\text{up-parabola flow } (-)}.$$

(II) In phase space, for $x_{j_2}^* = a_{j_1 j_2 s_1} \neq a_{j_1 j_2 s_2}$ and $\overline{x}_{j_1} \neq a_{j_2 j_1 1}$,

$$\frac{dx_{j_1}}{dx_{j_2}}\bigg|_{x_{j_2}^* = a_{j_1 j_2 s_1}} = \frac{a_{j_1 j_2 0}(x_{j_2} - a_{j_1 j_2 s_1})(x_{j_2} - a_{j_1 j_2 s_2})^2}{a_{j_2 j_1 0}(x_{j_1} - a_{j_2 j_1 1})}\bigg|_{x_{j_2}^* = a_{j_1 j_2 s_1}} = 0.$$

If

$$\frac{d^2 x_{j_1}}{dx_{j_2}^2}\bigg|_{x_{j_2}^* = a_{j_1 j_2 s_1}} = \frac{a_{j_1 j_2 0}(a_{j_1 j_2 s_1} - a_{j_1 j_2 s_2})^2}{a_{j_2 j_1 0}(\overline{x}_{j_1} - a_{j_2 j_1 1})} > 0,$$

the flow at $x_{j_2}^* = a_{j_1 j_2 s_1}$ is an up-parabola flow in the x_{j_1}-direction. If

$$\frac{d^2 x_{j_1}}{dx_{j_2}^2}\bigg|_{x_{j_2}^* = a_{j_1 j_2 s_1}} = \frac{a_{j_1 j_2 0}(a_{j_1 j_2 s_1} - a_{j_1 j_2 s_2})^2}{a_{j_2 j_1 0}(\overline{x}_{j_1} - a_{j_2 j_1 1})} < 0,$$

the flow at $x_{j_2}^* = a_{j_1 j_2 s_1}$ is a down-parabola flow in the x_{j_1}-direction. However,

$$\dot{x}_{j_2} = a_{j_2 j_1 0}(\overline{x}_{j_1} - a_{j_2 j_1 1}).$$

For $a_{j_2 j_1 0}(\overline{x}_{j_1} - a_{j_2 j_1 1}) > 0$ and $a_{j_2 j_1 0}(\overline{x}_{j_1} - a_{j_2 j_1 1}) < 0$, the flows at $x_{j_2}^* = a_{j_1 j_2 s_1}$ are positive and negative in the x_{j_2}-direction, respectively.

Thus, the simple equilibrium of $x_{j_2}^* = a_{j_1 j_2 s_1}$ in the x_{j_2}-direction has the following properties:

- For $a_{j_1 j_2 0}(a_{j_1 j_2 s_1} - a_{j_1 j_2 s_2})^2 > 0$ and $a_{j_2 j_1 0}(\overline{x}_{j_1} - a_{j_2 j_1 1}) > 0$,

$$(a_{j_1 j_2 s_1}, \dot{x}_{j_2}) = \underbrace{(\mathrm{UP}, \mathrm{pF})}_{\text{up-parabola flow }(+)}.$$

- For $a_{j_1 j_2 0}(a_{j_1 j_2 s_1} - a_{j_1 j_2 s_2})^2 < 0$ and $a_{j_2 j_1 0}(\overline{x}_{j_1} - a_{j_2 j_1 1}) > 0$,

$$(a_{j_1 j_2 s_1}, \dot{x}_{j_2}) = \underbrace{(\mathrm{DP}, \mathrm{pF})}_{\text{down-parabola flow }(+)}.$$

- For $a_{j_1 j_2 0}(a_{j_1 j_2 s_1} - a_{j_1 j_2 s_2})^2 > 0$ and $a_{j_2 j_1 0}(\overline{x}_{j_1} - a_{j_2 j_1 1}) < 0$,

$$(a_{j_1 j_2 s_1}, \dot{x}_{j_2}) = \underbrace{(\mathrm{DP}, \mathrm{nF})}_{\text{down-parabola flow }(-)}.$$

- For $a_{j_1 j_2 0}(a_{j_1 j_2 s_1} - a_{j_1 j_2 s_2})^2 < 0$ and $a_{j_2 j_1 0}(\overline{x}_{j_1} - a_{j_2 j_1 1}) < 0$,

$$(a_{j_1 j_2 s_1}, \dot{x}_{j_2}) = \underbrace{(\mathrm{UP}, \mathrm{nF})}_{\text{up-parabola flow }(-)}.$$

Therefore, from cases (I) and (II), the equilibrium of $(x_{j_2}^*, x_{j_1}^*) = (a_{j_1j_2}1, a_{j_2j_1}1)$ has the following properties as in Eqs. (2.15)–(2.18):

- For $a_{j_1j_2}0(a_{j_1j_2s_1} - a_{j_1j_2s_2})^2 > 0$ and $a_{j_2j_1}0 > 0$,

$$(a_{j_1j_2s_1}, a_{j_2j_1}1) = \underbrace{(\mathrm{UP}_+, \mathrm{UP}_+)}_{\text{positive saddle}} .$$

  The equilibrium of $(x_{j_2}^*, x_{j_1}^*) = (a_{j_1j_2s_1}, a_{j_2j_1}1)$ is a $(\mathrm{UP}_+,\mathrm{UP}_+)$-positive saddle.
- For $a_{j_1j_2}0(a_{j_1j_2s_1} - a_{j_1j_2s_2})^2 < 0$ and $a_{j_2j_1}0 > 0$,

$$(a_{j_1j_2s_1}, a_{j_2j_1}1) = \underbrace{(\mathrm{DP}_+, \mathrm{DP}_-)}_{\text{counter-clockwise center}} .$$

  The equilibrium of $(x_{j_2}^*, x_{j_1}^*) = (a_{j_1j_2s_1}, a_{j_2j_1}1)$ is a $(\mathrm{DP}_+,\mathrm{DP}_-)$-counter-clockwise center.
- For $a_{j_1j_2}0(a_{j_1j_2s_1} - a_{j_1j_2s_2})^2 > 0$ and $a_{j_2j_1}0 < 0$,

$$(a_{j_1j_2s_1}, a_{j_2j_1}1) = \underbrace{(\mathrm{DP}_-, \mathrm{UP}_+)}_{\text{clockwise center}} .$$

  The equilibrium of $(x_{j_2}^*, x_{j_1}^*) = (a_{j_1j_2s_1}, a_{j_2j_1}1)$ is a $(\mathrm{DP}_-,\mathrm{DP}_+)$-clockwise center.
- For $a_{j_1j_2}0(a_{j_1j_2s_1} - a_{j_1j_2s_2})^2 < 0$ and $a_{j_2j_1}0 < 0$,

$$(a_{j_1j_2s_1}, a_{j_2j_1}1) = \underbrace{(\mathrm{UP}_-, \mathrm{UP}_-)}_{\text{negative saddle}} .$$

  The equilibrium of $(x_{j_2}^*, x_{j_1}^*) = (a_{j_1j_2s_1}, a_{j_2j_1}1)$ is a $(\mathrm{UP}_-,\mathrm{UP}_-)$-negative saddle.

(ii$_2$) For $x_{j_2}^* = a_{j_1j_2s_2} \neq a_{j_1j_2s_1}$ but $\overline{x}_{j_1} \neq a_{j_2j_1}1$,

$$\frac{dx_{j_1}}{dx_{j_2}}\Big|_{x_{j_2}^* = a_{j_1j_2s_2}} = \frac{a_{j_1j_2}0(x_{j_2} - a_{j_1j_2s_1})(x_{j_2} - a_{j_1j_2s_2})^2}{a_{j_2j_1}0(\overline{x}_{j_1} - a_{j_2j_1}1)}\Big|_{x_{j_2}^* = a_{j_1j_2s_2}} = 0,$$
$$\frac{d^2x_{j_1}}{dx_{j_2}^2}\Big|_{x_{j_2}^* - a_{j_1j_2s_2}} = 0.$$

If

$$\frac{d^3x_{j_1}}{dx_{j_2}^3}\Big|_{x_{j_2}^* = a_{j_1j_2s_2}} = \frac{2a_{j_1j_2}0(a_{j_1j_2s_2} - a_{j_1j_2s_1})}{a_{j_2j_1}0(\overline{x}_{j_1} - a_{j_2j_1}1)} > 0,$$

the flow at $x_{j_2}^* = a_{j_1j_2s_2}$ and $\overline{x}_{j_1} \neq a_{j_2j_1}1$ is an increasing-inflection flow. If

$$\frac{d^3x_{j_1}}{dx_{j_2}^3}\bigg|_{x_{j_2}^*=a_{j_1j_2s_2}} = \frac{2a_{j_1j_20}(a_{j_1j_2s_2}-a_{j_1j_2s_1})}{a_{j_2j_10}(\overline{x}_{j_1}-a_{j_2j_11})} < 0,$$

the flow at $x_{j_2}^*=a_{j_1j_2s_2}$ and $\overline{x}_{j_1}\neq a_{j_2j_11}$ is a decreasing-inflection flow. However,

$$\dot{x}_{j_2}=a_{j_2j_10}(\overline{x}_{j_1}-a_{j_2j_11}).$$

For $a_{j_2j_10}(\overline{x}_{j_1}-a_{j_2j_11})>0$ and $a_{j_2j_10}(\overline{x}_{j_1}-a_{j_2j_11})<0$, the flows at $x_{j_2}^*=a_{j_1j_2s_2}$ and $\overline{x}_{j_1}\neq a_{j_2j_11}$ are positive and negative in the x_{j_2}-direction, respectively. Thus, at $x_{j_2}^*=a_{j_1j_2s_2}$ and $\overline{x}_{j_1}\neq a_{j_2j_11}$, the following properties exist:

- For $a_{j_1j_20}(a_{j_1j_2s_2}-a_{j_1j_2s_1})>0$ and $a_{j_2j_10}(\overline{x}_{j_1}-a_{j_2j_11})>0$,

$$(a_{j_1j_2s_2},\dot{x}_{j_2}) = \underbrace{(\text{II, pF})}_{\text{increasing-inflection flow }(+)}.$$

- For $a_{j_1j_20}(a_{j_1j_2s_2}-a_{j_1j_2s_1})<0$ and $a_{j_2j_10}(\overline{x}_{j_1}-a_{j_2j_11})>0$,

$$(a_{j_1j_2s_2},\dot{x}_{j_2}) = \underbrace{(\text{DI, pF})}_{\text{decreasing-inflection flow }(+)}.$$

- For $a_{j_1j_20}(a_{j_1j_2s_2}-a_{j_1j_2s_1})>0$ and $a_{j_2j_10}(\overline{x}_{j_1}-a_{j_2j_11})<0$,

$$(a_{j_1j_2s_2},\dot{x}_{j_2}) = \underbrace{(\text{DI, nF})}_{\text{decreasing-inflection flow }(-)}.$$

- For $a_{j_1j_20}(a_{j_1j_2s_2}-a_{j_1j_2s_1})<0$ and $a_{j_2j_10}(\overline{x}_{j_1}-a_{j_2j_11})<0$,

$$(a_{j_1j_2s_2},\dot{x}_{j_2}) = \underbrace{(\text{II, nF})}_{\text{increasing-inflection flow }(-)}.$$

From the proof (I) of case (ii$_1$) and the above analysis, therefore, the equilibrium of $(x_{j_2}^*,x_{j_1}^*)=(a_{j_1j_2s_2},a_{j_2j_11})$ has the following properties as in Eqs. (2.19)–(2.22):

- For $a_{j_1j_20}(a_{j_1j_2s_2}-a_{j_1j_2s_1})>0$ and $a_{j_2j_10}>0$,

$$(a_{j_1j_2s_2},a_{j_2j_11}) = \underbrace{(\text{US, UP})}_{\text{up-parabola upper-saddle}}.$$

  The equilibrium of $(x_{j_1}^*,x_{j_2}^*)=(a_{j_1j_2s_2},a_{j_2j_11})$ is a (US,UP)-up-parabola upper-saddle.

- For $a_{j_1j_2 0}(a_{j_1j_2s_2} - a_{j_1j_2s_1}) < 0$ and $a_{j_2j_1 0} > 0$,

$$(a_{j_1j_2s_2}, a_{j_2j_1 1}) = \underbrace{(\mathrm{LS}, \mathrm{DP})}_{\text{down-parabola lower-saddle}}.$$

The equilibrium of $(x^*_{j_1}, x^*_{j_2}) = (a_{j_1j_2s_2}, a_{j_2j_1 1})$ is an (LS,DP)-down-parabola lower-saddle.

- For $a_{j_1j_2 0}(a_{j_1j_2s_2} - a_{j_1j_2s_1}) > 0$ and $a_{j_2j_1 0} < 0$,

$$(a_{j_1j_2s_2}, a_{j_2j_1 1}) = \underbrace{(\mathrm{US}, \mathrm{DP})}_{\text{down-parabola upper-saddle}}.$$

The equilibrium of $(x^*_{j_1}, x^*_{j_2}) = (a_{j_1j_2s_2}, a_{j_2j_1 1})$ is a (US,DP)-down-parabola upper-saddle.

- For $a_{j_1j_2 0}(a_{j_1j_2s_2} - a_{j_1j_2s_1}) < 0$ and $a_{j_2j_1 0} < 0$,

$$(a_{j_1j_2s_2}, a_{j_2j_1 1}) = \underbrace{(\mathrm{LS}, \mathrm{UP})}_{\text{up-parabola lower-saddle}}.$$

The equilibrium of $(x^*_{j_1}, x^*_{j_2}) = (a_{j_1j_2s_2}, a_{j_2j_1 1})$ is an (LS,UP)-up-parabola lower-saddle.

(iii) If

$$A_{j_2j_1 1} \neq 0, A_{j_2j_1 2} \neq 0 \text{ and } \Delta_{j_1j_2} = 0, a_{j_1j_2s_1} = a_{j_1j_2s_2} \equiv a_{j_1j_2 1},$$

the two-dimensional dynamical system becomes

$$\begin{aligned} \dot{x}_{j_1} &= a_{j_1j_2 0}(x_{j_2} - a_{j_1j_2 1})^3, \\ \dot{x}_{j_2} &= a_{j_2j_1 0}(x_{j_1} - a_{j_2j_1 1}). \end{aligned}$$

In phase space,

$$\frac{dx_{j_2}}{dx_{j_1}} = \frac{a_{j_2j_1 0}(x_{j_1} - a_{j_2j_1 1})}{a_{j_1j_2 0}(x_{j_2} - a_{j_1j_2 1})^3},$$

and

$$a_{j_2j_1 0}(x_{j_1} - a_{j_2j_1 1})dx_{j_1} = a_{j_1j_2 0}(x_{j_2} - a_{j_1j_2 1})^3 dx_{j_2}.$$

With an initial condition $(x_{j_1 0}, x_{j_2 0})$, the first integral manifold is

$$\begin{aligned}&\frac{1}{2}a_{j_2j_10}\left[(x_{j_1}-a_{j_2j_11})^2-(x_{j_10}-a_{j_2j_11})^2\right]\\&=\frac{1}{4}a_{j_1j_20}\left[(x_{j_2}-a_{j_1j_21})^4-(x_{j_20}-a_{j_1j_21})^4\right].\end{aligned}$$

Consider two cases (I) $x_{j_1}^*=a_{j_2j_11}$ and (II) $x_{j_2}^*=a_{j_1j_21}$.

(I) For $x_{j_1}^*=a_{j_2j_11}$ but $\overline{x}_{j_2}\neq a_{j_1j_21}$,

$$\frac{dx_{j_2}}{dx_{j_1}}\bigg|_{x_{j_1}^*=a_{j_2j_11}}=\frac{a_{j_2j_10}(x_{j_1}-a_{j_2j_11})}{a_{j_1j_20}(\overline{x}_{j_2}-a_{j_1j_21})^3}\bigg|_{x_{j_1}^*=a_{j_2j_11}}=0.$$

If

$$\frac{d^2x_{j_2}}{dx_{j_1}^2}\bigg|_{x_{j_1}^*=a_{j_2j_11}}=\frac{a_{j_2j_10}}{a_{j_1j_20}(\overline{x}_{j_2}-a_{j_1j_21})^3}>0,$$

the flow at $x_{j_1}^*=a_{j_2j_11}$ is an up-parabola flow in the x_{j_2}-direction. If

$$\frac{d^2x_{j_2}}{dx_{j_1}^2}\bigg|_{x_{j_1}^*=a_{j_2j_11}}=\frac{a_{j_2j_10}}{a_{j_1j_20}(\overline{x}_{j_2}-a_{j_1j_21})^3}<0,$$

the flow at $x_{j_1}^*=a_{j_2j_11}$ is a down-parabola flow in the x_{j_2}-direction. However,

$$\dot{x}_{j_1}=a_{j_1j_20}(\overline{x}_{j_2}-a_{j_1j_21})^3.$$

For $\dot{x}_{j_1}>0$ and $\dot{x}_{j_1}<0$, the flows at $x_{j_1}^*=a_{j_2j_11}$ are positive and negative in the x_{j_1}-direction, respectively. Thus, at $x_{j_1}^*=a_{j_2j_11}$ but $\overline{x}_{j_2}\neq a_{j_1j_21}$, the following properties exist:

- For $a_{j_1j_20}(\overline{x}_{j_2}-a_{j_1j_21})^3>0$ and $a_{j_2j_10}>0$,

$$(\dot{x}_{j_1},a_{j_2j_11})=\underbrace{(\text{pF},\text{UP})}_{\text{up-parabola flow }(+)}.$$

- For $a_{j_1j_20}(\overline{x}_{j_2}-a_{j_1j_21})^3<0$ and $a_{j_2j_10}>0$,

$$(\dot{x}_{j_1},a_{j_2j_11})=\underbrace{(\text{nF},\text{DP})}_{\text{down-parabola flow }(-)}.$$

- For $a_{j_1j_20}(\overline{x}_{j_2}-a_{j_1j_21})^3>0$ and $a_{j_2j_10}<0$,

$$(\dot{x}_{j_1},a_{j_2j_11})=\underbrace{(\text{pF},\text{DP})}_{\text{down-parabola flow }(+)}.$$

- For $a_{j_1j_20}(\overline{x}_{j_2}-a_{j_1j_21})^3<0$ and $a_{j_2j_10}<0$,

$$(\dot{x}_{j_1},a_{j_2j_11})=\underbrace{(\text{nF},\text{UP})}_{\text{up-parabola flow }(-)}.$$

(II) In phase space, for $x_{j_2}^*=a_{j_1j_21}$ and $\overline{x}_{j_1}\neq a_{j_2j_11}$,

$$\begin{aligned}\frac{dx_{j_1}}{dx_{j_2}}\bigg|_{x_{j_2}^*=a_{j_1j_21}}&=\frac{a_{j_1j_20}(x_{j_2}-a_{j_1j_21})^3}{a_{j_2j_10}(\overline{x}_{j_1}-a_{j_2j_11})}\bigg|_{x_{j_2}^*=a_{j_1j_2s_1}}=0,\\ \frac{d^2x_{j_1}}{dx_{j_2}^2}\bigg|_{x_{j_2}^*=a_{j_1j_21}}&=\frac{d^3x_{j_1}}{dx_{j_2}^3}\bigg|_{x_{j_2}^*=a_{j_1j_21}}=0.\end{aligned}$$

If

$$\frac{d^4x_{j_1}}{dx_{j_2}^4}\bigg|_{x_{j_2}^*=a_{j_1j_2s_1}}=\frac{6a_{j_1j_20}}{a_{j_2j_10}(\overline{x}_{j_1}-a_{j_2j_11})}>0,$$

the flow at $x_{j_2}^*=a_{j_1j_21}$ is a third-order up-parabola flow in the x_{j_1}-direction. If

$$\frac{d^4x_{j_1}}{dx_{j_2}^4}\bigg|_{x_{j_2}^*=a_{j_1j_2s_1}}=\frac{6a_{j_1j_20}}{a_{j_2j_10}(\overline{x}_{j_1}-a_{j_2j_11})}<0,$$

the flow at $x_{j_2}^*=a_{j_1j_21}$ is a third-order down-parabola flow in the x_{j_1}-direction. However,

$$\dot{x}_{j_2}=a_{j_2j_10}(\overline{x}_{j_1}-a_{j_2j_11}).$$

For $a_{j_2j_10}(\overline{x}_{j_1}-a_{j_2j_11})>0$ and $a_{j_2j_10}(\overline{x}_{j_1}-a_{j_2j_11})<0$, the flows at $x_{j_2}^*=a_{j_1j_2s_1}$ are positive and negative in the x_{j_2}-direction, respectively. Thus, the equilibrium of $x_{j_2}^*=a_{j_1j_21}$ in the x_{j_2}-direction has the following properties:

- For $a_{j_1j_20}>0$ and $a_{j_2j_10}(\overline{x}_{j_1}-a_{j_2j_11})>0$,

$$(a_{j_1j_21},\dot{x}_{j_2})=\underbrace{(3^{\text{rd}}\text{UP},\text{pF})}_{\text{third-order up-parabola flow }(+)}.$$

- For $a_{j_1j_20}<0$ and $a_{j_2j_10}(\overline{x}_{j_1}-a_{j_2j_11})>0$,

$$(a_{j_1j_21},\dot{x}_{j_2})=\underbrace{(3^{\text{rd}}\text{DP},\text{pF})}_{\text{third-order down-parabola flow }(+)}.$$

- For $a_{j_1j_2 0} > 0$ and $a_{j_2j_1 0}(\overline{x}_{j_1} - a_{j_2j_1 1}) < 0$,

$$(a_{j_1j_2 1}, \dot{x}_{j_2}) = \underbrace{(3^{\text{rd}}\text{DP}, \text{nF})}_{\text{third-order down-parabola flow } (-)}.$$

- For $a_{j_1j_2 0} < 0$ and $a_{j_2j_1 0}(\overline{x}_{j_1} - a_{j_2j_1 1}) < 0$,

$$(a_{j_1j_2 1}, \dot{x}_{j_2}) = \underbrace{(3^{\text{rd}}\text{UP}, \text{nF})}_{\text{third-order up-parabola flow } (-)}.$$

Therefore, from cases (I) and (II), the equilibrium of $(x^*_{j_2}, x^*_{j_1}) = (a_{j_1j_2 1}, a_{j_2j_1 1})$ has the following properties as in Eqs. (2.26)–(2.29):

- For $a_{j_1j_2 0} > 0$ and $a_{j_2j_1 0} > 0$,

$$(a_{j_1j_2 1}, a_{j_2j_1 1}) = \underbrace{(3^{\text{rd}}\text{UP}_+, \text{UP}_+)}_{\text{third-order positive saddle}}.$$

  The equilibrium of $(x^*_{j_2}, x^*_{j_1}) = (a_{j_1j_2 1}, a_{j_2j_1 1})$ is a $(3^{\text{rd}}\text{UP}_+, \text{UP}_+)$-third-order positive saddle.
- For $a_{j_1j_2 0} < 0$ and $a_{j_2j_1 0} > 0$,

$$(a_{j_1j_2 1}, a_{j_2j_1 1}) = \underbrace{(3^{\text{rd}}\text{DP}_+, \text{DP}_-)}_{\text{third-order CCW center}}.$$

  The equilibrium of $(x^*_{j_2}, x^*_{j_1}) = (a_{j_1j_2 1}, a_{j_2j_1 1})$ is a $(3^{\text{rd}}\text{DP}_+, \text{DP}_-)$-third-order counter-clockwise center.
- For $a_{j_1j_2 0} > 0$ and $a_{j_2j_1 0} < 0$,

$$(a_{j_1j_2 1}, a_{j_2j_1 1}) = \underbrace{(3^{\text{rd}}\text{DP}_-, \text{UP}_+)}_{\text{third-order CW center}}.$$

  The equilibrium of $(x^*_{j_2}, x^*_{j_1}) = (a_{j_1j_2 1}, a_{j_2j_1 1})$ is a $(3^{\text{rd}}\text{DP}_-, \text{DP}_+)$-third-order clockwise center.
- For $a_{j_1j_2 0} < 0$ and $a_{j_2j_1 0} < 0$,

$$(a_{j_1j_2 1}, a_{j_2j_1 1}) = \underbrace{(3^{\text{rd}}\text{UP}_-, \text{UP}_-)}_{\text{third-order negative saddle}}.$$

  The equilibrium of $(x^*_{j_2}, x^*_{j_1}) = (a_{j_1j_2 1}, a_{j_2j_1 1})$ is a $(3^{\text{rd}}\text{UP}_-, \text{UP}_-)$-third-order negative saddle.

(iv) If

$$A_{j_2 j_1 2} \neq 0 \text{ and } \Delta_{j_1 j_2} > 0,$$

the corresponding dynamical system becomes

$$\begin{aligned}
\dot{x}_{j_2} &= a_{j_2 j_1 0}(x_{j_1} - a_{j_2 j_1 1}),\\
\dot{x}_{j_1} &= a_{j_1 j_2 0}(x_{j_2} - a_{j_1 j_2 1})(x_{j_2} - a_{j_1 j_2 2})(x_{j_2} - a_{j_1 j_2 3})
\end{aligned}$$

where

$$\begin{aligned}
\{a_{j_1 j_2 1}, a_{j_1 j_2 2}, a_{j_1 j_2 2}\} &= \mathrm{sort}\{b_{j_1 j_2 1}, b_{j_1 j_2 2}, b_{j_1 j_2 3}\}, a_{j_1 j_2 1} < a_{j_1 j_2 2} < a_{j_1 j_2 3};\\
b_{j_1 j_2 2}, b_{j_1 j_2 3} &= -\frac{1}{2}(B_{j_1 j_2} \pm \sqrt{\Delta_{j_1 j_2}}), a_{j_1 j_2 0} = A_{j_1 j_2 0}.
\end{aligned}$$

In phase space,

$$\frac{dx_{j_2}}{dx_{j_1}} = \frac{a_{j_2 j_1 0}(x_{j_1} - a_{j_2 j_1 1})}{a_{j_1 j_2 0}\prod_{s_1=1}^{3}(x_{j_2} - a_{j_1 j_2 s_1})}$$

and

$$\begin{aligned}
a_{j_2 j_1 0}(x_{j_1} - a_{j_2 j_1 1})dx_{j_1} &= a_{j_1 j_2 0}(x_{j_2} - a_{j_1 j_2 1})(x_{j_2} - a_{j_1 j_2 2})(x_{j_2} - a_{j_1 j_2 1})dx_{j_2}\\
&= a_{j_1 j_2 0}\big[(x_{j_2} - a_{j_1 j_2 s_1})^3\\
&+ \sum_{s_2=1, s_2 \neq s_1}^{3}(a_{j_1 j_2 s_1} - a_{j_1 j_2 s_2})(x_{j_2} - a_{j_1 j_2 s_1})^2\\
&+ \prod_{s_2=1, s_2 \neq s_1}^{3}(a_{j_1 j_2 s_1} - a_{j_1 j_2 s_2})(x_{j_2} - a_{j_1 j_2 s_1})\big]dx_{j_2}
\end{aligned}$$

With an initial condition $(x_{j_1 0}, x_{j_2 0})$, the first integral manifold is

$$\begin{aligned}
&\frac{1}{2}a_{j_2 j_1 0}\big[(x_{j_1} - a_{j_2 j_1 1})^2 - (x_{j_1 0} - a_{j_2 j_1 1})^2\big]\\
&= a_{j_1 j_2 0}\Big\{\frac{1}{4}\big[(x_{j_2} - a_{j_1 j_2 s_1})^4 - (x_{j_2 0} - a_{j_1 j_2 s_1})^4\big]\\
&+ \frac{1}{3}\sum_{s_2=1, s_2 \neq s_1}^{3}(a_{j_1 j_2 s_1} - a_{j_1 j_2 s_2})\big[(x_{j_2} - a_{j_1 j_2 s_1})^3 - (x_{j_2 0} - a_{j_1 j_2 s_1})^3\big]\\
&+ \frac{1}{2}\prod_{s_2=1, s_2 \neq s_1}^{3}(a_{j_1 j_2 s_1} - a_{j_1 j_2 s_2})\big[(x_{j_2} - a_{j_1 j_2 s_1})^2 - (x_{j_2 0} - a_{j_1 j_2 s_1})^2\big]\Big\}.
\end{aligned}$$

The three equilibriums are

$$\begin{aligned}
x^*_{j_2} &= a_{j_1 j_2 s_1}\ (s_1 = 1, 2, 3) \text{ for the } x_{j_1}\text{-direction},\\
x^*_{j_1} &= a_{j_2 j_1 1} \text{ for the } x_{j_2}\text{-direction}.
\end{aligned}$$

Consider two cases (I) $x^*_{j_1} = a_{j_2 j_1 1}$ and (II) $x^*_{j_2} = a_{j_1 j_2 s_1}$.

(I) For $x_{j_1}^* = a_{j_2j_11}$ but $\overline{x}_{j_2} \neq a_{j_1j_2s_1}, a_{j_1j_2s_2}$ $(s_1, s_2 \in \{1,2,3\},\ s_1 \neq s_2)$,

$$\frac{dx_{j_2}}{dx_{j_1}}\bigg|_{x_{j_1}^* = a_{j_2j_11}} = \frac{a_{j_2j_10}(x_{j_1} - a_{j_2j_11})}{a_{j_1j_20}\prod_{s_1=1}^{3}(\overline{x}_{j_2} - a_{j_1j_2s_1})}\bigg|_{x_{j_1}^* = a_{j_2j_11}} = 0.$$

If

$$\frac{d^2x_{j_2}}{dx_{j_1}^2}\bigg|_{x_{j_1}^* = a_{j_2j_11}} = \frac{a_{j_2j_10}}{a_{j_1j_20}\prod_{s_1=1}^{3}(\overline{x}_{j_2} - a_{j_1j_2s_1})} > 0,$$

the flow at $x_{j_1}^* = a_{j_2j_11}$ and $\overline{x}_{j_2} \neq a_{j_1j_2s_1}(s_1 \in \{1,2,3\})$ is an up-parabola flow in the x_{j_2}-direction. If

$$\frac{d^2x_{j_2}}{dx_{j_1}^2}\bigg|_{x_{j_1}^* = a_{j_2j_11}} = \frac{a_{j_2j_10}}{a_{j_1j_20}\prod_{s_1=1}^{3}(\overline{x}_{j_2} - a_{j_1j_2s_1})} < 0,$$

the flow at $x_{j_1}^* = a_{j_2j_11}$ and $\overline{x}_{j_2} \neq a_{j_1j_2s_1}, a_{j_1j_2s_2}$ is a down-parabola flow in the x_{j_2}-direction. However,

$$\dot{x}_{j_1} = a_{j_1j_20}\prod_{s_1=1}^{3}(\overline{x}_{j_2} - a_{j_1j_2s_1}).$$

For $\dot{x}_{j_1} > 0$ and $\dot{x}_{j_1} < 0$, the flows at $x_{j_1}^* = a_{j_2j_11}$ and $\overline{x}_{j_2} \neq a_{j_1j_2s_1}$ $(s_1 \in \{1,2,3\})$ are positive and negative in the x_{j_1}-direction, respectively. Thus, at $x_{j_1}^* = a_{j_2j_11}$ and $\overline{x}_{j_2} \neq a_{j_1j_2s_1}, a_{j_1j_2s_2}$, the following properties exist:

- For $a_{j_1j_20}\prod_{s_1=1}^{3}(\overline{x}_{j_2} - a_{j_1j_2s_1}) > 0$ and $a_{j_2j_10} > 0$,

$$(\dot{x}_{j_1}, a_{j_2j_11}) = \underbrace{(\text{pF}, \text{UP})}_{\text{up-parabola flow }(+)}$$

- For $a_{j_1j_20}\prod_{s_1=1}^{3}(\overline{x}_{j_2} - a_{j_1j_2s_1}) < 0$ and $a_{j_2j_10} > 0$,

$$(\dot{x}_{j_1}, a_{j_2j_11}) = \underbrace{(\text{nF}, \text{DP})}_{\text{down-paraboa flow }(-)}.$$

- For $a_{j_1j_20}\prod_{s_1=1}^{3}(\overline{x}_{j_2} - a_{j_1j_2s_1}) > 0$ and $a_{j_2j_10} < 0$,

$$(\dot{x}_{j_1}, a_{j_2j_11}) = \underbrace{(\text{pF}, \text{DP})}_{\text{down-paraboa flow }(+)}.$$

- For $a_{j_1j_20}\prod_{s_1=1}^{3}(\overline{x}_{j_2}-a_{j_1j_2s_1})<0$ and $a_{j_2j_10}<0$,

$$(\dot{x}_{j_1},a_{j_2j_11})=\underbrace{(\text{nF},\text{UP})}_{\text{up-paraboa flow }(-)}.$$

(II) For $x_{j_2}^*=a_{j_1j_2s_1}$ but $\overline{x}_{j_1}\neq a_{j_2j_11}$ (s_1, $s_2\in\{1,2,3\}$, $s_1\neq s_2$),

$$\frac{dx_{j_1}}{dx_{j_2}}\bigg|_{x_{j_2}^*=a_{j_1j_2s_1}}=\frac{a_{j_1j_20}(x_{j_2}-a_{j_1j_2s_1})\prod_{s_2=1,s_2\neq s_1}^{3}(x_{j_2}-a_{j_1j_2s_2})}{a_{j_2j_10}(\overline{x}_{j_1}-a_{j_2j_11})}\bigg|_{x_{j_2}^*=a_{j_1j_2s_1}}=0.$$

If

$$\frac{d^2x_{j_1}}{dx_{j_2}^2}\bigg|_{x_{j_2}^*=a_{j_1j_2s_1}}=\frac{a_{j_1j_20}\prod_{s_2=1,s_2\neq s_1}^{3}(a_{j_1j_2s_1}-a_{j_1j_2s_2})}{a_{j_2j_10}(\overline{x}_{j_1}-a_{j_2j_11})}\bigg|_{x_{j_2}^*=a_{j_1j_2s_1}}>0,$$

the flow $x_{j_2}^*=a_{j_1j_2s_1}$ but $\overline{x}_{j_1}\neq a_{j_2j_11}$ is an up-parabola flow in the x_{j_1}-direction. If

$$\frac{d^2x_{j_1}}{dx_{j_2}^2}\bigg|_{x_{j_2}^*=a_{j_1j_2s_1}}=\frac{a_{j_1j_20}\prod_{s_2=1,s_2\neq s_1}^{3}(a_{j_1j_2s_1}-a_{j_1j_2s_2})}{a_{j_2j_10}(\overline{x}_{j_1}-a_{j_2j_11})}\bigg|_{x_{j_2}^*=a_{j_1j_2s_1}}<0,$$

the flow at $x_{j_2}^*=a_{j_1j_21}$ and $\overline{x}_{j_1}\neq a_{j_2j_11}$ is a down-parabola flow in the x_{j_1}-direction. However,

$$\dot{x}_{j_1}=a_{j_2j_10}(\overline{x}_{j_1}-a_{j_2j_11}).$$

For $a_{j_2j_10}(\overline{x}_{j_1}-a_{j_2j_11})>0$ and $a_{j_2j_10}(\overline{x}_{j_1}-a_{j_2j_11})<0$, the flows at $x_{j_2}^*=a_{j_1j_21}$ and $\overline{x}_{j_1}\neq a_{j_2j_11}$ are positive and negative in the x_{j_2}-direction, respectively. Thus, at $x_{j_2}^*=a_{j_1j_21}$ and $\overline{x}_{j_1}\neq a_{j_2j_11}$, the following properties exist:

- For $a_{j_1j_20}\prod_{s_2=1,s_2\neq s_1}^{3}(a_{j_1j_2s_1}-a_{j_1j_2s_2})>0$ and $a_{j_2j_10}(\overline{x}_{j_1}-a_{j_2j_11})>0$,

$$(a_{j_1j_2s_1},\dot{x}_{j_2})=\underbrace{(\text{UP},\text{pF})}_{\text{up-parabola flow }(+)}.$$

- For $a_{j_1j_20}\prod_{s_2=1,s_2\neq s_1}^{3}(a_{j_1j_2s_1}-a_{j_1j_2s_2})<0$ and $a_{j_2j_10}(\overline{x}_{j_1}-a_{j_2j_11})>0$,

$$(a_{j_1j_2s_1},\dot{x}_{j_2})=\underbrace{(\text{DP},\text{pF})}_{\text{down-parabola flow }(+)}.$$

- For $a_{j_1j_20}\prod_{s_2=1,s_2\neq s_1}^{3}(a_{j_1j_2s_1}-a_{j_1j_2s_2})>0$ and $a_{j_2j_10}(\overline{x}_{j_1}-a_{j_2j_11})<0$,

$$(a_{j_1j_2s_1},\dot{x}_{j_2})=\underbrace{(\text{DP},\text{nF})}_{\text{down-parabola flow }(-)}.$$

- For $a_{j_1j_20}\prod_{s_2=1,s_2\neq s_1}^{3}(a_{j_1j_2s_1}-a_{j_1j_2s_2})<0$ and $a_{j_2j_10}(\overline{x}_{j_1}-a_{j_2j_11})<0$,

$$(a_{j_1j_2s_1},\dot{x}_{j_2})=\underbrace{(\text{UP},\text{nF})}_{\text{up-parabola flow }(-)}.$$

Therefore, from cases (I) and (II), the equilibrium of $(x_{j_2}^*,x_{j_1}^*)=(a_{j_1j_2s_1},a_{j_2j_11})$ (s_1, $s_2\in\{1,2\}$, $s_1\neq s_2$) has the following properties as in Eq. (2.35)–(2.38):

- For $a_{j_1j_20}\prod_{s_2=1,s_2\neq s_1}^{3}(a_{j_1j_2s_1}-a_{j_1j_2s_2})>0$ and $a_{j_2j_10}>0$,

$$(a_{j_1j_2s_1},a_{j_2j_11})=\underbrace{(\text{UP}_+,\text{UP}_+)}_{\text{positive saddle}}.$$

  The equilibrium of $(x_{j_2}^*,x_{j_1}^*)=(a_{j_1j_2s_1},a_{j_2j_11})$ is a $(\text{UP}_+,\text{UP}_+)$-positive-saddle.
- For $a_{j_1j_20}\prod_{s_2=1,s_2\neq s_1}^{3}(a_{j_1j_2s_1}-a_{j_1j_2s_2})<0$ and $a_{j_2j_10}>0$,

$$(a_{j_1j_2s_1},a_{j_2j_11},)=\underbrace{(\text{DP}_+,\text{DP}_-)}_{\text{CCW center}}.$$

  The equilibrium of $(x_{j_2}^*,x_{j_1}^*)=(a_{j_1j_2s_1},a_{j_2j_11})$ is a $(\text{DP}_+,\text{DP}_-)$-counter-clockwise-center.
- For $a_{j_1j_20}\prod_{s_2=1,s_2\neq s_1}^{3}(a_{j_1j_2s_1}-a_{j_1j_2s_2})>0$ and $a_{j_2j_10}<0$,

$$(a_{j_1j_2s_1},a_{j_2j_11})=\underbrace{(\text{DP}_-,\text{DP}_+)}_{\text{CW center}}.$$

  The equilibrium of $(x_{j_2}^*,x_{j_1}^*)=(a_{j_1j_2s_1},a_{j_2j_11})$ is a $(\text{DP}_-,\text{DP}_+)$-clockwise-center.
- For $a_{j_1j_20}\prod_{s_2=1,s_2\neq s_1}^{3}(a_{j_1j_2s_1}-a_{j_1j_2s_2})<0$ and $a_{j_2j_10}<0$,

$$(a_{j_1j_2s_1},a_{j_2j_11})=\underbrace{(\text{UP}_-,\text{UP}_-)}_{\text{negative saddle}}.$$

  The equilibrium of $(x_{j_2}^*,x_{j_1}^*)=(a_{j_1j_2s_1},a_{j_2j_11})$ is a $(\text{UP}_-,\text{UP}_-)$-negative-saddle.

(v) Under

$$A_{j_2 j_1 2} \neq 0 \text{ and } \Delta_{j_1 j_2} = 0,$$

the two-dimensional dynamical system is

$$\begin{aligned}\dot{x}_{j_1} &= a_{j_1 j_2 0}(x_{j_2} - a_{j_1 j_2 s_1})(x_{j_2} - a_{j_1 j_2 s_2})^2,\\ \dot{x}_{j_2} &= a_{j_2 j_1 0}(x_{j_1} - a_{j_2 j_1 1}).\end{aligned}$$

From cases (ii$_2$) and (iv), the equilibrium of $(x^*_{j_2}, x^*_{j_1}) = (a_{j_1 j_2 s_2}, a_{j_2 j_1 1})$ has the following appearing bifurcations as in Eqs. (2.41)–(2.44):

- For $a_{j_1 j_2 0}(a_{j_1 j_2 s_2} - a_{j_1 j_2 s_1}) > 0$ and $a_{j_2 j_1 0} > 0$,

$$\underbrace{(a_{j_1 j_2 s_2}, a_{j_2 j_1 1})}_{\substack{\text{(US, UP)-parabola-saddle}\\ \text{appearing bifurcation}}} \rightleftharpoons \left\{ \begin{array}{c} \underbrace{(a_{j_1 j_2 (s_2)_1}, a_{j_2 j_1 1})}_{(\text{DP}_+, \text{DP}_-)\text{-CCW center}} \\ \underbrace{(a_{j_1 j_2 (s_2)_2}, a_{j_2 j_1 1})}_{(\text{UP}_+, \text{UP}_+)\text{-positive saddle}} \end{array} \right\}.$$

  A (US,UP)-parabola-saddle of $(x^*_{j_2}, x^*_{j_1}) = (a_{j_1 j_2 s_2}, a_{j_2 j_1 1})$ is a (US,UP)-appearing bifurcation for a (DP$_+$,DP$_-$)-counter-clockwise-center and a (UP$_+$,UP$_+$)-positive-saddle.
- For $a_{j_1 j_2 0}(a_{j_1 j_2 s_2} - a_{j_1 j_2 s_1}) < 0$ and $a_{j_2 j_1 0} > 0$,

$$\underbrace{(a_{j_1 j_2 s_2}, a_{j_2 j_1 1})}_{\substack{\text{(LS, DP)-saddle-parabola}\\ \text{appearing bifurcation}}} \rightleftharpoons \left\{ \begin{array}{c} \underbrace{(a_{j_1 j_2 (s_2)_1}, a_{j_2 j_1 1})}_{(\text{UP}_+, \text{UP}_+)\text{-positive saddle}} \\ \underbrace{(a_{j_1 j_2 (s_2)_2}, a_{j_2 j_1 1})}_{(\text{DP}_+, \text{DP}_-)\text{-CCW center}} \end{array} \right\}.$$

  An (LS,DP)-parabola-saddle of $(x^*_{j_2}, x^*_{j_1}) = (a_{j_1 j_2 s_2}, a_{j_2 j_1 1})$ is an (LS,DP)-appearing bifurcation for a (UP$_+$,UP$_+$)-positive saddle and a (DP$_+$,DP$_-$)-counter-clockwise center.
- For $a_{j_1 j_2 0}(a_{j_1 j_2 s_2} - a_{j_1 j_2 s_1}) > 0$ and $a_{j_2 j_1 0} < 0$,

$$\underbrace{(a_{j_1 j_2 s_2}, a_{j_2 j_1 1})}_{\substack{\text{(US, DP)-parabola-saddle}\\ \text{appearing bifurcation}}} \rightleftharpoons \left\{ \begin{array}{c} \underbrace{(a_{j_1 j_2 (s_2)_1}, a_{j_2 j_1 1})}_{(\text{UP}_-, \text{UP}_-)\text{-negative saddle}} \\ \underbrace{(a_{j_1 j_2 (s_2)_2}, a_{j_2 j_1 1})}_{(\text{DP}_-, \text{DP}_+)\text{-CW center}} \end{array} \right\}.$$

A (US,DP)-parabola-saddle of $(x_{j_2}^*, x_{j_1}^*) = (a_{j_1j_2s_2}, a_{j_2j_1}1)$ is a (US,DP)-appearing bifurcation for a $(\text{UP}_-,\text{UP}_-)$-negative saddle and a $(\text{DP}_-,\text{DP}_+)$-clockwise center.

- For $a_{j_1j_2}0(a_{j_1j_2s_2} - a_{j_1j_2s_1}) < 0$ and $a_{j_2j_1}0 < 0$,

$$\underbrace{(a_{j_1j_2s_2}, a_{j_2j_1}1)}_{\substack{\text{(LS, UP)-saddle-parabola}\\ \text{appearing bifurcation}}} \rightleftharpoons \left\{ \begin{array}{c} \underbrace{(a_{j_1j_2(s_2)_1}, a_{j_2j_1}1)}_{(\text{DP}_-,\text{DP}_+)\text{-CW center}} \\ \underbrace{(a_{j_1j_2(s_2)_2}, a_{j_2j_1}1)}_{(\text{UP}_-,\text{UP}_-)\text{-negative saddle}} \end{array} \right\}.$$

An (LS,UP)-parabola-saddle of $(x_{j_2}^*, x_{j_1}^*) = (a_{j_1j_2s_2}, a_{j_2j_1}1)$ is an (LS,DP)-appearing bifurcation for a $(\text{DP}_-,\text{DP}_+)$-clockwise center and a $(\text{UP}_-,\text{UP}_-)$-negative saddle.

(vi) Consider

$$\begin{aligned} \dot{x}_{j_1} &= a_{j_1j_2}0(x_{j_2} - a_{j_1j_2s_1})(x_{j_2} - a_{j_1j_2(s_2)_1})(x_{j_2} - a_{j_1j_2(s_2)_2}) \\ \dot{x}_{j_2} &= a_{j_2j_1}0(x_{j_2} - a_{j_2j_1}1) \end{aligned}$$

with

$$a_{j_1j_2(s_2)_1}^- \rightleftharpoons a_{j_1j_2(s_2)_2}^+ \text{ and } a_{j_1j_2(s_2)_2}^- \rightleftharpoons a_{j_1j_2(s_2)_1}^+.$$

Under

$$A_{j_2j_1}2 \neq 0, \text{and } \Delta_{j_1j_2} = (a_{j_1j_2(s_2)_1} - a_{j_1j_2(s_2)_2})^2 = 0,$$

from cases (ii) and (iv), the equilibrium of $(x_{j_2}^*, x_{j_1}^*) = (a_{j_1j_2s_2}, a_{j_2j_1}1)$ has the following appearing bifurcations as in Eqs. (2.48)–(2.51):

- For $a_{j_1j_2}0(a_{j_1j_2s_2} - a_{j_1j_2}1) > 0$ and $a_{j_2j_1}0 > 0$,

$$\left\{ \begin{array}{c} \underbrace{(a_{j_1j_2(s_2)_1}^-, a_{j_2j_1}^-1)}_{(\text{UP}_+,\text{DP}_-)\text{-CCW center}} \\ \underbrace{(a_{j_1j_2(s_2)_2}^-, a_{j_2j_1}^-1)}_{(\text{UP}_+,\text{UP}_+)\text{-positive saddle}} \end{array} \right\} \rightleftharpoons \underbrace{(a_{j_1j_2s_2}, a_{j_2j_1}1)}_{\substack{\text{(US, UP)-parabola-saddle}\\ \text{switching bifurcation}}} \rightleftharpoons \left\{ \begin{array}{c} \underbrace{(a_{j_1j_2(s_2)_1}^+, a_{j_2j_1}^+1)}_{(\text{DP}_+,\text{DP}_-)\text{-CCW center}} \\ \underbrace{(a_{j_1j_2(s_2)_2}^+, a_{j_2j_1}^+1)}_{(\text{UP}_+,\text{UP}_+)\text{-positive saddle}} \end{array} \right\}.$$

A (US,UP)-parabola-saddle of $(x_{j_2}^*, x_{j_1}^*) = (a_{j_1j_2s_2}, a_{j_1j_2}1)$ is a (US, UP)-switching bifurcation for a $(\text{DP}_+,\text{DP}_-)$-counter-clockwise-center and a $(\text{UP}_+,\text{UP}_+)$-positive-saddle.

- For $a_{j_1j_20}(a_{j_1j_2s_2}-a_{j_1j_21})<0$ and $a_{j_2j_10}>0$,

$$\left\{\begin{array}{c}\underbrace{(a^{-}_{j_1j_2(s_2)_1}, a^{-}_{j_2j_11})}_{(\mathrm{UP}_+,\mathrm{UP}_+)\text{-positive saddle}}\\ \underbrace{(a^{-}_{j_1j_2(s_2)_2}, a^{-}_{j_2j_11})}_{(\mathrm{UP}_+,\mathrm{DP}_-)\text{-CCW center}}\end{array}\right\} \rightleftharpoons \underbrace{(a_{j_1j_2s_2}, a_{j_2j_11})}_{\substack{(\mathrm{LS},\,\mathrm{DP})\text{-parabola-saddle}\\ \text{switching bifurcation}}} \rightleftharpoons \left\{\begin{array}{c}\underbrace{(a^{+}_{j_1j_2(s_2)_1}, a^{+}_{j_2j_11})}_{(\mathrm{UP}_+,\mathrm{UP}_+)\text{-positive saddle}}\\ \underbrace{(a^{+}_{j_1j_2(s_2)_2}, a^{+}_{j_2j_11})}_{(\mathrm{DP}_+,\mathrm{DP}_-)\text{-CCW center}}\end{array}\right\}.$$

An (LS,DP)-parabola-saddle of $(x^*_{j_2},x^*_{j_1})=(a_{j_1j_2s_2},a_{j_1j_21})$ is an (LS, DP)-switching bifurcation for a (UP_+,UP_+)-positive saddle and a (DP_-,DP_+)-counter-clockwise center .

- For $a_{j_1j_20}(a_{j_1j_2s_2}-a_{j_1j_21})>0$ and $a_{j_2j_10}<0$,

$$\left\{\begin{array}{c}\underbrace{(a^{-}_{j_1j_2(s_2)_1}, a^{-}_{j_2j_11})}_{(\mathrm{UP}_-,\mathrm{UP}_-)\text{-negative saddle}}\\ \underbrace{(a^{-}_{j_1j_2(s_2)_2}, a^{-}_{j_2j_11})}_{(\mathrm{DP}_-,\mathrm{DP}_+)\text{-CW center}}\end{array}\right\} \rightleftharpoons \underbrace{(a_{j_1j_2s_2}, a_{j_2j_11})}_{\substack{(\mathrm{US},\,\mathrm{DP})\text{-parabola-saddle}\\ \text{switching bifurcation}}} \rightleftharpoons \left\{\begin{array}{c}\underbrace{(a^{+}_{j_1j_2(s_2)_1}, a^{+}_{j_2j_11})}_{(\mathrm{UP}_-,\mathrm{UP}_-)\text{-negative saddle}}\\ \underbrace{(a^{+}_{j_1j_2(s_2)_2}, a^{+}_{j_2j_11})}_{(\mathrm{DP}_-,\mathrm{DP}_+)\text{-CW center}}\end{array}\right\}.$$

A (US,DP)-parabola-saddle of $(x^*_{j_2},x^*_{j_1})=(a_{j_1j_2s_2},a_{j_1j_21})$ is a (US,DP)-switching bifurcation for a (UP_-,UP_-)-negative saddle and a (DP_-,DP_+)-clockwise center.

- For $a_{j_1j_20}(a_{j_1j_2s_2}-a_{j_1j_21})<0$ and $a_{j_2j_10}<0$

$$\left\{\begin{array}{c}\underbrace{(a^{-}_{j_1j_2(s_2)_1}, a^{-}_{j_2j_11})}_{(\mathrm{DP}_-,\mathrm{DP}_+)\text{-CW center}}\\ \underbrace{(a^{-}_{j_1j_2(s_2)_2}, a^{-}_{j_2j_11})}_{(\mathrm{UP}_-,\mathrm{UP}_-)\text{-negative saddle}}\end{array}\right\} \rightleftharpoons \underbrace{(a_{j_1j_2s_2}, a_{j_2j_11})}_{\substack{(\mathrm{LS},\,\mathrm{UP})\text{-parabola-saddle}\\ \text{switching bifurcation}}} \rightleftharpoons \left\{\begin{array}{c}\underbrace{(a^{+}_{j_1j_2(s_2)_1}, a^{+}_{j_2j_11})}_{(\mathrm{DP}_-,\mathrm{DP}_+)\text{-CW center}}\\ \underbrace{(a^{-}_{j_1j_2(s_2)_2}, a^{+}_{j_2j_11})}_{(\mathrm{UP}_-,\mathrm{UP}_-)\text{-negative saddle}}\end{array}\right\}.$$

An (LS,UP) -parabola-saddle of $(x^*_{j_2},x^*_{j_1})=(a_{j_1j_2s_2},a_{j_1j_21})$ is an (LS, UP)-switching bifurcation for a (DP_-,DP_+)-clockwise center and a (UP_-,UP_-)-negative saddle.

(vii) Under

$$A_{j_2j_12}\neq 0 \text{ and } \Delta_{j_1j_2}=0, a_{j_1j_2s_1}=a_{j_1j_2s_2}\equiv a_{j_1j_21},$$

the two-dimensional dynamical system is

$$\begin{aligned}\dot{x}_{j_1}&=a_{j_1j_20}(x_{j_2}-a_{j_1j_21})^3,\\ \dot{x}_{j_2}&=a_{j_2j_10}(x_{j_1}-a_{j_2j_11}).\end{aligned}$$

From cases (iii) and (iv), the equilibrium of $(x^*_{j_2},x^*_{j_1})=(a_{j_1j_21},a_{j_2j_11})$ has the following switching bifurcations as in Eqs. (2.54)–(2.57):

- For $a_{j_1j_20} > 0$ and $a_{j_2j_10} > 0$,

$$\underbrace{(a_{j_1j_21}, a_{j_2j_11})}_{(\text{UP}_+,\text{UP}_+)\text{-positive saddle}} \rightleftharpoons \underbrace{(a_{j_1j_21}, a_{j_2j_11})}_{\substack{(3^{\text{rd}}\text{UP}_+,\text{UP}_+)\text{-positive saddle}\\ \text{appearing bifurcation}}} \rightleftharpoons \left\{\begin{array}{c} \underbrace{(a_{j_1j_21}, a_{j_2j_11})}_{(\text{UP}_+,\text{UP}_+)\text{-positive saddle}} \\ \underbrace{(a_{j_1j_22}, a_{j_2j_11})}_{(\text{DP}_+,\text{DP}_-)\text{-CCW center}} \\ \underbrace{(a_{j_1j_23}, a_{j_2j_11})}_{(\text{UP}_+,\text{UP}_+)\text{-positive saddle}} \end{array}\right\}.$$

A (3^{rd}UP$_+$,UP$_+$)-positive saddle of $(x_{j_2}^*, x_{j_1}^*) = (a_{j_1j_21}, a_{j_2j_11})$ is a (3^{rd}UP$_+$, UP$_+$)-appearing bifurcation from a (UP$_+$,UP$_+$)-positive-saddle to a (UP$_+$, UP$_+$)-positive-saddle, a (DP$_+$,DP$_-$)-counter-clockwise-center, and a (UP$_+$, UP$_+$)-positive-saddle.

- For $a_{j_1j_20} < 0$ and $a_{j_2j_10} > 0$

$$\underbrace{(a_{j_1j_21}, a_{j_2j_11})}_{(\text{DP}_+,\text{DP}_-)\text{-CCW center}} \rightleftharpoons \underbrace{(a_{j_1j_21}, a_{j_2j_11})}_{\substack{(3^{\text{rd}}\text{DP}_+,\text{DP}_-)\text{-CCW center}\\ \text{appearing bifurcation}}} \rightleftharpoons \left\{\begin{array}{c} \underbrace{(a_{j_1j_21}, a_{j_2j_11})}_{(\text{DP}_+,\text{DP}_-)\text{-CCW center}} \\ \underbrace{(a_{j_1j_22}, a_{j_2j_11})}_{(\text{UP}_+,\text{UP}_+)\text{-positive saddle}} \\ \underbrace{(a_{j_1j_23}, a_{j_2j_11})}_{(\text{DP}_+,\text{DP}_-)\text{-CCW center}} \end{array}\right\}.$$

A (3^{rd}DP$_+$,UP$_-$)-third-order counter-clockwise center of $(x_{j_2}^*, x_{j_1}^*) = (a_{j_1j_21}, a_{j_2j_11})$ is a (3^{rd}DP$_+$,UP$_-$)-appearing bifurcation from a (DP$_+$, UP$_-$)-counter-clockwise center to a (DP$_+$,UP$_-$)-counter-clockwise center, a (UP$_+$,UP$_+$)-positive saddle, and a (DP$_+$,UP$_-$)-counter-clockwise center.

- For $a_{j_1j_20} > 0$ and $a_{j_2j_10} < 0$,

$$\underbrace{(a_{j_1j_21}, a_{j_2j_11})}_{(\text{DP}_-,\text{DP}_+)\text{-CW center}} \rightleftharpoons \underbrace{(a_{j_1j_21}, a_{j_2j_11})}_{\substack{(3^{\text{rd}}\text{DP}_-,\text{DP}_+)\text{-CW center}\\ \text{appearing bifurcation}}} \rightleftharpoons \left\{\begin{array}{c} \underbrace{(a_{j_1j_21}, a_{j_2j_11})}_{(\text{DP}_-,\text{DP}_+)\text{-CW center}} \\ \underbrace{(a_{j_1j_22}, a_{j_2j_11})}_{(\text{UP}_-,\text{UP}_-)\text{-negative saddle}} \\ \underbrace{(a_{j_1j_23}, a_{j_2j_11})}_{(\text{DP}_-,\text{DP}_+)\text{-CW center}} \end{array}\right\}.$$

A (3^{rd}DP$_-$,UP$_+$)-third-order clockwise center of $(x_{j_2}^*, x_{j_1}^*) = (a_{j_1j_21}, a_{j_2j_11})$ is a (3^{rd}DP$_-$,UP$_+$)-appearing bifurcation from a (DP$_-$,UP$_+$)-clockwise center to a (DP$_-$,UP$_+$)-clockwise center, a (UP$_-$,UP$_-$)-negative saddle, and a (DP$_-$, UP$_+$)-clockwise center.

- For $a_{j_1j_20} < 0$ and $a_{j_2j_10} < 0$,

$$\underbrace{(a_{j_1j_21},a_{j_2j_11})}_{(\mathrm{UP}_-,\mathrm{UP}_-)\text{-negative saddle}} \rightleftharpoons \underbrace{(a_{j_1j_21},a_{j_2j_11})}_{\substack{(3^{\mathrm{rd}}\mathrm{UP}_-,\mathrm{UP}_-)\text{-negative saddle} \\ \text{appearing bifurcation}}} \rightleftharpoons \left\{ \begin{array}{c} \underbrace{(a_{j_1j_21},a_{j_2j_11})}_{(\mathrm{UP}_-,\mathrm{UP}_-)\text{-negative saddle}} \\ \underbrace{(a_{j_1j_22},a_{j_2j_11})}_{(\mathrm{DP}_-,\mathrm{DP}_+)\text{-CW center}} \\ \underbrace{(a_{j_1j_23},a_{j_2j_11})}_{(\mathrm{UP}_-,\mathrm{UP}_-)\text{-negative saddle}} \end{array} \right\}.$$

A $(3^{\mathrm{rd}}\mathrm{UP}_-,\mathrm{UP}_-)$-negative saddle of $(x_{j_2}^*,x_{j_1}^*)=(a_{j_1j_21},a_{j_2j_11})$ is a $(3^{\mathrm{rd}}\mathrm{UP}_-$, $\mathrm{UP}_-)$-appearing bifurcation from a $(\mathrm{UP}_-,\mathrm{UP}_-)$-negative-saddle to a $(\mathrm{UP}_-$, $\mathrm{UP}_-)$-negative-saddle, a $(\mathrm{DP}_-,\mathrm{DP}_+)$-clockwise-center, and a $(\mathrm{UP}_-,\mathrm{UP}_-)$-negative saddle.

In the end, from the above local analysis and the first integral of the crossing-linear and cubic systems, this theorem is proved. ■

2.3 Bifurcations and Limit Cycles

In this section, the appearing and switching bifurcations for the saddle and center will be presented and the limit cycles and homoclinic orbits will be discussed.

2.3.1 *Third-Order Saddles and Centers*

Consider a two-dimensional system as

$$\begin{aligned} \dot{x}_1 &= a_{120}(x_2 - a_{121})^3, \\ \dot{x}_2 &= a_{210}(x_1 - a_{211}). \end{aligned} \tag{2.58}$$

The first integral manifold is determined by

$$\frac{1}{4}a_{120}\left[(x_2 - a_{121})^4 - (x_{20} - a_{121})^4\right] = \frac{1}{2}a_{210}\left[(x_1 - a_{211})^2 - (x_{10} - a_{211})^2\right]. \tag{2.59}$$

For such a system, its phase portraits near third-order saddle and center equilibriums of $(x_2^*,x_1^*)=(a_{121},a_{211})$ for a two-dimensional system are presented in Fig. 2.1. In Fig. 2.1a, a phase portrait near a $(3^{\mathrm{rd}}\mathrm{UP}_+,\mathrm{UP}_+)$-third-order positive saddle of $(x_2^*,x_1^*)=(a_{121},a_{211})$ for $a_{120} > 0$ and $a_{210} > 0$ is presented. At $x_1^*=a_{211}$, the equilibrium in the x_2 -direction is a positive and negative positive up-parabola on the right and left sides. At $x_1^* = a_{121}$, the equilibrium in the x_1-direction is a third-order positive and negative positive up-parabola on the upper and lower sides. Thus, $(x_2^*, x_1^*) = (a_{121}, a_{211})$ is a (3^{rd}UP$_+$,UP$_+$)-third-order positive saddle. The phase portraits near a (3^{rd}DP$_+$,DP$_-$)-third-order counter-clockwise center ($a_{120} < 0$, $a_{210} > 0$), a (3^{rd}DP$_-$,DP$_+$)-third-order clockwise center ($a_{120} > 0$, $a_{210} < 0$), and a (3^{rd}UP$_-$,UP$_-$)-third-order negative saddle ($a_{120} < 0$, $a_{210} < 0$) are sketched in Fig. 2.1b–d, respectively.

$$
(a_{121}, a_{211}) = \begin{cases} \underbrace{(3^{\text{rd}}\text{UP}_+, \text{UP}_+)}_{\text{third-order positive saddle}} & \text{for } a_{120} > 0, a_{210} > 0; \\ \underbrace{(3^{\text{rd}}\text{DP}_+, \text{DP}_-)}_{\text{third-order CCW center}} & \text{for } a_{120} < 0, a_{210} > 0; \\ \underbrace{(3^{\text{rd}}\text{DP}_-, \text{DP}_+)}_{\text{third-order CW center}} & \text{for } a_{120} > 0, a_{210} < 0; \\ \underbrace{(3^{\text{rd}}\text{UP}_-, \text{UP}_-)}_{\text{third-order negative saddle}} & \text{for } a_{120} < 0, a_{210} < 0. \end{cases} \tag{2.60}
$$

Consider a two-dimensional system as

$$
\begin{aligned} \dot{x}_1 &= a_{120}(x_2 - a_{121}), \\ \dot{x}_2 &= a_{210}(x_1 - a_{211})^3 \end{aligned} \tag{2.61}
$$

and the first integral manifold is

$$
\begin{aligned} &\frac{1}{2} a_{120} \left[(x_2 - a_{121})^2 - (x_{20} - a_{121})^2 \right] \\ &= \frac{1}{4} a_{210} \left[(x_1 - a_{211})^4 - (x_{10} - a_{211})^4 \right]. \end{aligned} \tag{2.62}
$$

The singular equilibrium possesses four cases summarized as follows:

$$
(a_{121}, a_{211}) = \begin{cases} \underbrace{(\text{UP}_+, 3^{\text{rd}}\text{UP}_+)}_{\text{third-order psoitive saddle}} & \text{for } a_{120} > 0, a_{210} > 0; \\ \underbrace{(\text{DP}_+, 3^{\text{rd}}\text{DP}_-)}_{\text{third-order CCW center}} & \text{for } a_{120} < 0, a_{210} > 0, \\ \underbrace{(\text{DP}_-, 3^{\text{rd}}\text{DP}_+)}_{\text{third-order CW center}} & \text{for } a_{120} > 0, a_{210} < 0; \\ \underbrace{(\text{UP}_-, 3^{\text{rd}}\text{UP}_-)}_{\text{third-order negative saddle}} & \text{for } a_{120} < 0, a_{210} < 0. \end{cases} \tag{2.63}
$$

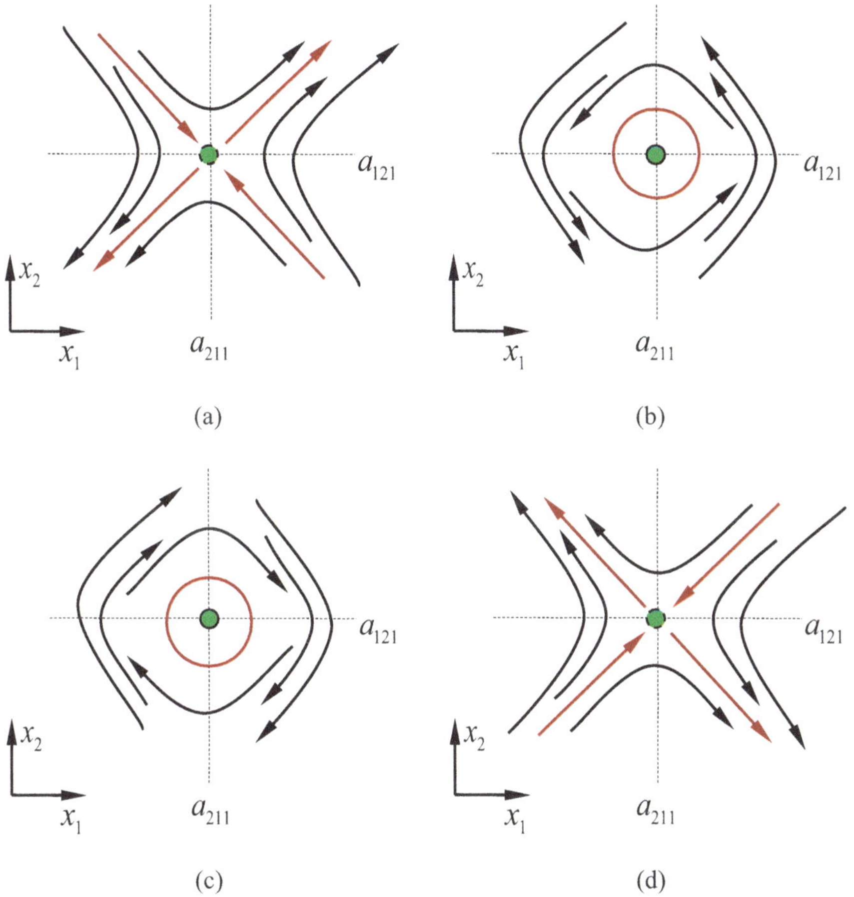

Fig. 2.1 Phase portraits for two-dimensional systems in the x_1-direction with $x_1^* = a_{211}$ and in the x_2-direction with $x_2^* = a_{121}$: the equilibrium of $(x_2^*, x_1^*) = (a_{121}, a_{211})$: (**a**) ($3^{\text{rd}}\text{UP}_+$,$\text{UP}_+$)-third-order positive-saddle ($a_{120} > 0$, $a_{210} > 0$), (**b**) (3^{rd}DP_+,DP_-)-third-order counter-clockwise center ($a_{120} < 0$, $a_{210} > 0$), (**c**) (3^{rd}DP_-,DP_+)-third-order clockwise center ($a_{120} > 0$, $a_{210} < 0$), (**d**) (3^{rd}UP_-,UP_-)-third-order negative-saddle ($a_{120} < 0$, $a_{210} < 0$)

For such a system, its phase portraits near third-order saddle and center equilibriums of $(x_2^*, x_1^*) = (a_{121}, a_{211})$ for a two-dimensional system are presented in Fig. 2.2. In Fig. 2.2a, a phase portrait near a (UP_+,3rd UP_+)-third-order positive saddle of $(x_2^*, x_1^*) = (a_{121}, a_{211})$ for $a_{120} > 0$ and $a_{210} > 0$ is presented. At $x_1^* = a_{211}$, the equilibrium in thex_2-direction is a third-order positive and negative positive up-parabola on the right and left sides, respectively. At $x_1^* = a_{121}$, the equilibrium in the x_1-direction is a positive and negative positive up-parabola on the upper and lower sides, respectively. Thus, $(x_2^*, x_1^*) = (a_{121}, a_{211})$ is a (UP_+,3^{rd}UP_+)-third-order positive saddle. The phase portraits near a (DP_+,3^{rd}DP_-)-third-order counter-clockwise center ($a_{120} < 0$, $a_{210} > 0$), a (DP_-,3^{rd}DP_+)-third-order clockwise center ($a_{120} > 0$, $a_{210} < 0$), and a (UP_-,3^{rd}UP_-)-third-order negative saddle ($a_{120} < 0$, $a_{210} < 0$) are sketched in Fig. 2.2b–d, respectively.

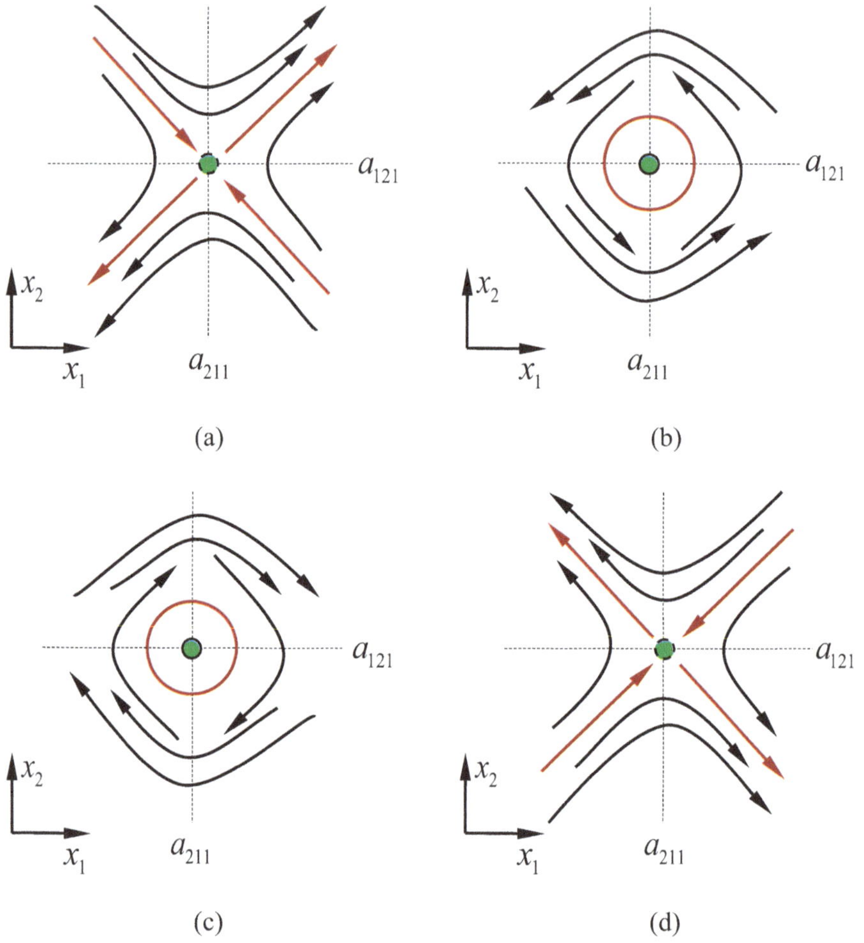

Fig. 2.2 Phase portraits for two-dimensional systems in the x_1-direction with $x_1^* = a_{211}$ and in the x_2-direction with$x_2^* = a_{121}$: the equilibrium of $(x_2^*, x_1^*) = (a_{121}, a_{211})$: (**a**) ($UP_+$,$3^{rd}UP_+$)-third-order positive saddle ($a_{120} > 0$, $a_{210} > 0$), (**b**) (DP_+,$3^{rd}DP_-$)-third-order counter-clockwise center ($a_{120} < 0$, $a_{210} > 0$), (**c**) (DP_-,$3^{rd}DP_+$)-third-order clockwise center ($a_{120} > 0$, $a_{210} < 0$), (**d**) (UP_-,$3^{rd}UP_-$)-third-order negative-saddle ($a_{120} < 0$, $a_{210} < 0$)

2.3.2 Limit Cycles and Homoclinic Orbits

Consider a two-dimensional system as

$$\begin{aligned}\dot{x}_1 &= a_{120}(x_2 - a_{121}),\\ \dot{x}_2 &= a_{210}(x_1 - a_{211})(x_1 - a_{212})(x_1 - a_{213})\end{aligned} \tag{2.64}$$

and the first integral manifold is

$$
\begin{aligned}
&\frac{1}{2}a_{120}\left[(x_2-a_{121})^2-(x_{20}-a_{121})^2\right]\\
&=a_{210}\Big\{\frac{1}{4}\left[(x_1-a_{21s_1})^4-(x_{10}-a_{21s_1})^4\right]\\
&+\frac{1}{3}\sum\nolimits_{s_2=1,s_2\neq s_1}^{3}(a_{21s_1}-a_{21s_2})\left[(x_1-a_{21s_1})^3-(x_{10}-a_{21s_1})^3\right]\\
&+\frac{1}{2}\prod\nolimits_{s_2=1,s_2\neq s_1}^{3}(a_{21s_1}-a_{21s_2})\left[(x_1-a_{21s_1})^2-(x_{10}-a_{21s_1})^2\right]\Big\}.
\end{aligned}
\tag{2.65}
$$

The phase portraits near center and saddle equilibriums of $(x_2^*,x_1^*)=(a_{121},a_{211})$, (a_{121},a_{212}), (a_{121},a_{213}) for a two-dimensional system are presented in Fig. 2.3. In Fig. 2.3a, a phase portrait near equilibriums of $(x_2^*,x_1^*)=(a_{121},a_{211})$, (a_{121},a_{212}) (a_{121},a_{213}) for $a_{120}>0$ and $a_{210}>0$ are presented. The equilibrium of $(x_2^*,x_1^*)=(a_{121},a_{211})$ is a (UP_+,UP_+)-positive saddle, the equilibrium of $(x_2^*,x_1^*)=(a_{121},a_{212})$ is a (DP_-,DP_+)-clockwise center, and the equilibrium of $(x_2^*,x_1^*)=(a_{121},a_{213})$ is a (UP_+,UP_+)-positive saddle. Similarly, (i) the phase portraits for the (DP_+,DP_-)-counter-clockwise center, (UP_-,UP_-)-negative saddle, and (DP_+,DP_-)-counter-clockwise center ($a_{120}<0$ and $a_{210}>0$); (ii) the (DP_-,DP_+)-clockwise center, (UP_+,UP_+)-positive saddle, and (DP_-,DP_+)-clockwise center ($a_{120}>0$ and $a_{210}<0$); and (iii) the (UP_-,UP_-)-negative saddle, (DP_+,DP_-)-counter-clockwise center, and (UP_-,UP_-)-negative saddle and ($a_{120}<0$ and $a_{210}<0$) are presented in Fig. 2.3b–d, respectively. Thus, for such two-dimensional crossing-cubic nonlinear systems, there are two local saddles and one local center, or one local saddle and two local centers, which are separated by the separatrix (or homoclinic orbit). The saddles and centers are in the four cases as follows:

(i) For $a_{120}>0$ and $a_{210}>0$,

$$
\left\{\begin{array}{c}(a_{121},a_{211})\\(a_{121},a_{212})\\(a_{121},a_{213})\end{array}\right\}=\left\{\begin{array}{c}\underbrace{(\mathrm{UP}_+,\mathrm{UP}_+)}_{\text{positive saddle}}\\\underbrace{(\mathrm{DP}_-,\mathrm{DP}_+)}_{\text{CW center}}\\\underbrace{(\mathrm{UP}_+,\mathrm{UP}_+)}_{\text{positive saddle}}\end{array}\right\}. \tag{2.66}
$$

(ii) For $a_{120}<0$ and $a_{210}>0$,

$$
\left\{\begin{array}{c}(a_{121},a_{211})\\(a_{121},a_{212})\\(a_{121},a_{213})\end{array}\right\}=\left\{\begin{array}{c}\underbrace{(\mathrm{DP}_+,\mathrm{DP}_-)}_{\text{CCW center}}\\\underbrace{(\mathrm{UP}_-,\mathrm{UP}_-)}_{\text{negative saddle}}\\\underbrace{(\mathrm{DP}_+,\mathrm{DP}_-)}_{\text{CCW center}}\end{array}\right\}. \tag{2.67}
$$

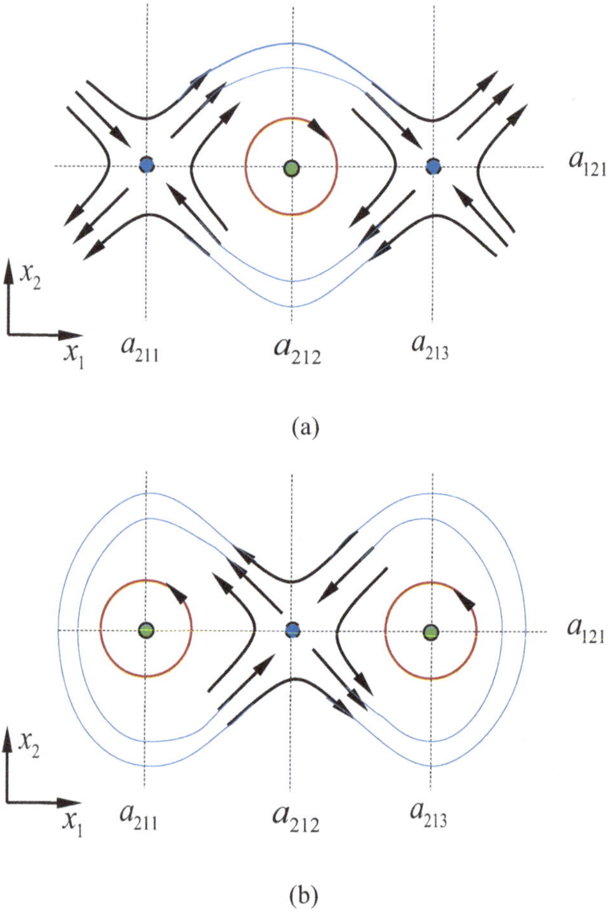

Fig. 2.3 Phase portraits for two-dimensional systems in the x_1-direction with $x_1^* = a_{211}, a_{212}, a_{213}$ and in the x_2-direction with $x_2^* = a_{121}$. The equilibriums of $(x_1^*, x_2^*) = (a_{211}, a_{121})$, (a_{212}, a_{121}): (**a**) $(\mathrm{UP}_+,\mathrm{UP}_+)$-positive saddle, $(\mathrm{DP}_-,\mathrm{DP}_+)$-CW center, and $(\mathrm{UP}_+,\mathrm{UP}_+)$-positive saddle equilibriums ($a_{120} > 0$, $a_{210} > 0$); (**b**) $(\mathrm{DP}_+,\mathrm{DP}_-)$-CCW center, $(\mathrm{UP}_-,\mathrm{UP}_-)$-negative saddle, and $(\mathrm{DP}_+,\mathrm{DP}_-)$-CCW center equilibriums ($a_{120} < 0$, $a_{210} > 0$); (**c**) $(\mathrm{DP}_-,\mathrm{DP}_+)$-CW center, $(\mathrm{DP}_+,\mathrm{UP}_+)$-positive saddle, and $(\mathrm{DP}_-,\mathrm{DP}_+)$-CW center ($a_{120} > 0$, $a_{210} < 0$); (**d**) $(\mathrm{UP}_-,\mathrm{UP}_-)$-negative saddle, $(\mathrm{DP}_+,\mathrm{DP}_-)$-CCW center, and $(\mathrm{UP}_-,\mathrm{UP}_-)$-negative saddle equilibriums ($a_{120} < 0$, $a_{210} < 0$)

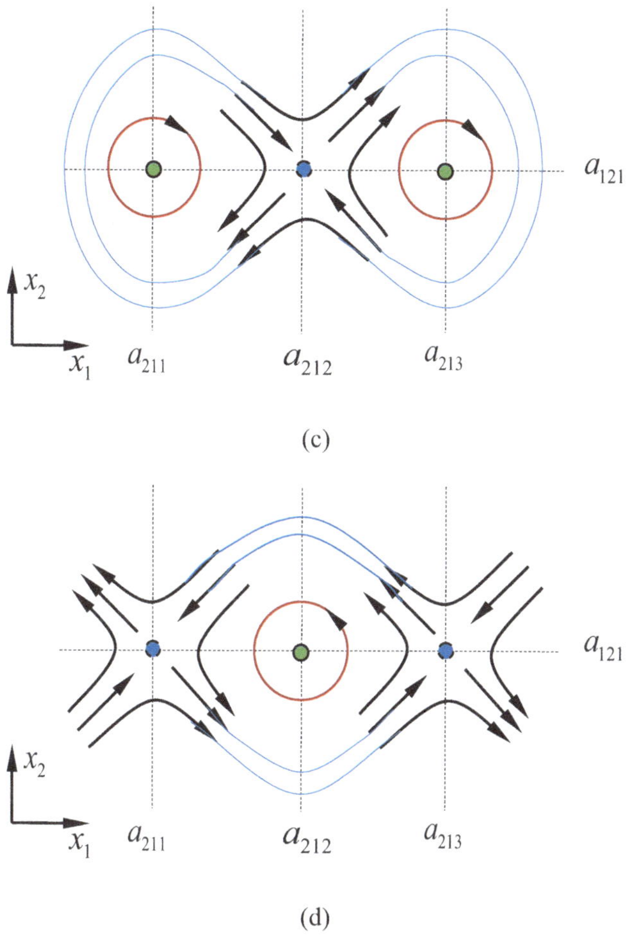

Fig. 2.3 (continued)

(iii) For $a_{120} > 0$ and $a_{210} < 0$,

$$\left\{\begin{matrix}(a_{121}, a_{211})\\(a_{121}, a_{212})\\(a_{121}, a_{213})\end{matrix}\right\} = \left\{\begin{matrix}\underbrace{(\mathrm{DP}_-, \mathrm{DP}_+)}_{\text{CW center}}\\\underbrace{(\mathrm{UP}_+, \mathrm{UP}_+)}_{\text{positive saddle}}\\\underbrace{(\mathrm{DP}_-, \mathrm{DP}_+)}_{\text{CW center}}\end{matrix}\right\}. \tag{2.68}$$

(iv) For $a_{120} < 0$ and $a_{210} < 0$,

$$\left\{\begin{array}{c}(a_{121}, a_{211}) \\ (a_{121}, a_{212}) \\ (a_{121}, a_{213})\end{array}\right\} = \left\{\begin{array}{c}\underbrace{(\mathrm{UP}_-, \mathrm{UP}_-)}_{\text{negative saddle}} \\ \underbrace{(\mathrm{DP}_+, \mathrm{DP}_-)}_{\text{CCW center}} \\ \underbrace{(\mathrm{UP}_-, \mathrm{UP}_-)}_{\text{negative saddle}}\end{array}\right\}. \tag{2.69}$$

Consider a two-dimensional system as

$$\begin{aligned}\dot{x}_1 &= a_{120}(x_2 - a_{121})(x_2 - a_{122})(x_2 - a_{123}) \\ \dot{x}_2 &= a_{210}(x_1 - a_{211}),\end{aligned} \tag{2.70}$$

and the first integral manifold is

$$\begin{aligned}&\frac{1}{2}a_{210}\left[(x_1 - a_{211})^2 - (x_{10} - a_{211})^2\right] \\ &= a_{120}\Big\{\frac{1}{4}\left[(x_2 - a_{12s_1})^4 - (x_{20} - a_{12s_1})^4\right] \\ &+ \frac{1}{3}\sum\nolimits_{s_2 = 1, s_2 \neq s_1}^{3} (a_{12s_1} - a_{12s_2})\left[(x_2 - a_{12s_1})^3 - (x_{20} - a_{12s_1})^3\right] \\ &+ \frac{1}{2}\prod\nolimits_{s_2 = 1, s_2 \neq s_1}^{3} (a_{12s_1} - a_{12s_2})\left[(x_2 - a_{12s_1})^2 - (x_{20} - a_{12s_1})^2\right]\Big\}.\end{aligned} \tag{2.71}$$

Thus, the phase portraits near sink-saddle and source-saddle equilibriums of $(x_2^*, x_1^*) = (a_{121}, a_{211}), (a_{122}, a_{211}), (a_{123}, a_{211})$ for a two-dimensional system are presented in Fig. 2.4. In Fig. 2.4a, a phase portrait near equilibriums of $(x_2^*, x_1^*) = (a_{121}, a_{211})$, (a_{122}, a_{211}), (a_{123}, a_{211}) for $a_{120} > 0$ and $a_{210} > 0$ are presented. The equilibrium of $(x_2^*, x_1^*) = (a_{121}, a_{211})$ is a (UP_+,UP_+)-positive saddle, the equilibrium of $(x_2^*, x_1^*) = (a_{122}, a_{211})$ is a (DP_+,DP_-)-counter clockwise center, and the equilibrium of $(x_2^*, x_1^*) = (a_{122}, a_{211})$ is a (UP_+,UP_+)-positive saddle. Similarly, presented in Fig. 2.4b–d are the phase portraits for (i) the (DP_+,DP_-)-counter-clockwise center, (UP_+,UP_+)-positive saddle, and (DP_+,DP_-)-counter-clockwise center ($a_{120} < 0$ and $a_{210} > 0$); (ii) the (DP_-,DP_+)-clockwise center, (UP_-,UP_-)-negative saddle, and (DP_-,DP_+)-clockwise center ($a_{120} > 0$ and $a_{210} < 0$); and (iii) the (UP_-,UP_-)-negative saddle, the (DP_-,DP_+)-clockwise center, and the (UP_-,UP_-)-negative saddle and ($a_{120} < 0$ and $a_{210} < 0$). The saddles and centers are in the four cases as follows:

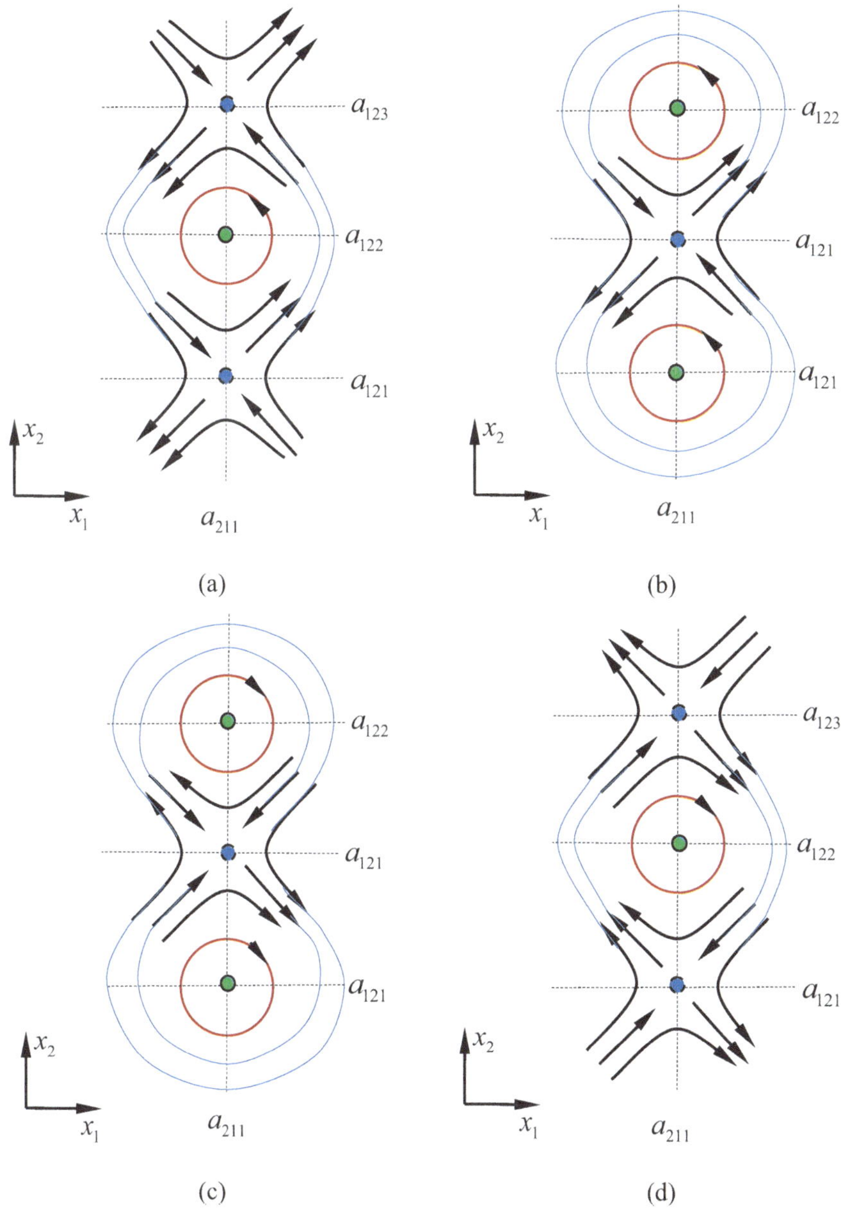

Fig. 2.4 Phase portraits for two-dimensional systems in the x_1-direction with $x_1^* = a_{211}, a_{212}, a_{213}$ and in the x_2-direction with $x_2^* = a_{121}$. The equilibriums of $(x_1^*, x_2^*) = (a_{211}, a_{121})$, (a_{212}, a_{121}): (**a**) $(a_{120} > 0, a_{210} > 0)$, (**b**) $(a_{120} < 0, a_{210} > 0)$, (**c**) $(a_{120} > 0, a_{210} < 0)$, (**d**) $(a_{120} < 0, a_{210} < 0)$

(i) For $a_{120} > 0$ and $a_{210} > 0$,

$$\begin{Bmatrix} (a_{121}, a_{211}) \\ (a_{122}, a_{211}) \\ (a_{123}, a_{211}) \end{Bmatrix} = \begin{Bmatrix} \underbrace{(\mathrm{UP}_+, \mathrm{UP}_+)}_{\text{positive saddle}} \\ \underbrace{(\mathrm{DP}_+, \mathrm{DP}_-)}_{\text{CCW center}} \\ \underbrace{(\mathrm{UP}_+, \mathrm{UP}_+)}_{\text{positive saddle}} \end{Bmatrix}. \tag{2.72}$$

(ii) For $a_{120} < 0$ and $a_{210} > 0$,

$$\begin{Bmatrix} (a_{121}, a_{211}) \\ (a_{122}, a_{211}) \\ (a_{123}, a_{211}) \end{Bmatrix} = \begin{Bmatrix} \underbrace{(\mathrm{DP}_+, \mathrm{DP}_-)}_{\text{CCW center}} \\ \underbrace{(\mathrm{UP}_+, \mathrm{UP}_+)}_{\text{positive saddle}} \\ \underbrace{(\mathrm{DP}_+, \mathrm{DP}_-)}_{\text{CCW center}} \end{Bmatrix}. \tag{2.73}$$

(iii) For $a_{120} > 0$ and $a_{210} < 0$,

$$\begin{Bmatrix} (a_{121}, a_{211}) \\ (a_{122}, a_{211}) \\ (a_{123}, a_{211}) \end{Bmatrix} = \begin{Bmatrix} \underbrace{(\mathrm{DP}_-, \mathrm{DP}_+)}_{\text{CW center}} \\ \underbrace{(\mathrm{UP}_-, \mathrm{UP}_-)}_{\text{negative saddle}} \\ \underbrace{(\mathrm{DP}_-, \mathrm{DP}_+)}_{\text{CW center}} \end{Bmatrix}. \tag{2.74}$$

(iv) For $a_{120} < 0$ and $a_{210} < 0$,

$$\begin{Bmatrix} (a_{121}, a_{211}) \\ (a_{122}, a_{211}) \\ (a_{123}, a_{211}) \end{Bmatrix} = \begin{Bmatrix} \underbrace{(\mathrm{UP}_-, \mathrm{UP}_-)}_{\text{negative saddle}} \\ \underbrace{(\mathrm{DP}_-, \mathrm{DP}_+)}_{\text{CW center}} \\ \underbrace{(\mathrm{UP}_-, \mathrm{UP}_-)}_{\text{negative saddle}} \end{Bmatrix}. \tag{2.75}$$

2.4 Switching and Appearing Bifurcations

Consider a dynamical system as

$$\begin{aligned}\dot{x}_{j_1} &= a_{j_1 j_2 0}(x_{j_2} - a_{j_1 j_2 1})^3,\\ \dot{x}_{j_2} &= a_{j_2 j_1 0}(x_{j_1} - a_{j_2 j_1 1}).\end{aligned} \tag{2.76}$$

The two appearing bifurcations from a saddle to saddle, center, and saddle in the x_{j_1}-direction are discussed, and the two appearing bifurcations from a center to center, saddle, and center are presented. The corresponding properties are listed as flows.

(i) For $a_{j_1 j_2 0} > 0$ and $a_{j_2 j_1 0} > 0$, the ($3^{\mathrm{rd}}\mathrm{UP}_+$,$\mathrm{UP}_+$)-third-order positive saddle appearing bifurcation is

$$\underbrace{(\mathrm{UP}_+,\mathrm{UP}_+)}_{\text{positive saddle}} \rightleftharpoons \underbrace{(3^{\mathrm{rd}}\mathrm{UP}_+,\mathrm{UP}_+)}_{\text{appearing bifurcation}} \rightleftharpoons \left\{\begin{array}{c}\underbrace{(\mathrm{UP}_+,\mathrm{UP}_+)}_{\text{positive saddle}}\\ \underbrace{(\mathrm{DP}_+,\mathrm{DP}_-)}_{\text{CCW center}}\\ \underbrace{(\mathrm{UP}_+,\mathrm{UP}_+)}_{\text{positive saddle}}\end{array}\right\}. \tag{2.77}$$

(ii) For $a_{j_1 j_2 0} < 0$ and $a_{j_2 j_1 0} > 0$, the ($3^{\mathrm{rd}}\mathrm{DP}_+$,$\mathrm{DP}_-$)-third-order counter-clockwise appearing bifurcation is

$$\underbrace{(\mathrm{DP}_+,\mathrm{DP}_-)}_{\text{CCW center}} \rightleftharpoons \underbrace{(3^{\mathrm{rd}}\mathrm{DP}_+,\mathrm{DP}_-)}_{\text{appearing bifurcation}} \rightleftharpoons \left\{\begin{array}{c}\underbrace{(\mathrm{DP}_+,\mathrm{DP}_-)}_{\text{CCW center}}\\ \underbrace{(\mathrm{UP}_+,\mathrm{UP}_+)}_{\text{positive saddle}}\\ \underbrace{(\mathrm{DP}_+,\mathrm{DP}_-)}_{\text{CCW center}}\end{array}\right\}. \tag{2.78}$$

(iii) For $a_{j_1 j_2 0} > 0$ and $a_{j_2 j_1 0} < 0$, the ($3^{\mathrm{rd}}\mathrm{DP}_-$,$\mathrm{DP}_+$)-third-order clockwise center appearing bifurcation is

$$\underbrace{(\mathrm{DP}_-,\mathrm{DP}_+)}_{\text{CW center}} \rightleftharpoons \underbrace{(3^{\mathrm{rd}}\mathrm{DP}_-,\mathrm{DP}_+)}_{\text{appearing bifurcation}} \rightleftharpoons \left\{\begin{array}{c}\underbrace{(\mathrm{DP}_-,\mathrm{DP}_+)}_{\text{CW center}}\\ \underbrace{(\mathrm{UP}_-,\mathrm{UP}_-)}_{\text{negairve saddle}}\\ \underbrace{(\mathrm{DP}_-,\mathrm{DP}_+)}_{\text{CW center}}\end{array}\right\}. \tag{2.79}$$

(iv) For $a_{j_2 j_1 0} < 0$ and $a_{j_1 j_2 0} < 0$, the (3^{rd}UP$_-$,UP$_-$)-third-order negative saddle appearing bifurcation is

$$\underbrace{(\text{UP}_-,\text{UP}_-)}_{\text{negative saddle}} \rightleftharpoons \underbrace{(3^{\text{rd}}\text{UP}_-,\text{UP}_-)}_{\text{appearing bifurcation}} \rightleftharpoons \left\{ \begin{array}{c} \underbrace{(\text{UP}_-,\text{UP}_-)}_{\text{negative saddle}} \\ \underbrace{(\text{DP}_-,\text{DP}_+)}_{\text{CW center}} \\ \underbrace{(\text{UP}_-,\text{UP}_-)}_{\text{negative saddle}} \end{array} \right\} \tag{2.80}$$

The appearing bifurcations of the third-order saddles and centers for $x_2^* = a_{j_1 j_2 1}$, $a_{j_1 j_2 2}, a_{j_1 j_2 3}$ and $x_{j_1}^* = a_{j_2 j_1 1}$ are sketched in Fig. 2.5. A (3^{rd}UP$_+$,UP$_+$)-third-order positive saddle switching bifurcation ($a_{j_1 j_2 0} > 0, a_{j_2 j_1 0} > 0$), a ($3^{\text{rd}}DP_+$, DP$_-$)-third-order counter-clockwise switching bifurcation ($a_{j_1 j_2 0} < 0, a_{j_2 j_1 0} > 0$), a (3^{rd}DP$_-$, DP$_+$)-third-order clockwise center switching bifurcation ($a_{j_1 j_2 0} > 0$, $a_{j_2 j_1 0} < 0$), and a (3^{rd}UP$_-$,UP$_-$)-third-order positive saddle bifurcation ($a_{j_1 j_2 0} < 0$, $a_{j_2 j_1 0} < 0$) are presented in Fig. 2.5a–d, respectively.

Consider a dynamical system as

$$\begin{aligned} \dot{x}_{j_1} &= a_{j_1 j_2 0}(x_{j_2} - a_{j_1 j_2 1}), \\ \dot{x}_{j_2} &= a_{j_2 j_1 0}(x_{j_1} - a_{j_2 j_1 1})^3. \end{aligned} \tag{2.81}$$

The two appearing bifurcations from a saddle to saddle, center, and saddle in the x_{j_2}-direction are discussed, and the two appearing bifurcations from a center to center, saddle, and center are discussed as well, which are presented as follows:

(i) For $a_{j_1 j_2 0} > 0$ and $a_{j_2 j_1 0} > 0$, the (UP$_+$,3^{rd}UP$_+$)-saddle-parabola appearing bifurcation is

$$\underbrace{(\text{UP}_+,\text{UP}_+)}_{\text{positive saddle}} \rightleftharpoons \underbrace{(\text{UP}_+,3^{\text{rd}}\text{UP}_+)}_{\text{appearing bifurcation}} \rightleftharpoons \left\{ \begin{array}{c} \underbrace{(\text{UP}_+,\text{UP}_+)}_{\text{positive saddle}} \\ \underbrace{(\text{DP}_-,\text{DP}_+)}_{\text{CW center}} \\ \underbrace{(\text{UP}_+,\text{UP}_+)}_{\text{positive saddle}} \end{array} \right\}. \tag{2.82}$$

(ii) For $a_{j_1 j_2 0} < 0$ and $a_{j_2 j_1 0} > 0$, the (DP$_+$,3^{rd}DP$_-$)-third-order counter-clockwise appearing bifurcation is

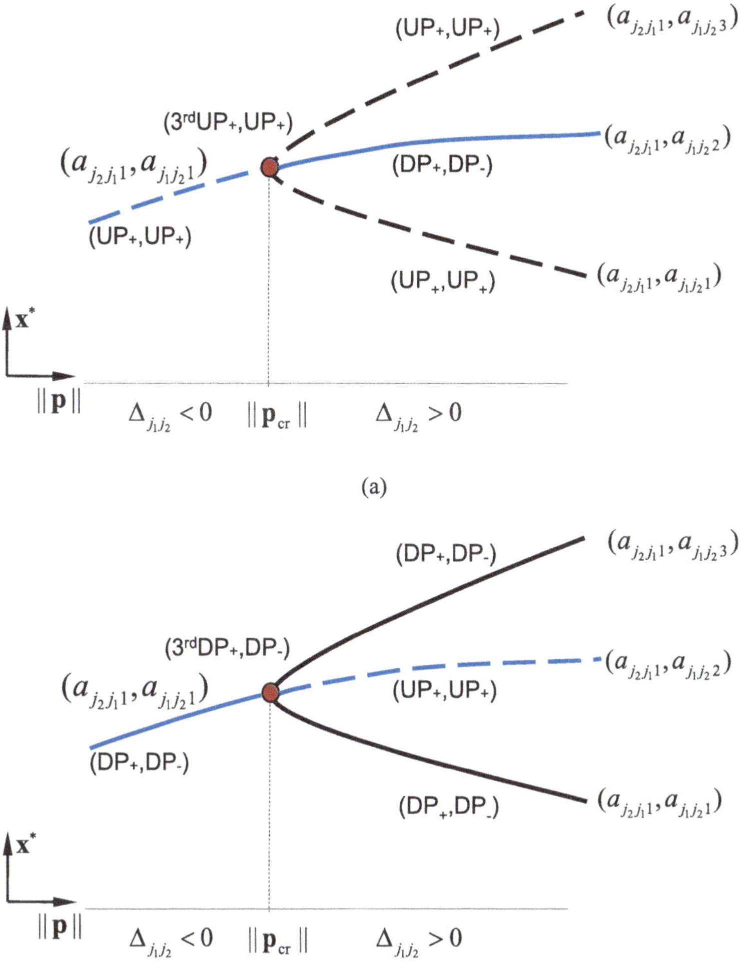

(b)

Fig. 2.5 The appearing bifurcations of parabola-saddles for $x^*_{j_2} = a_{j_1j_2 1}$ and $x^*_{j_1} = a_{j_2j_1 1}, a_{j_2j_1 2}$: (**a**) ($3^{\text{rd}}UP_+$,UP$_+$)-third-order positive saddle ($a_{j_1j_2 0} > 0, a_{j_2j_1 0} > 0$), (**b**) ($3^{\text{rd}}DP_+$,DP$_-$)-third-order counter-clockwise center ($a_{j_1j_2 0} < 0$, $a_{j_2j_1 0} > 0$), (**c**) (3^{rd}DP$_-$,DP$_+$)-third-order clockwise center ($a_{j_1j_2 0} > 0, a_{j_2j_1 0} < 0$), (**d**) ($3^{\text{rd}}UP_-$,UP$_-$)-third-order negative saddle ($a_{j_1j_2 0} < 0, a_{j_2j_1 0} < 0$). UP$_+$: positive up-parabola, DP$_+$: positive down-parabola, UP$_-$: negative up-parabola, DP$_-$: negative down-parabola

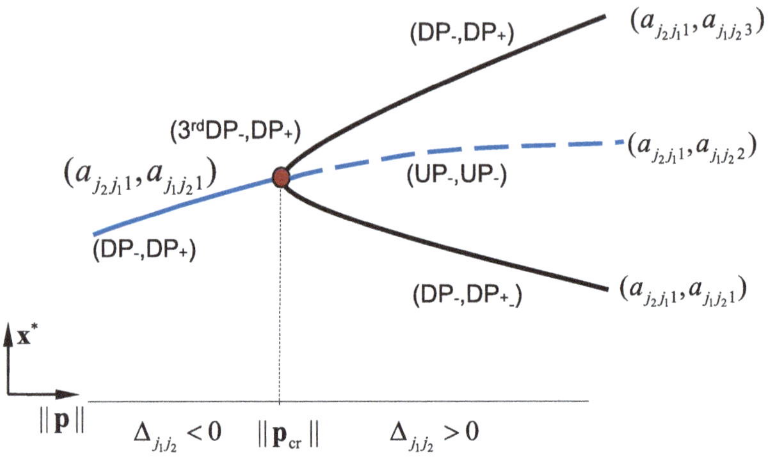

(c)

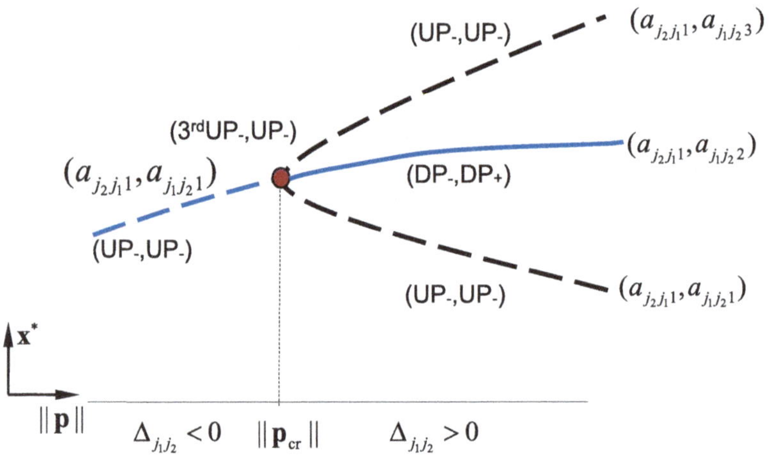

(d)

Fig. 2.5 (continued)

$$\underbrace{(\mathrm{DP}_{+},\mathrm{DP}_{-})}_{\text{CCW center}} \rightleftharpoons \underbrace{(\mathrm{DP}_{+},3^{\mathrm{rd}}\mathrm{DP}_{-})}_{\text{appearing bifurcation}} \rightleftharpoons \left\{ \begin{array}{c} \underbrace{(\mathrm{DP}_{+},\mathrm{DP}_{-})}_{\text{CCW center}} \\ \underbrace{(\mathrm{UP}_{-},\mathrm{UP}_{-})}_{\text{negative saddle}} \\ \underbrace{(\mathrm{DP}_{+},\mathrm{DP}_{-})}_{\text{CCW center}} \end{array} \right\}. \tag{2.83}$$

(iii) For $a_{j_1j_20} > 0$ and $a_{j_2j_10} < 0$, the (DP_{-},$3^{\mathrm{rd}}\mathrm{DP}_{+}$)-third-order clockwise center appearing bifurcation is

$$\underbrace{(\mathrm{DP}_{-},\mathrm{DP}_{+})}_{\text{CW center}} \rightleftharpoons \underbrace{(\mathrm{DP}_{-},3^{\mathrm{rd}}\mathrm{DP}_{+})}_{\text{appearing bifurcation}} \rightleftharpoons \left\{ \begin{array}{c} \underbrace{(\mathrm{DP}_{-},\mathrm{DP}_{+})}_{\text{CW center}} \\ \underbrace{(\mathrm{UP}_{+},\mathrm{UP}_{+})}_{\text{positive saddle}} \\ \underbrace{(\mathrm{DP}_{-},\mathrm{DP}_{+})}_{\text{CW center}} \end{array} \right\}. \tag{2.84}$$

(iv) For $a_{j_2j_10} < 0$ and $a_{j_1j_20} < 0$, the (3rd UP_{-},UP_{-})-third-order negative saddle appearing bifurcation is

$$\underbrace{(\mathrm{UP}_{-},\mathrm{UP}_{-})}_{\text{negative saddle}} \rightleftharpoons \underbrace{(\mathrm{UP}_{-},3^{\mathrm{rd}}\mathrm{UP}_{-})}_{\text{appearing bifurcation}} \rightleftharpoons \left\{ \begin{array}{c} \underbrace{(\mathrm{UP}_{-},\mathrm{UP}_{-})}_{\text{negative saddle}} \\ \underbrace{(\mathrm{DP}_{+},\mathrm{DP}_{-})}_{\text{CW center}} \\ \underbrace{(\mathrm{UP}_{-},\mathrm{UP}_{-})}_{\text{negative saddle}} \end{array} \right\}. \tag{2.85}$$

The above third-order saddles and center appearing bifurcations in the x_{j_2}-direction can be similarly illustrated as in the x_{j_1}-direction.

The two third-order saddle switching bifurcations of saddle, center, and saddle will not be presented herein, and the two third-order center switching bifurcations of center, saddle, center will not be presented. Readers can complete illustrations as an exercise.

Chapter 3
Crossing-Quadratic and Crossing-Cubic Vector Fields

In this chapter, a cubic dynamical system possessing crossing-quadratic and crossing-cubic vector fields is considered, and the corresponding dynamical behaviors are discussed. As in constant and crossing-cubic system, the third-order parabola flows exist for the appearing bifurcation from a parabola flow to a series of three parabola flows, and the inflection flows are the appearing and switching bifurcations of up-parabola and down parabola flows, respectively. In addition to parabola saddles and third-order saddle sand centers, the third-order parabola-saddle and double inflection-saddles exist. The (2,3)-third-order parabola-saddle are the appearing bifurcations of a 2×3 network of saddle and center. The (2,2)-double inflection-saddles are the appearing bifurcations of a 2×2 network of saddle and center. The (2,3)-third-order parabola-saddles are also for the switching bifurcations of a double inflection-saddle with saddle (center). The networks of saddles and centers with homoclinic orbits are presented through the first integral manifolds.

3.1 Crossing-Quadratic and Crossing-Cubic Systems

Consider a dynamical system possessing variable-crossing univariate quadratic and cubic vector fields. The corresponding dynamical behaviors will be presented through the following theorem:

Theorem 3.1 *Consider a two-dimensional, dynamical system with variable-crossing univariate quadratic and cubic vector fields as*

$$\begin{aligned} \dot{x}_{j_1} &= f_{j_1}(\mathbf{x}, \mathbf{p}) = A_{j_1 j_2 3} + A_{j_1 j_2 2} x_{j_2} + A_{j_1 j_2 1} x_{j_2}^2 + A_{j_1 j_2 0} x_{j_2}^3, \\ \dot{x}_{j_2} &= f_{j_2}(\mathbf{x}, \mathbf{p}) = A_{j_2 j_1 3} + A_{j_2 j_1 2} x_{j_1} + A_{j_2 j_1 1} x_{j_1}^2 + A_{j_2 j_1 0} x_{j_1}^3. \end{aligned} \tag{3.1}$$

A. C. J. Luo, *Two-dimensional Crossing-Variable Cubic Nonlinear Systems*,
https://doi.org/10.1007/978-3-031-62810-8_3

For $A_{j_2j_10}=0, A_{j_2j_11}\neq 0$, *the two-dimensional dynamical system possesses variable-crossing univariate quadratic and cubic vector fields as*

$$\begin{aligned}\dot{x}_{j_2} &= a_{j_2j_10}(x_{j_1}^2 + B_{j_2j_1}x_{j_1} + C_{j_2j_1}),\\ \dot{x}_{j_1} &= a_{j_1j_20}(x_{j_2} - b_{j_1j_21})(x_{j_2}^2 + B_{j_1j_2}x_{j_2} + C_{j_1j_2})\end{aligned} \tag{3.2}$$

with

$$\begin{aligned}&\Delta_{j_2j_1} = (B_{j_2j_1})^2 - 4C_{j_2j_1}, \Delta_{j_1j_2} = (B_{j_1j_2})^2 - 4C_{j_1j_2}\\ &\text{for } j_1, j_2 \in \{1,2\}, j_1 \neq j_2.\end{aligned} \tag{3.3}$$

(i) If

$$\Delta_{j_1j_2} < 0 \text{ and } \Delta_{j_2j_1} < 0, \tag{3.4}$$

then

$$\begin{aligned}&\dot{x}_{j_1} = a_{j_1j_20}(x_{j_2} - a_{j_1j_21})\left[(x_{j_2} - a_{j_1j_2})^2 + b_{j_1j_2}\right],\\ &\dot{x}_{j_2} = a_{j_2j_10}\left[(x_{j_1} - a_{j_2j_1})^2 + b_{j_2j_1}\right]\\ &j_1, j_2 \in \{1,2\}, j_2 \neq j_1.\end{aligned} \tag{3.5}$$

The first integral manifold is

$$\begin{aligned}&a_{j_2j_10}\Big\{\frac{1}{3}\left[(x_{j_1} - a_{j_2j_1})^3 - (x_{j_10} - a_{j_2j_1})^3\right]\\ &+b_{j_2j_1}\left[(x_{j_1} - a_{j_2j_1}) - (x_{j_10} - a_{j_2j_1})\right]\Big\}\\ &= a_{j_1j_20}\Big\{\frac{1}{4}\left[(x_{j_2} - a_{j_1j_21})^4 - (x_{j_20} - a_{j_1j_21})^4\right]\\ &+\frac{2}{3}(a_{j_1j_21} - a_{j_1j_2})\left[(x_{j_2} - a_{j_1j_21})^3 - (x_{j_20} - a_{j_1j_21})^3\right]\\ &+\frac{1}{2}\left[(a_{j_1j_21} - a_{j_1j_2})^2 + b_{j_1j_2}\right]\left[(x_{j_2} - a_{j_1j_21})^2 - (x_{j_20} - a_{j_1j_21})^2\right]\Big\}.\end{aligned} \tag{3.6}$$

The equilibrium of $x_{j_2}^* = a_{j_1j_21}$ *in the* x_{j_2}*-direction has the following properties:*

- *For* $a_{j_1j_20} > 0$ *and* $a_{j_2j_10} > 0$,

$$(a_{j_1j_21}, \dot{x}_{j_2}) = \underbrace{(\text{UP}, \text{pF})}_{\text{up-parabola flow }(+)}. \tag{3.7}$$

  The flow at $x_{j_2}^* = a_{j_1j_21}$ *is a* (UP,pF)*-positive up-parabola flow.*

- *For* $a_{j_1j_20} < 0$ *and* $a_{j_2j_10} > 0$,

$$(a_{j_1j_21}, \dot{x}_{j_2}) = \underbrace{(\text{DP}, \text{pF})}_{\text{down-parabola flow }(+)}. \tag{3.8}$$

  The flow at $x_{j_2}^* = a_{j_1j_21}$ *is a* (DP,pF)*-positive down-parabola flow.*

- *For* $a_{j_1 j_2 0} > 0$ *and* $a_{j_2 j_1 0} < 0$,

$$(a_{j_1 j_2 1}, \dot{x}_{j_2}) = \underbrace{(\text{DP}, \text{nF})}_{\text{down-parabola flow } (-)}. \tag{3.9}$$

The flow at $x_{j_2}^* = a_{j_1 j_2 1}$ *is a* (DP,nF)*-negative down-parabola flow.*

- *For* $a_{j_1 j_2 0} < 0$ *and* $a_{j_2 j_1 0} < 0$,

$$(a_{j_1 j_2 1}, \dot{x}_{j_2}) = \underbrace{(\text{UP}, \text{nF})}_{\text{up-parabola flow } (-)}. \tag{3.10}$$

The flow at $x_{j_2}^* = a_{j_1 j_2 1}$ *is a* (UP,nF)*-negative up-parabola flow.*

(ii) If

$$\Delta_{j_1 j_2} < 0 \text{ and } \Delta_{j_2 j_1} = 0, \tag{3.11}$$

then

$$\begin{aligned} &\dot{x}_{j_2} = a_{j_2 j_1 0}(x_{j_1} - a_{j_2 j_1 1})^2, \\ &\dot{x}_{j_1} = a_{j_1 j_2 0}(x_{j_2} - a_{j_1 j_2 1})\left[(x_{j_2} - a_{j_1 j_2})^2 + b_{j_1 j_2}\right], \\ &j_1, j_2 \in \{1,2\}; j_1 \neq j_2. \end{aligned} \tag{3.12}$$

The first integral manifold is

$$\begin{aligned} &\frac{1}{3} a_{j_2 j_1 0}\left[(x_{j_1} - a_{j_2 j_1 1})^3 - (x_{j_1 0} - a_{j_2 j_1 1})^3\right] \\ &= a_{j_1 j_2 0}\Big\{\frac{1}{4}\left[(x_{j_1} - a_{j_1 j_2 1})^4 - (x_{j_1 0} - a_{j_1 j_2 1})^4\right] \\ &+ \frac{2}{3}(a_{j_1 j_2 1} - a_{j_1 j_2})\left[(x_{j_2} - a_{j_1 j_2 1})^3 - (x_{j_2 0} - a_{j_1 j_2 1})^3\right] \\ &+ \frac{1}{2}\left[(a_{j_1 j_2 1} - a_{j_1 j_2})^2 + b_{j_1 j_2}\right]\left[(x_{j_2} - a_{j_1 j_2 1})^2 - (x_{j_2 0} - a_{j_1 j_2 1})^2\right]\Big\}. \end{aligned} \tag{3.13}$$

The equilibrium of $(x_{j_2}^*, x_{j_1}^*) = (a_{j_1 j_2 1}, a_{j_2 j_1 1})$ *has the following properties:*

- *For* $a_{j_1 j_2 0} > 0$ *and* $a_{j_2 j_1 0} > 0$,

$$(a_{j_1 j_2 1}, a_{j_2 j_1 1}) = \underbrace{(\text{UP}, \text{US})}_{\text{up-parabola upper-saddle}}. \tag{3.14}$$

The equilibrium of $(x_{j_2}^*, x_{j_1}^*) = (a_{j_1 j_2 1}, a_{j_2 j_1 1})$ *is a* (UP,US)*-up-parabola upper-saddle.*

- *For* $a_{j_1 j_2 0} < 0$ *and* $a_{j_2 j_1 0} > 0$,

$$(a_{j_1j_2}1, a_{j_2j_1}1) = \underbrace{(\mathrm{DP}, \mathrm{US})}_{\text{down-parabola upper-saddle}}. \tag{3.15}$$

The equilibrium of $(x^*_{j_2}, x^*_{j_1}) = (a_{j_1j_2}1, a_{j_2j_1}1)$ *is a* (DP,US)*-down-parabola upper-saddle.*

- *For* $a_{j_1j_2}0 > 0$ *and* $a_{j_2j_1}0 < 0$,

$$(a_{j_1j_2}1, a_{j_2j_1}1) = \underbrace{(\mathrm{DP}, \mathrm{LS})}_{\text{down-parabola lower-saddle}}. \tag{3.16}$$

The equilibrium of $(x^*_{j_2}, x^*_{j_1}) = (a_{j_1j_2}1, a_{j_2j_1}1)$ *is a* (DP,LS)*-down-parabola lower-saddle.*

- *For* $a_{j_1j_2}0 < 0$ *and* $a_{j_2j_1}0 < 0$,

$$(a_{j_1j_2}1, a_{j_2j_1}1) = \underbrace{(\mathrm{UP}, \mathrm{LS})}_{\text{up-parabola lower-saddle}}. \tag{3.17}$$

The equilibrium of $(x^*_{j_2}, x^*_{j_1}) = (a_{j_1j_2}1, a_{j_2j_1}1)$ *is a* (UP,LS)*-up-parabola lower-saddle.*

(iii) If

$$\Delta_{j_1j_2} < 0 \text{ and } \Delta_{j_2j_1} > 0, \tag{3.18}$$

then

$$\begin{aligned}
\dot{x}_{j_2} &= a_{j_1j_2}0(x_{j_1} - a_{j_2j_1}1)(x_{j_1} - a_{j_2j_1}2),\\
\dot{x}_{j_1} &= a_{j_2j_1}0(x_{j_2} - a_{j_1j_2}1)\left[(x_{j_2} - a_{j_1j_2})^2 + b_{j_1j_2}\right],\\
j_1, j_2 &\in \{1,2\}; j_1 \neq j_2
\end{aligned} \tag{3.19}$$

where

$$\begin{aligned}
&\{a_{j_2j_1}1, a_{j_2j_1}2\} = \mathrm{sort}\{b_{j_2j_1}1, b_{j_2j_1}2\}, a_{j_2j_1}1 < a_{j_2j_1}2;\\
&b_{j_2j_1}1, b_{j_2j_1}2 = -\frac{1}{2}(B_{j_2j_1} \pm \sqrt{\Delta_{j_2j_1}}), a_{j_2j_1}0 = A_{j_2j_1}0;\\
&a_{j_1j_2} = -\frac{1}{2}B_{j_1j_2}, b_{j_1j_2} = \frac{1}{4}(-\Delta_{j_1j_2}).
\end{aligned} \tag{3.20}$$

The first integral manifold is

$$\begin{aligned}
&a_{j_2j_1}0\Big\{\frac{1}{3}\left[(x_{j_1} - a_{j_2j_1}1)^3 - (x_{j_1}0 - a_{j_2j_1}1)^3\right]\\
&+\frac{1}{2}(a_{j_2j_1}1 - a_{j_2j_1}2)\left[(x_{j_1} - a_{j_2j_1}1)^2 - (x_{j_1}0 - a_{j_2j_1}1)^2\right]\Big\}\\
&= a_{j_1j_2}0\Big\{\frac{1}{4}\left[(x_{j_1} - a_{j_1j_2}1)^4 - (x_{j_1}0 - a_{j_1j_2}1)^4\right]\\
&+\frac{2}{3}(a_{j_1j_2}1 - a_{j_1j_2})\left[(x_{j_2} - a_{j_1j_2}1)^3 - (x_{j_2}0 - a_{j_1j_2}1)^3\right]\\
&+\frac{1}{2}\left[(a_{j_1j_2}1 - a_{j_1j_2})^2 + b_{j_1j_2}\right]\left[(x_{j_2} - a_{j_1j_2}1)^2 - (x_{j_2}0 - a_{j_1j_2}1)^2\right]\Big\}.
\end{aligned} \tag{3.21}$$

The equilibrium of $(x^*_{j_2}, x^*_{j_1}) = (a_{j_1 j_2 1}, a_{j_2 j_1 l_1})$ $(l_1, l_2 \in \{1,2\}, l_1 \neq l_2)$ *has the following properties:*

- *For* $a_{j_1 j_2 0} > 0$ *and* $a_{j_2 j_1 0}(a_{j_2 j_1 l_1} - a_{j_2 j_1 l_2}) > 0$,

$$(a_{j_1 j_2 1}, a_{j_2 j_1 l_1}) = \underbrace{(\mathrm{UP}_+, \mathrm{UP}_+)}_{\text{positive saddle}}. \tag{3.22}$$

  The equilibrium of $(x^*_{j_2}, x^*_{j_1}) = (a_{j_1 j_2 1}, a_{j_2 j_1 l_1})$ *is a* $(\mathrm{UP}_+,\mathrm{UP}_+)$*-positive saddle.*

- *For* $a_{j_1 j_2 0} < 0$ *and* $a_{j_2 j_1 0}(a_{j_2 j_1 l_1} - a_{j_2 j_1 l_2}) > 0$,

$$(a_{j_1 j_2 1}, a_{j_2 j_1 l_1}) = \underbrace{(\mathrm{DP}_+, \mathrm{DP}_-)}_{\text{CCW center}}. \tag{3.23}$$

  The equilibrium of $(x^*_{j_2}, x^*_{j_1}) = (a_{j_1 j_2 1}, a_{j_2 j_1 l_1})$ *is a* $(\mathrm{DP}_+,\mathrm{DP}_-)$*-counter-clockwise center.*

- *For* $a_{j_1 j_2 0} > 0$ *and* $a_{j_2 j_1 0}(a_{j_2 j_1 l_1} - a_{j_2 j_1 l_2}) < 0$,

$$(a_{j_1 j_2 1}, a_{j_2 j_1 l_1}) = \underbrace{(\mathrm{DP}_-, \mathrm{DP}_+)}_{\text{CW center}}. \tag{3.24}$$

  The equilibrium of $(x^*_{j_2}, x^*_{j_1}) = (a_{j_1 j_2 1}, a_{j_2 j_1 l_1})$ *is a* $(\mathrm{DP}_-,\mathrm{DP}_+)$*-clockwise center.*

- *For* $a_{j_1 j_2 0} < 0$ *and* $a_{j_2 j_1 0}(a_{j_2 j_1 l_1} - a_{j_2 j_1 l_2}) < 0$,

$$(a_{j_1 j_2 1}, a_{j_2 j_1 l_1}) = \underbrace{(\mathrm{UP}_-, \mathrm{UP}_-)}_{\text{negative saddle}}. \tag{3.25}$$

  The equilibrium of $(x^*_{j_2}, x^*_{j_1}) = (a_{j_1 j_2 1}, a_{j_2 j_1 l_1})$ *is a* $(\mathrm{UP}_-,\mathrm{UP}_-)$*-negative saddle.*

(iv) If

$$\Delta_{j_1 j_2} = 0 \text{ and } \Delta_{j_2 j_1} < 0, a_{j_1 j_2 s_1} \neq a_{j_1 j_2 s_2} \tag{3.26}$$

then

$$\begin{aligned} &\dot{x}_{j_1} = a_{j_1 j_2 0}(x_{j_2} - a_{j_1 j_2 s_1})(x_{j_2} - a_{j_1 j_2 s_2})^2, \\ &\dot{x}_{j_2} = a_{j_2 j_1 0}\left[(x_{j_1} - a_{j_2 j_1})^2 + b_{j_2 j_1}\right] \\ &j_1, j_2 \in \{1,2\}, j_2 \neq j_1. \end{aligned} \tag{3.27}$$

The first integral manifold is

$$
\begin{aligned}
& a_{j_2 j_1 0}\Big\{\frac{1}{3}\left[(x_{j_1}-a_{j_2 j_1})^3-(x_{j_1 0}-a_{j_2 j_1})^3\right] \\
& +b_{j_2 j_1}\left[(x_{j_1}-a_{j_2 j_1})-(x_{j_1 0}-a_{j_2 j_1})\right]\Big\} \\
& =a_{j_1 j_2 0}\Big\{\frac{1}{4}\left[(x_{j_2}-a_{j_1 j_2 s_1})^4-(x_{j_2 0}-a_{j_1 j_2 s_1})^4\right] \\
& +\frac{2}{3}(a_{j_1 j_2 s_1}-a_{j_1 j_2 s_2})\left[(x_{j_2}-a_{j_1 j_2 s_1})^3-(x_{j_2 0}-a_{j_1 j_2 s_1})^3\right] \\
& +\frac{1}{2}(a_{j_1 j_2 s_1}-a_{j_1 j_2 s_2})^2\left[(x_{j_2}-a_{j_1 j_2 s_1})^2-(x_{j_2 0}-a_{j_1 j_2 s_1})^2\right]\Big\}.
\end{aligned}
\tag{3.28}
$$

(iv$_1$) The equilibrium of $x^*_{j_2}=a_{j_1 j_2 s_1}$ (s_1, $s_2\in\{1,2\}$, $s_1\neq s_2$) in the x_{j_2}-direction has the following properties:

- *For $a_{j_1 j_2 0}(a_{j_1 j_2 s_1}-a_{j_1 j_2 s_2})^2>0$ and $a_{j_2 j_1 0}>0$,*

$$
(a_{j_1 j_2 s_1},\dot{x}_{j_2})=\underbrace{(\mathrm{UP},\mathrm{pF})}_{\text{up-parabola flow }(+)}. \tag{3.29}
$$

The flow at $x^*_{j_2}=a_{j_1 j_2 s_1}$ is a (UP,pF)*-positive up-parabola flow.*

- *For $a_{j_1 j_2 0}(a_{j_1 j_2 s_1}-a_{j_1 j_2 s_2})^2<0$ and $a_{j_2 j_1 0}>0$,*

$$
(a_{j_1 j_2 s_1},\dot{x}_{j_2})=\underbrace{(\mathrm{DP},\mathrm{pF})}_{\text{down-parabola flow }(+)}. \tag{3.30}
$$

The flow at $x^*_{j_2}=a_{j_1 j_2 s_1}$ is a (DP,pF)*-positive down-parabola flow.*

- *For $a_{j_1 j_2 0}(a_{j_1 j_2 s_1}-a_{j_1 j_2 s_2})^2>0$ and $a_{j_2 j_1 0}<0$,*

$$
(a_{j_1 j_2 s_1},\dot{x}_{j_2})=\underbrace{(\mathrm{DP},\mathrm{nF})}_{\text{down-parabola flow }(-)}. \tag{3.31}
$$

The flow at $x^*_{j_2}=a_{j_1 j_2 s_1}$ is a (DP,nF)*-negative down-parabola flow.*

- *For $a_{j_1 j_2 0}(a_{j_1 j_2 s_1}-a_{j_1 j_2 s_2})^2<0$ and $a_{j_2 j_1 0}<0$,*

$$
(a_{j_1 j_2 s_1},\dot{x}_{j_2})=\underbrace{(\mathrm{UP},\mathrm{nF})}_{\text{up-parabola flow }(-)}. \tag{3.32}
$$

The flow at $x^*_{j_2}=a_{j_1 j_2 s_1}$ is a (UP,nF)*-negative up-parabola flow.*

(iv$_2$) The equilibrium of $x^*_{j_2}=a_{j_1 j_2 s_2}$ (s_1, $s_2\in\{1,2\}$, $s_1\neq s_2$) in the x_{j_2}-direction has the following properties:

- *For $a_{j_1 j_2 0}(a_{j_1 j_2 s_2}-a_{j_1 j_2 s_1})>0$ and $a_{j_2 j_1 0}>0$,*

$$(a_{j_1j_2s_2},\dot{x}_{j_2}) = \underbrace{(\text{II},\text{pF})}_{\text{increasing-inflection flow }(+)}. \tag{3.33}$$

The flow at $x_{j_2}^* = a_{j_1j_2s_2}$ *is an* (II,pF)*-positive increasing-inflection flow.*

- *For* $a_{j_1j_2 0}(a_{j_1j_2s_2} - a_{j_1j_2s_1}) < 0$ *and* $a_{j_2j_1 0} > 0$,

$$(a_{j_1j_2s_2},\dot{x}_{j_2}) = \underbrace{(\text{DI},\text{pF})}_{\text{decreasing-inflection flow }(+)}. \tag{3.34}$$

The flow at $x_{j_2}^* = a_{j_1j_2s_2}$ *is a* (DI,pF)*-positive decreasing-inflection flow.*

- *For* $a_{j_1j_2 0}(a_{j_1j_2s_2} - a_{j_1j_2s_1}) > 0$ *and* $a_{j_2j_1 0} < 0$,

$$(a_{j_1j_2s_2},\dot{x}_{j_2}) = \underbrace{(\text{DI},\text{nF})}_{\text{decreasing-inflection flow }(-)}. \tag{3.35}$$

The flow at $x_{j_2}^* = a_{j_1j_2s_2}$ *is a* (DI,nF)*-negative decreasing-inflection flow.*

- *For* $a_{j_1j_2 0}(a_{j_1j_2s_2} - a_{j_1j_2s_1}) < 0$ *and* $a_{j_2j_1 0} < 0$,

$$(a_{j_1j_2s_2},\dot{x}_{j_2}) = \underbrace{(\text{II},\text{nF})}_{\text{increasing-inflection flow }(-)}. \tag{3.36}$$

The flow at $x_{j_2}^* = a_{j_1j_2s_2}$ *is a* (II,nF)*-negative increasing-inflection flow.*

(v) If

$$\Delta_{j_1j_2} = 0 \text{ and } \Delta_{j_2j_1} < 0, a_{j_1j_2s_1} = a_{j_1j_2s_2} = a_{j_1j_2 1}, \tag{3.37}$$

then

$$\begin{aligned}
&\dot{x}_{j_1} = a_{j_1j_2 0}(x_{j_2} - a_{j_1j_2 1})^3,\\
&\dot{x}_{j_2} = a_{j_2j_1 0}\left[(x_{j_1} - a_{j_2j_1})^2 + b_{j_2j_1}\right]\\
&j_1, j_2 \in \{1,2\}, j_2 \neq j_1.
\end{aligned} \tag{3.38}$$

The first integral manifold is

$$\begin{aligned}
&a_{j_2j_1 0}\Big\{\frac{1}{3}\left[(x_{j_1} - a_{j_2j_1})^3 - (x_{j_1 0} - a_{j_2j_1})^3\right]\\
&+ b_{j_2j_1}\left[(x_{j_1} - a_{j_2j_1}) - (x_{j_1 0} - a_{j_2j_1})\right]\Big\}\\
&= \frac{1}{4}a_{j_1j_2 0}\left[(x_{j_2} - a_{j_1j_2 1})^4 - (x_{j_2 0} - a_{j_1j_2 1})^4\right].
\end{aligned} \tag{3.39}$$

The equilibrium of $x_{j_2}^* = a_{j_1j_21}$ *in the* x_{j_2}*-direction has the following properties:*

- *For* $a_{j_1j_20} > 0$ *and* $a_{j_2j_10} > 0$,

$$(a_{j_1j_21}, \dot{x}_{j_2}) = \underbrace{(3^{\text{rd}}\text{UP}, \text{pF})}_{\text{third-order up-parabola flow } (+)}. \tag{3.40}$$

The flow at $x_{j_2}^* = a_{j_1j_21}$ *is a* (3^{rd}UP,pF)*-positive third-order up-parabola flow.*

- *For* $a_{j_1j_20} < 0$ *and* $a_{j_2j_10} > 0$,

$$(a_{j_1j_21}, \dot{x}_{j_2}) = \underbrace{(3^{\text{rd}}\text{DP}, \text{pF})}_{\text{third-order down-parabola flow } (+)}. \tag{3.41}$$

The flow at $x_{j_2}^* = a_{j_1j_21}$ *is a* (3^{rd}DP,pF)*-positive third-order down-parabola flow.*

- *For* $a_{j_1j_20} > 0$ *and* $a_{j_2j_10} < 0$,

$$(a_{j_1j_21}, \dot{x}_{j_2}) = \underbrace{(3^{\text{rd}}\text{DP}, \text{nF})}_{\text{third-order down-parabola flow } (-)}. \tag{3.42}$$

The flow at $x_{j_2}^* = a_{j_1j_21}$ *is a* (3^{rd}DP,nF)*-negative third-order down-parabola flow.*

- *For* $a_{j_1j_20} < 0$ *and* $a_{j_2j_10} < 0$,

$$(a_{j_1j_21}, \dot{x}_{j_2}) = \underbrace{(3^{\text{rd}}\text{UP}, \text{nF})}_{\text{third-order up-parabola flow } (-)}. \tag{3.43}$$

The flow at $x_{j_2}^* = a_{j_1j_21}$ *is a* (3^{rd}UP,nF)*-negative third-order up-parabola flow.*

(vi) If

$$\Delta_{j_1j_2} = 0 \text{ and } \Delta_{j_2j_1} = 0, a_{j_2j_1s_1} \neq a_{j_2j_1s_2} \tag{3.44}$$

then

$$\begin{aligned} &\dot{x}_{j_1} = a_{j_1j_20}(x_{j_2} - a_{j_1j_2s_1})(x_{j_2} - a_{j_1j_2s_2})^2, \\ &\dot{x}_{j_2} = a_{j_2j_10}(x_{j_1} - a_{j_2j_11})^2, \\ &j_1, j_2 \in \{1,2\}; j_1 \neq j_2; \end{aligned} \tag{3.45}$$

and the first integral manifold is

$$\begin{aligned}&\frac{1}{3}a_{j_2j_10}\left[(x_{j_1}-a_{j_2j_11})^3-(x_{j_10}-a_{j_2j_11})^3\right]\\&=a_{j_1j_20}\Big\{\frac{1}{4}\left[(x_{j_1}-a_{j_1j_2s_1})^4-(x_{j_10}-a_{j_1j_2s_1})^4\right]\\&+\frac{2}{3}(a_{j_1j_2s_1}-a_{j_1j_2s_2})\left[(x_{j_2}-a_{j_1j_2s_1})^3-(x_{j_20}-a_{j_1j_2s_1})^3\right]\\&+\frac{1}{2}(a_{j_1j_2s_1}-a_{j_1j_2s_2})^2\left[(x_{j_2}-a_{j_1j_2s_1})^2-(x_{j_20}-a_{j_1j_2s_1})^2\right]\Big\}.\end{aligned}\tag{3.46}$$

(vi_1) The equilibrium of $(x^*_{j_2},x^*_{j_1})=(a_{j_1j_2s_1},a_{j_2j_11})$ has the following properties:

- *For $a_{j_1j_20}(a_{j_1j_2s_1}-a_{j_1j_2s_2})^2>0$ and $a_{j_2j_10}>0$,*

$$(a_{j_1j_2s_1},a_{j_2j_11})=\underbrace{(\mathrm{UP},\mathrm{US})}_{\text{up-parabola upper-saddle}}.\tag{3.47}$$

The equilibrium of $(x^*_{j_2},x^*_{j_1})=(a_{j_1j_2s_1},a_{j_2j_11})$ is a (UP,US)-*up-parabola upper-saddle.*

- *For $a_{j_1j_20}(a_{j_1j_2s_1}-a_{j_1j_2s_2})^2<0$ and $a_{j_2j_10}>0$,*

$$(a_{j_1j_2s_1},a_{j_2j_11})=\underbrace{(\mathrm{DP},\mathrm{US})}_{\text{down-parabola upper-saddle}}.\tag{3.48}$$

The equilibrium of $(x^*_{j_2},x^*_{j_1})=(a_{j_1j_2s_1},a_{j_2j_11})$ is a (DP,US)-down-*parabola upper-saddle.*

- *For $a_{j_1j_20}(a_{j_1j_2s_1}-a_{j_1j_2s_2})^2>0$ and $a_{j_2j_10}<0$,*

$$(a_{j_1j_2s_1},a_{j_2j_11})=\underbrace{(\mathrm{DP},\mathrm{LS})}_{\text{down-parabola lower-saddle}}.\tag{3.49}$$

The equilibrium of $(x^*_{j_2},x^*_{j_1})=(a_{j_1j_2s_1},a_{j_2j_11})$ is a (DP,LS)-*down-parabola lower-saddle.*

- *For $a_{j_1j_20}(a_{j_1j_2s_1}-a_{j_1j_2s_2})^2<0$ and $a_{j_2j_10}<0$,*

$$(a_{j_1j_2s_1},a_{j_2j_11})=\underbrace{(\mathrm{UP},\mathrm{LS})}_{\text{up-parabola lower-saddle}}.\tag{3.50}$$

The equilibrium of $(x^*_{j_2},x^*_{j_1})=(a_{j_1j_2s_1},a_{j_2j_11})$ is a (UP,LS)-*up-parabola lower-saddle.*

(vi_2) The equilibrium of $(x^*_{j_2},x^*_{j_1})=(a_{j_1j_2s_2},a_{j_2j_11})$ has the following properties:

- *For $a_{j_1j_20}(a_{j_1j_2s_2}-a_{j_1j_2s_1})>0$ and $a_{j_2j_10}>0$,*

$$(a_{j_1j_2s_2}, a_{j_2j_1 1}) = \underbrace{(\mathrm{II}_+, \mathrm{II}_+)}_{\text{double-inflection saddle}} . \tag{3.51}$$

The equilibrium of $(x_{j_2}^*, x_{j_1}^*) = (a_{j_1j_2s_2}, a_{j_2j_1 1})$ *is an* (II$_+$,II$_+$)*-double increasing-inflection saddle.*

- *For* $a_{j_1j_2 0}(a_{j_1j_2s_2} - a_{j_1j_2s_1}) < 0$ *and* $a_{j_2j_1 0} > 0$,

$$(a_{j_1j_2s_2}, a_{j_2j_1 1}) = \underbrace{(\mathrm{DI}_+, \mathrm{DI}_-)}_{\text{double-inflection saddle}} . \tag{3.52}$$

The equilibrium of $(x_{j_2}^*, x_{j_1}^*) = (a_{j_1j_2s_2}, a_{j_2j_1 1})$ *is a* (DI$_+$,DI$_-$)*-double decreasing-inflection saddle.*

- *For* $a_{j_1j_2 0}(a_{j_1j_2s_2} - a_{j_1j_2s_1}) > 0$ *and* $a_{j_2j_1 0} < 0$,

$$(a_{j_1j_2s_2}, a_{j_2j_1 1}) = \underbrace{(\mathrm{DI}_-, \mathrm{DI}_+)}_{\text{double-inflection saddle}} . \tag{3.53}$$

The equilibrium of $(x_{j_2}^*, x_{j_1}^*) = (a_{j_1j_2s_2}, a_{j_2j_1 1})$ *is a* (DI$_-$,DI$_+$)*-double decreasing-inflection saddle.*

- *For* $a_{j_1j_2 0}(a_{j_1j_2s_2} - a_{j_1j_2s_1}) < 0$ *and* $a_{j_2j_1 0} < 0$,

$$(a_{j_1j_2s_2}, a_{j_2j_1 1}) = \underbrace{(\mathrm{II}_-, \mathrm{II}_-)}_{\text{double-inflection saddle}} . \tag{3.54}$$

The equilibrium of $(x_{j_2}^*, x_{j_1}^*) = (a_{j_1j_2s_2}, a_{j_2j_1 1})$ *is an* (II$_-$,II$_-$)*-double-increasing -inflection saddle.*

(vii) If

$$\Delta_{j_1j_2} = 0 \text{ and } \Delta_{j_2j_1} = 0, a_{j_2j_1s_1} = a_{j_2j_1s_2} = a_{j_2j_1 1}, \tag{3.55}$$

then

$$\begin{aligned} \dot{x}_{j_1} &= a_{j_1j_2 0}(x_{j_2} - a_{j_1j_2 1})^3, \\ \dot{x}_{j_2} &= a_{j_2j_1 0}(x_{j_1} - a_{j_2j_1 1})^2, \\ j_1, j_2 &\in \{1, 2\}; j_1 \neq j_2; \end{aligned} \tag{3.56}$$

and the first integral manifold is

$$\begin{aligned}&\frac{1}{3}a_{j_2j_10}\left[(x_{j_1}-a_{j_2j_11})^3-(x_{j_10}-a_{j_2j_11})^3\right]\\&=\frac{1}{4}a_{j_1j_20}\left[(x_{j_2}-a_{j_1j_21})^4-(x_{j_20}-a_{j_1j_21})^4\right].\end{aligned}\tag{3.57}$$

The equilibrium of $(x^*_{j_2},x^*_{j_1})=(a_{j_1j_21},a_{j_2j_11})$ *has the following properties:*

- *For* $a_{j_1j_20}>0$ *and* $a_{j_2j_10}>0$,

$$(a_{j_1j_21},a_{j_2j_11})=\underbrace{(3^{\text{rd}}\text{UP},\text{US})}_{\text{third-order up-parabola upper-saddle}}.\tag{3.58}$$

  The equilibrium of $(x^*_{j_2},x^*_{j_1})=(a_{j_1j_21},a_{j_2j_11})$ *is a* (3rdUP,US)-*third-order-up-parabola upper-saddle.*
- *For* $a_{j_1j_20}<0$ *and* $a_{j_2j_10}>0$,

$$(a_{j_1j_21},a_{j_2j_11})=\underbrace{(3^{\text{rd}}\text{DP},\text{US})}_{\text{third-order down-parabola upper-saddle}}.\tag{3.59}$$

  The equilibrium of $(x^*_{j_1},x^*_{j_2})=(a_{j_2j_11},a_{j_1j_21})$ *is a* (3rdDP,US)-*third-order down-parabola upper-saddle.*
- *For* $a_{j_1j_20}>0$ *and* $a_{j_2j_10}<0$,

$$(a_{j_1j_21},a_{j_2j_11})=\underbrace{(3^{\text{rd}}\text{DP},\text{LS})}_{\text{third-order down-parabola lower-saddle}}.\tag{3.60}$$

  The equilibrium of $(x^*_{j_1},x^*_{j_2})=(a_{j_2j_11},a_{j_1j_21})$ *is a* (3rdDP,LS)-*third-order down-parabola lower-saddle.*
- *For* $a_{j_1j_20}<0$ *and* $a_{j_2j_10}<0$,

$$(a_{j_1j_21},a_{j_2j_11})=\underbrace{(3^{\text{rd}}\text{UP},\text{LS})}_{\text{third-order up-parabola lower-saddle}}.\tag{3.61}$$

  The equilibrium of $(x^*_{j_1},x^*_{j_2})=(a_{j_2j_11},a_{j_1j_21})$ *is a* (3rdUP,LS)-*third-order-up-parabola lower-saddle.*

(viii) If

$$\Delta_{j_1j_2}=0\text{ and }\Delta_{j_2j_1}>0,\tag{3.62}$$

then

$$\begin{aligned}
&\dot{x}_{j_2} = a_{j_2 j_1 0}(x_{j_1} - a_{j_2 j_1 1})(x_{j_1} - a_{j_2 j_1 2}),\\
&\dot{x}_{j_1} = a_{j_1 j_2 0}(x_{j_2} - a_{j_1 j_2 s_1})(x_{j_2} - a_{j_1 j_2 s_2})^2,\\
&j_1, j_2 \in \{1,2\}; j_1 \neq j_2.
\end{aligned} \tag{3.63}$$

The first integral manifold (s_1, s_2, l_1, $l_2 \in \{1,2\}$; $s_1 \neq s_2$, $l_1 \neq l_2$) is

$$\begin{aligned}
&a_{j_2 j_1 0}\Big\{\frac{1}{3}\left[(x_{j_1} - a_{j_2 j_1 l_1})^3 - (x_{j_1 0} - a_{j_2 j_1 l_1})^3\right]\\
&+\frac{1}{2}(a_{j_2 j_1 l_1} - a_{j_2 j_1 l_2})\left[(x_{j_1} - a_{j_2 j_1 l_1})^3 - (x_{j_1 0} - a_{j_2 j_1 l_1})^3\right]\Big\}\\
&= a_{j_1 j_2 0}\Big\{\frac{1}{4}\left[(x_{j_1} - a_{j_1 j_2 s_1})^4 - (x_{j_1 0} - a_{j_1 j_2 s_1})^4\right]\\
&+\frac{2}{3}(a_{j_1 j_2 s_1} - a_{j_1 j_2 s_2})\left[(x_{j_2} - a_{j_1 j_2 s_1})^3 - (x_{j_2 0} - a_{j_1 j_2 s_1})^3\right]\\
&+\frac{1}{2}(a_{j_1 j_2 s_1} - a_{j_1 j_2 s_2})^2\left[(x_{j_2} - a_{j_1 j_2 s_1})^2 - (x_{j_2 0} - a_{j_1 j_2 s_1})^2\right]\Big\}.
\end{aligned} \tag{3.64}$$

(viii$_l$) The equilibrium of $(x^*_{j_2}, x^*_{j_1}) = (a_{j_1 j_2 s_1}, a_{j_2 j_1 l_1})$ (s_1, s_2, l_1, $l_2 \in \{1,2\}$; $s_1 \neq s_2$, $l_1 \neq l_2$) has the following properties:

- *For $a_{j_1 j_2 0}(a_{j_1 j_2 s_1} - a_{j_1 j_2 s_2})^2 > 0$ and $a_{j_2 j_1 0}(a_{j_2 j_1 l_1} - a_{j_2 j_1 l_2}) > 0$,*

$$(a_{j_1 j_2 s_1}, a_{j_2 j_1 l_1}) = \underbrace{(\mathrm{UP}_+, \mathrm{UP}_+)}_{\text{positive saddle}}. \tag{3.65}$$

  The equilibrium of $(x^*_{j_2}, x^*_{j_1}) = (a_{j_1 j_2 s_1}, a_{j_2 j_1 l_1})$ is a (UP$_+$,UP$_+$)*-positive saddle.*

- *For $a_{j_1 j_2 0}(a_{j_1 j_2 s_1} - a_{j_1 j_2 s_2})^2 < 0$ and $a_{j_2 j_1 0}(a_{j_2 j_1 l_1} - a_{j_2 j_1 l_2}) > 0$,*

$$(a_{j_1 j_2 s_1}, a_{j_2 j_1 l_1}) = \underbrace{(\mathrm{DP}_+, \mathrm{DP}_-)}_{\text{CCW center}}. \tag{3.66}$$

  The equilibrium of $(x^*_{j_2}, x^*_{j_1}) = (a_{j_1 j_2 s_1}, a_{j_2 j_1 l_1})$ is a (DP$_+$,DP$_-$)*-counter-clockwise center.*

- *For $a_{j_1 j_2 0}(a_{j_1 j_2 s_1} - a_{j_1 j_2 s_2})^2 > 0$ and $a_{j_2 j_1 0}(a_{j_2 j_1 l_1} - a_{j_2 j_1 l_2}) < 0$,*

$$(a_{j_1 j_2 s_1}, a_{j_2 j_1 l_1}) = \underbrace{(\mathrm{DP}_-, \mathrm{DP}_+)}_{\text{CW center}}. \tag{3.67}$$

  The equilibrium of $(x^*_{j_2}, x^*_{j_1}) = (a_{j_1 j_2 s_1}, a_{j_2 j_1 l_1})$ is a (DP$_-$,DP$_+$)*-clockwise center.*

- *For $a_{j_1 j_2 0}(a_{j_1 j_2 s_1} - a_{j_1 j_2 s_2})^2 < 0$ and $a_{j_2 j_1 0}(a_{j_2 j_1 l_1} - a_{j_2 j_1 l_2}) < 0$,*

$$(a_{j_1j_2s_1}, a_{j_2j_1l_1}) = \underbrace{(\text{UP}_-, \text{UP}_-)}_{\text{negative saddle}}. \tag{3.68}$$

The equilibrium of $(x_{j_2}^*, x_{j_1}^*) = (a_{j_1j_2s_1}, a_{j_2j_1l_1})$ *is a* (DP$_-$,UP$_-$)*-negative saddle.*

(viii$_2$) The equilibrium of $(x_{j_2}^*, x_{j_1}^*) = (a_{j_1j_2s_2}, a_{j_2j_1l_1})$ $(s_1, s_2, l_1, l_2 \in \{1,2\}, s_1 \neq s_2, l_1 \neq l_2)$ *has the following properties:*

- *For* $a_{j_1j_20}(a_{j_1j_2s_2} - a_{j_1j_2s_1}) > 0$ *and* $a_{j_2j_10}(a_{j_2j_1l_1} - a_{j_2j_1l_2}) > 0$,

$$(a_{j_1j_2s_2}, a_{j_2j_1l_1}) = \underbrace{(\text{US}, \text{UP})}_{\text{up-parabola upper-saddle}}. \tag{3.69}$$

  The equilibrium of $(x_{j_2}^*, x_{j_1}^*) = (a_{j_1j_2s_2}, a_{j_2j_1l_1})$ *is a* (US,UP)*-up-parabola upper-saddle.*
- *For* $a_{j_1j_20}(a_{j_1j_2s_2} - a_{j_1j_2s_1}) < 0$ *and* $a_{j_2j_10}(a_{j_2j_1l_1} - a_{j_2j_1l_2}) > 0$,

$$(a_{j_1j_2s_2}, a_{j_2j_1l_1}) = \underbrace{(\text{LS}, \text{DP})}_{\text{down-parabola lower-saddle}}. \tag{3.70}$$

  The equilibrium of $(x_{j_2}^*, x_{j_1}^*) = (a_{j_1j_2s_2}, a_{j_2j_1l_1})$ *is an* (LS,DP)*-down-parabola lower-saddle.*
- *For* $a_{j_1j_20}(a_{j_1j_2s_2} - a_{j_1j_2s_1}) > 0$ *and* $a_{j_2j_10}(a_{j_2j_1l_1} - a_{j_2j_1l_2}) < 0$,

$$(a_{j_1j_2s_2}, a_{j_2j_1l_1}) = \underbrace{(\text{US}, \text{DP})}_{\text{down-parabola upper-saddle}}. \tag{3.71}$$

  The equilibrium of $(x_{j_2}^*, x_{j_1}^*) = (a_{j_1j_2s_2}, a_{j_2j_1l_1})$ *is a* (US,DP)*-down-parabola upper-saddle.*
- *For* $a_{j_1j_20}(a_{j_1j_2s_2} - a_{j_1j_2s_1}) < 0$ *and* $a_{j_2j_10}(a_{j_2j_1l_1} - a_{j_2j_1l_2}) < 0$,

$$(a_{j_1j_2s_2}, a_{j_2j_1l_1}) = \underbrace{(\text{LS}, \text{UP})}_{\text{up-parabola lower-saddle}}. \tag{3.72}$$

  The equilibrium of $(x_{j_2}^*, x_{j_1}^*) = (a_{j_1j_2s_2}, a_{j_2j_1l_1})$ *is an* (LS,UP)*-up-parabola lower-saddle.*

(ix) If

$$\Delta_{j_1j_2} = 0 \text{ and } \Delta_{j_2j_1} > 0, a_{j_1j_2s_1} = a_{j_1j_2s_2} = a_{j_1j_21}, \tag{3.73}$$

then

$$\begin{aligned}
&\dot{x}_{j_2} = a_{j_2j_10}(x_{j_1} - a_{j_2j_11})(x_{j_1} - a_{j_2j_12}),\\
&\dot{x}_{j_1} = a_{j_1j_20}(x_{j_2} - a_{j_1j_21})^3,\\
&j_1, j_2 \in \{1,2\}; j_1 \neq j_2.
\end{aligned} \tag{3.74}$$

The first integral manifold is

$$\begin{aligned}
&a_{j_2j_10}\{\frac{1}{3}\left[(x_{j_1} - a_{j_2j_1l_1})^3 - (x_{j_10} - a_{j_2j_1l_1})^3\right]\\
&+\frac{1}{2}(a_{j_2j_1l_1} - a_{j_2j_1l_2})\left[(x_{j_1} - a_{j_2j_1l_1})^3 - (x_{j_10} - a_{j_2j_1l_1})^3\right]\}\\
&= \frac{1}{4}a_{j_1j_20}\left[(x_{j_1} - a_{j_1j_2s_1})^4 - (x_{j_10} - a_{j_1j_2s_1})^4\right].
\end{aligned} \tag{3.75}$$

The equilibrium of $(x^*_{j_2}, x^*_{j_1}) = (a_{j_1j_21}, a_{j_2j_1l_1})$ $(l_1, l_2 \in \{1,2\}, l_1 \neq l_2)$ *has the following properties:*

- *For* $a_{j_1j_20} > 0$ *and* $a_{j_2j_10}(a_{j_2j_1l_1} - a_{j_2j_1l_2}) > 0$,

$$(a_{j_1j_21}, a_{j_2j_1l_1}) = \underbrace{(3^{\text{rd}}\text{UP}_+, \text{UP}_+)}_{\text{third-order positive saddle}}. \tag{3.76}$$

  The equilibrium of $(x^*_{j_2}, x^*_{j_1}) = (a_{j_1j_21}, a_{j_2j_1l_1})$ *is a* (3^{rd}UP_+,UP_+)*-third-order positive saddle.*
- *For* $a_{j_1j_20} < 0$ *and* $a_{j_2j_10}(a_{j_2j_1l_1} - a_{j_2j_1l_2}) > 0$,

$$(a_{j_1j_21}, a_{j_2j_1l_1}) = \underbrace{(3^{\text{rd}}\text{DP}_+, \text{DP}_-)}_{\text{third-order CCW center}}. \tag{3.77}$$

  The equilibrium of $(x^*_{j_2}, x^*_{j_1}) = (a_{j_1j_21}, a_{j_2j_1l_1})$ *is a* (3^{rd}DP_+,DP_-)*-third-order-counter-clockwise center.*
- *For* $a_{j_1j_20} > 0$ *and* $a_{j_2j_10}(a_{j_2j_1l_1} - a_{j_2j_1l_2}) < 0$,

$$(a_{j_1j_21}, a_{j_2j_1l_1}) = \underbrace{(3^{\text{rd}}\text{DP}_-, \text{DP}_+)}_{\text{third-order CW center}}. \tag{3.78}$$

  The equilibrium of $(x^*_{j_2}, x^*_{j_1}) = (a_{j_1j_21}, a_{j_2j_1l_1})$ *is a* (3^{rd}DP_-,DP_+)*-third-order-clockwise center.*
- *For* $a_{j_1j_20} < 0$ *and* $a_{j_2j_10}(a_{j_2j_1l_1} - a_{j_2j_1l_2}) < 0$,

$$(a_{j_1j_21}, a_{j_2j_1l_1}) = \underbrace{(3^{\text{rd}}\text{UP}_-, \text{UP}_-)}_{\text{third-order negative saddle}}. \tag{3.79}$$

The equilibrium of $(x_{j_2}^*, x_{j_1}^*) = (a_{j_1 j_2 1}, a_{j_2 j_1 l_1})$ *is a* (3^{rd}UP$_-$,UP$_-$)*-third-order negative saddle.*

(x) If

$$\Delta_{j_1 j_2} > 0 \text{ and } \Delta_{j_2 j_1} < 0, \tag{3.80}$$

then

$$\begin{aligned} &\dot{x}_{j_1} = a_{j_1 j_2 0}(x_{j_2} - a_{j_1 j_2 1})(x_{j_2} - a_{j_1 j_2 2})(x_{j_2} - a_{j_1 j_2 3}), \\ &\dot{x}_{j_2} = a_{j_2 j_1 0}\left[(x_{j_1} - a_{j_2 j_1})^2 + b_{j_2 j_1}\right] \\ &j_1, j_2 \in \{1,2\}, j_2 \neq j_1 \end{aligned} \tag{3.81}$$

where

$$\begin{aligned} &\{a_{j_1 j_2 1}, a_{j_1 j_2 2}, a_{j_1 j_2 3}\} = \text{sort}\{b_{j_1 j_2 1}, b_{j_1 j_2 2}, b_{j_1 j_2 3}\}, a_{j_1 j_2 1} < a_{j_1 j_2 2} < a_{j_1 j_2 3}; \\ &b_{j_1 j_2 2}, b_{j_1 j_2 3} = -\frac{1}{2}(B_{j_1 j_2} \pm \sqrt{\Delta_{j_1 j_2}}), a_{j_1 j_2 0} = A_{j_1 j_2 0}; \\ &a_{j_2 j_1} = -\frac{1}{2} B_{j_2 j_1}, b_{j_2 j_1} = \frac{1}{4}(-\Delta_{j_2 j_1}). \end{aligned} \tag{3.82}$$

The first integral manifold is

$$\begin{aligned} &a_{j_2 j_1 0}\{\frac{1}{3}\left[(x_{j_1} - a_{j_2 j_1})^3 - (x_{j_1 0} - a_{j_2 j_1})^3\right] + b_{j_2 j_1}\left[(x_{j_1} - a_{j_2 j_1}) - (x_{j_1 0} - a_{j_2 j_1})\right]\} \\ &= a_{j_1 j_2 0}\{\frac{1}{4}\left[(x_{j_2} - a_{j_1 j_2 s_1})^4 - (x_{j_2 0} - a_{j_1 j_2 s_1})^4\right] \\ &+ \frac{1}{3}\sum\nolimits_{s_2=1, s_2 \neq s_1}^{3} (a_{j_1 j_2 s_1} - a_{j_1 j_2 s_2})\left[(x_{j_2} - a_{j_1 j_2 s_1})^3 - (x_{j_2 0} - a_{j_1 j_2 s_1})^3\right] \\ &+ \frac{1}{2}\prod\nolimits_{s_2=1, s_2 \neq s_1}^{3} (a_{j_1 j_2 s_1} - a_{j_1 j_2 s_2})\left[(x_{j_2} - a_{j_1 j_2 s_1})^2 - (x_{j_2 0} - a_{j_1 j_2 s_1})^2\right]\}. \end{aligned} \tag{3.83}$$

The equilibrium of $x_{j_2}^* = a_{j_1 j_2 s_1}$ $(s_1, s_2 \in \{1,2,3\}, s_1 \neq s_2)$ in the x_{j_2}-direction *has the following properties:*

- *For* $a_{j_1 j_2 0}\prod_{s_2=1, s_2 \neq s_1}^{3}(a_{j_1 j_2 s_1} - a_{j_1 j_2 s_2}) > 0$ *and* $a_{j_2 j_1 0} > 0$,

$$(a_{j_1 j_2 s_1}, \dot{x}_{j_2}) = \underbrace{(\text{UP}, \text{pF})}_{\text{up-parabola flow } (+)}. \tag{3.84}$$

The flow at $x_{j_2}^* = a_{j_1 j_2 s_1}$ *is a* (UP,pF)*-positive up-parabola flow.*

- *For* $a_{j_1j_20}\prod_{s_2=1,s_2\neq s_1}^{3}(a_{j_1j_2s_1}-a_{j_1j_2s_2})<0$ *and* $a_{j_2j_10}>0$,

$$(a_{j_1j_2s_1},\dot{x}_{j_2})=\underbrace{(\text{DP},\text{pF})}_{\text{down-parabola flow }(+)}. \tag{3.85}$$

The flow at $x_{j_2}^*=a_{j_1j_2s_1}$ *is a* (DP,pF)-*positive down-parabola flow.*

- *For* $a_{j_1j_20}\prod_{s_2=1,s_2\neq s_1}^{3}(a_{j_1j_2s_1}-a_{j_1j_2s_2})>0$ *and* $a_{j_2j_10}<0$,

$$(a_{j_1j_2s_1},\dot{x}_{j_2})=\underbrace{(\text{DP},\text{nF})}_{\text{down-parabola flow }(-)}. \tag{3.86}$$

The flow at $x_{j_2}^*=a_{j_1j_2s_1}$ *is a* (DP,nF)-*negative down-parabola flow.*

- *For* $a_{j_1j_20}\prod_{s_2=1,s_2\neq s_1}^{3}(a_{j_1j_2s_1}-a_{j_1j_2s_2})<0$ *and* $a_{j_2j_10}<0$,

$$(a_{j_1j_2s_1},\dot{x}_{j_2})=\underbrace{(\text{UP},\text{nF})}_{\text{up-parabola flow }(-)}. \tag{3.87}$$

The flow at $x_{j_2}^*=a_{j_1j_2s_1}$ *is a* (UP,nF)-*negative up-parabola flow.*

(xi) If

$$\Delta_{j_1j_2}>0 \text{ and } \Delta_{j_2j_1}=0, \tag{3.88}$$

then

$$\begin{aligned}
&\dot{x}_{j_1}=a_{j_1j_20}(x_{j_2}-a_{j_1j_21})(x_{j_2}-a_{j_1j_22})(x_{j_2}-a_{j_1j_23}),\\
&\dot{x}_{j_2}=a_{j_2j_10}(x_{j_1}-a_{j_2j_11})^2\\
&j_1,j_2\in\{1,2\},j_2\neq j_1.
\end{aligned} \tag{3.89}$$

The first integral manifold is

$$\begin{aligned}
&\frac{1}{3}a_{j_2j_10}\left[(x_{j_1}-a_{j_2j_11})^3-(x_{j_10}-a_{j_2j_11})^3\right]\\
&=a_{j_1j_20}\Big\{\frac{1}{4}\left[(x_{j_2}-a_{j_1j_2s_1})^4-(x_{j_20}-a_{j_1j_2s_1})^4\right]\\
&+\frac{1}{3}\sum\nolimits_{s_2=1,s_2\neq s_1}^{3}(a_{j_1j_2s_1}-a_{j_1j_2s_2})\left[(x_{j_2}-a_{j_1j_2s_1})^3-(x_{j_20}-a_{j_1j_2s_1})^3\right]\\
&+\frac{1}{2}\prod\nolimits_{s_2=1,s_2\neq s_1}^{3}(a_{j_1j_2s_1}-a_{j_1j_2s_2})\left[(x_{j_2}-a_{j_1j_2s_1})^2-(x_{j_20}-a_{j_1j_2s_1})^2\right]\Big\}.
\end{aligned} \tag{3.90}$$

The equilibrium of $(x_{j_2}^*, x_{j_1}^*) = (a_{j_1 j_2 s_1}, a_{j_2 j_1 1})$ $(s_1,\ s_2 \in \{1,2,3\},\ s_1 \neq s_2)$ *has the following properties:*

- *For* $a_{j_1 j_2 0} \prod_{s_2=1, s_2 \neq s_1}^{3} (a_{j_1 j_2 s_1} - a_{j_1 j_2 s_2}) > 0$ *and* $a_{j_2 j_1 0} > 0$,

$$(a_{j_1 j_2 s_1}, a_{j_2 j_1 1}) = \underbrace{(\mathrm{UP}, \mathrm{US})}_{\text{up-parabola upper-saddle}}. \tag{3.91}$$

  The equilibrium of $(x_{j_2}^*, x_{j_1}^*) = (a_{j_1 j_2 s_1}, a_{j_2 j_1 1})$ *is a* (UP,US)*-up-parabola upper-saddle.*

- *For* $a_{j_1 j_2 0} \prod_{s_2=1, s_2 \neq s_1}^{3} (a_{j_1 j_2 s_1} - a_{j_1 j_2 s_2}) < 0$ *and* $a_{j_2 j_1 0} > 0$,

$$(a_{j_1 j_2 s_1}, a_{j_2 j_1 1}) = \underbrace{(\mathrm{DP}, \mathrm{US})}_{\text{down-parabola upper-saddle}}. \tag{3.92}$$

  The equilibrium of $(x_{j_2}^*, x_{j_1}^*) = (a_{j_1 j_2 s_1}, a_{j_2 j_1 1})$ *is a* (DP,US)*-down-parabola upper-saddle.*

- *For* $a_{j_1 j_2 0} \prod_{s_2=1, s_2 \neq s_1}^{3} (a_{j_1 j_2 s_1} - a_{j_1 j_2 s_2}) > 0$ *and* $a_{j_2 j_1 0} < 0$,

$$(a_{j_1 j_2 s_1}, a_{j_2 j_1 1}) = \underbrace{(\mathrm{DP}, \mathrm{LS})}_{\text{down-parabola lower-saddle}}. \tag{3.93}$$

  The equilibrium of $(x_{j_2}^*, x_{j_1}^*) = (a_{j_1 j_2 s_1}, a_{j_2 j_1 1})$ *is a* (DP,LS)*-down-parabola lower-saddle.*

- *For* $a_{j_1 j_2 0} \prod_{s_2=1, s_2 \neq s_1}^{3} (a_{j_1 j_2 s_1} - a_{j_1 j_2 s_2}) < 0$ *and* $a_{j_2 j_1 0} < 0$,

$$(a_{j_1 j_2 s_1}, a_{j_2 j_1 1}) = \underbrace{(\mathrm{UP}, \mathrm{LS})}_{\text{up-parabola lower-saddle}}. \tag{3.94}$$

  The equilibrium of $(x_{j_2}^*, x_{j_1}^*) = (a_{j_1 j_2 s_1}, a_{j_2 j_1 1})$ *is a* (UP,LS)*-up-parabola lower-saddle.*

(xii) If

$$\Delta_{j_1 j_2} > 0 \text{ and } \Delta_{j_2 j_1} > 0, \tag{3.95}$$

then

$$\begin{aligned}
&\dot{x}_{j_1} = a_{j_1 j_2 0}(x_{j_2} - a_{j_1 j_2 1})(x_{j_2} - a_{j_1 j_2 2})(x_{j_2} - a_{j_1 j_2 3}),\\
&\dot{x}_{j_2} = a_{j_2 j_1 0}(x_{j_1} - a_{j_2 j_1 1})(x_{j_1} - a_{j_2 j_1 2}),\\
&j_1, j_2 \in \{1,2\}; j_1 \neq j_2.
\end{aligned} \tag{3.96}$$

The first integral manifold is

$$\begin{aligned}
&a_{j_2 j_1 0}\{\tfrac{1}{3}\left[(x_{j_1} - a_{j_2 j_1 l_1})^3 - (x_{j_1 0} - a_{j_2 j_1 l_1})^3\right]\\
&+\tfrac{1}{2}(a_{j_2 j_1 l_1} - a_{j_2 j_1 l_2})\left[(x_{j_1} - a_{j_2 j_1 l_1})^2 - (x_{j_1 0} - a_{j_2 j_1 l_1})^2\right]\}\\
&= a_{j_1 j_2 0}\{\tfrac{1}{4}\left[(x_{j_2} - a_{j_1 j_2 s_1})^4 - (x_{j_2 0} - a_{j_1 j_2 s_1})^4\right]\\
&+\tfrac{1}{3}\sum\nolimits_{s_2=1, s_2\neq s_1}^{3}(a_{j_1 j_2 s_1} - a_{j_1 j_2 s_2})\left[(x_{j_2} - a_{j_1 j_2 s_1})^3 - (x_{j_2 0} - a_{j_1 j_2 s_1})^3\right]\\
&+\tfrac{1}{2}\prod\nolimits_{s_2=1, s_2\neq s_1}^{3}(a_{j_1 j_2 s_1} - a_{j_1 j_2 s_2})\left[(x_{j_2} - a_{j_1 j_2 s_1})^2 - (x_{j_2 0} - a_{j_1 j_2 s_1})^2\right]\}.
\end{aligned} \tag{3.97}$$

(xii$_1$) The equilibrium of $(x^*_{j_2}, x^*_{j_1}) = (a_{j_1 j_2 s_1}, a_{j_2 j_1 l_1})$ $(s_1,\ s_2 \in \{1,2,3\},\ s_1 \neq s_2,\ l_1,\ l_2 \in \{1,2\},\ l_1 \neq l_2)$ *has the following properties:*

- *For* $a_{j_1 j_2 0}\prod_{s_2=1, s_2\neq s_1}^{3}(a_{j_1 j_2 s_1} - a_{j_1 j_2 s_2}) > 0$ *and* $a_{j_2 j_1 0}(a_{j_2 j_1 l_1} - a_{j_2 j_1 l_2}) > 0$,

$$(a_{j_1 j_2 s_1}, a_{j_2 j_1 l_1}) = \underbrace{(\mathrm{UP}_+, \mathrm{UP}_+)}_{\text{positive saddle}}. \tag{3.98}$$

  The equilibrium of $(x^*_{j_1}, x^*_{j_2}) = (a_{j_1 j_2 s_1}, a_{j_2 j_1 l_1})$ *is a* $(\mathrm{UP}_+,\mathrm{UP}_+)$*-positive saddle.*

- *For* $a_{j_1 j_2 0}\prod_{s_2=1, s_2\neq s_1}^{3}(a_{j_1 j_2 s_1} - a_{j_1 j_2 s_2}) < 0$ *and* $a_{j_2 j_1 0}(a_{j_2 j_1 l_1} - a_{j_2 j_1 l_2}) > 0$,

$$(a_{j_1 j_2 s_1}, a_{j_2 j_1 l_1}) = \underbrace{(\mathrm{DP}_+, \mathrm{DP}_-)}_{\text{CCW center}}. \tag{3.99}$$

  The equilibrium of $(x^*_{j_1}, x^*_{j_2}) = (a_{j_1 j_2 s_1}, a_{j_2 j_1 l_1})$ *is a* $(\mathrm{DP}_+,\mathrm{DP}_-)$*-counter-clockwise center.*

- *For* $a_{j_1 j_2 0}\prod_{s_2=1, s_2\neq s_1}^{3}(a_{j_1 j_2 s_1} - a_{j_1 j_2 s_2}) > 0$ *and* $a_{j_2 j_1 0}(a_{j_2 j_1 l_1} - a_{j_2 j_1 l_2}) < 0$,

$$(a_{j_1 j_2 s_1}, a_{j_2 j_1 l_1}) = \underbrace{(\mathrm{DP}_-, \mathrm{DP}_+)}_{\text{CW center}}. \tag{3.100}$$

  The equilibrium of $(x^*_{j_1}, x^*_{j_2}) = (a_{j_1 j_2 s_1}, a_{j_2 j_1 l_1})$ *is a* $(\mathrm{DP}_-,\mathrm{DP}_+)$*-clockwise center.*

- *For* $a_{j_1j_20}\prod_{s_2=1,s_2\neq s_1}^{3}(a_{j_1j_2s_1}-a_{j_1j_2s_2})<0$ *and* $a_{j_2j_10}(a_{j_2j_1l_1}-a_{j_2j_1l_2})<0$,

$$(a_{j_1j_2s_1},a_{j_2j_1l_1})=\underbrace{(\mathrm{UP}_-,\mathrm{UP}_-)}_{\text{negative saddle}}. \tag{3.101}$$

The equilibrium of $(x^*_{j_1},x^*_{j_2})=(a_{j_1j_2s_1},a_{j_2j_1l_1})$ *is a* $(\mathrm{UP}_-,\mathrm{UP}_-)$*-negative saddle.*

(xii$_2$*) For* $(x^*_{j_2},x^*_{j_1})=(a_{j_1j_2s_1},a_{j_2j_1l_1})$ $(s_1=1,2,3;\ l_1=1,2)$*, the corresponding equilibrium networks have the following properties:*

- *For* $a_{j_1j_20}>0$ *and* $a_{j_2j_10}>0$,

$$\left\{\begin{matrix}(a_{j_1j_23},a_{j_2j_11}) & (a_{j_1j_23},a_{j_2j_12})\\ (a_{j_1j_22},a_{j_2j_11}) & (a_{j_1j_22},a_{j_2j_12})\\ (a_{j_1j_21},a_{j_2j_11}) & (a_{j_1j_21},a_{j_2j_12})\end{matrix}\right\}=\left\{\begin{matrix}\underbrace{(\mathrm{DP}_-,\mathrm{DP}_+)}_{\text{CW center}} & \underbrace{(\mathrm{UP}_+,\mathrm{UP}_+)}_{\text{positive saddle}}\\ \underbrace{(\mathrm{UP}_-,\mathrm{UP}_-)}_{\text{negative saddle}} & \underbrace{(\mathrm{DP}_+,\mathrm{DP}_-)}_{\text{CCW center}}\\ \underbrace{(\mathrm{DP}_-,\mathrm{DP}_+)}_{\text{CW center}} & \underbrace{(\mathrm{UP}_+,\mathrm{UP}_+)}_{\text{positive saddle}}\end{matrix}\right\}. \tag{3.102}$$

- *For* $a_{j_1j_20}<0$ *and* $a_{j_2j_10}>0$,

$$\left\{\begin{matrix}(a_{j_1j_23},a_{j_2j_11}) & (a_{j_1j_23},a_{j_2j_12})\\ (a_{j_1j_22},a_{j_2j_11}) & (a_{j_1j_22},a_{j_2j_12})\\ (a_{j_1j_21},a_{j_2j_11}) & (a_{j_1j_21},a_{j_2j_12})\end{matrix}\right\}=\left\{\begin{matrix}\underbrace{(\mathrm{UP}_-,\mathrm{UP}_-)}_{\text{negative saddle}} & \underbrace{(\mathrm{DP}_+,\mathrm{DP}_-)}_{\text{CCW center}}\\ \underbrace{(\mathrm{DP}_-,\mathrm{DP}_+)}_{\text{CW center}} & \underbrace{(\mathrm{UP}_+,\mathrm{UP}_+)}_{\text{positive saddle}}\\ \underbrace{(\mathrm{UP}_-,\mathrm{UP}_-)}_{\text{negative saddle}} & \underbrace{(\mathrm{DP}_+,\mathrm{DP}_-)}_{\text{CCW center}}\end{matrix}\right\}. \tag{3.103}$$

- *For* $a_{j_1j_20}>0$ *and* $a_{j_2j_10}<0$,

$$\left\{\begin{matrix}(a_{j_1j_23},a_{j_2j_11}) & (a_{j_1j_23},a_{j_2j_12})\\ (a_{j_1j_22},a_{j_2j_11}) & (a_{j_1j_22},a_{j_2j_12})\\ (a_{j_1j_21},a_{j_2j_11}) & (a_{j_1j_21},a_{j_2j_12})\end{matrix}\right\}=\left\{\begin{matrix}\underbrace{(\mathrm{UP}_+,\mathrm{UP}_+)}_{\text{positive saddle}} & \underbrace{(\mathrm{DP}_-,\mathrm{DP}_+)}_{\text{CW center}}\\ \underbrace{(\mathrm{DP}_+,\mathrm{DP}_-)}_{\text{CCW center}} & \underbrace{(\mathrm{UP}_-,\mathrm{UP}_-)}_{\text{negative saddle}}\\ \underbrace{(\mathrm{UP}_+,\mathrm{UP}_+)}_{\text{positive saddle}} & \underbrace{(\mathrm{DP}_-,\mathrm{DP}_+)}_{\text{CW center}}\end{matrix}\right\}. \tag{3.104}$$

- *For* $a_{j_1j_20} < 0$ *and* $a_{j_2j_10} < 0$,

$$\begin{Bmatrix} (a_{j_1j_2}3, a_{j_2j_1}1) & (a_{j_1j_2}3, a_{j_2j_1}2) \\ (a_{j_1j_2}2, a_{j_2j_1}1) & (a_{j_1j_2}2, a_{j_2j_1}2) \\ (a_{j_1j_2}1, a_{j_2j_1}1) & (a_{j_1j_2}1, a_{j_2j_1}2) \end{Bmatrix} = \begin{Bmatrix} \underbrace{(\mathrm{DP}_+, \mathrm{DP}_-)}_{\text{CCW center}} & \underbrace{(\mathrm{UP}_-, \mathrm{UP}_-)}_{\text{negative saddle}} \\ \underbrace{(\mathrm{UP}_+, \mathrm{UP}_+)}_{\text{positive saddle}} & \underbrace{(\mathrm{DP}_-, \mathrm{DP}_+)}_{\text{CW center}} \\ \underbrace{(\mathrm{DP}_+, \mathrm{DP}_-)}_{\text{CCW center}} & \underbrace{(\mathrm{UP}_-, \mathrm{UP}_-)}_{\text{negative saddle}} \end{Bmatrix}. \tag{3.105}$$

(xiii) Under

$$\Delta_{j_1j_2} < 0 \text{ and } \Delta_{j_2j_1} = 0, \tag{3.106}$$

the standard form is

$$\begin{aligned} \dot{x}_{j_2} &= a_{j_2j_10}(x_{j_1} - a_{j_2j_1}1)^2, \\ \dot{x}_{j_1} &= a_{j_1j_20}(x_{j_2} - a_{j_1j_2}1)\left[(x_{j_2} - a_{j_1j_2})^2 + b_{j_1j_2}\right], \\ & j_1, j_2 \in \{1,2\}; j_1 \neq j_2. \end{aligned} \tag{3.107}$$

The equilibrium of $(x^*_{j_2}, x^*_{j_1}) = (a_{j_1j_2}1, a_{j_2j_1}1)$ *possesses the following appearing bifurcations:*

- *For* $a_{j_1j_20} > 0$ *and* $a_{j_2j_10} > 0$,

$$\underbrace{(a_{j_1j_2}1, a_{j_2j_1}1)}_{(\mathrm{UP},\mathrm{US})\text{-appearing bifurcation}} \rightleftharpoons \begin{Bmatrix} \underbrace{(a_{j_1j_2}1, a_{j_2j_1}2)}_{(\mathrm{UP}_+,\mathrm{UP}_+)\text{-positive saddle}} \\ \underbrace{(a_{j_1j_2}1, a_{j_2j_1}1)}_{(\mathrm{DP}_-,\mathrm{DP}_+)\text{-CW center}} \end{Bmatrix}. \tag{3.108}$$

  The (UP,US)*-up-parabola upper-saddle at* $(x^*_{j_2}, x^*_{j_1}) = (a_{j_1j_2}1, a_{j_2j_1}1)$ *is a* (UP,US)*-appearing bifurcation for* (UP_+,UP_+)*-positive saddle and* (DP_-,DP_+)*-clockwise center.*

- *For* $a_{j_1j_20} < 0$ *and* $a_{j_2j_10} > 0$,

$$\underbrace{(a_{j_1j_2}1, a_{j_2j_1}1)}_{(\mathrm{DP},\mathrm{US})\text{-appearing bifurcation}} \rightleftharpoons \begin{Bmatrix} \underbrace{(a_{j_1j_2}1, a_{j_2j_1}2)}_{(\mathrm{DP}_+,\mathrm{DP}_-)\text{-CCW center}} \\ \underbrace{(a_{j_1j_2}1, a_{j_2j_1}1)}_{(\mathrm{UP}_-,\mathrm{UP}_-)\text{-negative saddle}} \end{Bmatrix}. \tag{3.109}$$

The (DP,US)-*down-parabola upper-saddle at* $(x^*_{j_2}, x^*_{j_1}) = (a_{j_1 j_2 1}, a_{j_2 j_1 1})$ *is a* (DP, US)-*appearing bifurcation for* (DP$_+$,DP$_-$)-*counter-clockwise center and* (UP$_-$,UP$_-$)-*negative saddle.*

- *For* $a_{j_1 j_2 0} > 0$ *and* $a_{j_2 j_1 0} < 0$,

$$\underbrace{(a_{j_1 j_2 1}, a_{j_2 j_1 1})}_{(\text{DP},\text{LS})\text{-appearing bifurcation}} \rightleftharpoons \left\{ \begin{matrix} \underbrace{(a_{j_1 j_2 1}, a_{j_2 j_1 2})}_{(\text{DP}_-,\text{DP}_+)\text{-CW center}} \\ \underbrace{(a_{j_1 j_2 1}, a_{j_2 j_1 1})}_{(\text{UP}_+,\text{UP}_+)\text{-positive saddle}} \end{matrix} \right\}. \tag{3.110}$$

The (DP,LS)-*down-parabola lower-saddle at* $(x^*_{j_2}, x^*_{j_1}) = (a_{j_1 j_2 1}, a_{j_2 j_1 1})$ *is an* (DP,LS)-*appearing bifurcation for* (DP$_-$,DP$_+$)-*clockwise center and* (UP$_+$, UP$_+$)-positive saddle.

- *For* $a_{j_1 j_2 0} < 0$ *and* $a_{j_2 j_1 0} < 0$,

$$\underbrace{(a_{j_1 j_2 1}, a_{j_2 j_1 1})}_{(\text{UP},\text{LS})\text{-appearing bifurcation}} \rightleftharpoons \left\{ \begin{matrix} \underbrace{(a_{j_1 j_2 1}, a_{j_2 j_1 2})}_{(\text{UP}_-,\text{UP}_-)\text{-negative saddle}} \\ \underbrace{(a_{j_1 j_2 1}, a_{j_2 j_1 1})}_{(\text{DP}_+,\text{DP}_-)\text{-CCW center}} \end{matrix} \right\}. \tag{3.111}$$

The (UP,LS)-*up-parabola lower-saddle at* $(x^*_{j_2}, x^*_{j_1}) = (a_{j_1 j_2 1}, a_{j_2 j_1 1})$ *is a* (UP, LS)-*appearing bifurcation for* (UP$_-$,UP$_-$)-*negative saddle and* (DP$_+$, DP$_-$)-counter-*clockwise center.*

(xiv) Under

$$\Delta_{j_1 j_2} = 0 \text{ and } \Delta_{j_2 j_1} < 0, a_{j_1 j_2 s_1} \neq a_{j_1 j_2 s_2}, \tag{3.112}$$

then

$$\begin{aligned} \dot{x}_{j_1} &= a_{j_1 j_2 0}(x_{j_2} - a_{j_1 j_2 s_1})(x_{j_2} - a_{j_1 j_2 s_2})^2, \\ \dot{x}_{j_2} &= a_{j_2 j_1 0}\left[(x_{j_1} - a_{j_2 j_1})^2 + b_{j_2 j_1}\right] \\ j_1, j_2 &\in \{1,2\}, j_2 \neq j_1. \end{aligned} \tag{3.113}$$

The equilibrium state of $x^*_{j_2} = a_{j_1 j_2 s_2}$ $(s_1, s_2 \in \{1,2\}, s_1 \neq s_2)$ *in the* x_{j_2}-*direction is for the following appearing bifurcations:*

- *For* $a_{j_1j_20}(a_{j_1j_2s_2}-a_{j_1j_2s_1})>0$ *and* $a_{j_2j_10}>0$,

$$\underbrace{(a_{j_1j_2s_2},\dot{x}_{j_2})}_{\text{(II, pF)-flow appearing bifurcation}} \rightleftharpoons \left\{\begin{array}{c}\underbrace{(a_{j_1j_2(s_2)_1},\dot{x}_{j_2})}_{\text{(DP, pF)-flow}}\\ \underbrace{(a_{j_1j_2(s_2)_2},\dot{x}_{j_2})}_{\text{(UP, pF)-flow}}\end{array}\right\}. \tag{3.114}$$

  The (II,pF)*-positive increasing-inflection flow is the* (II,pF)*-appearing bifurcations for up-parabola and down-parabola flows.*
- *For* $a_{j_1j_20}(a_{j_1j_2s_2}-a_{j_1j_2s_1})<0$ *and* $a_{j_2j_10}>0$,

$$\underbrace{(a_{j_1j_2s_2},\dot{x}_{j_2})}_{\text{(DI, pF)-flow appearing bifurcation}} \rightleftharpoons \left\{\begin{array}{c}\underbrace{(a_{j_1j_2(s_2)_1},\dot{x}_{j_2})}_{\text{(UP, pF)-flow}}\\ \underbrace{(a_{j_1j_2(s_2)_2},\dot{x}_{j_2})}_{\text{(DP, pF)-flow}}\end{array}\right\}. \tag{3.115}$$

  The (DI,pF)*-positive decreasing-inflection flow is the* (DI,pF)*-appearing bifurcations for up-parabola and down-parabola flows.*
- *For* $a_{j_1j_20}(a_{j_1j_2s_2}-a_{j_1j_2s_1})>0$ *and* $a_{j_2j_10}<0$,

$$\underbrace{(a_{j_1j_2s_2},\dot{x}_{j_2})}_{\text{(DI, nF)-flow appearing bifurcation}} \rightleftharpoons \left\{\begin{array}{c}\underbrace{(a_{j_1j_2(s_2)_1},\dot{x}_{j_2})}_{\text{(UP, nF)-flow}}\\ \underbrace{(a_{j_1j_2(s_2)_2},\dot{x}_{j_2})}_{\text{(DP, nF)-flow}}\end{array}\right\}. \tag{3.116}$$

  The (DI,nF)*-negative decreasing-inflection flow is the* (DI,nF)*-appearing bifurcations for up-parabola and down-parabola flows.*
- *For* $a_{j_1j_20}(a_{j_1j_2s_2}-a_{j_1j_2s_1})<0$ *and* $a_{j_2j_10}<0$,

$$\underbrace{(a_{j_1j_2s_2},\dot{x}_{j_2})}_{\text{(II, nF)-flow appearing bifurcation}} \rightleftharpoons \left\{\begin{array}{c}\underbrace{(a_{j_1j_2(s_2)_1},\dot{x}_{j_2})}_{\text{(DP, nF)-flow}}\\ \underbrace{(a_{j_1j_2(s_2)_2},\dot{x}_{j_2})}_{\text{(UP, nF)-flow}}\end{array}\right\}. \tag{3.117}$$

  The (II,nF)*-negative increasing-inflection flow is the* (II,nF)*-appearing bifurcations for up-parabola and down-parabola flows.*

– *Such inflection bifurcations are the switching bifurcations for up-parabola and down-parabola flows.*

(xv) Under

$$\Delta_{j_1j_2} = 0 \text{ and } \Delta_{j_2j_1} < 0, a_{j_1j_2s_1} = a_{j_1j_2s_2} = a_{j_1j_2 1}, \tag{3.118}$$

the standard form is

$$\begin{aligned} \dot{x}_{j_1} &= a_{j_1j_2 0}(x_{j_2} - a_{j_1j_2 1})^3, \\ \dot{x}_{j_2} &= a_{j_2j_1 0}\left[(x_{j_1} - a_{j_2j_1})^2 + b_{j_2j_1}\right] \\ j_1, j_2 &\in \{1,2\}, j_2 \neq j_1. \end{aligned} \tag{3.119}$$

The equilibrium state of $x_{j_2}^* = a_{j_1j_2 1}$ in the x_{j_2}-direction *is for the following appearing (switching) bifurcations:*

- *For* $a_{j_1j_2 0} > 0$ *and* $a_{j_2j_1 0} > 0$,

$$\underbrace{(a_{j_1j_2 1}, \dot{x}_{j_2})}_{(\text{UP, pF})\text{-flow}} \rightleftharpoons \underbrace{(a_{j_1j_2 1}, \dot{x}_{j_2})}_{\substack{(3^{\text{rd}}\text{UP, pF})\text{-flow} \\ \text{appearing bifurcation}}} \rightleftharpoons \left\{ \begin{array}{c} \underbrace{(a_{j_1j_2 3}, \dot{x}_{j_2})}_{(\text{UP, pF})\text{-flow}} \\ \underbrace{(a_{j_1j_2 2}, \dot{x}_{j_2})}_{(\text{DP, pF})\text{-flow}} \\ \underbrace{(a_{j_1j_2 3}, \dot{x}_{j_2})}_{(\text{UP, pF})\text{-flow}} \end{array} \right\}. \tag{3.120}$$

  The (3^{rd}UP,pF)-*third-order positive up-parabola flow is a* (3^{rd}UP,pF)-*appearing bifurcation from an up-parabola flow to up-parabola, down-parabola, and up-parabola flows.*
- *For* $a_{j_1j_2 0} < 0$ *and* $a_{j_2j_1 0} > 0$,

$$\underbrace{(a_{j_1j_2 1}, \dot{x}_{j_2})}_{(\text{DP, pF})\text{-flow}} \rightleftharpoons \underbrace{(a_{j_1j_2 1}, \dot{x}_{j_2})}_{\substack{(3^{\text{rd}}\text{DP, pF})\text{-flow} \\ \text{appearing bifurcation}}} \rightleftharpoons \left\{ \begin{array}{c} \underbrace{(a_{j_1j_2 3}, \dot{x}_{j_2})}_{(\text{DP, pF})\text{-flow}} \\ \underbrace{(a_{j_1j_2 2}, \dot{x}_{j_2})}_{(\text{UP, pF})\text{-flow}} \\ \underbrace{(a_{j_1j_2 3}, \dot{x}_{j_2})}_{(\text{DP, pF})\text{-flow}} \end{array} \right\}. \tag{3.121}$$

  The (3^{rd}DP,pF)-*third-order positive down-parabola flow is a* (3^{rd}DP,pF)-*appearing bifurcation from a down-parabola flow to down-parabola, up-parabola, and down-parabola flows.*
- *For* $a_{j_1j_2 0} > 0$ *and* $a_{j_2j_1 0} < 0$,

$$\underbrace{(a_{j_1j_21},\dot{x}_{j_2})}_{(\mathrm{DP,nF})\text{-flow}} \rightleftharpoons \underbrace{(a_{j_1j_21},\dot{x}_{j_2})}_{\substack{(3^{\mathrm{rd}}\mathrm{DP,nF})\text{-flow}\\ \text{appearing bifurcation}}} \rightleftharpoons \left\{\begin{array}{c} \underbrace{(a_{j_1j_23},\dot{x}_{j_2})}_{(\mathrm{DP,nF})\text{-flow}} \\ \underbrace{(a_{j_1j_22},\dot{x}_{j_2})}_{(\mathrm{UP,nF})\text{-flow}} \\ \underbrace{(a_{j_1j_23},\dot{x}_{j_2})}_{(\mathrm{DP,nF})\text{-flow}} \end{array}\right\}. \tag{3.122}$$

The (3^{rd}DP,nF)*-third-order negative down-parabola flow is a* (3^{rd}DP,nF)*-appearing bifurcation from a down-parabola flow to down-parabola, up-parabola, and down-parabola flows.*

- *For* $a_{j_1j_20}<0$ *and* $a_{j_2j_10}<0$,

$$\underbrace{(a_{j_1j_21},\dot{x}_{j_2})}_{(\mathrm{UP,pF})\text{-flow}} \rightleftharpoons \underbrace{(a_{j_1j_21},\dot{x}_{j_2})}_{\substack{(3^{\mathrm{rd}}\mathrm{UP,pF})\text{-flow}\\ \text{appearing bifucation}}} \rightleftharpoons \left\{\begin{array}{c} \underbrace{(a_{j_1j_23},\dot{x}_{j_2})}_{(\mathrm{UP,pF})\text{-flow}} \\ \underbrace{(a_{j_1j_22},\dot{x}_{j_2})}_{(\mathrm{DP,pF})\text{-flow}} \\ \underbrace{(a_{j_1j_23},\dot{x}_{j_2})}_{(\mathrm{UP,pF})\text{-flow}} \end{array}\right\}. \tag{3.123}$$

The (3^{rd}UP,nF)*-third-order negative up-parabola flow is a* (3^{rd}UP,nF)*-appearing bifurcation from an up-parabola flow to up-parabola, down-parabola, and up-parabola flows.*

(xvi) Under

$$\Delta_{j_1j_2}=0 \text{ and } \Delta_{j_2j_1}=0, a_{j_2j_1s_1}\neq a_{j_2j_1s_2}, \tag{3.124}$$

the standard form is

$$\begin{aligned} &\dot{x}_{j_1}=a_{j_1j_20}(x_{j_2}-a_{j_1j_2s_1})(x_{j_2}-a_{j_1j_2s_2})^2,\\ &\dot{x}_{j_2}=a_{j_2j_10}(x_{j_1}-a_{j_2j_11})^2\\ &j_1,j_2\in\{1,2\};j_1\neq j_2. \end{aligned} \tag{3.125}$$

(xvi$_1$) The equilibrium of $(x^*_{j_2},x^*_{j_1})=(a_{j_1j_2s_1},a_{j_2j_11})$ *has the following appearing bifurcations:*

- *For* $a_{j_1j_20}>0$ *and* $a_{j_2j_10}>0$,

$$\underbrace{(a_{j_1j_2s_1}, a_{j_2j_1}1)}_{(\mathrm{UP},\mathrm{US})\text{-appearing bifurcation}} \rightleftharpoons \left\{ \begin{array}{c} \underbrace{(a_{j_1j_2s_1}, a_{j_2j_1}2)}_{(\mathrm{UP}_+,\mathrm{UP}_+)\text{-positive saddle}} \\ \underbrace{(a_{j_1j_2s_1}, a_{j_2j_1}1)}_{(\mathrm{DP}_-,\mathrm{DP}_+)\text{-CW center}} \end{array} \right\}. \tag{3.126}$$

The (UP,US)-*up-parabola upper-saddle is a* (UP,US)-*appearing bifurcation for center and saddle.*

- *For $a_{j_1j_2 0} < 0$ and $a_{j_2j_1 0} > 0$,*

$$\underbrace{(a_{j_1j_2s_1}, a_{j_2j_1}1)}_{(\mathrm{DP},\mathrm{US})\text{-appearing bifurcation}} \rightleftharpoons \left\{ \begin{array}{c} \underbrace{(a_{j_1j_2s_1}, a_{j_2j_1}2)}_{(\mathrm{DP}_+,\mathrm{DP}_-)\text{-CCW center}} \\ \underbrace{(a_{j_1j_2s_1}, a_{j_2j_1}1)}_{(\mathrm{UP}_-,\mathrm{UP}_-)\text{-negative saddle}} \end{array} \right\}. \tag{3.127}$$

The (DP,US)-*down-parabola upper-saddle is a* (DP,US)-*appearing bifurcation for center and saddle.*

- *For $a_{j_1j_2 0} > 0$ and $a_{j_2j_1 0} < 0$,*

$$\underbrace{(a_{j_1j_2s_1}, a_{j_2j_1}1)}_{(\mathrm{DP},\mathrm{LS})\text{-appearing bifurcation}} \rightleftharpoons \left\{ \begin{array}{c} \underbrace{(a_{j_1j_2s_1}, a_{j_2j_1}2)}_{(\mathrm{DP}_-,\mathrm{DP}_+)\text{-CW center}} \\ \underbrace{(a_{j_1j_2s_1}, a_{j_2j_1}1)}_{(\mathrm{UP}_+,\mathrm{UP}_+)\text{-positive saddle}} \end{array} \right\}. \tag{3.128}$$

The (DP,LS)-*down-parabola lower-saddle is a* (DP,LS)-*appearing bifurcation for center and saddle.*

- *For $a_{j_1j_2 0} < 0$ and $a_{j_2j_1 0} < 0$,*

$$\underbrace{(a_{j_1j_2s_1}, a_{j_2j_1}1)}_{(\mathrm{UP},\mathrm{LS})\text{-appearing bifurcation}} \rightleftharpoons \left\{ \begin{array}{c} \underbrace{(a_{j_1j_2s_1}, a_{j_2j_1}2)}_{(\mathrm{UP}_-,\mathrm{UP}_-)\text{-negative saddle}} \\ \underbrace{(a_{j_1j_2s_1}, a_{j_2j_1}1)}_{(\mathrm{DP}_+,\mathrm{DP}_-)\text{-CCW center}} \end{array} \right\}. \tag{3.129}$$

The (UP,LS)-*up-parabola lower-saddle is a* (UP,LS)-*appearing bifurcation for center and saddle.*

(xvi$_2$) The equilibrium of $(x^*_{j_2}, x^*_{j_1}) = (a_{j_1j_2s_2}, a_{j_2j_1}1)$ has the following bifurcations:

- *For $a_{j_1j_2 0}(a_{j_1j_2s_2} - a_{j_1j_2s_1}) > 0$ and $a_{j_2j_1 0} > 0$,*

$$\underbrace{(a_{j_1j_2s_2}, a_{j_2j_1}1)}_{(\text{II}_+,\text{II}_+)\text{-appearing bifurcation}} \rightleftharpoons \left\{ \begin{array}{cc} \underbrace{(a_{j_1j_2(s_2)_2}, a_{j_2j_1}1)}_{(\text{DP}_-,\text{DP}_+)\text{-CW center}} & \underbrace{(a_{j_1j_2(s_2)_2}, a_{j_2j_1}2)}_{(\text{UP}_+,\text{UP}_+)\text{-positive saddle}} \\ \underbrace{(a_{j_1j_2(s_2)_1}, a_{j_2j_1}1)}_{(\text{UP}_-,\text{UP}_-)\text{-negative saddle}} & \underbrace{(a_{j_1j_2(s_2)_1}, a_{j_2j_1}2)}_{(\text{DP}_+,\text{DP}_-)\text{-CCW center}} \end{array} \right\}. \tag{3.130}$$

The (II$_+$,II$_+$)*-double increasing-inflection saddle is an* (II$_+$,II$_+$)*-appearing bifurcation for a network of saddles and centers.*

- *For* $a_{j_1j_20}(a_{j_1j_2s_2} - a_{j_1j_2s_1}) < 0$ *and* $a_{j_2j_10} > 0$,

$$\underbrace{(a_{j_1j_2s_2}, a_{j_2j_1}1)}_{(\text{DI}_+,\text{DI}_-)\text{-appearing bifurcation}} \rightleftharpoons \left\{ \begin{array}{cc} \underbrace{(a_{j_1j_2(s_2)_2}, a_{j_2j_1}1)}_{(\text{UP}_-,\text{UP}_-)\text{-negative saddle}} & \underbrace{(a_{j_1j_2(s_2)_2}, a_{j_2j_1}2)}_{(\text{DP}_+,\text{DP}_-)\text{-CCW center}} \\ \underbrace{(a_{j_1j_2(s_2)_1}, a_{j_2j_1}1)}_{(\text{DP}_-,\text{DP}_+)\text{-CW center}} & \underbrace{(a_{j_1j_2(s_2)_1}, a_{j_2j_1}2)}_{(\text{UP}_+,\text{UP}_+)\text{-positive saddle}} \end{array} \right\}. \tag{3.131}$$

The (DI$_+$,DI$_-$)*-double decreasing-inflection saddle is a* (DI$_+$,DI$_-$)*-appearing bifurcation for a network of saddles and centers.*

- *For* $a_{j_1j_20}(a_{j_1j_2s_2} - a_{j_1j_2s_1}) > 0$ *and* $a_{j_2j_10} < 0$,

$$\underbrace{(a_{j_1j_2s_2}, a_{j_2j_1}1)}_{(\text{DI}_-,\text{DI}_+)\text{-appearing bifurcation}} \rightleftharpoons \left\{ \begin{array}{cc} \underbrace{(a_{j_1j_2(s_2)_2}, a_{j_2j_1}1)}_{(\text{UP}_+,\text{UP}_+)\text{-positive saddle}} & \underbrace{(a_{j_1j_2(s_2)_2}, a_{j_2j_1}2)}_{(\text{DP}_-,\text{DP}_+)\text{-CW center}} \\ \underbrace{(a_{j_1j_2(s_2)_1}, a_{j_2j_1}1)}_{(\text{DP}_+,\text{DP}_-)\text{-CCW center}} & \underbrace{(a_{j_1j_2(s_2)_1}, a_{j_2j_1}2)}_{(\text{UP}_-,\text{UP}_-)\text{-negative saddle}} \end{array} \right\}. \tag{3.132}$$

The (DI$_-$,DI$_+$)*-double decreasing-inflection saddle is a* (DI$_-$,DI$_+$)*-appearing bifurcation for a network of saddles and centers.*

- *For* $a_{j_1j_20}(a_{j_1j_2s_2} - a_{j_1j_2s_1}) < 0$ *and* $a_{j_2j_10} < 0$,

$$\underbrace{(a_{j_1j_2s_2}, a_{j_2j_1}1)}_{(\text{II}_-,\text{II}_-)\text{-appearing bifurcation}} \rightleftharpoons \left\{ \begin{array}{cc} \underbrace{(a_{j_1j_2(s_2)_2}, a_{j_2j_1}1)}_{(\text{DP}_+,\text{DP}_-)\text{-CCW center}} & \underbrace{(a_{j_1j_2(s_2)_2}, a_{j_2j_1}2)}_{(\text{UP}_-,\text{UP}_-)\text{-negative saddle}} \\ \underbrace{(a_{j_1j_2(s_2)_1}, a_{j_2j_1}1)}_{(\text{UP}_+,\text{UP}_+)\text{-positive saddle}} & \underbrace{(a_{j_1j_2(s_2)_1}, a_{j_2j_1}2)}_{(\text{DP}_-,\text{DP}_+)\text{-CW center}} \end{array} \right\}. \tag{3.133}$$

The (II$_-$,II$_-$)*-double increasing-inflection saddle is an* (II$_-$,II$_-$)*-appearing bifurcation for a network of saddles and centers.*

(xvii) Under

$$\Delta_{j_1j_2}=0 \text{ and } \Delta_{j_2j_1}=0, a_{j_2j_1s_1}=a_{j_2j_1s_2}=a_{j_2j_11}, \tag{3.134}$$

the standard form is

$$\begin{aligned}
&\dot{x}_{j_1}=a_{j_1j_20}(x_{j_2}-a_{j_1j_21})^3,\\
&\dot{x}_{j_2}=a_{j_2j_10}(x_{j_1}-a_{j_2j_11})^2,\\
&j_1,j_2\in\{1,2\};j_1\neq j_2.
\end{aligned} \tag{3.135}$$

The equilibrium of $(x_{j_2}^*,x_{j_1}^*)=(a_{j_1j_21},a_{j_2j_11})$ *has the following appearing bifurcations:*

- *For* $a_{j_1j_20}>0$ *and* $a_{j_2j_10}>0$,

$$\underbrace{(a_{j_1j_21},a_{j_2j_11})}_{(3^{\text{rd}}\text{UP},\,\text{US})\text{-appearing bifurcation}} \rightleftharpoons \left\{\begin{array}{cc}
\underbrace{(a_{j_1j_23},a_{j_2j_11})}_{(\text{DP}_-,\,\text{DP}_+)\text{-CW center}} & \underbrace{(a_{j_1j_23},a_{j_2j_12})}_{(\text{UP}_+,\,\text{UP}_+)\text{-positive saddle}}\\
\underbrace{(a_{j_1j_22},a_{j_2j_11})}_{(\text{UP}_-,\,\text{UP}_-)\text{-negative saddle}} & \underbrace{(a_{j_1j_21},a_{j_2j_12})}_{(\text{DP}_+,\,\text{DP}_-)\text{-CCW center}}\\
\underbrace{(a_{j_1j_21},a_{j_2j_11})}_{(\text{DP}_-,\,\text{DP}_+)\text{-CW center}} & \underbrace{(a_{j_1j_21},a_{j_2j_12})}_{(\text{UP}_+,\,\text{UP}_+)\text{-positive saddle}}
\end{array}\right\}. \tag{3.136}$$

The (3^rd^UP,US)-*third-order-up-parabola upper-saddle is a* (3^rd^UP,US)-*appearing bifurcation for a network of saddles and centers.*

- *For* $a_{j_1j_20}<0$ *and* $a_{j_2j_10}>0$,

$$\underbrace{(a_{j_1j_21},a_{j_2j_11})}_{(3^{\text{rd}}\text{DP},\,\text{US})\text{-appearing bifucation}} \rightleftharpoons \left\{\begin{array}{cc}
\underbrace{(a_{j_1j_23},a_{j_2j_11})}_{(\text{UP}_-,\,\text{UP}_-)\text{-negative saddle}} & \underbrace{(a_{j_1j_23},a_{j_2j_12})}_{(\text{DP}_+,\,\text{DP}_-)\text{-CCW center}}\\
\underbrace{(a_{j_1j_22},a_{j_2j_11})}_{(\text{DP}_-,\,\text{DP}_+)\text{-CW center}} & \underbrace{(a_{j_1j_21},a_{j_2j_12})}_{(\text{UP}_+,\,\text{UP}_+)\text{-positive saddle}}\\
\underbrace{(a_{j_1j_21},a_{j_2j_11})}_{(\text{UP}_-,\,\text{UP}_-)\text{-negative saddle}} & \underbrace{(a_{j_1j_21},a_{j_2j_12})}_{(\text{DP}_+,\,\text{UP}_-)\text{-CCW center}}
\end{array}\right\}. \tag{3.137}$$

The (3^{rd}DP,US)-*third-order*-down-*parabola upper-saddle is a* (3^{rd}DP,US)-*appearing bifurcation for a network of saddles and centers.*

- *For* $a_{j_1j_20}>0$ *and* $a_{j_2j_10}<0$,

$$\underbrace{(a_{j_1j_2}1, a_{j_2j_1}1)}_{(3^{\text{rd}}\text{DP, LS})\text{-appearing bifucation}} \rightleftharpoons \left\{ \begin{array}{cc} \underbrace{(a_{j_1j_2}3, a_{j_2j_1}1)}_{(\text{UP}_+, \text{UP}_+)\text{-positive saddle}} & \underbrace{(a_{j_1j_2}3, a_{j_2j_1}2)}_{(\text{DP}_-, \text{DP}_+)\text{-CW center}} \\ \underbrace{(a_{j_1j_2}2, a_{j_2j_1}1)}_{(\text{DP}_+, \text{DP}_-)\text{-CCW center}} & \underbrace{(a_{j_1j_2}1, a_{j_2j_1}2)}_{(\text{UP}_-, \text{UP}_-)\text{-negative saddle}} \\ \underbrace{(a_{j_1j_2}1, a_{j_2j_1}1)}_{(\text{UP}_+, \text{UP}_+)\text{-positive saddle}} & \underbrace{(a_{j_1j_2}1, a_{j_2j_1}2)}_{(\text{DP}_-, \text{DP}_+)\text{-CW center}} \end{array} \right\}. \tag{3.138}$$

The $(3^{\text{rd}}\text{DP,LS})$*-third-order down-parabola lower-saddle is a* $(3^{\text{rd}}\text{DP,LS})$*-appearing bifurcation for a network of saddles and centers.*

- *For* $a_{j_1j_2}0 < 0$ *and* $a_{j_2j_1}0 < 0$,

$$\underbrace{(a_{j_1j_2}1, a_{j_2j_1}1)}_{(3^{\text{rd}}\text{UP, LS})\text{-appearing bifurcation}} \rightleftharpoons \left\{ \begin{array}{cc} \underbrace{(a_{j_1j_2}3, a_{j_2j_1}1)}_{(\text{DP}_+, \text{DP}_-)\text{-CCW center}} & \underbrace{(a_{j_1j_2}3, a_{j_2j_1}2)}_{(\text{UP}_-, \text{UP}_-)\text{-negative saddle}} \\ \underbrace{(a_{j_1j_2}2, a_{j_2j_1}1)}_{(\text{UP}_+, \text{UP}_+)\text{-positive saddle}} & \underbrace{(a_{j_1j_2}1, a_{j_2j_1}2)}_{(\text{DP}_-, \text{DP}_+)\text{-CW center}} \\ \underbrace{(a_{j_1j_2}1, a_{j_2j_1}1)}_{(\text{DP}_+, \text{DP}_-)\text{-CCW center}} & \underbrace{(a_{j_1j_2}1, a_{j_2j_1}2)}_{(\text{UP}_-, \text{UP}_-)\text{-negative saddle}} \end{array} \right\}. \tag{3.139}$$

The $(3^{\text{rd}}\text{UP,LS})$*-third-order-up-parabola lower-saddle is a* $(3^{\text{rd}}\text{UP,LS})$*-appearing bifurcation for a network of saddles and centers.*

(xix) Under

$$\Delta_{j_1j_2} = 0 \text{ and } \Delta_{j_2j_1} > 0, \tag{3.140}$$

the standard form is

$$\begin{aligned} \dot{x}_{j_2} &= a_{j_2j_1}0(x_{j_1} - a_{j_2j_1}1)(x_{j_1} - a_{j_2j_1}2), \\ \dot{x}_{j_1} &= a_{j_1j_2}0(x_{j_2} - a_{j_1j_2s_1})(x_{j_2} - a_{j_1j_2s_2})^2, \\ j_1, j_2 &\in \{1,2\}; j_1 \neq j_2. \end{aligned} \tag{3.141}$$

The equilibrium of $(x_{j_2}^*, x_{j_1}^*) = (a_{j_1j_2s_2}, a_{j_2j_1l_1})$ $(s_1, s_2, l_1, l_2 \in \{1,2\}, s_1 \neq s_2; l_1 \neq l_2)$ *has the following appearing bifurcation properties:*

- *For* $a_{j_1j_2}0(a_{j_1j_2s_2} - a_{j_1j_2s_1}) > 0$ *and* $a_{j_2j_1}0(a_{j_2j_1l_1} - a_{j_2j_1l_2}) > 0$,

$$\underbrace{(a_{j_1j_2s_2}, a_{j_2j_1l_1})}_{(\text{US},\,\text{UP})\text{-appearing bifurcation}} \rightleftharpoons \left\{ \begin{array}{c} \underbrace{(a_{j_1j_2(s_2)_2}, a_{j_2j_1l_1})}_{(\text{UP}_+,\,\text{UP}_+)\text{-positive saddle}} \\ \underbrace{(a_{j_1j_2(s_2)_1}, a_{j_2j_1l_1})}_{(\text{DP}_+,\,\text{DP}_-)\text{-CCW center}} \end{array} \right\}. \tag{3.142}$$

The (US,UP)*-up-parabola upper-saddle is a* (US,UP)*-appearing bifurcation for saddle and center.*

- *For* $a_{j_1j_2 0}(a_{j_1j_2s_2} - a_{j_1j_2s_1}) < 0$ *and* $a_{j_2j_1 0}(a_{j_2j_1l_1} - a_{j_2j_1l_2}) > 0$,

$$\underbrace{(a_{j_1j_2s_2}, a_{j_2j_1l_1})}_{(\text{LS},\,\text{DP})\text{-appearing bifurcation}} \rightleftharpoons \left\{ \begin{array}{c} \underbrace{(a_{j_1j_2(s_2)_2}, a_{j_2j_1l_1})}_{(\text{DP}_+,\,\text{DP}_-)\text{-CCW saddle}} \\ \underbrace{(a_{j_1j_2(s_2)_1}, a_{j_2j_1l_1})}_{(\text{UP}_+,\,\text{UP}_+)\text{-positive saddle}} \end{array} \right\}. \tag{3.143}$$

The (LS,DP)*-down-parabola lower-saddle is an* (LS,DP)*-appearing bifurcation for saddle and center.*

- *For* $a_{j_1j_2 0}(a_{j_1j_2s_2} - a_{j_1j_2s_1}) > 0$ *and* $a_{j_2j_1 0}(a_{j_2j_1l_1} - a_{j_2j_1l_2}) < 0$,

$$\underbrace{(a_{j_1j_2s_2}, a_{j_2j_1l_1})}_{(\text{US},\,\text{DP})\text{-appearing bifurcation}} \rightleftharpoons \left\{ \begin{array}{c} \underbrace{(a_{j_1j_2(s_2)_2}, a_{j_2j_1l_1})}_{(\text{DP}_-,\,\text{DP}_+)\text{-CW center}} \\ \underbrace{(a_{j_1j_2(s_2)_1}, a_{j_2j_1l_1})}_{(\text{UP}_-,\,\text{UP}_-)\text{-negative saddle}} \end{array} \right\}. \tag{3.144}$$

The (US,DP)*-down-parabola upper-saddle is a* (US,DP)*-appearing bifurcation for saddle and center.*

- *For* $a_{j_1j_2 0}(a_{j_1j_2s_2} - a_{j_1j_2s_1}) < 0$ *and* $a_{j_2j_1 0}(a_{j_2j_1l_1} - a_{j_2j_1l_2}) < 0$,

$$\underbrace{(a_{j_1j_2s_2}, a_{j_2j_1l_1})}_{(\text{LS},\,\text{UP})\text{-appearing bifurcation}} \rightleftharpoons \left\{ \begin{array}{c} \underbrace{(a_{j_1j_2(s_2)_2}, a_{j_2j_1l_1})}_{(\text{UP}_-,\,\text{UP}_-)\text{-negative saddle}} \\ \underbrace{(a_{j_1j_2(s_2)_1}, a_{j_2j_1l_1})}_{(\text{DP}_-,\,\text{DP}_+)\text{-CW center}} \end{array} \right\}. \tag{3.145}$$

The (LS,UP)*-up-parabola lower-saddle is an* (LS,UP)*-appearing bifurcation for saddle and center.*

(xx) Under

$$\Delta_{j_1j_2}=0 \text{ and } \Delta_{j_2j_1}>0, a_{j_1j_2s_1}=a_{j_1j_2s_2}=a_{j_1j_2}1, \tag{3.146}$$

the standard form is

$$\begin{aligned}
&\dot{x}_{j_2}=a_{j_2j_1}0(x_{j_1}-a_{j_2j_1}1)(x_{j_1}-a_{j_2j_1}2),\\
&\dot{x}_{j_1}=a_{j_1j_2}0(x_{j_2}-a_{j_1j_2}1)^3,\\
&j_1,j_2\in\{1,2\};j_1\neq j_2.
\end{aligned} \tag{3.147}$$

The equilibrium of $(x^*_{j_2},x^*_{j_1})=(a_{j_1j_2}1,a_{j_2j_1l_1})$ $(l_1, l_2\in\{1,2\}, l_1\neq l_2)$ *has the following appearing bifurcations:*

- *For* $a_{j_1j_2}0>0$ *and* $a_{j_2j_1}0(a_{j_2j_1l_1}-a_{j_2j_1l_2})>0$,

$$\underbrace{(a_{j_1j_2}1,a_{j_2j_1l_1})}_{(\mathrm{UP}_+,\mathrm{UP}_+)\text{-positive saddle}} \rightleftharpoons \underbrace{(a_{j_1j_2}1,a_{j_2j_1l_1})}_{(3^{\mathrm{rd}}\mathrm{UP}_+,\mathrm{UP}_+)\text{-appearing bifucation}} \rightleftharpoons \left\{\begin{array}{c}\underbrace{(a_{j_1j_2}3,a_{j_2j_1l_1})}_{(\mathrm{UP}_+,\mathrm{UP}_+)\text{-positive saddle}}\\ \underbrace{(a_{j_1j_2}2,a_{j_2j_1l_1})}_{(\mathrm{DP}_+,\mathrm{DP}_-)\text{-CCW center}}\\ \underbrace{(a_{j_1j_2}1,a_{j_2j_1l_1})}_{(\mathrm{UP}_+,\mathrm{UP}_+)\text{-positive saddle}}\end{array}\right\}. \tag{3.148}$$

The $(3^{\mathrm{rd}}\mathrm{UP}_+,\mathrm{UP}_+)$*-third-order positive saddle is a* $(3^{\mathrm{rd}}\mathrm{UP}_+,\mathrm{UP}_+)$*-appearing bifurcation from a positive saddle to positive saddle, counter-clockwise center, and positive saddle.*

- *For* $a_{j_1j_2}0<0$ *and* $a_{j_2j_1}0(a_{j_2j_1l_1}-a_{j_2j_1l_2})>0$,

$$\underbrace{(a_{j_1j_2}1,a_{j_2j_1l_1})}_{(\mathrm{DP}_+,\mathrm{DP}_-)\text{-CCW center}} \rightleftharpoons \underbrace{(a_{j_1j_2}1,a_{j_2j_1l_1})}_{(3^{\mathrm{rd}}\mathrm{DP}_+,\mathrm{DP}_-)\text{-appearing bifucation}} \rightleftharpoons \left\{\begin{array}{c}\underbrace{(a_{j_1j_2}3,a_{j_2j_1l_1})}_{(\mathrm{DP}_+,\mathrm{DP}_-)\text{-CCW center}}\\ \underbrace{(a_{j_1j_2}2,a_{j_2j_1l_1})}_{(\mathrm{UP}_+,\mathrm{UP}_+)\text{-positive saddle}}\\ \underbrace{(a_{j_1j_2}1,a_{j_2j_1l_1})}_{(\mathrm{DP}_+,\mathrm{DP}_-)\text{-CCW center}}\end{array}\right\}. \tag{3.149}$$

The $(3^{\mathrm{rd}}\mathrm{DP}_+,\mathrm{DP}_-)$*-third-order-counter-clockwise center is a* $(3^{\mathrm{rd}}\mathrm{DP}_+,\mathrm{DP}_-)$*-appearing bifurcation from a counter-clockwise center to counter-clockwise center, positive saddle, and counter-clockwise center.*

- *For* $a_{j_1j_2}0>0$ *and* $a_{j_2j_1}0(a_{j_2j_1l_1}-a_{j_2j_1l_2})<0$,

$$\underbrace{(a_{j_1j_2}1, a_{j_2j_1l_1})}_{(\text{DP}_-, \text{DP}_+)\text{-CW center}} \rightleftharpoons \underbrace{(a_{j_1j_2}1, a_{j_2j_1l_1})}_{(3^{\text{rd}}\text{DP}_-, \text{DP}_+)\text{-appearing bifucation}} \rightleftharpoons \left\{ \begin{array}{c} \underbrace{(a_{j_1j_2}3, a_{j_2j_1l_1})}_{(\text{DP}_-, \text{DP}_+)\text{-CW center}} \\ \underbrace{(a_{j_1j_2}2, a_{j_2j_1l_1})}_{(\text{UP}_-, \text{UP}_-)\text{-negative saddle}} \\ \underbrace{(a_{j_1j_2}1, a_{j_2j_1l_1})}_{(\text{DP}_-, \text{DP}_+)\text{-CW center}} \end{array} \right\}. \tag{3.150}$$

The $(3^{\text{rd}}\text{DP}_-,\text{DP}_+)$*-third-order clockwise center is a* $(3^{\text{rd}}\text{DP}_-,\text{DP}_+)$*-appearing bifurcation from a clockwise center to clockwise center, negative saddle, and clockwise center.*

- *For* $a_{j_1j_2}0 < 0$ *and* $a_{j_2j_1}0(a_{j_2j_1l_1} - a_{j_2j_1l_2}) < 0$,

$$\underbrace{(a_{j_1j_2}1, a_{j_2j_1l_1})}_{(\text{UP}_-, \text{UP}_-)\text{-negative saddle}} \rightleftharpoons \underbrace{(a_{j_1j_2}1, a_{j_2j_1l_1})}_{(3^{\text{rd}}\text{UP}_-, \text{UP}_-)\text{-appearing bifucation}} \rightleftharpoons \left\{ \begin{array}{c} \underbrace{(a_{j_1j_2}3, a_{j_2j_1l_1})}_{(\text{UP}_-, \text{UP}_-)\text{-negative saddle}} \\ \underbrace{(a_{j_1j_2}2, a_{j_2j_1l_1})}_{(\text{DP}_-, \text{DP}_+)\text{-CW center}} \\ \underbrace{(a_{j_1j_2}1, a_{j_2j_1l_1})}_{(\text{UP}_-, \text{UP}_-)\text{-negative saddle}} \end{array} \right\}. \tag{3.151}$$

The $(3^{\text{rd}}\text{UP}_-,\text{UP}_-)$*-third-order negative saddle is a* $(3^{\text{rd}}\text{UP}_-,\text{UP}_-)$*-appearing bifurcation from a negative saddle to negative saddle, clockwise center, and negative saddle.*

3.2 Proof of Theorem 3.1

Consider a two-dimensional, dynamical system as

$$\begin{aligned} \dot{x}_{j_1} &= f_{j_1}(\mathbf{x}, \mathbf{p}) = A_{j_1j_2}3 + A_{j_1j_2}2x_{j_2} + A_{j_1j_2}1x_{j_2}^2 + A_{j_1j_2}0x_{j_2}^3, \\ \dot{x}_{j_2} &= f_{j_2}(\mathbf{x}, \mathbf{p}) = A_{j_2j_1}3 + A_{j_2j_1}2x_{j_1} + A_{j_2j_1}1x_{j_1}^2 + A_{j_2j_1}0x_{j_1}^3. \end{aligned}$$

For $A_{j_2j_1}0 = 0, A_{j_2j_1}1 \neq 0,$ the two-dimensional dynamical system possesses variable-crossing univariate quadratic and cubic vector fields as

$$\begin{aligned}
\dot{x}_{j_2} &= a_{j_2 j_1 0}(x_{j_1}^2 + B_{j_2 j_1} x_{j_1} + C_{j_2 j_1}),\\
\dot{x}_{j_1} &= a_{j_1 j_2 0}(x_{j_2} - b_{j_1 j_2 1})(x_{j_2}^2 + B_{j_1 j_2} x_{j_2} + C_{j_1 j_2})
\end{aligned}$$

with

$$\begin{aligned}
&\Delta_{j_2 j_1} = (B_{j_2 j_1})^2 - 4C_{j_2 j_1}, \Delta_{j_1 j_2} = (B_{j_1 j_2})^2 - 4C_{j_1 j_2}\\
&\text{for } j_1, j_2 \in \{1,2\}, j_1 \neq j_2.
\end{aligned}$$

(i) If

$$\Delta_{j_1 j_2} < 0 \text{ and } \Delta_{j_2 j_1} < 0,$$

then

$$\begin{aligned}
&\dot{x}_{j_1} = a_{j_1 j_2 0}(x_{j_2} - a_{j_1 j_2 1})\left[(x_{j_2} - a_{j_1 j_2})^2 + b_{j_1 j_2}\right],\\
&\dot{x}_{j_2} = a_{j_2 j_1 0}\left[(x_{j_1} - a_{j_2 j_1})^2 + b_{j_2 j_1}\right]\\
&j_1, j_2 \in \{1,2\}, j_2 \neq j_1.
\end{aligned}$$

In phase space,

$$\frac{dx_{j_1}}{dx_{j_2}} = \frac{a_{j_1 j_2 0}}{a_{j_2 j_1 0}} \frac{(x_{j_2} - a_{j_1 j_2 1})\left[(x_{j_2} - a_{j_1 j_2})^2 + b_{j_1 j_2}\right]}{(x_{j_1} - a_{j_2 j_1})^2 + b_{j_2 j_1}}$$

and

$$\begin{aligned}
&a_{j_2 j_1 0}\left[(x_{j_1} - a_{j_2 j_1})^2 + b_{j_2 j_1}\right]dx_{j_1}\\
&= a_{j_1 j_2 0}\{(x_{j_2} - a_{j_1 j_2 1})^3 + 2(a_{j_1 j_2 1} - a_{j_1 j_2})(x_{j_2} - a_{j_1 j_2 1})^2\\
&+ \left[(a_{j_1 j_2 1} - a_{j_1 j_2})^2 + b_{j_1 j_2}\right](x_{j_2} - a_{j_1 j_2 1})\}dx_{j_2}.
\end{aligned}$$

With an initial condition $(x_{j_1 0}, x_{j_2 0})$, the first integral manifold is generated, i.e.,

$$\begin{aligned}
&a_{j_2 j_1 0}\{\frac{1}{3}\left[(x_{j_1} - a_{j_2 j_1})^3 - (x_{j_1 0} - a_{j_2 j_1})^3\right]\\
&+ b_{j_2 j_1}\left[(x_{j_1} - a_{j_2 j_1}) - (x_{j_1 0} - a_{j_2 j_1})\right]\}\\
&= a_{j_1 j_2 0}\{\frac{1}{4}\left[(x_{j_2} - a_{j_1 j_2 1})^4 - (x_{j_2 0} - a_{j_1 j_2 1})^4\right]\\
&+ \frac{2}{3}(a_{j_1 j_2 1} - a_{j_1 j_2})\left[(x_{j_2} - a_{j_1 j_2 1})^3 - (x_{j_2 0} - a_{j_1 j_2 1})^3\right]\\
&+ \frac{1}{2}\left[(a_{j_1 j_2 1} - a_{j_1 j_2})^2 + b_{j_1 j_2}\right]\left[(x_{j_2} - a_{j_1 j_2 1})^2 - (x_{j_2 0} - a_{j_1 j_2 1})^2\right]\}.
\end{aligned}$$

In phase space, at $x_{j_2}^* = a_{j_1 j_2 1}$ with $\overline{x}_{j_1} \in (-\infty, \infty)$, we have

$$\frac{dx_{j_1}}{dx_{j_2}}\bigg|_{x_{j_2}^*=a_{j_1j_21}} = \frac{a_{j_1j_20}}{a_{j_2j_10}} \frac{(x_{j_2}-a_{j_1j_21})\left[(x_{j_2}-a_{j_1j_2})^2+b_{j_1j_2}\right]}{(\bar{x}_{j_1}-a_{j_2j_1})^2+b_{j_2j_1}}\bigg|_{x_{j_2}^*=a_{j_1j_21}} = 0.$$

If

$$\frac{d^2x_{j_1}}{dx_{j_2}^2}\bigg|_{x_{j_2}^*=a_{j_1j_21}} = \frac{a_{j_1j_20}}{a_{j_2j_10}} \frac{\left[(a_{j_1j_21}-a_{j_1j_2})^2+b_{j_1j_2}\right]}{(\bar{x}_{j_1}-a_{j_2j_1})^2+b_{j_2j_1}} > 0,$$

there is an up-parabola flow at $x_{j_2}^* = a_{j_1j_21}$ in the x_{j_2}-direction. If

$$\frac{d^2x_{j_1}}{dx_{j_2}^2}\bigg|_{x_{j_2}^*=a_{j_1j_21}} = \frac{a_{j_1j_20}}{a_{j_2j_10}} \frac{\left[(a_{j_1j_21}-a_{j_1j_2})^2+b_{j_1j_2}\right]}{(\bar{x}_{j_1}-a_{j_2j_1})^2+b_{j_2j_1}} < 0,$$

there is a down-parabola flow at $x_{j_2}^* = a_{j_1j_21}$ in the x_{j_2}-direction.

Because

$$\dot{x}_{j_2} = a_{j_2j_10}\left[(\bar{x}_{j_1}-a_{j_2j_1})^2+b_{j_2j_1}\right]$$

and

$$(\bar{x}_{j_1}-a_{j_2j_1})^2+b_{j_2j_1} > 0 \text{ and } (a_{j_1j_21}-a_{j_1j_2})^2+b_{j_1j_2} > 0,$$

the flows at $x_{j_2}^* = a_{j_1j_21}$ are positive and negative in the x_{j_2}-direction for $a_{j_2j_10} > 0$ and $a_{j_2j_10} < 0$, respectively.

Therefore, the equilibrium of $x_{j_2}^* = a_{j_1j_21}$ in the x_{j_2}-direction has the following properties as in Eqs. (3.7)–(3.10):

- *For* $a_{j_1j_20} > 0$ *and* $a_{j_2j_10} > 0$,

$$(a_{j_1j_21}, \dot{x}_{j_2}) = \underbrace{(\text{UP}, \text{pF})}_{\text{up-parabola flow } (+)}.$$

  The flow at $x_{j_2}^* = a_{j_1j_21}$ is a (UP,pF)-positive up-parabola flow.

- *For* $a_{j_1j_20} < 0$ and $a_{j_2j_10} > 0$,

$$(a_{j_1j_21}, \dot{x}_{j_2}) = \underbrace{(\text{DP}, \text{pF})}_{\text{down-parabola flow } (+)}.$$

  The flow at $x_{j_2}^* = a_{j_1j_21}$ is a (DP,pF)-positive down-parabola flow.

- *For* $a_{j_1j_20} > 0$ and $a_{j_2j_10} < 0$,

$$(a_{j_1j_21}, \dot{x}_{j_2}) = \underbrace{(\text{DP}, \text{nF})}_{\text{down-parabola flow } (-)}.$$

The flow at $x^*_{j_2} = a_{j_1j_21}$ is a (DP,nF)-negative down-parabola flow.

- *For* $a_{j_1j_20} < 0$ and $a_{j_2j_10} < 0$,

$$(a_{j_1j_21}, \dot{x}_{j_2}) = \underbrace{(\text{UP}, \text{nF})}_{\text{up-parabola flow } (-)}.$$

The flow at $x^*_{j_2} = a_{j_1j_21}$ is a (UP,nF)-negative up-parabola flow.

(ii) If

$$\Delta_{j_1j_2} = 0 \text{ and } \Delta_{j_2j_1} < 0,$$

then

$$\begin{aligned}
&\dot{x}_{j_2} = a_{j_2j_10}(x_{j_1} - a_{j_2j_11})^2,\\
&\dot{x}_{j_1} = a_{j_1j_20}(x_{j_2} - a_{j_1j_21})\left[(x_{j_2} - a_{j_1j_2})^2 + b_{j_1j_2}\right],\\
&j_1, j_2 \in \{1, 2\}; j_1 \neq j_2.
\end{aligned}$$

In phase space,

$$\frac{dx_{j_2}}{dx_{j_1}} = \frac{a_{j_1j_20}(x_{j_2} - a_{j_1j_21})^2}{a_{j_2j_10}(x_{j_2} - a_{j_1j_21})\left[(x_{j_1} - a_{j_2j_11})^2 + b_{j_2j_1}\right]},$$

and

$$\begin{aligned}
&a_{j_2j_10}(x_{j_1} - a_{j_2j_11})^2 dx_{j_1}\\
&= a_{j_1j_20}\{(x_{j_2} - a_{j_1j_21})^3 + 2(a_{j_1j_21} - a_{j_1j_2})(x_{j_2} - a_{j_1j_21})^2\\
&+ \left[(a_{j_1j_21} - a_{j_1j_2})^2 + b_{j_1j_2}\right](x_{j_2} - a_{j_1j_21})\}dx_{j_2}.
\end{aligned}$$

With an initial condition (x_{j_10}, x_{j_20}), the first integral manifold is generated, i.e.,

$$\begin{aligned}
&\frac{1}{3}a_{j_2j_10}\left[(x_{j_1} - a_{j_2j_11})^3 - (x_{j_10} - a_{j_2j_11})^3\right]\\
&= a_{j_1j_20}\{\frac{1}{4}\left[(x_{j_1} - a_{j_1j_21})^4 - (x_{j_10} - a_{j_1j_21})^4\right]\\
&+ \frac{2}{3}(a_{j_1j_21} - a_{j_1j_2})\left[(x_{j_2} - a_{j_1j_21})^3 - (x_{j_20} - a_{j_1j_21})^3\right]\\
&+ \frac{1}{2}\left[(a_{j_1j_21} - a_{j_1j_2})^2 + b_{j_1j_2}\right]\left[(x_{j_2} - a_{j_1j_21})^2 - (x_{j_20} - a_{j_1j_21})^2\right]\}.
\end{aligned}$$

Consider two cases (I) $x^*_{j_2} = a_{j_1j_21}$ and (II) $x^*_{j_1} = a_{j_2j_11}$.

(I) For $x_{j_2}^* = a_{j_1 j_2 1}$ and $\overline{x}_{j_1} \neq a_{j_2 j_1 1}$, we have

$$\frac{dx_{j_1}}{dx_{j_2}}\bigg|_{x_{j_2}^* = a_{j_1 j_2 1}} = \frac{a_{j_1 j_2 0}(x_{j_2} - a_{j_1 j_2 1})\left[(x_{j_2} - a_{j_1 j_2})^2 + b_{j_1 j_2}\right]}{a_{j_1 j_2 0}(\overline{x}_{j_1} - a_{j_2 j_1 1})^2}\bigg|_{x_{j_2}^* = a_{j_1 j_2 1}} = 0.$$

If

$$\frac{d^2 x_{j_1}}{dx_{j_2}^2}\bigg|_{x_{j_2}^* = a_{j_1 j_2 1}} = \frac{a_{j_1 j_2 0}}{a_{j_2 j_1 0}} \frac{(a_{j_1 j_2 1} - a_{j_1 j_2})^2 + b_{j_1 j_2}}{(\overline{x}_{j_1} - a_{j_2 j_1 1})^2} > 0,$$

there is an up-parabola flow at $x_{j_2}^* = a_{j_1 j_2 1}$ in the x_{j_2}-direction. If

$$\frac{d^2 x_{j_1}}{dx_{j_2}^2}\bigg|_{x_{j_2}^* = a_{j_1 j_2 1}} = \frac{a_{j_1 j_2 0}}{a_{j_2 j_1 0}} \frac{(a_{j_1 j_2 1} - a_{j_1 j_2})^2 + b_{j_1 j_2}}{(\overline{x}_{j_1} - a_{j_2 j_1 1})^2} < 0,$$

there is a down-parabola flow at $x_{j_2}^* = a_{j_1 j_2 1}$ in the x_{j_2}-direction. Because

$$\dot{x}_{j_2} = a_{j_2 j_1 0}(\overline{x}_{j_1} - a_{j_2 j_1 1})^2$$

and

$$(\overline{x}_{j_1} - a_{j_2 j_1 1})^2 > 0 \text{ and } (a_{j_1 j_2 1} - a_{j_1 j_2})^2 + b_{j_1 j_2} > 0,$$

the flows at $x_{j_2}^* = a_{j_1 j_2 1}$ are positive and negative in the x_{j_2}-direction for $a_{j_2 j_1 0} > 0$ and $a_{j_2 j_1 0} < 0$, respectively.

Thus, the equilibrium of $x_{j_2}^* = a_{j_1 j_2 1}$ with $\overline{x}_{j_1} \neq a_{j_2 j_1 1}$ in the x_{j_2}-direction has the following properties:

- For $a_{j_1 j_2 0} > 0$ and $a_{j_2 j_1 0}(\overline{x}_{j_1} - a_{j_2 j_1 1})^2 > 0$,

$$(a_{j_1 j_2 1}, \dot{x}_{j_2}) = \underbrace{(\text{UP}, \text{pF})}_{\text{up-parabola flow } (+)}.$$

- For $a_{j_1 j_2 0} < 0$ and $a_{j_2 j_1 0}(\overline{x}_{j_1} - a_{j_2 j_1 1})^2 > 0$,

$$(a_{j_1 j_2 1}, \dot{x}_{j_2}) = \underbrace{(\text{DP}, \text{pF})}_{\text{down-parabola flow } (+)}.$$

- For $a_{j_1j_20} > 0$ and $a_{j_2j_10}(\bar{x}_{j_1} - a_{j_2j_11})^2 < 0$,

$$(a_{j_1j_21}, \dot{x}_{j_2}) = \underbrace{(\text{DP}, \text{nF})}_{\text{down-parabola flow } (-)}.$$

- For $a_{j_1j_20} < 0$ and $a_{j_2j_10}(\bar{x}_{j_1} - a_{j_2j_11})^2 < 0$,

$$(a_{j_1j_21}, \dot{x}_{j_2}) = \underbrace{(\text{UP}, \text{nF})}_{\text{up-parabola flow } (-)}.$$

(II) For $x_{j_2}^* = a_{j_2j_11}$ and $\bar{x}_{j_2} \neq a_{j_1j_21}$, we have

$$\begin{aligned}
&\frac{dx_{j_2}}{dx_{j_1}}\bigg|_{x_{j_1}^* = a_{j_2j_11}} = \frac{a_{j_1j_20}(x_{j_1} - a_{j_2j_11})^2}{a_{j_2j_10}(\bar{x}_{j_2} - a_{j_1j_21})\left[(\bar{x}_{j_2} - a_{j_1j_2})^2 + b_{j_1j_2}\right]}\bigg|_{x_{j_1}^* = a_{j_2j_11}} = 0,\\
&\frac{d^2x_{j_2}}{dx_{j_1}^2}\bigg|_{x_{j_1}^* = a_{j_2j_11}} = 0.
\end{aligned}$$

If

$$\frac{d^3x_{j_2}}{dx_{j_1}^3}\bigg|_{x_{j_1}^* = a_{j_2j_11}} = \frac{2a_{j_1j_20}}{a_{j_1j_20}(\bar{x}_{j_2} - a_{j_1j_21})\left[(\bar{x}_{j_2} - a_{j_1j_2})^2 + b_{j_1j_2}\right]} > 0,$$

the flow at $x_{j_2}^* = a_{j_2j_11}$ is an increasing-inflection flow. If

$$\frac{d^3x_{j_2}}{dx_{j_1}^3}\bigg|_{x_{j_1}^* = a_{j_2j_11}} = \frac{2a_{j_1j_20}}{a_{j_1j_20}(\bar{x}_{j_2} - a_{j_1j_21})\left[(\bar{x}_{j_2} - a_{j_1j_2})^2 + b_{j_1j_2}\right]} < 0,$$

the flow at $x_{j_1}^* = a_{j_2j_11}$ is a decreasing-inflection flow. Because

$$\dot{x}_{j_1} = a_{j_1j_20}(\bar{x}_{j_2} - a_{j_1j_21})\left[(\bar{x}_{j_2} - a_{j_1j_2})^2 + b_{j_1j_2}\right],$$

the inflection flows at $x_{j_1}^* = a_{j_2j_11}$ are positive and negative in the x_{j_1}-direction for $a_{j_1j_20}(\bar{x}_{j_2} - a_{j_1j_21}) > 0$ and $a_{j_1j_20}(\bar{x}_{j_2} - a_{j_1j_21}) < 0$, respectively.

Thus, the equilibrium of $x_{j_1}^* = a_{j_2j_11}$ in the x_{j_2}-direction for the x_{j_1}-direction has the following properties:

- For $a_{j_1j_20}(\overline{x}_{j_2} - a_{j_1j_21}) > 0$ and $a_{j_2j_10} > 0$,

$$(\dot{x}_{j_2}, a_{j_2j_11}) = \underbrace{(\mathrm{pF}, \mathrm{II})}_{\text{increasing-inflection flow } (+)}.$$

- For $a_{j_1j_20}(\overline{x}_{j_2} - a_{j_1j_21}) < 0$ and $a_{j_2j_10} > 0$,

$$(\dot{x}_{j_2}, a_{j_2j_11}) = \underbrace{(\mathrm{nF}, \mathrm{DI})}_{\text{decreasing-inflection flow } (-)}.$$

- For $a_{j_1j_20}(\overline{x}_{j_2} - a_{j_1j_21}) > 0$ and $a_{j_2j_10} < 0$,

$$(\dot{x}_{j_2}, a_{j_2j_11}) = \underbrace{(\mathrm{pF}, \mathrm{DI})}_{\text{decreasing-inflection flow } (+)}.$$

- For $a_{j_1j_20}(\overline{x}_{j_2} - a_{j_1j_21}) < 0$ and $a_{j_2j_10} < 0$,

$$(\dot{x}_{j_2}, a_{j_2j_11}) = \underbrace{(\mathrm{nF}, \mathrm{II})}_{\text{increasing-inflection flow } (-)}.$$

Therefore, from cases (I) and (II), the equilibrium of $(x^*_{j_2}, x^*_{j_1}) = (a_{j_1j_21}, a_{j_2j_11})$ has the following properties as in Eqs. (3.14)–(3.17):

- For $a_{j_1j_20} > 0$ and $a_{j_2j_10} > 0$,

$$(a_{j_1j_21}, a_{j_2j_11}) = \underbrace{(\mathrm{UP}, \mathrm{US})}_{\text{up-parabola upper-saddle}}.$$

  The equilibrium of $(x^*_{j_2}, x^*_{j_1}) = (a_{j_1j_21}, a_{j_2j_11})$ is a (UP,US)-up-parabola upper-saddle.
- For $a_{j_1j_20} < 0$ and $a_{j_2j_10} > 0$,

$$(a_{j_1j_21}, a_{j_2j_11}) = \underbrace{(\mathrm{DP}, \mathrm{US})}_{\text{down-parabola upper-saddle}}.$$

  The equilibrium of $(x^*_{j_2}, x^*_{j_1}) = (a_{j_1j_21}, a_{j_2j_11})$ is a (DP,US)-down-parabola upper-saddle.

- For $a_{j_1j_2 0} > 0$ and $a_{j_2j_1 0} < 0$,

$$(a_{j_1j_2 1}, a_{j_2j_1 1}) = \underbrace{(\text{DP}, \text{LS})}_{\text{down-parabola lower-saddle}}.$$

  The equilibrium of $(x^*_{j_2}, x^*_{j_1}) = (a_{j_1j_2 1}, a_{j_2j_1 1})$ is a (DP,LS)-down-parabola lower-saddle.
- For $a_{j_1j_2 0} < 0$ and $a_{j_2j_1 0} < 0$,

$$(a_{j_1j_2 1}, a_{j_2j_1 1}) = \underbrace{(\text{UP}, \text{LS})}_{\text{up-parabola lower-saddle}}.$$

  The equilibrium of $(x^*_{j_2}, x^*_{j_1}) = (a_{j_1j_2 1}, a_{j_2j_1 1})$ is a (UP,LS)-up-parabola lower-saddle.

(iii) If

$$\Delta_{j_1j_2} < 0 \text{ and } \Delta_{j_2j_1} > 0,$$

then

$$\begin{aligned}
\dot{x}_{j_2} &= a_{j_1j_2 0}(x_{j_1} - a_{j_2j_1 1})(x_{j_1} - a_{j_2j_1 2}),\\
\dot{x}_{j_1} &= a_{j_2j_1 0}(x_{j_2} - a_{j_1j_2 1})\left[(x_{j_2} - a_{j_1j_2})^2 + b_{j_1j_2}\right],\\
j_1, j_2 &\in \{1, 2\}; j_1 \neq j_2
\end{aligned}$$

where

$$\begin{aligned}
&\{a_{j_2j_1 1}, a_{j_2j_1 2}\} = \text{sort}\{b_{j_2j_1 1}, b_{j_2j_1 2}\}, a_{j_2j_1 1} < a_{j_2j_1 2};\\
&b_{j_2j_1 1}, b_{j_2j_1 2} = -\frac{1}{2}(B_{j_2j_1} \pm \sqrt{\Delta_{j_2j_1}}), a_{j_2j_1 0} = A_{j_2j_1 0};\\
&a_{j_1j_2} = -\frac{1}{2}B_{j_1j_2}, b_{j_1j_2} = \frac{1}{4}(-\Delta_{j_1j_2}).
\end{aligned}$$

In phase space,

$$\frac{dx_{j_2}}{dx_{j_1}} = \frac{a_{j_1j_2 0}(x_{j_2} - a_{j_1j_2 1})(x_{j_2} - a_{j_1j_2 2})}{a_{j_2j_1 0}(x_{j_2} - a_{j_1j_2 1})\left[(x_{j_1} - a_{j_2j_1 1})^2 + b_{j_2j_1}\right]},$$

and

$$a_{j_2j_10}\left[(x_{j_1}-a_{j_2j_11})^2+(a_{j_2j_11}-a_{j_2j_12})(x_{j_1}-a_{j_2j_11})\right]dx_{j_1}$$
$$=a_{j_1j_20}\{(x_{j_2}-a_{j_1j_21})^3+2(a_{j_1j_21}-a_{j_1j_2})(x_{j_2}-a_{j_1j_21})^2$$
$$+\left[(a_{j_1j_21}-a_{j_1j_2})^2+b_{j_1j_2}\right](x_{j_2}-a_{j_1j_21})\}dx_{j_2}.$$

With an initial condition (x_{j_10},x_{j_20}), the first integral manifold is generated, i.e.,

$$a_{j_2j_10}\{\frac{1}{3}\left[(x_{j_1}-a_{j_2j_11})^3-(x_{j_10}-a_{j_2j_11})^3\right]$$
$$+\frac{1}{2}(a_{j_2j_11}-a_{j_2j_12})\left[(x_{j_1}-a_{j_2j_11})^2-(x_{j_10}-a_{j_2j_11})^2\right]\}$$
$$=a_{j_1j_20}\{\frac{1}{4}\left[(x_{j_1}-a_{j_1j_21})^4-(x_{j_10}-a_{j_1j_21})^4\right]$$
$$+\frac{2}{3}(a_{j_1j_21}-a_{j_1j_2})\left[(x_{j_2}-a_{j_1j_21})^3-(x_{j_20}-a_{j_1j_21})^3\right]$$
$$+\frac{1}{2}\left[(a_{j_1j_21}-a_{j_1j_2})^2+b_{j_1j_2}\right]\left[(x_{j_2}-a_{j_1j_21})^2-(x_{j_20}-a_{j_1j_21})^2\right]\}.$$

Consider two cases (I) $x_{j_2}^*=a_{j_1j_21}$ and (II) $x_{j_1}^*=a_{j_2j_1s_1}$.

(I) For $x_{j_2}^*=a_{j_1j_21}$ and $\overline{x}_{j_1}\neq a_{j_2j_11},a_{j_2j_12}$, we have

$$\frac{dx_{j_1}}{dx_{j_2}}\Big|_{x_{j_2}^*=a_{j_1j_21}}=\frac{a_{j_1j_20}(x_{j_2}-a_{j_1j_21})\left[(x_{j_2}-a_{j_1j_2})^2+b_{j_1j_2}\right]}{a_{j_1j_20}(\overline{x}_{j_1}-a_{j_2j_11})(\overline{x}_{j_1}-a_{j_2j_12})}\Big|_{x_{j_2}^*=a_{j_1j_21}}=0.$$

If

$$\frac{d^2x_{j_1}}{dx_{j_2}^2}\Big|_{x_{j_2}^*=a_{j_1j_21}}=\frac{a_{j_1j_20}}{a_{j_2j_10}}\frac{\left[(a_{j_1j_21}-a_{j_1j_2})^2+b_{j_1j_2}\right]}{(\overline{x}_{j_1}-a_{j_2j_11})(\overline{x}_{j_1}-a_{j_2j_12})}>0,$$

there is an up-parabola flow at $x_{j_2}^*=a_{j_1j_21}$ in the x_{j_2}-direction. If

$$\frac{d^2x_{j_1}}{dx_{j_2}^2}\Big|_{x_{j_2}^*=a_{j_1j_21}}=\frac{a_{j_1j_20}}{a_{j_2j_10}}\frac{\left[(a_{j_1j_21}-a_{j_1j_2})^2+b_{j_1j_2}\right]}{(\overline{x}_{j_1}-a_{j_2j_11})(\overline{x}_{j_1}-a_{j_2j_12})}<0,$$

there is a down-parabola flow at $x_{j_2}^*=a_{j_1j_21}$ in the x_{j_2}-direction. Let

$$\dot{x}_{j_2}=a_{j_2j_10}(\overline{x}_{j_1}-a_{j_2j_11})(\overline{x}_{j_1}-a_{j_2j_12})$$

and

$$(a_{j_1j_21} - a_{j_1j_2})^2 + b_{j_1j_2} > 0.$$

The flows at $x_{j_2}^* = a_{j_1j_21}$ are positive and negative in the x_{j_2}-direction for $\dot{x}_{j_2} > 0$ and $\dot{x}_{j_2} < 0$, respectively.

Thus, the equilibrium of $x_{j_2}^* = a_{j_1j_21}$ in the x_{j_2}-direction has the following properties:

- For $a_{j_1j_20} > 0$ and $a_{j_2j_10}\prod_{l_1=1}^{2}(\overline{x}_{j_1} - a_{j_2j_1l_1}) > 0$,

$$(a_{j_1j_21}, \dot{x}_{j_2}) = \underbrace{(\text{UP}, \text{pF})}_{\text{up-parabola flow }(+)}.$$

- For $a_{j_1j_20} < 0$ and $a_{j_2j_10}\prod_{l_1=1}^{2}(\overline{x}_{j_1} - a_{j_2j_1l_1}) > 0$,

$$(a_{j_1j_21}, \dot{x}_{j_2}) = \underbrace{(\text{DP}, \text{pF})}_{\text{down-parabola flow }(+)}.$$

- For $a_{j_1j_20} > 0$ and $a_{j_2j_10}\prod_{l_1=1}^{2}(\overline{x}_{j_1} - a_{j_2j_1l_1}) < 0$,

$$(a_{j_1j_21}, \dot{x}_{j_2}) = \underbrace{(\text{DP}, \text{nF})}_{\text{down-parabola flow }(-)}.$$

- For $a_{j_1j_20} < 0$ and $a_{j_2j_10}\prod_{l_1=1}^{2}(\overline{x}_{j_1} - a_{j_2j_1l_1}) < 0$,

$$(a_{j_1j_21}, \dot{x}_{j_2}) = \underbrace{(\text{UP}, \text{nF})}_{\text{up-parabola flow }(-)}.$$

(II) For $x_{j_1}^* = a_{j_2j_1l_1}$ ($l_1, l_2 \in \{1,2\}, l_1 \neq l_2$) and $\overline{x}_{j_2} \neq a_{j_1j_21}$, we have

$$\frac{dx_{j_2}}{dx_{j_1}}\bigg|_{x_{j_1}^* = a_{j_2j_1l_1}} = \frac{a_{j_2j_10}(x_{j_1} - a_{j_2j_1l_1})(x_{j_1} - a_{j_2j_1l_2})}{a_{j_1j_20}(\overline{x}_{j_2} - a_{j_1j_21})\left[(\overline{x}_{j_2} - a_{j_1j_2})^2 + b_{j_1j_2}\right]}\bigg|_{x_{j_1}^* = a_{j_2j_1l_1}} = 0.$$

If

$$\frac{d^2x_{j_2}}{dx_{j_1}^2}\bigg|_{x_{j_1}^* = a_{j_2j_1l_1}} = \frac{a_{j_2j_10}(a_{j_2j_1l_1} - a_{j_2j_1l_2})}{a_{j_1j_20}(\overline{x}_{j_2} - a_{j_1j_21})\left[(\overline{x}_{j_2} - a_{j_1j_2})^2 + b_{j_1j_2}\right]} > 0,$$

the flow at $x_{j_1}^* = a_{j_2j_1l_1}$ is an up-parabola flow. If

$$\left.\frac{d^2 x_{j_2}}{dx_{j_1}^2}\right|_{x_{j_1}^*=a_{j_2 j_1 l_1}} = \frac{a_{j_2 j_1 0}(a_{j_2 j_1 l_1} - a_{j_2 j_1 l_2})}{a_{j_1 j_2 0}(\bar{x}_{j_2} - a_{j_1 j_2 1})\left[(\bar{x}_{j_2} - a_{j_1 j_2})^2 + b_{j_1 j_2}\right]} < 0,$$

the flow at $x_{j_1}^* = a_{j_2 j_1 l_1}$ is a down-parabola flow. Because

$$\dot{x}_{j_1} = a_{j_1 j_2 0}(\bar{x}_{j_2} - a_{j_1 j_2 1})\left[(\bar{x}_{j_2} - a_{j_1 j_2})^2 + b_{j_1 j_2}\right],$$

the parabola flows at $x_{j_1}^* = a_{j_2 j_1 l_1}$ are positive and negative in the x_{j_1}-direction for $a_{j_1 j_2 0}(\bar{x}_{j_2} - a_{j_1 j_2 1}) > 0$ and $a_{j_1 j_2 0}(\bar{x}_{j_2} - a_{j_1 j_2 1}) < 0$, respectively.

Thus, the equilibrium of $x_{j_1}^* = a_{j_2 j_1 l_1}$ in the x_{j_2}-direction for the x_{j_1}-direction has the following properties:

- For $a_{j_1 j_2 0}(\bar{x}_{j_2} - a_{j_1 j_2 1}) > 0$ and $a_{j_2 j_1 0}(a_{j_2 j_1 l_1} - a_{j_2 j_1 l_2}) > 0$,

$$(\dot{x}_{j_1}, a_{j_2 j_1 l_1}) = \underbrace{(\mathrm{pF}, \mathrm{UP})}_{\text{up-parabola flow } (+)}.$$

- For $a_{j_1 j_2 0}(\bar{x}_{j_2} - a_{j_1 j_2 1}) < 0$ and $a_{j_2 j_1 0}(a_{j_2 j_1 l_1} - a_{j_2 j_1 l_2}) > 0$,

$$(\dot{x}_{j_1}, a_{j_2 j_1 l_1}) = \underbrace{(\mathrm{nF}, \mathrm{DP})}_{\text{down-parabola flow } (-)}.$$

- For $a_{j_1 j_2 0}(\bar{x}_{j_2} - a_{j_1 j_2 1}) > 0$ and $a_{j_2 j_1 0}(a_{j_2 j_1 l_1} - a_{j_2 j_1 l_2}) < 0$,

$$(\dot{x}_{j_1}, a_{j_2 j_1 l_1}) = \underbrace{(\mathrm{pF}, \mathrm{DP})}_{\text{down-parabola flow } (+)}.$$

- For $a_{j_1 j_2 0}(\bar{x}_{j_2} - a_{j_1 j_2 1}) < 0$ and $a_{j_2 j_1 0}(a_{j_2 j_1 l_1} - a_{j_2 j_1 l_2}) < 0$,

$$(\dot{x}_{j_1}, a_{j_2 j_1 l_1}) = \underbrace{(\mathrm{nF}, \mathrm{UP})}_{\text{up-parabola flow } (-)}.$$

Therefore, from cases (I) and (II), the equilibrium of $(x_{j_2}^*, x_{j_1}^*) = (a_{j_1 j_2 1}, a_{j_2 j_1 l_1})$ $(l_1, l_2 \in \{1, 2\}, l_1 \neq l_2)$ has the following properties as in Eqs. (3.22)–(3.25):

- For $a_{j_1 j_2 0} > 0$ and $a_{j_2 j_1 0}(a_{j_2 j_1 l_1} - a_{j_2 j_1 l_2}) > 0$,

$$(a_{j_1 j_2 1}, a_{j_2 j_1 l_1}) = \underbrace{(\mathrm{UP}_+, \mathrm{UP}_+)}_{\text{positive saddle}}.$$

The equilibrium of $(x_{j_2}^*, x_{j_1}^*) = (a_{j_1 j_2 1}, a_{j_2 j_1 l_1})$ is a (UP_+,UP_+)-positive saddle.

- For $a_{j_1j_20}<0$ and $a_{j_2j_10}(a_{j_2j_1l_1}-a_{j_2j_1l_2})>0$,

$$(a_{j_1j_21},a_{j_2j_1l_1})=\underbrace{(\mathrm{DP}_+,\mathrm{DP}_-)}_{\text{CCW center}}.$$

The equilibrium of $(x^*_{j_2},x^*_{j_1})=(a_{j_1j_21},a_{j_2j_1l_1})$ is a (DP$_+$,DP$_-$)-counter-clockwise center.

- For $a_{j_1j_20}>0$ and $a_{j_2j_10}(a_{j_2j_1l_1}-a_{j_2j_1l_2})<0$,

$$(a_{j_1j_21},a_{j_2j_1l_1})=\underbrace{(\mathrm{DP}_-,\mathrm{DP}_+)}_{\text{CW center}}.$$

The equilibrium of $(x^*_{j_2},x^*_{j_1})=(a_{j_1j_21},a_{j_2j_1l_1})$ is a (DP$_-$,DP$_+$)-clockwise center.

- For $a_{j_1j_20}<0$ and $a_{j_2j_10}(a_{j_2j_1l_1}-a_{j_2j_1l_2})<0$,

$$(a_{j_1j_21},a_{j_2j_1l_1})=\underbrace{(\mathrm{UP}_-,\mathrm{UP}_-)}_{\text{negative saddle}}.$$

The equilibrium of $(x^*_{j_2},x^*_{j_1})=(a_{j_1j_21},a_{j_2j_1l_1})$ is a (UP$_-$,UP$_-$)-negative saddle.

(iv) If

$$\Delta_{j_1j_2}=0 \text{ and } \Delta_{j_2j_1}<0, a_{j_1j_2s_1}\neq a_{j_1j_2s_2},$$

then

$$\begin{aligned}&\dot{x}_{j_1}=a_{j_1j_20}(x_{j_2}-a_{j_1j_2s_1})(x_{j_2}-a_{j_1j_2s_2})^2,\\&\dot{x}_{j_2}=a_{j_2j_10}\left[(x_{j_1}-a_{j_2j_1})^2+b_{j_2j_1}\right]\\&j_1,j_2\in\{1,2\},j_2\neq j_1.\end{aligned}$$

In phase space,

$$\frac{dx_{j_1}}{dx_{j_2}}=\frac{a_{j_1j_20}}{a_{j_2j_10}}\frac{(x_{j_2}-a_{j_1j_2s_1})(x_{j_2}-a_{j_1j_2s_2})^2}{(x_{j_1}-a_{j_2j_1})^2+b_{j_2j_1}},$$

and

$$\begin{aligned}&a_{j_2j_10}\left[(x_{j_1}-a_{j_2j_1})^2+b_{j_2j_1}\right]dx_{j_1}\\&=a_{j_1j_20}\left[(x_{j_2}-a_{j_1j_2s_1})^3+2(a_{j_1j_2s_1}-a_{j_1j_2s_2})(x_{j_2}-a_{j_1j_2s_1})^2\right.\\&\left.+(a_{j_1j_2s_1}-a_{j_1j_2s_2})^2(x_{j_2}-a_{j_1j_2s_1})\right]dx_{j_2}.\end{aligned}$$

With an initial condition (x_{j_10},x_{j_20}), the first integral manifold is generated, i.e.,

$$
\begin{aligned}
& a_{j_2j_10}\Big\{\frac{1}{3}\left[(x_{j_1}-a_{j_2j_1})^3-(x_{j_10}-a_{j_2j_1})^3\right] \\
& +b_{j_2j_1}\left[(x_{j_1}-a_{j_2j_1})-(x_{j_10}-a_{j_2j_1})\right]\Big\} \\
& =a_{j_1j_20}\Big\{\frac{1}{4}\left[(x_{j_2}-a_{j_1j_2s_1})^4-(x_{j_20}-a_{j_1j_2s_1})^4\right] \\
& +\frac{2}{3}(a_{j_1j_2s_1}-a_{j_1j_2s_2})\left[(x_{j_2}-a_{j_1j_2s_1})^3-(x_{j_20}-a_{j_1j_2s_1})^3\right] \\
& +\frac{1}{2}(a_{j_1j_2s_1}-a_{j_1j_2s_2})^2\left[(x_{j_2}-a_{j_1j_2s_1})^2-(x_{j_20}-a_{j_1j_2s_1})^2\right]\Big\}.
\end{aligned}
$$

(iv$_1$) In phase space, at $x^*_{j_2}=a_{j_1j_2s_1}$ (s_1, $s_2\in\{1,2\}$, $s_1\neq s_2$), we have

$$
\frac{dx_{j_1}}{dx_{j_2}}\Big|_{x^*_{j_2}=a_{j_1j_2s_1}}=\frac{a_{j_1j_20}}{a_{j_2j_10}}\frac{(x_{j_2}-a_{j_1j_2s_1})(x_{j_2}-a_{j_1j_2s_2})^2}{(\overline{x}_{j_1}-a_{j_2j_1})^2+b_{j_2j_1}}\Big|_{x^*_{j_2}=a_{j_1j_2s_1}}=0.
$$

If

$$
\frac{d^2x_{j_1}}{dx^2_{j_2}}\Big|_{x^*_{j_2}=a_{j_1j_2s_1}}=\frac{a_{j_1j_20}}{a_{j_2j_10}}\frac{(a_{j_1j_2s_1}-a_{j_1j_2s_2})^2}{(\overline{x}_{j_1}-a_{j_2j_1})^2+b_{j_2j_1}}>0,
$$

there is an up-parabola flow at $x^*_{j_2}=a_{j_1j_2s_1}$ in the x_{j_1}-direction. If

$$
\frac{d^2x_{j_1}}{dx^2_{j_2}}\Big|_{x^*_{j_2}=a_{j_1j_2s_1}}=\frac{a_{j_1j_20}}{a_{j_2j_10}}\frac{(a_{j_1j_2s_1}-a_{j_1j_2s_2})^2}{(\overline{x}_{j_1}-a_{j_2j_1})^2+b_{j_2j_1}}<0,
$$

there is a down-parabola flow at $x^*_{j_2}=a_{j_1j_2s_1}$ in the x_{j_1}-direction. Because

$$
\dot{x}_{j_2}=a_{j_2j_10}\left[(\overline{x}_{j_1}-a_{j_2j_1})^2+b_{j_2j_1}\right],
$$

the flows at $x^*_{j_2}=a_{j_1j_2s_1}$ in the x_{j_2}-direction are positive and negative for $a_{j_2j_10}>0$ and $a_{j_2j_10}<0$, respectively.

Therefore, the equilibrium of $x^*_{j_2}=a_{j_1j_2s_1}$ (s_1, $s_2\in\{1,2\}$, $s_1\neq s_2$) in the x_{j_2}-direction has the following properties as in Eqs. (3.29)–(3.32):

- For $a_{j_1j_20}(a_{j_1j_2s_1}-a_{j_1j_2s_2})^2>0$ and $a_{j_2j_10}>0$,

$$(a_{j_1j_2s_1}, \dot{x}_{j_2}) = \underbrace{(\mathrm{UP}, \mathrm{pF})}_{\text{up-parabola flow } (+)}.$$

The flow at $x^*_{j_2} = a_{j_1j_2s_1}$ is a (UP,pF)-positive up-parabola flow.

- For $a_{j_1j_20}(a_{j_1j_2s_1} - a_{j_1j_2s_2})^2 < 0$ and $a_{j_2j_10} > 0$,

$$(a_{j_1j_2s_1}, \dot{x}_{j_2}) = \underbrace{(\mathrm{DP}, \mathrm{pF})}_{\text{down-parabola flow } (+)}.$$

The flow at $x^*_{j_2} = a_{j_1j_2s_1}$ is a (DP,pF)-positive down-parabola flow.

- For $a_{j_1j_20}(a_{j_1j_2s_1} - a_{j_1j_2s_2})^2 > 0$ and $a_{j_2j_10} < 0$,

$$(a_{j_1j_2s_1}, \dot{x}_{j_2}) = \underbrace{(\mathrm{DP}, \mathrm{nF})}_{\text{down-parabola flow } (-)}.$$

The flow at $x^*_{j_2} = a_{j_1j_2s_1}$ is a (DP,nF)-negative down-parabola flow.

- For $a_{j_1j_20}(a_{j_1j_2s_1} - a_{j_1j_2s_2})^2 < 0$ and $a_{j_2j_10} < 0$,

$$(a_{j_1j_2s_1}, \dot{x}_{j_2}) = \underbrace{(\mathrm{UP}, \mathrm{nF})}_{\text{up-parabola flow } (-)}.$$

The flow at $x^*_{j_2} = a_{j_1j_2s_1}$ is a (UP,nF)-negative up-parabola flow.

(iv$_2$) In phase space, at $x^*_{j_2} = a_{j_1j_2s_2}$ (s_1, $s_2 \in \{1, 2\}$, $s_1 \neq s_2$), we have

$$\begin{aligned}
&\frac{dx_{j_1}}{dx_{j_2}}\bigg|_{x^*_{j_2} = a_{j_1j_2s_2}} = \frac{a_{j_1j_20}}{a_{j_2j_10}} \frac{(x_{j_2} - a_{j_1j_2s_1})(x_{j_2} - a_{j_1j_2s_2})^2}{(\bar{x}_{j_1} - a_{j_2j_1})^2 + b_{j_2j_1}}\bigg|_{x^*_{j_2} = a_{j_1j_2s_2}} = 0,\\
&\frac{d^2x_{j_1}}{dx^2_{j_2}}\bigg|_{x^*_{j_2} = a_{j_1j_2s_2}} = 0.
\end{aligned}$$

If

$$\frac{d^3x_{j_1}}{dx^3_{j_2}}\bigg|_{x^*_{j_2} = a_{j_1j_2s_2}} = \frac{a_{j_1j_20}}{a_{j_2j_10}} \frac{2(a_{j_1j_2s_2} - a_{j_1j_2s_1})}{(\bar{x}_{j_1} - a_{j_2j_1})^2 + b_{j_2j_1}} > 0,$$

there is an increasing-inflection flow at $x^*_{j_2} = a_{j_1j_2s_2}$ in the x_{j_1}-direction. If

$$\left.\frac{d^3x_{j_1}}{dx_{j_2}^3}\right|_{x_{j_2}^*=a_{j_1j_2s_2}} = \frac{a_{j_1j_20}}{a_{j_2j_10}}\frac{2(a_{j_1j_2s_2}-a_{j_1j_2s_1})}{(\overline{x}_{j_1}-a_{j_2j_1})^2+b_{j_2j_1}} < 0,$$

there is a decreasing-inflection flow at $x_{j_2}^* = a_{j_1j_2s_2}$ in the x_{j_1}-direction. Because

$$\dot{x}_{j_2} = a_{j_2j_10}\left[(\overline{x}_{j_1}-a_{j_2j_1})^2 + b_{j_2j_1}\right],$$

the flows at $x_{j_2}^* = a_{j_1j_2s_2}$ in the x_{j_2}-direction are positive and negative for $a_{j_2j_10} > 0$ and $a_{j_2j_10} < 0$, respectively.

Therefore, the equilibrium of $x_{j_2}^* = a_{j_1j_2s_2}$ (s_1, $s_2 \in \{1,2\}$, $s_1 \neq s_2$) in the x_{j_2}-direction has the following properties as in Eqs. (3.33)–(3.36):

- For $a_{j_1j_20}(a_{j_1j_2s_2} - a_{j_1j_2s_1}) > 0$ and $a_{j_2j_10} > 0$,

$$(a_{j_1j_2s_2}, \dot{x}_{j_2}) = \underbrace{(\text{II}, \text{pF})}_{\text{increasing-inflection flow }(+)}.$$

  The flow at $x_{j_2}^* = a_{j_1j_2s_2}$ is an (II,pF)-positive increasing-inflection flow.
- For $a_{j_1j_20}(a_{j_1j_2s_2} - a_{j_1j_2s_1}) < 0$ and $a_{j_2j_10} > 0$,

$$(a_{j_1j_2s_2}, \dot{x}_{j_2}) = \underbrace{(\text{DI}, \text{pF})}_{\text{decreasing-inflection flow }(+)}.$$

  The flow at $x_{j_2}^* = a_{j_1j_2s_2}$ is a (DI,pF)-positive decreasing-inflection flow.
- For $a_{j_1j_20}(a_{j_1j_2s_2} - a_{j_1j_2s_1}) > 0$ and $a_{j_2j_10} < 0$,

$$(a_{j_1j_2s_2}, \dot{x}_{j_2}) = \underbrace{(\text{DI}, \text{nF})}_{\text{decreasing-inflection flow }(-)}.$$

  The flow at $x_{j_2}^* = a_{j_1j_2s_2}$ is a (DI,nF)-negative decreasing-inflection flow.
- For $a_{j_1j_20}(a_{j_1j_2s_2} - a_{j_1j_2s_1}) < 0$ and $a_{j_2j_10} < 0$,

$$(a_{j_1j_2s_2}, \dot{x}_{j_2}) = \underbrace{(\text{II}, \text{nF})}_{\text{increasing-inflection flow }(-)}.$$

  The flow at $x_{j_2}^* = a_{j_1j_2s_2}$ is a (II,nF)-negative increasing-inflection flow.

(v) If

$$\Delta_{j_1j_2} = 0 \text{ and } \Delta_{j_2j_1} < 0, a_{j_1j_2s_1} = a_{j_1j_2s_2} = a_{j_1j_21},$$

then

$$\begin{aligned}
&\dot{x}_{j_1} = a_{j_1 j_2 0}(x_{j_2} - a_{j_1 j_2 1})^3,\\
&\dot{x}_{j_2} = a_{j_2 j_1 0}\left[(x_{j_1} - a_{j_2 j_1})^2 + b_{j_2 j_1}\right]\\
&j_1, j_2 \in \{1,2\}, j_2 \neq j_1.
\end{aligned}$$

In phase space,

$$\frac{dx_{j_1}}{dx_{j_2}} = \frac{a_{j_1 j_2 0}}{a_{j_2 j_1 0}} \frac{(x_{j_2} - a_{j_1 j_2 1})^3}{(x_{j_1} - a_{j_2 j_1})^2 + b_{j_2 j_1}},$$

and

$$\begin{aligned}
&a_{j_2 j_1 0}\left[(x_{j_1} - a_{j_2 j_1})^2 + b_{j_2 j_1}\right] dx_{j_1}\\
&= a_{j_1 j_2 0}(x_{j_2} - a_{j_1 j_2 1})^3 dx_{j_2}.
\end{aligned}$$

With an initial condition $(x_{j_1 0}, x_{j_2 0})$, the first integral manifold is obtained, i.e.,

$$\begin{aligned}
&a_{j_2 j_1 0}\{\frac{1}{3}\left[(x_{j_1} - a_{j_2 j_1})^3 - (x_{j_1 0} - a_{j_2 j_1})^3\right]\\
&+ b_{j_2 j_1}\left[(x_{j_1} - a_{j_2 j_1}) - (x_{j_1 0} - a_{j_2 j_1})\right]\}\\
&= \frac{1}{4} a_{j_1 j_2 0}\left[(x_{j_2} - a_{j_1 j_2 1})^4 - (x_{j_2 0} - a_{j_1 j_2 1})^4\right].
\end{aligned}$$

In phase space, at $x^*_{j_2} = a_{j_1 j_2 1}$,

$$\begin{aligned}
&\frac{dx_{j_1}}{dx_{j_2}}\Big|_{x^*_{j_2} = a_{j_1 j_2 1}} = \frac{a_{j_1 j_2 0}}{a_{j_2 j_1 0}} \frac{(x_{j_2} - a_{j_1 j_2 1})^3}{(\overline{x}_{j_1} - a_{j_2 j_1})^2 + b_{j_2 j_1}}\Big|_{x^*_{j_2} = a_{j_1 j_2 1}} = 0,\\
&\frac{d^2 x_{j_1}}{dx^2_{j_2}}\Big|_{x^*_{j_2} = a_{j_1 j_2 1}} = \frac{d^3 x_{j_1}}{dx^3_{j_2}}\Big|_{x^*_{j_2} = a_{j_1 j_2 1}} = 0.
\end{aligned}$$

If

$$\frac{d^4 x_{j_1}}{dx^4_{j_2}}\Big|_{x^*_{j_2} = a_{j_1 j_2 1}} = \frac{a_{j_1 j_2 0}}{a_{j_2 j_1 0}} \frac{6}{(\overline{x}_{j_1} - a_{j_2 j_1})^2 + b_{j_2 j_1}} > 0,$$

there is a third-order up-parabola flow at $x^*_{j_2} = a_{j_1 j_2 1}$ in the x_{j_1}-direction. If

$$\frac{d^4 x_{j_1}}{dx^4_{j_2}}\Big|_{x^*_{j_2} = a_{j_1 j_2 1}} = \frac{a_{j_1 j_2 0}}{a_{j_2 j_1 0}} \frac{6}{(\overline{x}_{j_1} - a_{j_2 j_1})^2 + b_{j_2 j_1}} < 0,$$

there is a third-order down-parabola flow at $x^*_{j_2} = a_{j_1 j_2 1}$ in the x_{j_1}-direction. Because

$$\dot{x}_{j_2} = a_{j_2j_10}\left[(\overline{x}_{j_1} - a_{j_2j_1})^2 + b_{j_2j_1}\right],$$

the flows at $x^*_{j_2} = a_{j_1j_21}$ in the x_{j_2}-direction are positive and negative for $a_{j_2j_10} > 0$ and $a_{j_2j_10} < 0$, respectively.

Therefore, the equilibrium of $x^*_{j_2} = a_{j_1j_21}$ in the x_{j_2}-direction has the following properties as in Eqs. (3.40)–(3.43):

- For $a_{j_1j_20} > 0$ and $a_{j_2j_10} > 0$,

$$(a_{j_1j_21}, \dot{x}_{j_2}) = \underbrace{(3^{\text{rd}}\text{UP, pF})}_{\text{third-order up-parabola flow } (+)}.$$

  The flow at $x^*_{j_2} = a_{j_1j_21}$ is a (3^{rd}UP,pF)-positive third-order up-parabola flow.

- For $a_{j_1j_20} < 0$ and $a_{j_2j_10} > 0$,

$$(a_{j_1j_21}, \dot{x}_{j_2}) = \underbrace{(3^{\text{rd}}\text{DP, pF})}_{\text{third-order down-parabola flow } (+)}.$$

  The flow at $x^*_{j_2} = a_{j_1j_21}$ is a (3^{rd}DP,pF)-positive third-order down-parabola flow.

- For $a_{j_1j_20} > 0$ and $a_{j_2j_10} < 0$,

$$(a_{j_1j_21}, \dot{x}_{j_2}) = \underbrace{(3^{\text{rd}}\text{DP, nF})}_{\text{third-order down-parabola flow } (-)}.$$

  The flow at $x^*_{j_2} = a_{j_1j_21}$ is a (3^{rd}DP,nF)-negative third-order down-parabola flow.

- For $a_{j_1j_20} < 0$ and $a_{j_2j_10} < 0$,

$$(a_{j_1j_21}, \dot{x}_{j_2}) = \underbrace{(3^{\text{rd}}\text{UP, nF})}_{\text{third-order up-parabola flow } (-)}.$$

  The flow at $x^*_{j_2} \doteq a_{j_1j_21}$ is a (3^{rd}UP,nF)-negative third-order up-parabola flow.

(vi) If

$$\Delta_{j_1j_2} = 0 \text{ and } \Delta_{j_2j_1} = 0, a_{j_2j_1s_1} \neq a_{j_2j_1s_2},$$

then

$$\begin{aligned}
&\dot{x}_{j_1} = a_{j_1 j_2 0}(x_{j_2} - a_{j_1 j_2 s_1})(x_{j_2} - a_{j_1 j_2 s_2})^2,\\
&\dot{x}_{j_2} = a_{j_2 j_1 0}(x_{j_1} - a_{j_2 j_1 1})^2,\\
&j_1, j_2 \in \{1,2\}; j_1 \neq j_2.
\end{aligned}$$

In phase space,

$$\frac{dx_{j_1}}{dx_{j_2}} = \frac{a_{j_1 j_2 0}}{a_{j_2 j_1 0}} \frac{(x_{j_2} - a_{j_1 j_2 s_1})(x_{j_2} - a_{j_1 j_2 s_2})^2}{(x_{j_1} - a_{j_2 j_1 1})^2},$$

and

$$\begin{aligned}
&a_{j_2 j_1 0}(x_{j_1} - a_{j_2 j_1 1})^2 dx_{j_1}\\
&= a_{j_1 j_2 0}\Big[(x_{j_2} - a_{j_1 j_2 s_1})^3 + 2(a_{j_1 j_2 s_1} - a_{j_1 j_2 s_2})(x_{j_2} - a_{j_1 j_2 s_1})^2\\
&+ (a_{j_1 j_2 s_1} - a_{j_1 j_2 s_2})^2 (x_{j_2} - a_{j_1 j_2 s_1})\Big] dx_{j_2}.
\end{aligned}$$

With an initial condition $(x_{j_1 0}, x_{j_2 0})$, the first integral manifold is obtained, i.e.,

$$\begin{aligned}
&\frac{1}{3} a_{j_2 j_1 0}\left[(x_{j_1} - a_{j_2 j_1 1})^3 - (x_{j_1 0} - a_{j_2 j_1 1})^3\right]\\
&= a_{j_1 j_2 0}\Big\{\frac{1}{4}\left[(x_{j_2} - a_{j_1 j_2 s_1})^4 - (x_{j_2 0} - a_{j_1 j_2 s_1})^4\right]\\
&+ \frac{2}{3}(a_{j_1 j_2 s_1} - a_{j_1 j_2 s_2})\left[(x_{j_2} - a_{j_1 j_2 s_1})^3 - (x_{j_2 0} - a_{j_1 j_2 s_1})^3\right]\\
&+ \frac{1}{2}(a_{j_1 j_2 s_1} - a_{j_1 j_2 s_2})^2\left[(x_{j_2} - a_{j_1 j_2 s_1})^2 - (x_{j_2 0} - a_{j_1 j_2 s_1})^2\right]\Big\}.
\end{aligned}$$

(vi$_1$) Consider two cases (I) $x_{j_2}^* = a_{j_1 j_2 s_1}$ and (II) $x_{j_1}^* = a_{j_2 j_1 1}$.

(I) In phase space, at $x_{j_2}^* = a_{j_1 j_2 s_1}$ (s_1, $s_2 \in \{1,2\}$, $s_1 \neq s_2$) and $\overline{x}_{j_1} \neq a_{j_2 j_1 1}$, we have

$$\frac{dx_{j_1}}{dx_{j_2}}\bigg|_{x_{j_2}^* = a_{j_1 j_2 s_1}} = \frac{a_{j_1 j_2 0}}{a_{j_2 j_1 0}} \frac{(x_{j_2} - a_{j_1 j_2 s_1})(x_{j_2} - a_{j_1 j_2 s_2})^2}{(\overline{x}_{j_1} - a_{j_2 j_1 1})^2}\bigg|_{x_{j_2}^* = a_{j_1 j_2 s_1}} = 0.$$

If

$$\frac{d^2 x_{j_1}}{dx_{j_2}^2}\bigg|_{x_{j_2}^* = a_{j_1 j_2 s_1}} = \frac{a_{j_1 j_2 0}}{a_{j_2 j_1 0}} \frac{(a_{j_1 j_2 s_1} - a_{j_1 j_2 s_2})^2}{(\overline{x}_{j_1} - a_{j_2 j_1 1})^2} > 0,$$

there is an up-parabola flow at $x_{j_2}^* = a_{j_1 j_2 s_1}$ in the x_{j_1}-direction. If

$$\frac{d^2 x_{j_1}}{dx_{j_2}^2}\bigg|_{x_{j_2}^*=a_{j_1j_2s_1}} = \frac{a_{j_1j_20}}{a_{j_2j_10}} \frac{(a_{j_1j_2s_1} - a_{j_1j_2s_2})^2}{(\overline{x}_{j_1} - a_{j_2j_1})^2} < 0,$$

there is a down-parabola flow at $x_{j_2}^* = a_{j_1j_2s_1}$ in the x_{j_1}-direction. Because

$$\dot{x}_{j_2} = a_{j_2j_10}(\overline{x}_{j_1} - a_{j_2j_11})^2,$$

the flows at $x_{j_2}^* = a_{j_1j_2s_1}$ in the x_{j_2}-direction are positive and negative for $a_{j_2j_10} > 0$ and $a_{j_2j_10} < 0$, respectively.

Thus, the equilibrium of $x_{j_2}^* = a_{j_1j_2s_1}$ (s_1, $s_2 \in \{1,2\}$, $s_1 \neq s_2$) with $\overline{x}_{j_1} \neq a_{j_2j_11}$ in the x_{j_2}-direction has the following properties:

- For $a_{j_1j_20}(a_{j_1j_2s_1} - a_{j_1j_2s_2})^2 > 0$ and $a_{j_2j_10}(\overline{x}_{j_1} - a_{j_2j_11})^2 > 0$,

$$(a_{j_1j_2s_1}, \dot{x}_{j_2}) = \underbrace{(\text{UP}, \text{pF})}_{\text{up-parabola flow } (+)}.$$

- For $a_{j_1j_20}(a_{j_1j_2s_1} - a_{j_1j_2s_2})^2 < 0$ and $a_{j_2j_10}(\overline{x}_{j_1} - a_{j_2j_11})^2 > 0$,

$$(a_{j_1j_2s_1}, \dot{x}_{j_2}) = \underbrace{(\text{DP}, \text{pF})}_{\text{down-parabola flow } (+)}.$$

- For $a_{j_1j_20}(a_{j_1j_2s_1} - a_{j_1j_2s_2})^2 > 0$ and $a_{j_2j_10}(\overline{x}_{j_1} - a_{j_2j_11})^2 < 0$,

$$(a_{j_1j_2s_1}, \dot{x}_{j_2}) = \underbrace{(\text{DP}, \text{nF})}_{\text{down-parabola flow } (-)}.$$

- For $a_{j_1j_20}(a_{j_1j_2s_1} - a_{j_1j_2s_2})^2 < 0$ and $a_{j_2j_10}(\overline{x}_{j_1} - a_{j_2j_11})^2 < 0$,

$$(a_{j_1j_2s_1}, \dot{x}_{j_2}) = \underbrace{(\text{UP}, \text{nF})}_{\text{up-parabola flow } (-)}.$$

(II) In phase space, at $x_{j_1}^* = a_{j_2j_11}$ and $\overline{x}_{j_2} \neq a_{j_1j_2s_1}, a_{j_1j_2s_2}$ (s_1, $s_2 \in \{1,2\}$, $s_1 \neq s_2$), we have

$$\frac{dx_{j_2}}{dx_{j_1}}\Big|_{x_{j_1}^*=a_{j_2j_1 1}} = \frac{a_{j_2j_1 0}}{a_{j_1j_2 0}}\frac{(x_{j_1}-a_{j_2j_1 1})^2}{(\overline{x}_{j_2}-a_{j_1j_2s_1})(\overline{x}_{j_2}-a_{j_1j_2s_2})^2}\Big|_{x_{j_1}^*=a_{j_2j_1 1}} = 0,$$
$$\frac{d^2x_{j_2}}{dx_{j_1}^2}\Big|_{x_{j_1}^*=a_{j_2j_1 1}} = 0.$$

If

$$\frac{d^3x_{j_2}}{dx_{j_1}^3}\Big|_{x_{j_1}^*=a_{j_2j_1 1}} = \frac{a_{j_1j_2 0}}{a_{j_2j_1 0}}\frac{2}{(\overline{x}_{j_2}-a_{j_1j_2s_1})(\overline{x}_{j_2}-a_{j_1j_2s_2})^2} > 0,$$

there is an increasing-inflection flow at $x_{j_1}^*=a_{j_2j_1 1}$ in the x_{j_2}-direction. If

$$\frac{d^3x_{j_2}}{dx_{j_1}^3}\Big|_{x_{j_1}^*=a_{j_2j_1 1}} = \frac{a_{j_1j_2 0}}{a_{j_2j_1 0}}\frac{2}{(\overline{x}_{j_2}-a_{j_1j_2s_1})(\overline{x}_{j_2}-a_{j_1j_2s_2})^2} < 0,$$

there is a decreasing-inflection flow at $x_{j_1}^*=a_{j_2j_1 1}$ in the x_{j_1}-direction. Because

$$\dot{x}_{j_1} = a_{j_1j_2 0}(\overline{x}_{j_2}-a_{j_1j_2s_1})(\overline{x}_{j_2}-a_{j_1j_2s_2})^2,$$

the flows at $x_{j_1}^*=a_{j_2j_1 1}$ in the x_{j_1}-direction are positive and negative for $\dot{x}_{j_1}>0$ and $\dot{x}_{j_1}<0$, respectively.

Thus, the equilibrium of $x_{j_1}^*=a_{j_2j_1 1}$ and $\overline{x}_{j_2}\neq a_{j_1j_2s_1}, a_{j_1j_2s_2}$ $(s_1, s_2\in\{1,2\}, s_1\neq s_2)$ in the x_{j_2}-direction has the following properties:

- For $a_{j_1j_2 0}(\overline{x}_{j_2}-a_{j_1j_2s_1})(\overline{x}_{j_2}-a_{j_1j_2s_2})^2>0$ and $a_{j_2j_1 0}>0$,

$$(\dot{x}_{j_1}, a_{j_2j_1 1}) = \underbrace{(\text{pF, II})}_{\text{increasing-inflection flow }(+)}.$$

- For $a_{j_1j_2 0}(\overline{x}_{j_2}-a_{j_1j_2s_1})(\overline{x}_{j_2}-a_{j_1j_2s_2})^2<0$ and $a_{j_2j_1 0}>0$,

$$(\dot{x}_{j_1}, a_{j_2j_1 1}) = \underbrace{(\text{nF, DI})}_{\text{decreasing-infleciton flow }(-)}.$$

- For $a_{j_1j_2 0}(\overline{x}_{j_2}-a_{j_1j_2s_1})(\overline{x}_{j_2}-a_{j_1j_2s_2})^2>0$ and $a_{j_2j_1 0}<0$,

$$(\dot{x}_{j_1}, a_{j_2j_1 1}) = \underbrace{(\text{pF, DI})}_{\text{decresing-inflection flow }(+)}.$$

- For $a_{j_1j_20}(\overline{x}_{j_2} - a_{j_1j_2s_1})(\overline{x}_{j_2} - a_{j_1j_2s_2})^2 < 0$ and $a_{j_2j_10} < 0$,

$$(\dot{x}_{j_1}, a_{j_2j_11}) = \underbrace{(\text{nF}, \text{II})}_{\text{increasing-inflection flow } (-)}.$$

Therefore, from cases (I) and (II), the equilibrium of $(x^*_{j_2}, x^*_{j_1}) = (a_{j_1j_2s_1}, a_{j_2j_11})$ has the following properties as in Eqs. (3.47)–(3.50):

- For $a_{j_1j_20}(a_{j_1j_2s_1} - a_{j_1j_2s_2})^2 > 0$ and $a_{j_2j_10} > 0$,

$$(a_{j_1j_2s_1}, a_{j_2j_11}) = \underbrace{(\text{UP}, \text{US})}_{\text{up-parabola upper-saddle}}.$$

  The equilibrium of $(x^*_{j_2}, x^*_{j_1}) = (a_{j_1j_2s_1}, a_{j_2j_11})$ is a (UP,US)-up-parabola upper-saddle.
- For $a_{j_1j_20}(a_{j_1j_2s_1} - a_{j_1j_2s_2})^2 < 0$ and $a_{j_2j_10} > 0$,

$$(a_{j_1j_2s_1}, a_{j_2j_11}) = \underbrace{(\text{DP}, \text{US})}_{\text{down-parabola upper-saddle}}.$$

  The equilibrium of $(x^*_{j_2}, x^*_{j_1}) = (a_{j_1j_2s_1}, a_{j_2j_11})$ is a (DP,US)-down-parabola upper-saddle.
- For $a_{j_1j_20}(a_{j_1j_2s_1} - a_{j_1j_2s_2})^2 > 0$ and $a_{j_2j_10} < 0$,

$$(a_{j_1j_2s_1}, a_{j_2j_11}) = \underbrace{(\text{DP}, \text{LS})}_{\text{down-parabola lower-saddle}}.$$

  The equilibrium of $(x^*_{j_2}, x^*_{j_1}) = (a_{j_1j_2s_1}, a_{j_2j_11})$ is a (DP,LS)-down-parabola lower-saddle.
- For $a_{j_1j_20}(a_{j_1j_2s_1} - a_{j_1j_2s_2})^2 < 0$ and $a_{j_2j_10} < 0$,

$$(a_{j_1j_2s_1}, a_{j_2j_11}) = \underbrace{(\text{UP}, \text{LS})}_{\text{up-parabola lower-saddle}}.$$

  The equilibrium of $(x^*_{j_2}, x^*_{j_1}) = (a_{j_1j_2s_1}, a_{j_2j_11})$ is a (UP,LS)-up-parabola lower-saddle.

(vi$_2$) In phase space, at $x^*_{j_2} = a_{j_1j_2s_2}$ ($s_1, s_2 \in \{1, 2\}, s_1 \neq s_2$) and $\overline{x}_{j_1} \neq a_{j_2j_11}$, we have

$$\begin{aligned}
&\frac{dx_{j_1}}{dx_{j_2}}\Big|_{x_{j_2}^*=a_{j_1j_2s_2}} = \frac{a_{j_1j_20}}{a_{j_2j_10}}\frac{(x_{j_2}-a_{j_1j_2s_1})(x_{j_2}-a_{j_1j_2s_2})^2}{(\overline{x}_{j_1}-a_{j_2j_11})^2}\Big|_{x_{j_2}^*=a_{j_1j_2s_2}}=0,\\
&\frac{d^2x_{j_1}}{dx_{j_2}^2}\Big|_{x_{j_2}^*=a_{j_1j_2s_2}}=0.
\end{aligned}$$

If

$$\frac{d^3x_{j_1}}{dx_{j_2}^3}\Big|_{x_{j_2}^*=a_{j_1j_2s_2}} = \frac{a_{j_1j_20}}{a_{j_2j_10}}\frac{2(a_{j_1j_2s_2}-a_{j_1j_2s_1})}{(\overline{x}_{j_1}-a_{j_2j_11})^2}>0,$$

there is an increasing-inflection flow at $x_{j_2}^*=a_{j_1j_2s_2}$ in the x_{j_1}-direction. If

$$\frac{d^3x_{j_1}}{dx_{j_2}^3}\Big|_{x_{j_2}^*=a_{j_1j_2s_2}} = \frac{a_{j_1j_20}}{a_{j_2j_10}}\frac{2(a_{j_1j_2s_2}-a_{j_1j_2s_1})}{(\overline{x}_{j_1}-a_{j_2j_11})^2}<0,$$

there is a decreasing-inflection flow at $x_{j_2}^*=a_{j_1j_2s_2}$ in the x_{j_1}-direction. Because

$$\dot{x}_{j_2}=a_{j_2j_10}(\overline{x}_{j_1}-a_{j_2j_1})^2,$$

the inflection flows at $x_{j_2}^*=a_{j_1j_2s_2}$ in the x_{j_2}-direction are positive and negative for $a_{j_2j_10}(\overline{x}_{j_1}-a_{j_2j_1})^2>0$ and $a_{j_2j_10}(\overline{x}_{j_1}-a_{j_2j_1})^2<0$, respectively.

Thus, the equilibrium of $x_{j_2}^*=a_{j_1j_2s_2}$ ($s_1, s_2\in\{1,2\}, s_1\neq s_2$) and $\overline{x}_{j_1}\neq a_{j_2j_11}$ in the x_{j_2}-direction has the following properties:

- For $a_{j_1j_20}(a_{j_1j_2s_2}-a_{j_1j_2s_1})>0$ and $a_{j_2j_10}(\overline{x}_{j_1}-a_{j_2j_1})^2>0$,

$$(a_{j_1j_2s_2},\dot{x}_{j_2})=\underbrace{(\text{II},\text{pF})}_{\text{increasing-inflection flow }(+)}.$$

- For $a_{j_1j_20}(a_{j_1j_2s_2}-a_{j_1j_2s_1})<0$ and $a_{j_2j_10}(\overline{x}_{j_1}-a_{j_2j_1})^2>0$,

$$(a_{j_1j_2s_2},\dot{x}_{j_2})=\underbrace{(\text{DI},\text{pF})}_{\text{deceasing-inflection flow }(+)}.$$

- For $a_{j_1j_20}(a_{j_1j_2s_2}-a_{j_1j_2s_1})>0$ and $a_{j_2j_10}(\overline{x}_{j_1}-a_{j_2j_1})^2<0$,

$$(a_{j_1j_2s_2},\dot{x}_{j_2})=\underbrace{(\text{DI},\text{nF})}_{\text{decreasing-inflection flow }(-)}.$$

- For $a_{j_1j_20}(a_{j_1j_2s_2} - a_{j_1j_2s_1}) < 0$ and $a_{j_2j_10}(\bar{x}_{j_1} - a_{j_2j_1})^2 < 0$,

$$(a_{j_1j_2s_2}, \dot{x}_{j_2}) = \underbrace{(\text{II}, \text{nF})}_{\text{increasing-inflection flow } (-)}.$$

Therefore, from the proof (II) of case (vi$_1$) and the above analysis, the equilibrium of $(x^*_{j_2}, x^*_{j_1}) = (a_{j_1j_2s_2}, a_{j_2j_11})$ has the following properties as in Eqs. (3.51)–(3.54):

- For $a_{j_1j_20}(a_{j_1j_2s_2} - a_{j_1j_2s_1}) > 0$ and $a_{j_2j_10} > 0$,

$$(a_{j_1j_2s_2}, a_{j_2j_11}) = \underbrace{(\text{II}_+, \text{II}_+)}_{\text{double-inflection saddle}}.$$

  The equilibrium of $(x^*_{j_2}, x^*_{j_1}) = (a_{j_1j_2s_2}, a_{j_2j_11})$ is an (II$_+$,II$_+$)-double increasing-inflection saddle.
- For $a_{j_1j_20}(a_{j_1j_2s_2} - a_{j_1j_2s_1}) < 0$ and $a_{j_2j_10} > 0$,

$$(a_{j_1j_2s_2}, a_{j_2j_11}) = \underbrace{(\text{DI}_+, \text{DI}_-)}_{\text{double-inflection saddle}}.$$

  The equilibrium of $(x^*_{j_2}, x^*_{j_1}) = (a_{j_1j_2s_2}, a_{j_2j_11})$ is a (DI$_+$,DI$_-$)-double decreasing-inflection saddle.
- For $a_{j_1j_20}(a_{j_1j_2s_2} - a_{j_1j_2s_1}) > 0$ and $a_{j_2j_10} < 0$,

$$(a_{j_1j_2s_2}, a_{j_2j_11}) = \underbrace{(\text{DI}_-, \text{DI}_+)}_{\text{double-inflection saddle}}.$$

  The equilibrium of $(x^*_{j_2}, x^*_{j_1}) = (a_{j_1j_2s_2}, a_{j_2j_11})$ is a (DI$_-$,DI$_+$)-double decreasing-inflection saddle.
- For $a_{j_1j_20}(a_{j_1j_2s_2} - a_{j_1j_2s_1}) < 0$ and $a_{j_2j_10} < 0$,

$$(a_{j_1j_2s_2}, a_{j_2j_11}) = \underbrace{(\text{II}_-, \text{II}_-)}_{\text{double-inflection saddle}}.$$

  The equilibrium of $(x^*_{j_2}, x^*_{j_1}) = (a_{j_1j_2s_2}, a_{j_2j_11})$ is an (II$_-$,II$_-$)-double increasing -inflection saddle.

(vii) If

$$\Delta_{j_1j_2} = 0 \text{ and } \Delta_{j_2j_1} = 0, a_{j_2j_1s_1} = a_{j_2j_1s_2} = a_{j_2j_11},$$

then

$$\begin{aligned}
&\dot{x}_{j_1} = a_{j_1j_20}(x_{j_2} - a_{j_1j_21})^3,\\
&\dot{x}_{j_2} = a_{j_2j_10}(x_{j_1} - a_{j_2j_11})^2,\\
&j_1, j_2 \in \{1,2\}; j_1 \neq j_2.
\end{aligned}$$

In phase space,

$$\frac{dx_{j_1}}{dx_{j_2}} = \frac{a_{j_1j_20}}{a_{j_2j_10}} \frac{(x_{j_2} - a_{j_1j_21})^3}{(x_{j_1} - a_{j_2j_11})^2}$$

and

$$a_{j_2j_10}(x_{j_1} - a_{j_2j_11})^2 dx_{j_1} = a_{j_1j_20}(x_{j_2} - a_{j_1j_21})^3 dx_{j_2}.$$

With an initial condition (x_{j_10}, x_{j_20}), the first integral manifold is obtained, i.e.,

$$\begin{aligned}
&\frac{1}{3} a_{j_2j_10}\left[(x_{j_1} - a_{j_2j_11})^3 - (x_{j_10} - a_{j_2j_11})^3\right]\\
&= \frac{1}{4} a_{j_1j_20}\left[(x_{j_2} - a_{j_1j_2s_1})^4 - (x_{j_20} - a_{j_1j_2s_1})^4\right].
\end{aligned}$$

(vii$_1$) Consider two cases (I) $x^*_{j_2} = a_{j_1j_21}$ and (II) $x^*_{j_1} = a_{j_2j_11}$.

(I) In phase space, at $x^*_{j_2} = a_{j_1j_21}$ and $\bar{x}_{j_1} \neq a_{j_2j_11}$, we have

$$\begin{aligned}
&\left.\frac{dx_{j_1}}{dx_{j_2}}\right|_{x^*_{j_2} = a_{j_1j_21}} = \left.\frac{a_{j_1j_20}}{a_{j_2j_10}} \frac{(x_{j_2} - a_{j_1j_21})^3}{(\bar{x}_{j_1} - a_{j_2j_11})^2}\right|_{x^*_{j_2} = a_{j_1j_21}} = 0,\\
&\left.\frac{d^2x_{j_1}}{dx^2_{j_2}}\right|_{x^*_{j_2} = a_{j_1j_21}} = \left.\frac{d^3x_{j_1}}{dx^3_{j_2}}\right|_{x^*_{j_2} = a_{j_1j_21}} = 0.
\end{aligned}$$

If

$$\left.\frac{d^4x_{j_1}}{dx^4_{j_2}}\right|_{x^*_{j_2} = a_{j_1j_21}} = \frac{a_{j_1j_20}}{a_{j_2j_10}} \frac{6}{(\bar{x}_{j_1} - a_{j_2j_11})^2} > 0,$$

there is a third-order up-parabola flow at $x^*_{j_2} = a_{j_1j_21}$ in the x_{j_1}-direction. If

$$\left.\frac{d^4x_{j_1}}{dx^4_{j_2}}\right|_{x^*_{j_2} = a_{j_1j_21}} = \frac{a_{j_1j_20}}{a_{j_2j_10}} \frac{6}{(\bar{x}_{j_1} - a_{j_2j_11})^2} < 0,$$

there is a third-order down-parabola flow at $x^*_{j_2} = a_{j_1j_21}$ in the x_{j_1}-direction. Because

$$\dot{x}_{j_2} = a_{j_2j_10}(\overline{x}_{j_1} - a_{j_2j_11})^2,$$

the flows at $x^*_{j_2} = a_{j_1j_21}$ in the x_{j_2}-direction are positive and negative for $a_{j_2j_10} > 0$ and $a_{j_2j_10} < 0$, respectively.

Thus, the equilibrium of $x^*_{j_2} = a_{j_1j_21}$ in the x_{j_2}-direction has the following properties:

- For $a_{j_1j_20} > 0$ and $a_{j_2j_10}(\overline{x}_{j_1} - a_{j_2j_11})^2 > 0$,

$$(a_{j_1j_21}, \dot{x}_{j_2}) = \underbrace{(3^{\text{rd}}\text{UP, pF})}_{\text{third-order up-parabola flow } (+)}.$$

- For $a_{j_1j_20} < 0$ and $a_{j_2j_10}(\overline{x}_{j_1} - a_{j_2j_11})^2 > 0$,

$$(a_{j_1j_21}, \dot{x}_{j_2}) = \underbrace{(3^{\text{rd}}\text{DP, pF})}_{\text{third-order down-parabola flow } (+)}.$$

- For $a_{j_1j_20} > 0$ and $a_{j_2j_10}(\overline{x}_{j_1} - a_{j_2j_11})^2 < 0$,

$$(a_{j_1j_21}, \dot{x}_{j_2}) = \underbrace{(3^{\text{rd}}\text{DP, nF})}_{\text{third-order down-parabola flow } (-)}.$$

- For $a_{j_1j_20} < 0$ and $a_{j_2j_10}(\overline{x}_{j_1} - a_{j_2j_11})^2 < 0$,

$$(a_{j_1j_21}, \dot{x}_{j_2}) = \underbrace{(3^{\text{rd}}\text{UP, nF})}_{\text{third-order up-parabola flow } (-)}.$$

(II) In phase space, at $x^*_{j_1} = a_{j_2j_11}$ and $\overline{x}_{j_2} \neq a_{j_1j_21}$, we have

$$\begin{aligned}
&\frac{dx_{j_2}}{dx_{j_1}}\bigg|_{x^*_{j_1} = a_{j_2j_11}} = \frac{a_{j_2j_10}\,(x_{j_1} - a_{j_2j_11})^2}{a_{j_1j_20}\,(\overline{x}_{j_2} - a_{j_1j_21})^3}\bigg|_{x^*_{j_1} = a_{j_2j_11}} = 0,\\
&\frac{d^2x_{j_2}}{dx^2_{j_1}}\bigg|_{x^*_{j_1} = a_{j_2j_11}} = 0.
\end{aligned}$$

If

$$\frac{d^3x_{j_2}}{dx_{j_1}^3}\Big|_{x_{j_1}^*=a_{j_2j_1 1}} = \frac{a_{j_1j_2 0}}{a_{j_2j_1 0}} \frac{2}{(\overline{x}_{j_2} - a_{j_1j_2s_2})^3} > 0,$$

there is an increasing-inflection flow at $x_{j_1}^* = a_{j_2j_1 1}$ in the x_{j_2}-direction. If

$$\frac{d^3x_{j_2}}{dx_{j_1}^3}\Big|_{x_{j_1}^*=a_{j_2j_1 1}} = \frac{a_{j_1j_2 0}}{a_{j_2j_1 0}} \frac{2}{(\overline{x}_{j_2} - a_{j_1j_2 1})^3} < 0,$$

there is a decreasing-inflection flow at $x_{j_1}^* = a_{j_2j_1 1}$ in the x_{j_1}-direction. Because

$$\dot{x}_{j_1} = a_{j_1j_2 0}(\overline{x}_{j_2} - a_{j_1j_2 1})^3,$$

the flows at $x_{j_1}^* = a_{j_2j_1 1}$ in the x_{j_1}-direction are positive and negative for $\dot{x}_{j_1} > 0$ and $\dot{x}_{j_1} < 0$, respectively.

Thus, the equilibrium of $x_{j_1}^* = a_{j_2j_1 1}$ and $\overline{x}_{j_2} \neq a_{j_1j_2 1}$ in the x_{j_2}-direction has the following properties:

- For $a_{j_1j_2 0}(\overline{x}_{j_2} - a_{j_1j_2 1})^3 > 0$ and $a_{j_2j_1 0} > 0$,

$$(\dot{x}_{j_1}, a_{j_2j_1 1}) = \underbrace{(\text{pF}, \text{II})}_{\text{increasing-inflection flow }(+)}.$$

- For $a_{j_1j_2 0}(\overline{x}_{j_2} - a_{j_1j_2 1})^3 < 0$ and $a_{j_2j_1 0} > 0$,

$$(\dot{x}_{j_1}, a_{j_2j_1 1}) = \underbrace{(\text{nF}, \text{DI})}_{\text{decreasing-inflection flow }(-)}.$$

- For $a_{j_1j_2 0}(\overline{x}_{j_2} - a_{j_1j_2 1})^3 > 0$ and $a_{j_2j_1 0} < 0$,

$$(\dot{x}_{j_1}, a_{j_2j_1 1}) = \underbrace{(\text{pF}, \text{DI})}_{\text{decreasing-inflection flow }(+)}.$$

- For $a_{j_1j_2 0}(\overline{x}_{j_2} - a_{j_1j_2 1})^3 < 0$ and $a_{j_2j_1 0} < 0$,

$$(\dot{x}_{j_1}, a_{j_2j_1 1}) = \underbrace{(\text{nF}, \text{II})}_{\text{increasing-inflection flow }(-)}.$$

Therefore, from cases (I) and (II), the equilibrium of $(x_{j_2}^*, x_{j_1}^*) = (a_{j_1j_2 1}, a_{j_2j_1 1})$ has the following properties as in Eqs. (3.58)–(3.61):

- For $a_{j_1j_2 0} > 0$ and $a_{j_2j_1 0} > 0$,

$$(a_{j_1j_2 1}, a_{j_2j_1 1}) = \underbrace{(3^{\text{rd}}\text{UP}, \text{US})}_{\text{third-order up-parabola upper-saddle}}.$$

The equilibrium of $(x^*_{j_2}, x^*_{j_1}) = (a_{j_1j_2 1}, a_{j_2j_1 1})$ is a (3^{rd}UP,US)-third-order-up-parabola upper-saddle.

- For $a_{j_1j_2 0} < 0$ and $a_{j_2j_1 0} > 0$,

$$(a_{j_1j_2 1}, a_{j_2j_1 1}) = \underbrace{(3^{\text{rd}}\text{DP}, \text{US})}_{\text{third-order down-parabola upper-saddle}}.$$

The equilibrium of $(x^*_{j_1}, x^*_{j_2}) = (a_{j_2j_1 1}, a_{j_1j_2 1})$ is a (3^{rd}DP,US)-third-order down-parabola upper-saddle.

- For $a_{j_1j_2 0} > 0$ and $a_{j_2j_1 0} < 0$,

$$(a_{j_1j_2 1}, a_{j_2j_1 1}) = \underbrace{(3^{\text{rd}}\text{DP}, \text{LS})}_{\text{third-order down-parabola lower-saddle}}.$$

The equilibrium of $(x^*_{j_1}, x^*_{j_2}) = (a_{j_2j_1 1}, a_{j_1j_2 1})$ is a (3^{rd}DP,LS)-third-order down-parabola lower-saddle.

- For $a_{j_1j_2 0} < 0$ and $a_{j_2j_1 0} < 0$,

$$(a_{j_1j_2 1}, a_{j_2j_1 1}) = \underbrace{(3^{\text{rd}}\text{UP}, \text{LS})}_{\text{third-order up-parabola lower-saddle}}.$$

The equilibrium of $(x^*_{j_1}, x^*_{j_2}) = (a_{j_2j_1 1}, a_{j_1j_2 1})$ is a (3^{rd}UP,LS)-third-order-up-parabola lower-saddle.

(viii) If

$$\Delta_{j_1j_2} = 0 \text{ and } \Delta_{j_2j_1} > 0, a_{j_1j_2s_1} \neq a_{j_1j_2s_2},$$

then

$$\begin{aligned}
&\dot{x}_{j_2} = a_{j_2j_1 0}(x_{j_1} - a_{j_2j_1 1})(x_{j_1} - a_{j_2j_1 2}),\\
&\dot{x}_{j_1} = a_{j_1j_2 0}(x_{j_2} - a_{j_1j_2s_1})(x_{j_2} - a_{j_1j_2s_2})^2,\\
&j_1, j_2 \in \{1, 2\}; j_1 \neq j_2.
\end{aligned}$$

In phase space,

$$\frac{dx_{j_1}}{dx_{j_2}} = \frac{a_{j_1j_20}}{a_{j_2j_10}} \frac{(x_{j_2} - a_{j_1j_2s_1})(x_{j_2} - a_{j_1j_2s_2})^2}{(x_{j_1} - a_{j_2j_1l_1})(x_{j_1} - a_{j_2j_1l_1})}$$

and

$$\begin{aligned}
&a_{j_2j_10}\left[(x_{j_1} - a_{j_2j_1l_1})^2 + (a_{j_2j_1l_1} - a_{j_2j_1l_2})(x_{j_1} - a_{j_2j_1l_1})\right]dx_{j_1} \\
&= a_{j_1j_20}\Big[(x_{j_2} - a_{j_1j_2s_1})^3 + 2(a_{j_1j_2s_1} - a_{j_1j_2s_2})(x_{j_2} - a_{j_1j_2s_1})^2 \\
&+ (a_{j_1j_2s_1} - a_{j_1j_2s_2})^2(x_{j_2} - a_{j_1j_2s_1})\Big]dx_{j_2}.
\end{aligned}$$

With an initial condition $(x_{j_1 0}, x_{j_2 0})$, the first integral manifold is generated, i.e.,

$$\begin{aligned}
&a_{j_2j_10}\Big\{\frac{1}{3}\left[(x_{j_1} - a_{j_2j_1l_1})^3 - (x_{j_10} - a_{j_2j_1l_1})^3\right] \\
&+ \frac{1}{2}(a_{j_2j_1l_1} - a_{j_2j_1l_2})\left[(x_{j_1} - a_{j_2j_1l_1})^3 - (x_{j_10} - a_{j_2j_1l_1})^3\right]\Big\} \\
&= a_{j_1j_20}\Big\{\frac{1}{4}\left[(x_{j_1} - a_{j_1j_2s_1})^4 - (x_{j_10} - a_{j_1j_2s_1})^4\right] \\
&+ \frac{2}{3}(a_{j_1j_2s_1} - a_{j_1j_2s_2})\left[(x_{j_2} - a_{j_1j_2s_1})^3 - (x_{j_20} - a_{j_1j_2s_1})^3\right] \\
&+ \frac{1}{2}\left[(a_{j_1j_2s_1} - a_{j_1j_2s_2})^2\right]\left[(x_{j_2} - a_{j_1j_2s_1})^2 - (x_{j_20} - a_{j_1j_2s_1})^2\right]\Big\}.
\end{aligned}$$

(viii$_1$) Consider two cases (I) $x^*_{j_2} = a_{j_1j_2s_1}$ and (II) $x^*_{j_1} = a_{j_2j_1l_1}$.

(I) In phase space, at $x^*_{j_2} = a_{j_1j_2s_1}$ (s_1, $s_2 \in \{1,2\}$, $s_1 \neq s_2$) and $\bar{x}_{j_1} \neq a_{j_2j_1l_1}, a_{j_2j_1l_2}$ (l_1, $l_2 \in \{1,2\}$, $l_1 \neq l_2$), we have

$$\left.\frac{dx_{j_1}}{dx_{j_2}}\right|_{x^*_{j_2} = a_{j_1j_2s_1}} = \left.\frac{a_{j_1j_20}}{a_{j_2j_10}} \frac{(x_{j_2} - a_{j_1j_2s_1})(x_{j_2} - a_{j_1j_2s_2})^2}{(\bar{x}_{j_1} - a_{j_2j_1l_1})(\bar{x}_{j_1} - a_{j_2j_1l_2})}\right|_{x^*_{j_2} = a_{j_1j_2s_1}} = 0.$$

If

$$\left.\frac{d^2x_{j_1}}{dx^2_{j_2}}\right|_{x^*_{j_2} = a_{j_1j_2s_1}} = \frac{a_{j_1j_20}}{a_{j_2j_10}} \frac{(a_{j_1j_2s_1} - a_{j_1j_2s_2})^2}{(\bar{x}_{j_1} - a_{j_2j_1l_1})(\bar{x}_{j_1} - a_{j_2j_1l_2})} > 0,$$

there is an up-parabola flow at $x^*_{j_2} = a_{j_1j_2s_1}$ in the x_{j_1}-direction. If

$$\left.\frac{d^2x_{j_1}}{dx^2_{j_2}}\right|_{x^*_{j_2} = a_{j_1j_2s_1}} = \frac{a_{j_1j_20}}{a_{j_2j_10}} \frac{(a_{j_1j_2s_1} - a_{j_1j_2s_2})^2}{(\bar{x}_{j_1} - a_{j_2j_1l_1})(\bar{x}_{j_1} - a_{j_2j_1l_2})} < 0,$$

there is a down-parabola flow at $x^*_{j_2} = a_{j_1j_2s_1}$ in the x_{j_1}-direction. Because

$$\dot{x}_{j_2} = a_{j_2 j_1 0}(\overline{x}_{j_1} - a_{j_2 j_1 l_1})(\overline{x}_{j_1} - a_{j_2 j_1 l_2}),$$

the flows at $x_{j_2}^* = a_{j_1 j_2 s_1}$ in the x_{j_2}-direction are positive and negative for $\dot{x}_{j_2} > 0$ and $\dot{x}_{j_2} < 0$, respectively.

Thus, the equilibrium of $x_{j_2}^* = a_{j_1 j_2 s_1}$ (s_1, $s_2 \in \{1,2\}$, $s_1 \neq s_2$) with $\overline{x}_{j_1} \neq a_{j_2 j_1 l_1}$ ($l_1 = 1, 2$) in the x_{j_2}-direction has the following properties:

- For $a_{j_1 j_2 0}(a_{j_1 j_2 s_1} - a_{j_1 j_2 s_2})^2 > 0$ and $a_{j_2 j_1 0}\prod_{l_1=1}^{2}(\overline{x}_{j_1} - a_{j_2 j_1 l_1}) > 0$,

$$(a_{j_1 j_2 s_1}, \dot{x}_{j_2}) = \underbrace{(\mathrm{UP}, \mathrm{pF})}_{\text{up-parabola flow } (+)}.$$

- For $a_{j_1 j_2 0}(a_{j_1 j_2 s_1} - a_{j_1 j_2 s_2})^2 < 0$ and $a_{j_2 j_1 0}\prod_{l_1=1}^{2}(\overline{x}_{j_1} - a_{j_2 j_1 l_1}) > 0$,

$$(a_{j_1 j_2 s_1}, \dot{x}_{j_2}) = \underbrace{(\mathrm{DP}, \mathrm{pF})}_{\text{down-parabola flow } (+)}.$$

- For $a_{j_1 j_2 0}(a_{j_1 j_2 s_1} - a_{j_1 j_2 s_2})^2 > 0$ and $a_{j_2 j_1 0}\prod_{l_1=1}^{2}(\overline{x}_{j_1} - a_{j_2 j_1 l_1}) < 0$,

$$(a_{j_1 j_2 s_1}, \dot{x}_{j_2}) = \underbrace{(\mathrm{DP}, \mathrm{nF})}_{\text{down-parabola flow } (-)}.$$

- For $a_{j_1 j_2 0}(a_{j_1 j_2 s_1} - a_{j_1 j_2 s_2})^2 < 0$ and $a_{j_2 j_1 0}\prod_{l_1=1}^{2}(\overline{x}_{j_1} - a_{j_2 j_1 l_1}) < 0$,

$$(a_{j_1 j_2 s_1}, \dot{x}_{j_2}) = \underbrace{(\mathrm{UP}, \mathrm{nF})}_{\text{up-parabola flow } (-)}.$$

(II) In phase space, at $x_{j_1}^* = a_{j_2 j_1 l_1}$ (l_1, $l_2 \in \{1,2\}$, $l_1 \neq l_2$) and $\overline{x}_{j_2} \neq a_{j_1 j_2 s_1}, a_{j_1 j_2 s_2}$ (s_1, $s_2 \in \{1,2\}$, $s_1 \neq s_2$), we have

$$\frac{dx_{j_2}}{dx_{j_1}}\bigg|_{x_{j_1}^* = a_{j_2 j_1 l_1}} = \frac{a_{j_2 j_1 0}}{a_{j_1 j_2 0}} \frac{(x_{j_1} - a_{j_2 j_1 l_1})(x_{j_1} - a_{j_2 j_1 l_2})}{(\overline{x}_{j_2} - a_{j_1 j_2 s_1})(\overline{x}_{j_2} - a_{j_1 j_2 s_2})^2}\bigg|_{x_{j_1}^* = a_{j_2 j_1 l_1}} = 0.$$

If

$$\frac{d^2 x_{j_2}}{dx_{j_1}^2}\bigg|_{x_{j_1}^* = a_{j_2 j_1 l_1}} = \frac{a_{j_2 j_1 0}}{a_{j_1 j_2 0}} \frac{(a_{j_2 j_1 l_1} - a_{j_2 j_1 l_2})}{(\overline{x}_{j_2} - a_{j_1 j_2 s_1})(\overline{x}_{j_2} - a_{j_1 j_2 s_2})^2} > 0,$$

there is an up-parabola flow at $x_{j_1}^* = a_{j_2 j_1 l_1}$ in the x_{j_2}-direction. If

$$\left.\frac{d^2x_{j_2}}{dx_{j_1}^2}\right|_{x_{j_1}^*=a_{j_2j_1l_1}} = \frac{a_{j_2j_10}}{a_{j_1j_20}} \frac{(a_{j_2j_1l_1} - a_{j_2j_1l_2})}{(\bar{x}_{j_2} - a_{j_1j_2s_1})(\bar{x}_{j_2} - a_{j_1j_2s_2})^2} < 0,$$

there is a down-parabola flow at $x_{j_1}^* = a_{j_2j_1l_1}$ in the x_{j_1}-direction. Because

$$\dot{x}_{j_1} = a_{j_1j_20}(\bar{x}_{j_2} - a_{j_1j_2s_1})(\bar{x}_{j_2} - a_{j_1j_2s_2})^2,$$

the flows at $x_{j_1}^* = a_{j_2j_1l_1}$ in the x_{j_1}-direction are positive and negative for $\dot{x}_{j_1} > 0$ and $\dot{x}_{j_1} < 0$, respectively.

Thus, the equilibrium of $x_{j_1}^* = a_{j_2j_1l_1}$ (l_1, $l_2 \in \{1,2\}$, $l_1 \neq l_2$) and $\bar{x}_{j_2} \neq a_{j_1j_2s_1}, a_{j_1j_2s_2}$ (s_1, $s_2 \in \{1,2\}$, $s_1 \neq s_2$) in the x_{j_2}-direction has the following properties:

- For $a_{j_1j_20}(\bar{x}_{j_2} - a_{j_1j_2s_1})(\bar{x}_{j_2} - a_{j_1j_2s_2})^2 > 0$ and $a_{j_2j_10}(a_{j_2j_1l_1} - a_{j_2j_1l_2}) > 0$,

$$(\dot{x}_{j_1}, a_{j_2j_1l_1}) = \underbrace{(\text{pF}, \text{UP})}_{\text{up-parabola flow }(+)}.$$

- For $a_{j_1j_20}(\bar{x}_{j_2} - a_{j_1j_2s_1})(\bar{x}_{j_2} - a_{j_1j_2s_2})^2 < 0$ and $a_{j_2j_10}(a_{j_2j_1l_1} - a_{j_2j_1l_2}) > 0$,

$$(\dot{x}_{j_1}, a_{j_2j_1l_1}) = \underbrace{(\text{nF}, \text{DP})}_{\text{down-parabola flow }(-)}.$$

- For $a_{j_1j_20}(\bar{x}_{j_2} - a_{j_1j_2s_1})(\bar{x}_{j_2} - a_{j_1j_2s_2})^2 > 0$ and $a_{j_2j_10}(a_{j_2j_1l_1} - a_{j_2j_1l_2}) < 0$,

$$(\dot{x}_{j_1}, a_{j_2j_1l_1}) = \underbrace{(\text{pF}, \text{DP})}_{\text{down-parabola flow }(+)}.$$

- For $a_{j_1j_20}(\bar{x}_{j_2} - a_{j_1j_2s_1})(\bar{x}_{j_2} - a_{j_1j_2s_2})^2 < 0$ and $a_{j_2j_10}(a_{j_2j_1l_1} - a_{j_2j_1l_2}) < 0$,

$$(\dot{x}_{j_1}, a_{j_2j_1l_1}) = \underbrace{(\text{nF}, \text{UP})}_{\text{up-parabola flow }(-)}.$$

Therefore, from cases (I) and (II), the equilibrium of $(x_{j_2}^*, x_{j_1}^*) = (a_{j_1j_2s_1}, a_{j_2j_1l_1})$ (s_1, s_2, l_1, $l_2 \in \{1,2\}$, $s_1 \neq s_2$; $l_1 \neq l_2$) has the following properties as in Eqs. (3.65)–(3.68):

- For $a_{j_1j_20}(a_{j_1j_2s_1} - a_{j_1j_2s_2})^2 > 0$ and $a_{j_2j_10}(a_{j_2j_1l_1} - a_{j_2j_1l_2}) > 0$,

$$(a_{j_1j_2s_1}, a_{j_2j_1l_1}) = \underbrace{(\mathrm{UP}_+, \mathrm{UP}_+)}_{\text{positive saddle}}.$$

The equilibrium of $(x_{j_2}^*, x_{j_1}^*) = (a_{j_1j_2s_1}, a_{j_2j_1l_1})$ is a (UP_+,UP_+)-positive saddle.

- For $a_{j_1j_20}(a_{j_1j_2s_1} - a_{j_1j_2s_2})^2 < 0$ and $a_{j_2j_10}(a_{j_2j_1l_1} - a_{j_2j_1l_2}) > 0$,

$$(a_{j_1j_2s_1}, a_{j_2j_1l_1}) = \underbrace{(\mathrm{DP}_+, \mathrm{DP}_-)}_{\text{CCW center}}.$$

The equilibrium of $(x_{j_2}^*, x_{j_1}^*) = (a_{j_1j_2s_1}, a_{j_2j_1l_1})$ is a (DP_+,DP_-)-counter-clockwise center.

- For $a_{j_1j_20}(a_{j_1j_2s_1} - a_{j_1j_2s_2})^2 > 0$ and $a_{j_2j_10}(a_{j_2j_1l_1} - a_{j_2j_1l_2}) < 0$,

$$(a_{j_1j_2s_1}, a_{j_2j_1l_1}) = \underbrace{(\mathrm{DP}_-, \mathrm{DP}_+)}_{\text{CW center}}.$$

The equilibrium of $(x_{j_2}^*, x_{j_1}^*) = (a_{j_1j_2s_1}, a_{j_2j_1l_1})$ is a (DP_-,DP_+)-clockwise center.

- For $a_{j_1j_20}(a_{j_1j_2s_1} - a_{j_1j_2s_2})^2 < 0$ and $a_{j_2j_10}(a_{j_2j_1l_1} - a_{j_2j_1l_2}) < 0$,

$$(a_{j_1j_2s_1}, a_{j_2j_1l_1}) = \underbrace{(\mathrm{UP}_-, \mathrm{UP}_-)}_{\text{negative saddle}}.$$

The equilibrium of $(x_{j_2}^*, x_{j_1}^*) = (a_{j_1j_2s_1}, a_{j_2j_1l_1})$ is a (DP_-,UP_-)-negative saddle.

(viii$_2$) In phase space, at $x_{j_2}^* = a_{j_1j_2s_2}$ ($s_1, s_2 \in \{1, 2\}$, $s_1 \neq s_2$) and $x_{j_1} \neq a_{j_2j_1}1$, we have

$$\begin{aligned}
&\frac{dx_{j_1}}{dx_{j_2}}\Big|_{x_{j_2}^* = a_{j_1j_2s_2}} = \frac{a_{j_1j_20}}{a_{j_2j_10}} \frac{(x_{j_2} - a_{j_1j_2s_1})(x_{j_2} - a_{j_1j_2s_2})^2}{(\bar{x}_{j_1} - a_{j_2j_1l_1})(\bar{x}_{j_1} - a_{j_2j_1l_2})}\Big|_{x_{j_2}^* = a_{j_1j_2s_2}} = 0,\\
&\frac{d^2x_{j_1}}{dx_{j_2}^2}\Big|_{x_{j_2}^* = a_{j_1j_2s_2}} = 0.
\end{aligned}$$

If

$$\frac{d^3x_{j_1}}{dx_{j_2}^3}\Big|_{x_{j_2}^* = a_{j_1j_2s_2}} = \frac{a_{j_1j_20}}{a_{j_2j_10}} \frac{2(a_{j_1j_2s_2} - a_{j_1j_2s_1})}{(\bar{x}_{j_1} - a_{j_2j_1l_1})(\bar{x}_{j_1} - a_{j_2j_1l_2})} > 0,$$

there is an increasing-inflection flow at $x_{j_2}^* = a_{j_1j_2s_2}$ in the x_{j_1}-direction. If

$$\left.\frac{d^3x_{j_1}}{dx_{j_2}^3}\right|_{x_{j_2}^*=a_{j_1j_2s_2}} = \frac{a_{j_1j_20}}{a_{j_2j_10}} \frac{2(a_{j_1j_2s_2} - a_{j_1j_2s_1})}{(\overline{x}_{j_1} - a_{j_2j_1l_1})(\overline{x}_{j_1} - a_{j_2j_1l_2})} < 0,$$

there is a decreasing-inflection flow at $x_{j_2}^* = a_{j_1j_2s_2}$ in the x_{j_1}-direction. Because

$$\dot{x}_{j_2} = a_{j_2j_10}(\overline{x}_{j_1} - a_{j_2j_1l_1})(\overline{x}_{j_1} - a_{j_2j_1l_2}),$$

the flows at $x_{j_2}^* = a_{j_1j_2s_2}$ in the x_{j_2}-direction are positive and negative for $a_{j_2j_10} > 0$ and $a_{j_2j_10} < 0$, respectively.

Thus, the equilibrium of $x_{j_2}^* = a_{j_1j_2s_2}$ (s_1, $s_2 \in \{1, 2\}$, $s_1 \neq s_2$) and $x_{j_1} \neq a_{j_2j_11}$ in the x_{j_2}-direction has the following properties:

- For $a_{j_1j_20}(a_{j_1j_2s_2} - a_{j_1j_2s_1}) > 0$ and $a_{j_2j_10}\prod_{l_1=1}^{2}(\overline{x}_{j_1} - a_{j_2j_1l_1}) > 0$,

$$(a_{j_1j_2s_2}, \dot{x}_{j_2}) = \underbrace{(\text{II}, \text{pF})}_{\text{increasing-inflection flow } (+)}.$$

- For $a_{j_1j_20}(a_{j_1j_2s_2} - a_{j_1j_2s_1}) < 0$ and $a_{j_2j_10}\prod_{l_1=1}^{2}(\overline{x}_{j_1} - a_{j_2j_1l_1}) > 0$,

$$(a_{j_1j_2s_2}, \dot{x}_{j_2}) = \underbrace{(\text{DI}, \text{pF})}_{\text{decreasing-inflection flow } (+)}.$$

- For $a_{j_1j_20}(a_{j_1j_2s_2} - a_{j_1j_2s_1}) > 0$ and $a_{j_2j_10}\prod_{l_1=1}^{2}(\overline{x}_{j_1} - a_{j_2j_1l_1}) < 0$,

$$(a_{j_1j_2s_2}, \dot{x}_{j_2}) = \underbrace{(\text{DI}, \text{nF})}_{\text{decreasing-inflection flow } (-)}.$$

- For $a_{j_1j_20}(a_{j_1j_2s_2} - a_{j_1j_2s_1}) < 0$ and $a_{j_2j_10}\prod_{l_1=1}^{2}(\overline{x}_{j_1} - a_{j_2j_1l_1}) < 0$,

$$(a_{j_1j_2s_2}, \dot{x}_{j_2}) = \underbrace{(\text{II}, \text{nF})}_{\text{increasing-inflection flow } (-)}.$$

Therefore, from the proof (II) of (viii$_1$) and the foregoing analysis, the equilibrium of $(x_{j_2}^*, x_{j_1}^*) = (a_{j_1j_2s_2}, a_{j_2j_1l_1})$ (s_1, s_2, l_1, $l_2 \in \{1, 2\}$, $s_1 \neq s_2$, $l_1 \neq l_2$) has the following properties as in Eqs. (3.69)–(3.72):

- For $a_{j_1j_20}(a_{j_1j_2s_2}-a_{j_1j_2s_1})>0$ and $a_{j_2j_10}(a_{j_2j_1l_1}-a_{j_2j_1l_2})>0$,

$$(a_{j_1j_2s_2},a_{j_2j_1l_1})=\underbrace{(\text{US},\text{UP})}_{\text{up-parabola upper-saddle}}.$$

The equilibrium of $(x^*_{j_2},x^*_{j_1})=(a_{j_1j_2s_2},a_{j_2j_1l_1})$ is a (US,UP)-up-parabola upper-saddle.

- For $a_{j_1j_20}(a_{j_1j_2s_2}-a_{j_1j_2s_1})<0$ and $a_{j_2j_10}(a_{j_2j_1l_1}-a_{j_2j_1l_2})>0$,

$$(a_{j_1j_2s_2},a_{j_2j_1l_1})=\underbrace{(\text{LS},\text{DP})}_{\text{down-parabola lower-saddle}}.$$

The equilibrium of $(x^*_{j_2},x^*_{j_1})=(a_{j_1j_2s_2},a_{j_2j_1l_1})$ is an (LS,DP)-down-parabola lower-saddle.

- For $a_{j_1j_20}(a_{j_1j_2s_2}-a_{j_1j_2s_1})>0$ and $a_{j_2j_10}(a_{j_2j_1l_1}-a_{j_2j_1l_2})<0$,

$$(a_{j_1j_2s_2},a_{j_2j_1l_1})=\underbrace{(\text{US},\text{DP})}_{\text{down-parabola upper-saddle}}.$$

The equilibrium of $(x^*_{j_2},x^*_{j_1})=(a_{j_1j_2s_2},a_{j_2j_1l_1})$ is a (US,DP)-down-parabola upper-saddle.

- For $a_{j_1j_20}(a_{j_1j_2s_2}-a_{j_1j_2s_1})<0$ and $a_{j_2j_10}(a_{j_2j_1l_1}-a_{j_2j_1l_2})<0$,

$$(a_{j_1j_2s_2},a_{j_2j_1l_1})=\underbrace{(\text{LS},\text{UP})}_{\text{up-parabola lower-saddle}}.$$

The equilibrium of $(x^*_{j_2},x^*_{j_1})=(a_{j_1j_2s_2},a_{j_2j_1l_1})$ is an (LS,UP)-up-parabola lower-saddle.

(ix) If

$$\Delta_{j_1j_2}=0 \text{ and } \Delta_{j_2j_1}>0, a_{j_1j_2s_1}=a_{j_1j_2s_2}=a_{j_1j_21},$$

then

$$\begin{aligned}
&\dot{x}_{j_2}=a_{j_2j_10}(x_{j_1}-a_{j_2j_11})(x_{j_1}-a_{j_2j_12}),\\
&\dot{x}_{j_1}=a_{j_1j_20}(x_{j_2}-a_{j_1j_21})^3,\\
&j_1,j_2\in\{1,2\};j_1\neq j_2.
\end{aligned}$$

In phase space,

$$\frac{dx_{j_1}}{dx_{j_2}} = \frac{a_{j_1j_20}}{a_{j_2j_10}} \frac{(x_{j_2} - a_{j_1j_21})^3}{(x_{j_1} - a_{j_2j_1l_1})(x_{j_1} - a_{j_2j_1l_1})}$$

for $a_{j_2j_1l_1} \neq a_{j_2j_1l_2}$ ($l_1, l_2 \in \{1,2\}, l_1 \neq l_2$), and

$$\begin{aligned}&a_{j_2j_10}\left[(x_{j_1} - a_{j_2j_1l_1})^2 + (a_{j_2j_1l_1} - a_{j_2j_1l_2})(x_{j_1} - a_{j_2j_1l_1})\right]dx_{j_1}\\&= a_{j_1j_20}(x_{j_2} - a_{j_1j_21})^3 dx_{j_2}.\end{aligned}$$

With an initial condition (x_{j_10}, x_{j_20}), the first integral manifold is generated, i.e.,

$$\begin{aligned}&a_{j_2j_10}\Big\{\frac{1}{3}\left[(x_{j_1} - a_{j_2j_1l_1})^3 - (x_{j_10} - a_{j_2j_1l_1})^3\right]\\&+\frac{1}{2}(a_{j_2j_1l_1} - a_{j_2j_1l_2})\left[(x_{j_1} - a_{j_2j_1l_1})^2 - (x_{j_10} - a_{j_2j_1l_1})^2\right]\Big\}\\&= \frac{1}{4}a_{j_1j_20}\left[(x_{j_1} - a_{j_1j_2s_1})^4 - (x_{j_10} - a_{j_1j_2s_1})^4\right].\end{aligned}$$

Consider two cases (I) $x^*_{j_2} = a_{j_1j_21}$ and (II) $x^*_{j_1} = a_{j_2j_1l_1}$.

(I) In phase space, at $x^*_{j_2} = a_{j_1j_21}$ and $\overline{x}_{j_1} \neq a_{j_2j_1l_1}$ ($l_1, l_2 \in \{1,2\}, l_1 \neq l_2$), we have

$$\begin{aligned}&\frac{dx_{j_1}}{dx_{j_2}}\Big|_{x^*_{j_2}=a_{j_1j_21}} = \frac{a_{j_1j_20}}{a_{j_2j_10}} \frac{(x_{j_2} - a_{j_1j_21})^3}{(\overline{x}_{j_1} - a_{j_2j_1l_1})(\overline{x}_{j_1} - a_{j_2j_1l_1})}\Big|_{x^*_{j_2}=a_{j_1j_21}} = 0,\\&\frac{d^2x_{j_1}}{dx^2_{j_2}}\Big|_{x^*_{j_2}=a_{j_1j_21}} = \frac{d^3x_{j_1}}{dx^3_{j_2}}\Big|_{x^*_{j_2}=a_{j_1j_21}} = 0.\end{aligned}$$

If

$$\frac{d^4x_{j_1}}{dx^4_{j_2}}\Big|_{x^*_{j_2}=a_{j_1j_21}} = \frac{a_{j_1j_20}}{a_{j_2j_10}} \frac{6}{(\overline{x}_{j_1} - a_{j_2j_1l_1})(\overline{x}_{j_1} - a_{j_2j_1l_1})} > 0,$$

there is a third-order up-parabola flow at $x^*_{j_2} = a_{j_1j_21}$ in the x_{j_1}-direction. If

$$\frac{d^4x_{j_1}}{dx^4_{j_2}}\Big|_{x^*_{j_2}=a_{j_1j_21}} = \frac{a_{j_1j_20}}{a_{j_2j_10}} \frac{6}{(\overline{x}_{j_1} - a_{j_2j_1l_1})(\overline{x}_{j_1} - a_{j_2j_1l_1})} < 0,$$

there is a third-order down-parabola flow at $x^*_{j_2} = a_{j_1j_21}$ in the x_{j_1}-direction. Because

$$\dot{x}_{j_2} = a_{j_2j_10}(\overline{x}_{j_1} - a_{j_2j_1l_1})(\overline{x}_{j_1} - a_{j_2j_1l_1}),$$

the flows at $x_{j_2}^* = a_{j_1j_21}$ in the x_{j_2}-direction are positive and negative for $\dot{x}_{j_2} > 0$ and $\dot{x}_{j_2} < 0$, respectively.

Thus, the equilibrium of $x_{j_2}^* = a_{j_1j_2s_1}$ ($s_1, s_2 \in \{1,2\}$, $s_1 \neq s_2$) in the x_{j_2}-direction has the following properties:

- For $a_{j_1j_20} > 0$ and $a_{j_2j_10}\prod_{l_1=1}^{2}(\bar{x}_{j_1} - a_{j_2j_1l_1}) > 0$,

$$(a_{j_1j_2s_1}, \dot{x}_{j_2}) = \underbrace{(3^{\text{rd}}\text{UP}, \text{pF})}_{\text{third-order up-parabola flow } (+)}.$$

- For $a_{j_1j_20} < 0$ and $a_{j_2j_10}\prod_{l_1=1}^{2}(\bar{x}_{j_1} - a_{j_2j_1l_1}) > 0$,

$$(a_{j_1j_2s_1}, \dot{x}_{j_2}) = \underbrace{(3^{\text{rd}}\text{DP}, \text{pF})}_{\text{third-order down-parabola flow } (+)}.$$

- For $a_{j_1j_20} > 0$ and $a_{j_2j_10}\prod_{l_1=1}^{2}(\bar{x}_{j_1} - a_{j_2j_1l_1}) < 0$,

$$(a_{j_1j_2s_1}, \dot{x}_{j_2}) = \underbrace{(3^{\text{rd}}\text{DP}, \text{nF})}_{\text{third-order down-parabola flow } (-)}.$$

- For $a_{j_1j_20} < 0$ and $a_{j_2j_10}\prod_{l_1=1}^{2}(\bar{x}_{j_1} - a_{j_2j_1l_1}) < 0$,

$$(a_{j_1j_2s_1}, \dot{x}_{j_2}) = \underbrace{(3^{\text{rd}}\text{UP}, \text{nF})}_{\text{third-order up-parabola flow } (-)}.$$

(II) In phase space, at $x_{j_1}^* = a_{j_2j_1l_1}$ ($l_1, l_2 \in \{1,2\}$, $l_1 \neq l_2$) and $\bar{x}_{j_2} \neq a_{j_1j_21}$, we have

$$\frac{dx_{j_2}}{dx_{j_1}}\bigg|_{x_{j_1}^* = a_{j_2j_1l_1}} = \frac{a_{j_2j_10}}{a_{j_1j_20}}\frac{(x_{j_1} - a_{j_2j_1l_1})(x_{j_1} - a_{j_2j_1l_2})}{(\bar{x}_{j_2} - a_{j_1j_21})^3}\bigg|_{x_{j_1}^* = a_{j_2j_1l_1}} = 0.$$

If

$$\frac{d^2x_{j_2}}{dx_{j_1}^2}\bigg|_{x_{j_1}^* = a_{j_2j_11}} = \frac{a_{j_2j_10}}{a_{j_1j_20}}\frac{(a_{j_2j_1l_1} - a_{j_2j_1l_2})}{(\bar{x}_{j_2} - a_{j_1j_21})^3} > 0,$$

there is an up-parabola flow at $x_{j_1}^* = a_{j_2j_1l_1}$ in the x_{j_2}-direction. If

$$\frac{d^2x_{j_2}}{dx_{j_1}^2}\Big|_{x_{j_1}^*=a_{j_2j_11}} = \frac{a_{j_2j_10}}{a_{j_1j_20}}\frac{(a_{j_2j_1l_1}-a_{j_2j_1l_2})}{(\overline{x}_{j_2}-a_{j_1j_21})^3} < 0,$$

there is a down-parabola flow at $x_{j_1}^* = a_{j_2j_1l_1}$ in the x_{j_1}-direction. Because

$$\dot{x}_{j_1} = a_{j_1j_20}(\overline{x}_{j_2} - a_{j_1j_21})^3,$$

the flows at $x_{j_1}^* = a_{j_2j_1l_1}$ in the x_{j_1}-direction are positive and negative for $\dot{x}_{j_1} > 0$ and $\dot{x}_{j_1} < 0$, respectively.

Thus, the equilibrium of $x_{j_1}^* = a_{j_2j_1l_1}$ ($l_1, l_2 \in \{1,2\}, l_1 \neq l_2$) and $\overline{x}_{j_2} \neq a_{j_1j_21}$ in the x_{j_2}-direction has the following properties:

- For $a_{j_1j_20}(\overline{x}_{j_2} - a_{j_1j_21})^3 > 0$ and $a_{j_2j_10}(a_{j_2j_1l_1} - a_{j_2j_1l_2}) > 0$,

$$(\dot{x}_{j_1}, a_{j_2j_1l_1}) = \underbrace{(\text{pF}, \text{UP})}_{\text{up-parabola flow }(+)}.$$

- For $a_{j_1j_20}(\overline{x}_{j_2} - a_{j_1j_21})^3 < 0$ and $a_{j_2j_10}(a_{j_2j_1l_1} - a_{j_2j_1l_2}) > 0$,

$$(\dot{x}_{j_1}, a_{j_2j_1l_1}) = \underbrace{(\text{nF}, \text{DP})}_{\text{down-parabola flow }(-)}.$$

- For $a_{j_1j_20}(\overline{x}_{j_2} - a_{j_1j_21})^3 > 0$ and $a_{j_2j_10}(a_{j_2j_1l_1} - a_{j_2j_1l_2}) < 0$,

$$(\dot{x}_{j_1}, a_{j_2j_1l_1}) = \underbrace{(\text{pF}, \text{DP})}_{\text{down-parabola flow }(+)}.$$

- For $a_{j_1j_20}(\overline{x}_{j_2} - a_{j_1j_21})^3 < 0$ and $a_{j_2j_10}(a_{j_2j_1l_1} - a_{j_2j_1l_2}) < 0$,

$$(\dot{x}_{j_1}, a_{j_2j_1l_1}) = \underbrace{(\text{nF}, \text{UP})}_{\text{up-parabola flow }(-)}.$$

Therefore, from cases (I) and (II), the equilibrium of $(x_{j_2}^*, x_{j_1}^*) = (a_{j_1j_21}, a_{j_2j_1l_1})$ ($l_1, l_2 \in \{1,2\}, l_1 \neq l_2$) has the following properties as in Eqs. (3.76)–(3.79):

- For $a_{j_1j_20} > 0$ and $a_{j_2j_10}(a_{j_2j_1l_1} - a_{j_2j_1l_2}) > 0$,

$$(a_{j_1j_2s_1}, a_{j_2j_1l_1}) = \underbrace{(3^{\rm rd}{\rm UP}_+, {\rm UP}_+)}_{\text{third-order positive saddle}}.$$

The equilibrium of $(x^*_{j_2}, x^*_{j_1}) = (a_{j_1j_2s_1}, a_{j_2j_1l_1})$ is a $(3^{\rm rd}{\rm UP}_+,{\rm UP}_+)$-third-order positive saddle.

- For $a_{j_1j_20} < 0$ and $a_{j_2j_10}(a_{j_2j_1l_1} - a_{j_2j_1l_2}) > 0$,

$$(a_{j_1j_2s_1}, a_{j_2j_1l_1}) = \underbrace{(3^{\rm rd}{\rm DP}_+, {\rm DP}_-)}_{\text{third-order CCW center}}.$$

The equilibrium of $(x^*_{j_2}, x^*_{j_1}) = (a_{j_1j_2s_1}, a_{j_2j_1l_1})$ is a $(3^{\rm rd}{\rm DP}_+,{\rm DP}_-)$-third-order-counter-clockwise center.

- For $a_{j_1j_20} > 0$ and $a_{j_2j_10}(a_{j_2j_1l_1} - a_{j_2j_1l_2}) < 0$,

$$(a_{j_1j_2s_1}, a_{j_2j_1l_1}) = \underbrace{(3^{\rm rd}{\rm DP}_-, {\rm DP}_+)}_{\text{third-order CW center}}.$$

The equilibrium of $(x^*_{j_2}, x^*_{j_1}) = (a_{j_1j_2s_1}, a_{j_2j_1l_1})$ is a $(3^{\rm rd}{\rm DP}_-,{\rm DP}_+)$-third-order-clockwise center.

- For $a_{j_1j_20} < 0$ and $a_{j_2j_10}(a_{j_2j_1l_1} - a_{j_2j_1l_2}) < 0$,

$$(a_{j_1j_2s_1}, a_{j_2j_1l_1}) = \underbrace{(3^{\rm rd}{\rm UP}_-, {\rm UP}_-)}_{\text{third-order negative saddle}}.$$

The equilibrium of $(x^*_{j_2}, x^*_{j_1}) = (a_{j_1j_2s_1}, a_{j_2j_1l_1})$ is a $(3^{\rm rd}{\rm UP}_-,{\rm UP}_-)$-third-order negative saddle.

(x) If

$$\Delta_{j_1j_2} > 0 \text{ and } \Delta_{j_2j_1} < 0,$$

then

$$\begin{aligned}
&\dot{x}_{j_1} = a_{j_1j_20}(x_{j_2} - a_{j_1j_21})(x_{j_2} - a_{j_1j_22})(x_{j_2} - a_{j_1j_23}),\\
&\dot{x}_{j_2} = a_{j_2j_10}\left[(x_{j_1} - a_{j_2j_1})^2 + b_{j_2j_1}\right]\\
&j_1, j_2 \in \{1, 2\}, j_2 \neq j_1
\end{aligned}$$

where

$$\begin{aligned}
&\{a_{j_1j_21}, a_{j_1j_22}, a_{j_1j_23}\} = \text{sort}\{b_{j_1j_21}, b_{j_1j_22}, b_{j_1j_23}\}, a_{j_1j_21} < a_{j_1j_22} < a_{j_1j_23};\\
&b_{j_1j_22}, b_{j_1j_23} = -\frac{1}{2}(B_{j_1j_2} \pm \sqrt{\Delta_{j_1j_2}}), a_{j_1j_20} = A_{j_1j_20};\\
&a_{j_2j_1} = -\frac{1}{2}B_{j_2j_1}, b_{j_2j_1} = \frac{1}{4}(-\Delta_{j_2j_1}).
\end{aligned}$$

In phase space,

$$\frac{dx_{j_1}}{dx_{j_2}} = \frac{a_{j_1j_20}}{a_{j_2j_10}} \frac{(x_{j_2} - a_{j_1j_21})(x_{j_2} - a_{j_1j_22})(x_{j_2} - a_{j_1j_23})}{(x_{j_1} - a_{j_2j_1})^2 + b_{j_2j_1}}$$

and for $a_{j_1j_2s_1} \neq a_{j_1j_2s_2}$ $(s_1, s_2 \in \{1,2,3\}, s_1 \neq s_2)$,

$$\begin{aligned}
&a_{j_2j_10}\left[(x_{j_1} - a_{j_2j_1})^2 + b_{j_2j_1}\right]dx_{j_1}\\
&= a_{j_1j_20}\left[(x_{j_2} - a_{j_1j_2s_1})^3 + \sum\nolimits_{s_2=1,s_2\neq s_1}^{3}(a_{j_1j_2s_1} - a_{j_1j_2s_2})(x_{j_2} - a_{j_1j_2s_1})^2\right.\\
&\left.+\prod\nolimits_{s_2=1,s_2\neq s_1}^{3}(a_{j_1j_2s_1} - a_{j_1j_2s_2})(x_{j_2} - a_{j_1j_2s_1})\right]dx_{j_2}.
\end{aligned}$$

With an initial condition (x_{j_10}, x_{j_20}), the first integral manifold is generated, i.e.,

$$\begin{aligned}
&a_{j_2j_10}\left\{\frac{1}{3}\left[(x_{j_1} - a_{j_2j_1})^3 - (x_{j_10} - a_{j_2j_1})^3\right] + b_{j_2j_1}\left[(x_{j_1} - a_{j_2j_1}) - (x_{j_10} - a_{j_2j_1})\right]\right\}\\
&= a_{j_1j_20}\left\{\frac{1}{4}\left[(x_{j_2} - a_{j_1j_2s_1})^4 - (x_{j_20} - a_{j_1j_2s_1})^4\right]\right.\\
&+\frac{1}{3}\sum\nolimits_{s_2=1,s_2\neq s_1}^{3}(a_{j_1j_2s_1} - a_{j_1j_2s_2})\left[(x_{j_2} - a_{j_1j_2s_1})^3 - (x_{j_20} - a_{j_1j_2s_1})^3\right]\\
&\left.+\frac{1}{2}\prod\nolimits_{s_2=1,s_2\neq s_1}^{3}(a_{j_1j_2s_1} - a_{j_1j_2s_2})\left[(x_{j_2} - a_{j_1j_2s_1})^2 - (x_{j_20} - a_{j_1j_2s_1})^2\right]\right\}.
\end{aligned}$$

In phase space, at $x_{j_2}^* = a_{j_1j_2s_1}$ $(s_1, s_2 \in \{1,2,3\}, s_1 \neq s_2)$ with $\bar{x}_{j_1} \in (-\infty, \infty)$, we have

$$\left.\frac{dx_{j_1}}{dx_{j_2}}\right|_{x_{j_2}^* = a_{j_1j_2s_1}} = \frac{a_{j_1j_20}}{a_{j_2j_10}} \left.\frac{(x_{j_2} - a_{j_1j_2s_1})\prod_{s_2=1,s_2\neq s_1}^{3}(x_{j_2} - a_{j_1j_2s_2})}{(\bar{x}_{j_1} - a_{j_2j_1})^2 + b_{j_2j_1}}\right|_{x_{j_2}^* = a_{j_1j_2s_1}} = 0.$$

If

$$\left.\frac{d^2x_{j_1}}{dx_{j_2}^2}\right|_{x_{j_2}^* = a_{j_1j_2s_1}} = \frac{a_{j_1j_20}}{a_{j_2j_10}} \frac{\prod_{s_2=1,s_2\neq s_1}^{3}(a_{j_1j_2s_1} - a_{j_1j_2s_2})}{(\bar{x}_{j_1} - a_{j_2j_1})^2 + b_{j_2j_1}} > 0,$$

there is an up-parabola flow at $x_{j_2}^* = a_{j_1j_2s_1}$ in the x_{j_1}-direction. If

$$\left.\frac{d^2 x_{j_1}}{dx_{j_2}^2}\right|_{x_{j_2}^* = a_{j_1 j_2 s_1}} = \frac{a_{j_1 j_2 0}}{a_{j_2 j_1 0}} \frac{\prod_{s_2=1, s_2 \neq s_1}^{3} (a_{j_1 j_2 s_1} - a_{j_1 j_2 s_2})}{(\overline{x}_{j_1} - a_{j_2 j_1})^2 + b_{j_2 j_1}} < 0,$$

there is a down-parabola flow at $x_{j_2}^* = a_{j_1 j_2 s_1}$ in the x_{j_1}-direction. Because

$$\dot{x}_{j_2} = a_{j_2 j_1 0}\left[(\overline{x}_{j_1} - a_{j_2 j_1})^2 + b_{j_2 j_1}\right],$$

the flows at $x_{j_2}^* = a_{j_1 j_2 s_1}$ in the x_{j_2}-direction are positive and negative for $a_{j_2 j_1 0} > 0$ and $a_{j_2 j_1 0} < 0$, respectively.

Therefore, the equilibrium of $x_{j_2}^* = a_{j_1 j_2 s_1}$ (s_1, $s_2 \in \{1, 2, 3\}$, $s_1 \neq s_2$) in the x_{j_2}-direction has the following properties as in Eqs. (3.84)–(3.87):

- For $a_{j_1 j_2 0} \prod_{s_2=1, s_2 \neq s_1}^{3} (a_{j_1 j_2 s_1} - a_{j_1 j_2 s_2}) > 0$ and $a_{j_2 j_1 0} > 0$,

$$(a_{j_1 j_2 s_1}, \dot{x}_{j_2}) = \underbrace{(\mathrm{UP}, \mathrm{pF})}_{\text{up-parabola flow } (+)}.$$

  The flow at $x_{j_2}^* = a_{j_1 j_2 s_1}$ is a (UP,pF)-positive up-parabola flow.
- For $a_{j_1 j_2 0} \prod_{s_2=1, s_2 \neq s_1}^{3} (a_{j_1 j_2 s_1} - a_{j_1 j_2 s_2}) < 0$ and $a_{j_2 j_1 0} > 0$,

$$(a_{j_1 j_2 s_1}, \dot{x}_{j_2}) = \underbrace{(\mathrm{DP}, \mathrm{pF})}_{\text{down-parabola flow } (+)}.$$

  The flow at $x_{j_2}^* = a_{j_1 j_2 s_1}$ is a (DP,pF)-positive down-parabola flow.
- For $a_{j_1 j_2 0} \prod_{s_2=1, s_2 \neq s_1}^{3} (a_{j_1 j_2 s_1} - a_{j_1 j_2 s_2}) > 0$ and $a_{j_2 j_1 0} < 0$,

$$(a_{j_1 j_2 s_1}, \dot{x}_{j_2}) = \underbrace{(\mathrm{DP}, \mathrm{nF})}_{\text{down-parabola flow } (-)}.$$

  The flow at $x_{j_2}^* = a_{j_1 j_2 s_1}$ is a (DP,nF)-negative down-parabola flow.
- For $a_{j_1 j_2 0} \prod_{s_2=1, s_2 \neq s_1}^{3} (a_{j_1 j_2 s_1} - a_{j_1 j_2 s_2}) < 0$ and $a_{j_2 j_1 0} < 0$,

$$(a_{j_1 j_2 s_1}, \dot{x}_{j_2}) = \underbrace{(\mathrm{UP}, \mathrm{nF})}_{\text{up-parabola flow } (-)}.$$

  The flow at $x_{j_2}^* = a_{j_1 j_2 s_1}$ is a (UP,nF)-negative up-parabola flow.

(xi) If

$$\Delta_{j_1j_2} > 0 \text{ and } \Delta_{j_2j_1} = 0,$$

then

$$\begin{aligned}
&\dot{x}_{j_1} = a_{j_1j_20}(x_{j_2} - a_{j_1j_21})(x_{j_2} - a_{j_1j_22})(x_{j_2} - a_{j_1j_23}),\\
&\dot{x}_{j_2} = a_{j_2j_10}(x_{j_1} - a_{j_2j_11})^2\\
&j_1, j_2 \in \{1,2\}, j_2 \neq j_1.
\end{aligned}$$

In phase space,

$$\frac{dx_{j_1}}{dx_{j_2}} = \frac{a_{j_1j_20}}{a_{j_2j_10}}\frac{(x_{j_2} - a_{j_1j_21})(x_{j_2} - a_{j_1j_22})(x_{j_2} - a_{j_1j_23})}{(x_{j_1} - a_{j_2j_11})^2}$$

and for $a_{j_1j_2s_1} \neq a_{j_1j_2s_2}$ (s_1, $s_2 \in \{1,2,3\}$, $s_1 \neq s_2$),

$$\begin{aligned}
&a_{j_2j_10}(x_{j_1} - a_{j_2j_11})^2 dx_{j_1}\\
&= a_{j_1j_20}\left[(x_{j_2} - a_{j_1j_2s_1})^3 + \sum\nolimits_{s_2=1, s_2\neq s_1}^{3}(a_{j_1j_2s_1} - a_{j_1j_2s_2})(x_{j_2} - a_{j_1j_2s_1})^2\right.\\
&\left.+\prod\nolimits_{s_2=1, s_2\neq s_1}^{3}(a_{j_1j_2s_1} - a_{j_1j_2s_2})(x_{j_2} - a_{j_1j_2s_1})\right]dx_{j_2}.
\end{aligned}$$

With an initial condition (x_{j_10}, x_{j_20}), the first integral manifold is generated, i.e.,

$$\begin{aligned}
&\frac{1}{3}a_{j_2j_10}\left[(x_{j_1} - a_{j_2j_11})^3 - (x_{j_10} - a_{j_2j_11})^3\right]\\
&= a_{j_1j_20}\left\{\frac{1}{4}\left[(x_{j_2} - a_{j_1j_2s_1})^4 - (x_{j_20} - a_{j_1j_2s_1})^4\right]\right.\\
&+\frac{1}{3}\sum\nolimits_{s_2=1, s_2\neq s_1}^{3}(a_{j_1j_2s_1} - a_{j_1j_2s_2})\left[(x_{j_2} - a_{j_1j_2s_1})^3 - (x_{j_20} - a_{j_1j_2s_1})^3\right]\\
&\left.+\frac{1}{2}\prod\nolimits_{s_2=1, s_2\neq s_1}^{3}(a_{j_1j_2s_1} - a_{j_1j_2s_2})\left[(x_{j_2} - a_{j_1j_2s_1})^2 - (x_{j_20} - a_{j_1j_2s_1})^2\right]\right\}.
\end{aligned}$$

Consider two cases (I) $x^*_{j_2} = a_{j_1j_2s_1}$ and (II) $x^*_{j_1} = a_{j_2j_11}$.

(I) In phase space, at $x^*_{j_2} = a_{j_1j_2s_1}$ (s_1, $s_2 \in \{1,2,3\}$, $s_1 \neq s_2$) and $\overline{x}_{j_1} \neq a_{j_2j_11}$, we have

$$\left.\frac{dx_{j_1}}{dx_{j_2}}\right|_{x^*_{j_2}=a_{j_1j_2s_1}} = \frac{a_{j_1j_20}}{a_{j_2j_10}}\left.\frac{(x_{j_2} - a_{j_1j_2s_1})\prod_{s_2=1, s_2\neq s_1}^{3}(x_{j_2} - a_{j_1j_2s_2})}{(\overline{x}_{j_1} - a_{j_2j_11})^2}\right|_{x^*_{j_2}=a_{j_1j_2s_1}} = 0.$$

If

$$\frac{d^2x_{j_1}}{dx_{j_2}^2}\Big|_{x_{j_2}^*=a_{j_1j_2s_1}} = \frac{a_{j_1j_20}}{a_{j_2j_10}} \frac{\prod_{s_2=1,s_2\neq s_1}^{3}(a_{j_1j_2s_1}-a_{j_1j_2s_2})}{(\overline{x}_{j_1}-a_{j_2j_11})^2} > 0,$$

there is an up-parabola flow at $x_{j_2}^* = a_{j_1j_2s_1}$ in the x_{j_1}-direction. If

$$\frac{d^2x_{j_1}}{dx_{j_2}^2}\Big|_{x_{j_2}^*=a_{j_1j_2s_1}} = \frac{a_{j_1j_20}}{a_{j_2j_10}} \frac{\prod_{s_2=1,s_2\neq s_1}^{3}(a_{j_1j_2s_1}-a_{j_1j_2s_2})}{(\overline{x}_{j_1}-a_{j_2j_11})^2} < 0,$$

there is a down-parabola flow at $x_{j_2}^* = a_{j_1j_2s_1}$ in the x_{j_1}-direction. Because

$$\dot{x}_{j_2} = a_{j_2j_10}(\overline{x}_{j_1}-a_{j_2j_11})^2,$$

the flows at $x_{j_2}^* = a_{j_1j_2s_1}$ in the x_{j_2}-direction are positive and negative for $a_{j_2j_10} > 0$ and $a_{j_2j_10} < 0$, respectively.

Thus, the equilibrium of $x_{j_2}^* = a_{j_1j_2s_1}$ (s_1, $s_2 \in \{1,2,3\}$, $s_1 \neq s_2$) with $\overline{x}_{j_1} \neq a_{j_2j_11}$ in the x_{j_2}-direction has the following properties:

- For $a_{j_1j_20}\prod_{s_2=1,s_2\neq s_1}^{3}(a_{j_1j_2s_1}-a_{j_1j_2s_2}) > 0$ and $a_{j_2j_10}(\overline{x}_{j_1}-a_{j_2j_11})^2 > 0$,

$$(a_{j_1j_2s_1}, \dot{x}_{j_2}) = \underbrace{(\text{UP}, \text{pF})}_{\text{up-parabola flow } (+)}.$$

- For $a_{j_1j_20}\prod_{s_2=1,s_2\neq s_1}^{3}(a_{j_1j_2s_1}-a_{j_1j_2s_2}) < 0$ and $a_{j_2j_10}(\overline{x}_{j_1}-a_{j_2j_11})^2 > 0$,

$$(a_{j_1j_2s_1}, \dot{x}_{j_2}) = \underbrace{(\text{DP}, \text{pF})}_{\text{down-parabola flow } (+)}.$$

- For $a_{j_1j_20}\prod_{s_2=1,s_2\neq s_1}^{3}(a_{j_1j_2s_1}-a_{j_1j_2s_2}) > 0$ and $a_{j_2j_10}(\overline{x}_{j_1}-a_{j_2j_11})^2 < 0$,

$$(a_{j_1j_2s_1}, \dot{x}_{j_2}) = \underbrace{(\text{DP}, \text{nF})}_{\text{down-parabola flow } (-)}.$$

- For $a_{j_1j_20}\prod_{s_2=1,s_2\neq s_1}^{3}(a_{j_1j_2s_1}-a_{j_1j_2s_2}) < 0$ and $a_{j_2j_10}(\overline{x}_{j_1}-a_{j_2j_11})^2 < 0$,

$$(a_{j_1j_2s_1}, \dot{x}_{j_2}) = \underbrace{(\text{UP}, \text{nF})}_{\text{up-parabola flow } (-)}.$$

(II) In phase space, at $x_{j_1}^* = a_{j_2 j_1 1}$ and $\overline{x}_{j_2} \neq a_{j_1 j_2 s_1}$ $(s_1 = 1, 2, 3)$, we have

$$\frac{dx_{j_2}}{dx_{j_1}}\Big|_{x_{j_1}^* = a_{j_2 j_1 1}} = \frac{a_{j_2 j_1 0}}{a_{j_1 j_2 0}} \frac{(x_{j_1} - a_{j_2 j_1 1})^2}{\prod_{s_1=1}^{3}(\overline{x}_{j_2} - a_{j_1 j_2 s_1})}\Big|_{x_{j_1}^* = a_{j_2 j_1 1}} = 0,$$
$$\frac{d^2 x_{j_2}}{dx_{j_1}^2}\Big|_{x_{j_1}^* = a_{j_2 j_1 1}} = 0.$$

If

$$\frac{d^3 x_{j_2}}{dx_{j_1}^3}\Big|_{x_{j_1}^* = a_{j_2 j_1 1}} = \frac{a_{j_1 j_2 0}}{a_{j_2 j_1 0}} \frac{2}{\prod_{s_1=1}^{3}(\overline{x}_{j_2} - a_{j_1 j_2 s_1})} > 0,$$

there is an increasing-inflection flow at $x_{j_1}^* = a_{j_2 j_1 1}$ in the x_{j_2}-direction. If

$$\frac{d^3 x_{j_2}}{dx_{j_1}^3}\Big|_{x_{j_1}^* = a_{j_2 j_1 1}} = \frac{a_{j_1 j_2 0}}{a_{j_2 j_1 0}} \frac{2}{\prod_{s_1=1}^{3}(\overline{x}_{j_2} - a_{j_1 j_2 s_1})} < 0,$$

there is a decreasing-inflection flow at $x_{j_1}^* = a_{j_2 j_1 1}$ in the x_{j_1}-direction. Because

$$\dot{x}_{j_1} = a_{j_1 j_2 0} \prod_{s_1=1}^{3}(\overline{x}_{j_2} - a_{j_1 j_2 s_1}),$$

the flows at $x_{j_1}^* = a_{j_2 j_1 1}$ in the x_{j_1}-direction are positive and negative for $\dot{x}_{j_1} > 0$ and $\dot{x}_{j_1} < 0$, respectively.

Thus, the equilibrium of $x_{j_1}^* = a_{j_2 j_1 1}$ and $\overline{x}_{j_2} \neq a_{j_1 j_2 1}$ in the x_{j_2}-direction has the following properties:

- For $a_{j_1 j_2 0} \prod_{s_1=1}^{3}(\overline{x}_{j_2} - a_{j_1 j_2 s_1}) > 0$ and $a_{j_2 j_1 0} > 0$,

$$(\dot{x}_{j_1}, a_{j_2 j_1 1}) = \underbrace{(\text{pF}, \text{II})}_{\text{increasing-inflection flow } (+)}.$$

- For $a_{j_1 j_2 0} \prod_{s_1=1}^{3}(\overline{x}_{j_2} - a_{j_1 j_2 s_1}) < 0$ and $a_{j_2 j_1 0} > 0$,

$$(\dot{x}_{j_1}, a_{j_2 j_1 1}) = \underbrace{(\text{nF}, \text{DI})}_{\text{decreasing-inflection flow } (-)}.$$

- For $a_{j_1 j_2 0} \prod_{s_1=1}^{3}(\overline{x}_{j_2} - a_{j_1 j_2 s_1}) > 0$ and $a_{j_2 j_1 0} < 0$,

$$(\dot{x}_{j_1}, a_{j_2j_1 1}) = \underbrace{(\text{pF}, \text{DI})}_{\text{decreasing-inflection flow } (+)}.$$

- For $a_{j_1j_20}\prod_{s_1=1}^{3}(\overline{x}_{j_2} - a_{j_1j_2s_1}) < 0$ and $a_{j_2j_10} < 0$,

$$(\dot{x}_{j_1}, a_{j_2j_1 1}) = \underbrace{(\text{nF}, \text{II})}_{\text{increasing-inflection flow } (-)}.$$

Therefore, from two cases (I) and (II), the equilibrium of $(x^*_{j_2}, x^*_{j_1}) = (a_{j_1j_2s_1}, a_{j_2j_1 1})$ (s_1, $s_2 \in \{1, 2, 3\}$, $s_1 \neq s_2$) has the following properties as in Eqs. (3.91)–(3.94):

- For $a_{j_1j_20}\prod_{s_2=1, s_2\neq s_1}^{3}(a_{j_1j_2s_1} - a_{j_1j_2s_2}) > 0$ and $a_{j_2j_10} > 0$,

$$(a_{j_1j_2 1}, a_{j_2j_1 1}) = \underbrace{(\text{UP}, \text{US})}_{\text{up-parabola upper-saddle}}.$$

  The equilibrium of $(x^*_{j_2}, x^*_{j_1}) = (a_{j_1j_2s_1}, a_{j_2j_1 1})$ is a (UP,US)-up-parabola upper-saddle.
- For $a_{j_1j_20}\prod_{s_2=1, s_2\neq s_1}^{3}(a_{j_1j_2s_1} - a_{j_1j_2s_2}) < 0$ and $a_{j_2j_10} > 0$,

$$(a_{j_1j_2 1}, a_{j_2j_1 1}) = \underbrace{(\text{DP}, \text{US})}_{\text{down-parabola upper-saddle}}.$$

  The equilibrium of $(x^*_{j_2}, x^*_{j_1}) = (a_{j_1j_2s_1}, a_{j_2j_1 1})$ is a (DP,US)-down-parabola upper-saddle.
- For $a_{j_1j_20}\prod_{s_2=1, s_2\neq s_1}^{3}(a_{j_1j_2s_1} - a_{j_1j_2s_2}) > 0$ and $a_{j_2j_10} < 0$,

$$(a_{j_1j_2 1}, a_{j_2j_1 1}) = \underbrace{(\text{DP}, \text{LS})}_{\text{down-parabola lower-saddle}}.$$

  The equilibrium of $(x^*_{j_2}, x^*_{j_1}) = (a_{j_1j_2s_1}, a_{j_2j_1 1})$ is a (DP,LS)-down-parabola lower-saddle.
- For $a_{j_1j_20}\prod_{s_2=1, s_2\neq s_1}^{3}(a_{j_1j_2s_1} - a_{j_1j_2s_2}) < 0$ and $a_{j_2j_10} < 0$,

$$(a_{j_1j_2 1}, a_{j_2j_1 1}) = \underbrace{(\text{UP}, \text{LS})}_{\text{up-parabola lower-saddle}}.$$

  The equilibrium of $(x^*_{j_2}, x^*_{j_1}) = (a_{j_1j_2s_1}, a_{j_2j_1 1})$ is a (UP,LS)-up-parabola lower-saddle.

(xii) If

$$\Delta_{j_1j_2} > 0 \text{ and } \Delta_{j_2j_1} > 0,$$

then

$$\begin{aligned}
&\dot{x}_{j_1} = a_{j_1j_20}(x_{j_2} - a_{j_1j_21})(x_{j_2} - a_{j_1j_22})(x_{j_2} - a_{j_1j_23}),\\
&\dot{x}_{j_2} = a_{j_2j_10}(x_{j_1} - a_{j_2j_11})(x_{j_1} - a_{j_2j_12}),\\
&j_1, j_2 \in \{1,2\}; j_1 \neq j_2.
\end{aligned}$$

In phase space,

$$\frac{dx_{j_1}}{dx_{j_2}} = \frac{a_{j_1j_20}}{a_{j_2j_10}}\frac{(x_{j_2} - a_{j_1j_21})(x_{j_2} - a_{j_1j_22})(x_{j_2} - a_{j_1j_23})}{(x_{j_1} - a_{j_2j_11})(x_{j_1} - a_{j_2j_12})}$$

and for $a_{j_1j_2s_1} \neq a_{j_1j_2s_2}$ (s_1, $s_2 \in \{1,2,3\}$, $s_1 \neq s_2$) and $a_{j_2j_1l_1} \neq a_{j_2j_1l_2}$ (l_1, $l_2 \in \{1,2\}$, $l_1 \neq l_2$),

$$\begin{aligned}
&a_{j_2j_10}\left[(x_{j_1} - a_{j_2j_1l_1})^2 + (a_{j_2j_1l_1} - a_{j_2j_1l_2})(x_{j_1} - a_{j_2j_1l_1})\right]dx_{j_1}\\
&= a_{j_1j_20}[(x_{j_2} - a_{j_1j_2s_1})^3 + \sum\nolimits_{s_2=1,s_2\neq s_1}^{3}(a_{j_1j_2s_1} - a_{j_1j_2s_2})(x_{j_2} - a_{j_1j_2s_1})^2\\
&+\prod\nolimits_{s_2=1,s_2\neq s_1}^{3}(a_{j_1j_2s_1} - a_{j_1j_2s_2})(x_{j_2} - a_{j_1j_2s_1})]dx_{j_2}.
\end{aligned}$$

With an initial condition (x_{j_10}, x_{j_20}), the first integral manifold is generated, i.e.,

$$\begin{aligned}
&a_{j_2j_10}\{\frac{1}{3}\left[(x_{j_1} - a_{j_2j_1l_1})^3 - (x_{j_10} - a_{j_2j_1l_1})^3\right]\\
&+\frac{1}{2}(a_{j_2j_1l_1} - a_{j_2j_1l_2})\left[(x_{j_1} - a_{j_2j_1l_1})^2 - (x_{j_10} - a_{j_2j_1l_1})^2\right\}\\
&= a_{j_1j_20}\{\frac{1}{4}\left[(x_{j_2} - a_{j_1j_2s_1})^4 - (x_{j_20} - a_{j_1j_2s_1})^4\right]\\
&+\frac{1}{3}\sum\nolimits_{s_2=1,s_2\neq s_1}^{3}(a_{j_1j_2s_1} - a_{j_1j_2s_2})\left[(x_{j_2} - a_{j_1j_2s_1})^3 - (x_{j_20} - a_{j_1j_2s_1})^3\right]\\
&+\frac{1}{2}\prod\nolimits_{s_2=1,s_2\neq s_1}^{3}(a_{j_1j_2s_1} - a_{j_1j_2s_2})\left[(x_{j_2} - a_{j_1j_2s_1})^2 - (x_{j_20} - a_{j_1j_2s_1})^2\right]\}.
\end{aligned}$$

(xii_1) Consider cases (I) $x_{j_2}^* = a_{j_1j_2s_1}$ and (II) $x_{j_1}^* = a_{j_2j_1s_1}$.

(I) In phase space, at $x_{j_2}^* = a_{j_1j_2s_1}$ (s_1, $s_2 \in \{1,2,3\}$, $s_1 \neq s_2$) and $\bar{x}_{j_1} \neq a_{j_2j_1l_1}$ ($l_1 = 1, 2$), we have

$$\frac{dx_{j_1}}{dx_{j_2}}\bigg|_{x_{j_2}^* = a_{j_1j_2s_1}} = \frac{a_{j_1j_20}}{a_{j_2j_10}}\frac{(x_{j_2} - a_{j_1j_2s_1})\prod_{s_2=1,s_2\neq s_1}^{3}(x_{j_2} - a_{j_1j_2s_2})}{\prod_{l_1=1}^{2}(\bar{x}_{j_1} - a_{j_2j_1l_1})}\bigg|_{x_{j_2}^* = a_{j_1j_2s_1}} = 0.$$

If

$$\frac{d^2 x_{j_1}}{dx_{j_2}^2}\bigg|_{x_{j_2}^*=a_{j_1j_2s_1}} = \frac{a_{j_1j_2 0}}{a_{j_2j_1 0}} \frac{\prod_{s_2=1,s_2\neq s_1}^{3}(a_{j_1j_2s_1}-a_{j_1j_2s_2})}{\prod_{l_1=1}^{2}(\bar{x}_{j_1}-a_{j_2j_1l_1})} > 0,$$

there is an up-parabola flow at $x_{j_2}^* = a_{j_1j_2s_1}$ in the x_{j_1}-direction. If

$$\frac{d^2 x_{j_1}}{dx_{j_2}^2}\bigg|_{x_{j_2}^*=a_{j_1j_2s_1}} = \frac{a_{j_1j_2 0}}{a_{j_2j_1 0}} \frac{\prod_{s_2=1,s_2\neq s_1}^{3}(a_{j_1j_2s_1}-a_{j_1j_2s_2})}{\prod_{l_1=1}^{2}(\bar{x}_{j_1}-a_{j_2j_1l_1})} < 0,$$

there is a down-parabola flow at $x_{j_2}^* = a_{j_1j_2s_1}$ in the x_{j_1}-direction. Because

$$\dot{x}_{j_2} = a_{j_2j_1 0}\prod_{l_1=1}^{2}(\bar{x}_{j_1}-a_{j_2j_1l_1}),$$

the flows at $x_{j_2}^* = a_{j_1j_2s_1}$ in the x_{j_2}-direction are positive and negative for $a_{j_2j_1 0} > 0$ and $a_{j_2j_1 0} < 0$, respectively.

Thus, the equilibrium of $x_{j_2}^* = a_{j_1j_2s_1}$ (s_1, $s_2 \in \{1,2,3\}$, $s_1 \neq s_2$) in the x_{j_2}-direction has the following properties:

- For $a_{j_1j_2 0}\prod_{s_2=1,s_2\neq s_1}^{3}(a_{j_1j_2s_1}-a_{j_1j_2s_2}) > 0$ and $a_{j_2j_1 0}\prod_{l_1=1}^{2}(\bar{x}_{j_1}-a_{j_2j_1l_1}) > 0$,

$$(a_{j_1j_2s_1}, \dot{x}_{j_2}) = \underbrace{(\text{UP}, \text{pF})}_{\text{up-parabola flow } (+)}.$$

- For $a_{j_1j_2 0}\prod_{s_2=1,s_2\neq s_1}^{3}(a_{j_1j_2s_1}-a_{j_1j_2s_2}) < 0$ and $a_{j_2j_1 0}\prod_{l_1=1}^{2}(\bar{x}_{j_1}-a_{j_2j_1l_1}) > 0$,

$$(a_{j_1j_2s_1}, \dot{x}_{j_2}) = \underbrace{(\text{DP}, \text{pF})}_{\text{down-parabola flow } (+)}.$$

- For $a_{j_1j_2 0}\prod_{s_2=1,s_2\neq s_1}^{3}(a_{j_1j_2s_1}-a_{j_1j_2s_2}) > 0$ and $a_{j_2j_1 0}\prod_{l_1=1}^{2}(\bar{x}_{j_1}-a_{j_2j_1l_1}) < 0$,

$$(a_{j_1j_2s_1}, \dot{x}_{j_2}) = \underbrace{(\text{DP}, \text{nF})}_{\text{down-parabola flow } (-)}.$$

- For $a_{j_1j_2 0}\prod_{s_2=1,s_2\neq s_1}^{3}(a_{j_1j_2s_1}-a_{j_1j_2s_2}) < 0$ and $a_{j_2j_1 0}\prod_{l_1=1}^{2}(\bar{x}_{j_1}-a_{j_2j_1l_1}) < 0$,

$$(a_{j_1j_2s_1}, \dot{x}_{j_2}) = \underbrace{(\text{UP}, \text{nF})}_{\text{up-parabola flow } (-)}.$$

(II) In phase space, at $x_{j_1}^* = a_{j_2 j_1 l_1}$ (l_1, $\mathrm{l}_2 \in \{1, 2\}$, $l_1 \neq l_2$) and $\overline{x}_{j_2} \neq a_{j_1 j_2 s_1}$ ($s_1 = 1, 2, 3$), we have

$$\frac{dx_{j_2}}{dx_{j_1}}\bigg|_{x_{j_1}^* = a_{j_2 j_1 l_1}} = \frac{a_{j_2 j_1 0}}{a_{j_1 j_2 0}} \frac{(x_{j_1} - a_{j_2 j_1 l_1})(x_{j_1} - a_{j_2 j_1 l_2})}{\prod_{s_1=1}^{3}(\overline{x}_{j_2} - a_{j_1 j_2 s_1})}\bigg|_{x_{j_1}^* = a_{j_2 j_1 l_1}} = 0.$$

If

$$\frac{d^2 x_{j_2}}{dx_{j_1}^2}\bigg|_{x_{j_1}^* = a_{j_2 j_1 l_1}} = \frac{a_{j_2 j_1 0}}{a_{j_1 j_2 0}} \frac{(a_{j_2 j_1 l_1} - a_{j_2 j_1 l_2})}{\prod_{s_1=1}^{3}(\overline{x}_{j_2} - a_{j_1 j_2 s_1})} > 0,$$

there is an up-parabola flow at $x_{j_1}^* = a_{j_2 j_1 l_1}$ in the x_{j_2}-direction. If

$$\frac{d^2 x_{j_2}}{dx_{j_1}^2}\bigg|_{x_{j_1}^* = a_{j_2 j_1 l_1}} = \frac{a_{j_2 j_1 0}}{a_{j_1 j_2 0}} \frac{(a_{j_2 j_1 l_1} - a_{j_2 j_1 l_2})}{\prod_{s_1=1}^{3}(\overline{x}_{j_2} - a_{j_1 j_2 s_1})} < 0,$$

there is a down-parabola flow at $x_{j_1}^* = a_{j_2 j_1 l_1}$ in the x_{j_1}-direction. Because

$$\dot{x}_{j_1} = a_{j_1 j_2 0} \prod_{s_1=1}^{3}(\overline{x}_{j_2} - a_{j_1 j_2 s_1}),$$

the flows at $x_{j_1}^* = a_{j_2 j_1 l_1}$ in the x_{j_1}-direction are positive and negative for $\dot{x}_{j_1} > 0$ and $\dot{x}_{j_1} < 0$, respectively.

Thus, the equilibrium of $x_{j_1}^* = a_{j_2 j_1 l_1}$ ($l_1, l_2 \in \{1, 2\}$, $l_1 \neq l_2$) and $\overline{x}_{j_2} \neq a_{j_1 j_2 s_1}$ ($s_1 = 1$, 2, 3) in the x_{j_2}-direction has the following properties:

- For $a_{j_1 j_2 0} \prod_{s_1=1}^{3}(\overline{x}_{j_2} - a_{j_1 j_2 s_1}) > 0$ and $a_{j_2 j_1 0}(a_{j_2 j_1 l_1} - a_{j_2 j_1 l_2}) > 0$,

$$(\dot{x}_{j_1}, a_{j_2 j_1 l_1}) = \underbrace{(\mathrm{pF}, \mathrm{UP})}_{\text{up-parabola flow } (+)}.$$

- For $a_{j_1 j_2 0} \prod_{s_1=1}^{3}(\overline{x}_{j_2} - a_{j_1 j_2 s_1}) < 0$ and $a_{j_2 j_1 0}(a_{j_2 j_1 l_1} - a_{j_2 j_1 l_2}) > 0$,

$$(\dot{x}_{j_1}, a_{j_2 j_1 l_1}) = \underbrace{(\mathrm{nF}, \mathrm{DP})}_{\text{down-parabola flow } (-)}.$$

- For $a_{j_1 j_2 0} \prod_{s_1=1}^{3}(\overline{x}_{j_2} - a_{j_1 j_2 s_1}) > 0$ and $a_{j_2 j_1 0}(a_{j_2 j_1 l_1} - a_{j_2 j_1 l_2}) < 0$,

$$(\dot{x}_{j_1}, a_{j_2 j_1 l_1}) = \underbrace{(\text{pF}, \text{DP})}_{\text{down-parabola flow } (+)}.$$

- For $a_{j_1 j_2 0} \prod_{s_1=1}^{3} (\overline{x}_{j_2} - a_{j_1 j_2 s_1}) < 0$ and $a_{j_2 j_1 0}(a_{j_2 j_1 l_1} - a_{j_2 j_1 l_2}) < 0$,

$$(\dot{x}_{j_1}, a_{j_2 j_1 l_1}) = \underbrace{(\text{nF}, \text{UP})}_{\text{up-parabola flow } (-)}.$$

Therefore, from cases (I) and (II), the equilibrium of $(x^*_{j_2}, x^*_{j_1}) = (a_{j_1 j_2 s_1}, a_{j_2 j_1 l_1})$ ($s_1, s_2 \in \{1, 2, 3\}$, $s_1 \neq s_2$, $l_1, l_2 \in \{1, 2\}$, $l_1 \neq l_2$) has the following properties as in Eqs. (3.98)–(3.101):

- For $a_{j_1 j_2 0} \prod_{s_2=1, s_2 \neq s_1}^{3} (a_{j_1 j_2 s_1} - a_{j_1 j_2 s_2}) > 0$ and $a_{j_2 j_1 0}(a_{j_2 j_1 l_1} - a_{j_2 j_1 l_2}) > 0$,

$$(a_{j_1 j_2 s_1}, a_{j_2 j_1 l_1}) = \underbrace{(\text{UP}_+, \text{UP}_+)}_{\text{positive saddle}}.$$

  The equilibrium of $(x^*_{j_1}, x^*_{j_2}) = (a_{j_1 j_2 s_1}, a_{j_2 j_1 l_1})$ is a $(\text{UP}_+,\text{UP}_+)$-positive saddle.
- For $a_{j_1 j_2 0} \prod_{s_2=1, s_2 \neq s_1}^{3} (a_{j_1 j_2 s_1} - a_{j_1 j_2 s_2}) < 0$ and $a_{j_2 j_1 0}(a_{j_2 j_1 l_1} - a_{j_2 j_1 l_2}) > 0$,

$$(a_{j_1 j_2 s_1}, a_{j_2 j_1 l_1}) = \underbrace{(\text{DP}_+, \text{DP}_-)}_{\text{CCW center}}.$$

  The equilibrium of $(x^*_{j_1}, x^*_{j_2}) = (a_{j_1 j_2 s_1}, a_{j_2 j_1 l_1})$ is a $(\text{DP}_+,\text{DP}_-)$-counter-clockwise center.
- For $a_{j_1 j_2 0} \prod_{s_2=1, s_2 \neq s_1}^{3} (a_{j_1 j_2 s_1} - a_{j_1 j_2 s_2}) > 0$ and $a_{j_2 j_1 0}(a_{j_2 j_1 l_1} - a_{j_2 j_1 l_2}) < 0$,

$$(a_{j_1 j_2 s_1}, a_{j_2 j_1 l_1}) = \underbrace{(\text{DP}_-, \text{DP}_+)}_{\text{CW center}}.$$

  The equilibrium of $(x^*_{j_1}, x^*_{j_2}) = (a_{j_1 j_2 s_1}, a_{j_2 j_1 l_1})$ is called a $(\text{DP}_-,\text{DP}_+)$-clockwise center.
- For $a_{j_1 j_2 0} \prod_{s_2=1, s_2 \neq s_1}^{3} (a_{j_1 j_2 s_1} - a_{j_1 j_2 s_2}) < 0$ and $a_{j_2 j_1 0}(a_{j_2 j_1 l_1} - a_{j_2 j_1 l_2}) < 0$,

$$(a_{j_1 j_2 s_1}, a_{j_2 j_1 l_1}) = \underbrace{(\text{UP}_-, \text{UP}_-)}_{\text{negative saddle}}.$$

  The equilibrium of $(x^*_{j_1}, x^*_{j_2}) = (a_{j_1 j_2 s_1}, a_{j_2 j_1 l_1})$ is a $(\text{UP}_-,\text{UP}_-)$-negative saddle.

(xii_2) From case (xii_1), the corresponding saddle and center networks are obtained. For $x_{j_1}^* = a_{j_2 j_1 1}, a_{j_2 j_1 2}$ in the x_{j_1}-direction and $x_{j_2}^* = a_{j_1 j_2 1}, a_{j_1 j_2 2}, a_{j_1 j_2 3}$ in the x_{j_2}-direction, the corresponding equilibrium networks have the following properties as in Eqs. (3.102)–(3.105):

- For $a_{j_1 j_2 0} > 0$ and $a_{j_2 j_1 0} > 0$,

$$\left\{\begin{array}{ll} (a_{j_1 j_2 3}, a_{j_2 j_1 1}) & (a_{j_1 j_2 3}, a_{j_2 j_1 2}) \\ (a_{j_1 j_2 2}, a_{j_2 j_1 1}) & (a_{j_1 j_2 2}, a_{j_2 j_1 2}) \\ (a_{j_1 j_2 1}, a_{j_2 j_1 1}) & (a_{j_1 j_2 1}, a_{j_2 j_1 2}) \end{array}\right\} = \left\{\begin{array}{ll} \underbrace{(\text{DP}_-, \text{DP}_+)}_{\text{CW center}} & \underbrace{(\text{UP}_+, \text{UP}_+)}_{\text{positive saddle}} \\ \underbrace{(\text{UP}_-, \text{UP}_-)}_{\text{negative saddle}} & \underbrace{(\text{DP}_+, \text{DP}_-)}_{\text{CCW center}} \\ \underbrace{(\text{DP}_-, \text{DP}_+)}_{\text{CW center}} & \underbrace{(\text{UP}_+, \text{UP}_+)}_{\text{positive saddle}} \end{array}\right\}.$$

- For $a_{j_1 j_2 0} < 0$ and $a_{j_2 j_1 0} > 0$,

$$\left\{\begin{array}{ll} (a_{j_1 j_2 3}, a_{j_2 j_1 1}) & (a_{j_1 j_2 3}, a_{j_2 j_1 2}) \\ (a_{j_1 j_2 2}, a_{j_2 j_1 1}) & (a_{j_1 j_2 2}, a_{j_2 j_1 2}) \\ (a_{j_1 j_2 1}, a_{j_2 j_1 1}) & (a_{j_1 j_2 1}, a_{j_2 j_1 2}) \end{array}\right\} = \left\{\begin{array}{ll} \underbrace{(\text{UP}_-, \text{UP}_-)}_{\text{negative saddle}} & \underbrace{(\text{DP}_+, \text{DP}_-)}_{\text{CCW center}} \\ \underbrace{(\text{DP}_-, \text{DP}_+)}_{\text{CW center}} & \underbrace{(\text{UP}_+, \text{UP}_+)}_{\text{positive saddle}} \\ \underbrace{(\text{UP}_-, \text{UP}_-)}_{\text{negative saddle}} & \underbrace{(\text{DP}_+, \text{DP}_-)}_{\text{CCW center}} \end{array}\right\}.$$

- For $a_{j_1 j_2 0} > 0$ and $a_{j_2 j_1 0} < 0$,

$$\left\{\begin{array}{ll} (a_{j_1 j_2 3}, a_{j_2 j_1 1}) & (a_{j_1 j_2 3}, a_{j_2 j_1 2}) \\ (a_{j_1 j_2 2}, a_{j_2 j_1 1}) & (a_{j_1 j_2 2}, a_{j_2 j_1 2}) \\ (a_{j_1 j_2 1}, a_{j_2 j_1 1}) & (a_{j_1 j_2 1}, a_{j_2 j_1 2}) \end{array}\right\} = \left\{\begin{array}{ll} \underbrace{(\text{UP}_+, \text{UP}_+)}_{\text{positive saddle}} & \underbrace{(\text{DP}_-, \text{DP}_+)}_{\text{CW center}} \\ \underbrace{(\text{DP}_+, \text{DP}_-)}_{\text{CCW center}} & \underbrace{(\text{UP}_-, \text{UP}_-)}_{\text{negative saddle}} \\ \underbrace{(\text{UP}_+, \text{UP}_+)}_{\text{positive saddle}} & \underbrace{(\text{DP}_-, \text{DP}_+)}_{\text{CW center}} \end{array}\right\}.$$

- For $a_{j_1 j_2 0} < 0$ and $a_{j_2 j_1 0} < 0$,

$$\left\{\begin{array}{ll}(a_{j_1j_2}3, a_{j_2j_1}1) & (a_{j_1j_2}3, a_{j_2j_1}2)\\(a_{j_1j_2}2, a_{j_2j_1}1) & (a_{j_1j_2}2, a_{j_2j_1}2)\\(a_{j_1j_2}1, a_{j_2j_1}1) & (a_{j_1j_2}1, a_{j_2j_1}2)\end{array}\right\} = \left\{\begin{array}{ll}\underbrace{(\mathrm{DP}_+, \mathrm{DP}_-)}_{\text{CCW center}} & \underbrace{(\mathrm{UP}_-, \mathrm{UP}_-)}_{\text{negative saddle}}\\\underbrace{(\mathrm{UP}_+, \mathrm{UP}_+)}_{\text{positive saddle}} & \underbrace{(\mathrm{DP}_-, \mathrm{DP}_+)}_{\text{CW center}}\\\underbrace{(\mathrm{DP}_+, \mathrm{DP}_-)}_{\text{CCW center}} & \underbrace{(\mathrm{UP}_-, \mathrm{UP}_-)}_{\text{negative saddle}}\end{array}\right\}.$$

(xiii) Under

$$\Delta_{j_1j_2} < 0 \text{ and } \Delta_{j_2j_1} = 0,$$

the standard form is

$$\begin{aligned}&\dot{x}_{j_2} = a_{j_2j_1 0}(x_{j_1} - a_{j_2j_1}1)^2,\\&\dot{x}_{j_1} = a_{j_1j_2 0}(x_{j_1} - a_{j_1j_2}1)\left[(x_{j_1} - a_{j_1j_2})^2 + b_{j_1j_2}\right],\\&j_1, j_2 \in \{1,2\}; j_1 \neq j_2.\end{aligned}$$

From cases (ii) and (iii), the equilibrium of $(x^*_{j_2}, x^*_{j_1}) = (a_{j_1j_2}1, a_{j_2j_1}1)$ possesses the following appearing bifurcations as in Eqs. (3.108)–(3.111):

- For $a_{j_1j_2 0} > 0$ and $a_{j_2j_1 0} > 0$,

$$\underbrace{(a_{j_1j_2}1, a_{j_2j_1}1)}_{(\mathrm{UP},\mathrm{US})\text{-appearing bifurcation}} \rightleftharpoons \left\{\begin{array}{c}\underbrace{(a_{j_1j_2}1, a_{j_2j_1}2)}_{(\mathrm{UP}_+, \mathrm{UP}_+)\text{-positive saddle}}\\\underbrace{(a_{j_1j_2}1, a_{j_2j_1}1)}_{(\mathrm{DP}_-, \mathrm{DP}_+)\text{-CW center}}\end{array}\right\}.$$

  The (UP,US)-up-parabola upper-saddle at $(x^*_{j_2}, x^*_{j_1}) = (a_{j_1j_2}1, a_{j_2j_1}1)$ is a (UP,US)-appearing bifurcation for (UP_+,UP_+)-positive saddle and (DP_-, DP_+)-clockwise center.
- For $a_{j_1j_2 0} < 0$ and $a_{j_2j_1 0} > 0$,

$$\underbrace{(a_{j_1j_2}1, a_{j_2j_1}1)}_{(\mathrm{DP},\mathrm{US})\text{-appearing bifurcation}} \rightleftharpoons \left\{\begin{array}{c}\underbrace{(a_{j_1j_2}1, a_{j_2j_1}2)}_{(\mathrm{DP}_+, \mathrm{DP}_-)\text{-CCW center}}\\\underbrace{(a_{j_1j_2}1, a_{j_2j_1}1)}_{(\mathrm{UP}_-, \mathrm{UP}_-)\text{-negative saddle}}\end{array}\right\}.$$

The (DP,US)-down-parabola upper-saddle at $(x_{j_2}^*, x_{j_1}^*) = (a_{j_1j_2 1}, a_{j_2j_1 1})$ is a (DP,US)-appearing bifurcation for $(\text{DP}_+,\text{DP}_-)$-counter-clockwise center and $(\text{UP}_-,\text{UP}_-)$-negative saddle.

- For $a_{j_1j_2 0} > 0$ and $a_{j_2j_1 0} < 0$,

$$\underbrace{(a_{j_1j_2 1}, a_{j_2j_1 1})}_{(\text{DP, LS})\text{-appearing bifurcation}} \rightleftharpoons \left\{ \begin{array}{c} \underbrace{(a_{j_1j_2 1}, a_{j_2j_1 2})}_{(\text{DP}_-, \text{DP}_+)\text{-CW center}} \\ \underbrace{(a_{j_1j_2 1}, a_{j_2j_1 1})}_{(\text{UP}_+, \text{UP}_+)\text{-positive saddle}} \end{array} \right\}.$$

The (DP,LS)-down-parabola lower-saddle at $(x_{j_2}^*, x_{j_1}^*) = (a_{j_1j_2 1}, a_{j_2j_1 1})$ is a (DP,LS)-appearing bifurcation for $(\text{DP}_-,\text{DP}_+)$-clockwise center and $(\text{UP}_+,\text{UP}_+)$-positive saddle.

- For $a_{j_1j_2 0} < 0$ and $a_{j_2j_1 0} < 0$,

$$\underbrace{(a_{j_1j_2 1}, a_{j_2j_1 1})}_{(\text{UP, LS})\text{-appearing bifurcation}} \rightleftharpoons \left\{ \begin{array}{c} \underbrace{(a_{j_1j_2 1}, a_{j_2j_1 2})}_{(\text{UP}_-, \text{UP}_-)\text{-negative saddle}} \\ \underbrace{(a_{j_1j_2 1}, a_{j_2j_1 1})}_{(\text{DP}_+, \text{DP}_-)\text{-CCW center}} \end{array} \right\}.$$

The (UP,LS)-up-parabola lower-saddle at $(x_{j_2}^*, x_{j_1}^*) = (a_{j_1j_2 1}, a_{j_2j_1 1})$ is a (UP, LS)-appearing bifurcation for $(\text{UP}_-,\text{UP}_-)$-negative saddle and $(\text{DP}_+,\text{DP}_-)$-counter-clockwise center.

(xiv) Under

$$\Delta_{j_1j_2} = 0 \text{ and } \Delta_{j_2j_1} < 0, a_{j_1j_2 s_1} \neq a_{j_1j_2 s_2},$$

the standard form is

$$\begin{aligned} \dot{x}_{j_1} &= a_{j_1j_2 0}(x_{j_2} - a_{j_1j_2 s_1})(x_{j_2} - a_{j_1j_2 s_2})^2, \\ \dot{x}_{j_2} &= a_{j_2j_1 0}\left[(x_{j_1} - a_{j_2j_1})^2 + b_{j_2j_1}\right] \\ j_1, j_2 &\in \{1, 2\}, j_2 \neq j_1. \end{aligned}$$

From cases (iv$_2$) and (x), the corresponding bifurcations are obtained. The equilibrium state of $x_{j_2}^* = a_{j_1j_2 s_2}$ (s_1, $s_2 \in \{1, 2\}$, $s_1 \neq s_2$) in the x_{j_2}-direction is for the following appearing bifurcations as in Eqs. (3.114)–(3.117);

- For $a_{j_1j_20}(a_{j_1j_2s_2} - a_{j_1j_2s_1}) > 0$ and $a_{j_2j_10} > 0$,

$$\underbrace{(a_{j_1j_2s_2}, \dot{x}_{j_2})}_{\text{(II, pF)-flow appearing bifurcation}} \rightleftharpoons \left\{ \begin{array}{c} \underbrace{(a_{j_1j_2(s_2)_1}, \dot{x}_{j_2})}_{\text{(DP, pF)-flow}} \\ \underbrace{(a_{j_1j_2(s_2)_2}, \dot{x}_{j_2})}_{\text{(UP, pF)-flow}} \end{array} \right\}.$$

The (II,pF)-positive increasing-inflection flow is the (II,pF)-appearing bifurcations for up-parabola and down-parabola flows.

- For $a_{j_1j_20}(a_{j_1j_2s_2} - a_{j_1j_2s_1}) < 0$ and $a_{j_2j_10} > 0$,

$$\underbrace{(a_{j_1j_2s_2}, \dot{x}_{j_2})}_{\text{(DI, pF)-flow appearing bifurcation}} \rightleftharpoons \left\{ \begin{array}{c} \underbrace{(a_{j_1j_2(s_2)_1}, \dot{x}_{j_2})}_{\text{(UP, pF)-flow}} \\ \underbrace{(a_{j_1j_2(s_2)_2}, \dot{x}_{j_2})}_{\text{(DP, pF)-flow}} \end{array} \right\}.$$

The (DI,pF)-positive decreasing-inflection flow is the (DI,pF)-appearing bifurcations for up-parabola and down-parabola flows.

- For $a_{j_1j_20}(a_{j_1j_2s_2} - a_{j_1j_2s_1}) > 0$ and $a_{j_2j_10} < 0$,

$$\underbrace{(a_{j_1j_2s_2}, \dot{x}_{j_2})}_{\text{(DI, nF)-flow appearing bifurcation}} \rightleftharpoons \left\{ \begin{array}{c} \underbrace{(a_{j_1j_2(s_2)_1}, \dot{x}_{j_2})}_{\text{(UP, nF)-flow}} \\ \underbrace{(a_{j_1j_2(s_2)_2}, \dot{x}_{j_2})}_{\text{(DP, nF)-flow}} \end{array} \right\}.$$

The (DI,nF)-negative decreasing-inflection flow is the (DI,nF)-appearing bifurcations for up-parabola and down-parabola flows.

- For $a_{j_1j_20}(a_{j_1j_2s_2} - a_{j_1j_2s_1}) < 0$ and $a_{j_2j_10} < 0$,

$$\underbrace{(a_{j_1j_2s_2}, \dot{x}_{j_2})}_{\text{(II, nF)-flow appearing bifurcation}} \rightleftharpoons \left\{ \begin{array}{c} \underbrace{(a_{j_1j_2(s_2)_1}, \dot{x}_{j_2})}_{\text{(DP, nF)-flow}} \\ \underbrace{(a_{j_1j_2(s_2)_2}, \dot{x}_{j_2})}_{\text{(UP, nF)-flow}} \end{array} \right\}.$$

The (II,nF)-negative increasing-inflection flow is the (II,nF)-appearing bifurcations for up-parabola and down-parabola flows.

– Such inflection bifurcations are the appearing bifurcations for up-parabola and down-parabola flows.

(xv) Under

$$\Delta_{j_1j_2} = 0 \text{ and } \Delta_{j_2j_1} < 0, a_{j_1j_2s_1} = a_{j_1j_2s_2} = a_{j_1j_21},$$

the standard form is

$$\begin{aligned}
&\dot{x}_{j_1} = a_{j_1j_20}(x_{j_2} - a_{j_1j_21})^3,\\
&\dot{x}_{j_2} = a_{j_2j_10}\left[(x_{j_1} - a_{j_2j_1})^2 + b_{j_2j_1}\right]\\
&j_1, j_2 \in \{1,2\}, j_2 \neq j_1.
\end{aligned}$$

From cases (v) and (x), the corresponding bifurcations are obtained. The equilibrium state of $x_{j_2}^* = a_{j_1j_21}$ in the x_{j_2}-direction is for the following appearing bifurcations as in Eqs. (3.120)–(3.123):

- For $a_{j_1j_20} > 0$ and $a_{j_2j_10} > 0$,

$$\underbrace{(a_{j_1j_21}, \dot{x}_{j_2})}_{(\text{UP, pF})\text{-flow}} \rightleftharpoons \underbrace{(a_{j_1j_21}, \dot{x}_{j_2})}_{\substack{(3^{\text{rd}}\text{UP, pF})\text{-flow}\\ \text{appearing bifurcation}}} \rightleftharpoons \left\{\begin{array}{c} \underbrace{(a_{j_1j_23}, \dot{x}_{j_2})}_{(\text{UP, pF})\text{-flow}} \\ \underbrace{(a_{j_1j_22}, \dot{x}_{j_2})}_{(\text{DP, pF})\text{-flow}} \\ \underbrace{(a_{j_1j_23}, \dot{x}_{j_2})}_{(\text{UP, pF})\text{-flow}} \end{array}\right\}.$$

  The (3^{rd}UP,pF)-third-order positive up-parabola flow is the (3^{rd}UP,pF)-appearing bifurcation from an up-parabola flow to up-parabola, down-parabola, and up-parabola flows.
- For $a_{j_1j_20} < 0$ and $a_{j_2j_10} > 0$,

$$\underbrace{(a_{j_1j_21}, \dot{x}_{j_2})}_{(\text{DP, pF})\text{-flow}} \rightleftharpoons \underbrace{(a_{j_1j_21}, \dot{x}_{j_2})}_{\substack{(3^{\text{rd}}\text{DP, pF})\text{-flow}\\ \text{appearing bifurcation}}} \rightleftharpoons \left\{\begin{array}{c} \underbrace{(a_{j_1j_23}, \dot{x}_{j_2})}_{(\text{DP, pF})\text{-flow}} \\ \underbrace{(a_{j_1j_22}, \dot{x}_{j_2})}_{(\text{UP, pF})\text{-flow}} \\ \underbrace{(a_{j_1j_23}, \dot{x}_{j_2})}_{(\text{DP, pF})\text{-flow}} \end{array}\right\}.$$

  The (3^{rd}DP,pF)-third-order positive down-parabola flow is the (3^{rd}DP,pF)-appearing bifurcation from a down-parabola flow to down-parabola, up-parabola, and down-parabola flows.
- For $a_{j_1j_20} > 0$ and $a_{j_2j_10} < 0$,

$$\underbrace{(a_{j_1j_2}1,\dot{x}_{j_2})}_{(\text{DP},\text{nF})\text{-flow}} \rightleftharpoons \underbrace{(a_{j_1j_2}1,\dot{x}_{j_2})}_{\substack{(3^{\text{rd}}\text{DP},\text{nF})\text{-flow}\\ \text{appearing bifurcation}}} \rightleftharpoons \left\{\begin{array}{c} \underbrace{(a_{j_1j_2}3,\dot{x}_{j_2})}_{(\text{DP},\text{nF})\text{-flow}} \\ \underbrace{(a_{j_1j_2}2,\dot{x}_{j_2})}_{(\text{UP},\text{nF})\text{-flow}} \\ \underbrace{(a_{j_1j_2}3,\dot{x}_{j_2})}_{(\text{DP},\text{nF})\text{-flow}} \end{array}\right\}.$$

The (3^{rd}DP,nF)-third-order negative down-parabola flow is the (3^{rd}DP,nF)-appearing bifurcation from a down-parabola flow to down-parabola, up-parabola, and down-parabola flows.

- For $a_{j_1j_2 0} < 0$ and $a_{j_2j_1 0} < 0$,

$$\underbrace{(a_{j_1j_2}1,\dot{x}_{j_2})}_{(\text{UP},\text{pF})\text{-flow}} \rightleftharpoons \underbrace{(a_{j_1j_2}1,\dot{x}_{j_2})}_{\substack{(3^{\text{rd}}\text{UP},\text{pF})\text{-flow}\\ \text{appearing bifurcation}}} \rightleftharpoons \left\{\begin{array}{c} \underbrace{(a_{j_1j_2}3,\dot{x}_{j_2})}_{(\text{UP},\text{pF})\text{-flow}} \\ \underbrace{(a_{j_1j_2}2,\dot{x}_{j_2})}_{(\text{DP},\text{pF})\text{-flow}} \\ \underbrace{(a_{j_1j_2}3,\dot{x}_{j_2})}_{(\text{UP},\text{pF})\text{-flow}} \end{array}\right\}.$$

The (3^{rd}UP,nF)-third-order negative up-parabola flow is the (3^{rd}UP,nF)-appearing bifurcation from an up-parabola flow to up-parabola, down-parabola, and up-parabola flows.

(xvi) Under

$$\Delta_{j_1j_2} = 0 \text{ and } \Delta_{j_2j_1} = 0, a_{j_2j_1s_1} \neq a_{j_2j_1s_2},$$

the standard form is

$$\begin{aligned}
&\dot{x}_{j_1} = a_{j_1j_2 0}(x_{j_2} - a_{j_1j_2s_1})(x_{j_2} - a_{j_1j_2s_2})^2,\\
&\dot{x}_{j_2} = a_{j_2j_1 0}(x_{j_1} - a_{j_2j_1 1})^2,\\
&j_1, j_2 \in \{1,2\}; j_1 \neq j_2.
\end{aligned}$$

(xvi$_1$) Form cases (vi$_1$) and (viii), the corresponding bifurcations are obtained. The equilibrium of $(x_{j_2}^*, x_{j_1}^*) = (a_{j_1j_2s_1}, a_{j_2j_1 1})$ has the following appearing bifurcations as in Eqs. (3.126)–(3.129):

- For $a_{j_1j_2 0} > 0$ and $a_{j_2j_1 0} > 0$,

$$\underbrace{(a_{j_1j_2s_1}, a_{j_2j_1}1)}_{(\mathrm{UP},\,\mathrm{US})\text{-appearing bifurcation}} \rightleftharpoons \left\{ \begin{array}{c} \underbrace{(a_{j_1j_2s_1}, a_{j_2j_1}2)}_{(\mathrm{UP}_+,\,\mathrm{UP}_+)\text{-positive saddle}} \\ \underbrace{(a_{j_1j_2s_1}, a_{j_2j_1}1)}_{(\mathrm{DP}_-,\,\mathrm{DP}_+)\text{-CW center}} \end{array} \right\}.$$

The (UP,US)-up-parabola upper-saddle is a (UP,US)-appearing bifurcation for center and saddle.

- For $a_{j_1j_2}0 < 0$ and $a_{j_2j_1}0 > 0$,

$$\underbrace{(a_{j_1j_2s_1}, a_{j_2j_1}1)}_{(\mathrm{DP},\,\mathrm{US})\text{-appearing bifurcation}} \rightleftharpoons \left\{ \begin{array}{c} \underbrace{(a_{j_1j_2s_1}, a_{j_2j_1}2)}_{(\mathrm{DP}_+,\,\mathrm{DP}_-)\text{-CCW center}} \\ \underbrace{(a_{j_1j_2s_1}, a_{j_2j_1}1)}_{(\mathrm{UP}_-,\,\mathrm{UP}_-)\text{-negative saddle}} \end{array} \right\}.$$

The (DP,US)-down-parabola upper-saddle is a (DP,US)-appearing bifurcation for center and saddle.

- For $a_{j_1j_2}0 > 0$ and $a_{j_2j_1}0 < 0$,

$$\underbrace{(a_{j_1j_2s_1}, a_{j_2j_1}1)}_{(\mathrm{DP},\,\mathrm{LS})\text{-appearing bifurcation}} \rightleftharpoons \left\{ \begin{array}{c} \underbrace{(a_{j_1j_2s_1}, a_{j_2j_1}2)}_{(\mathrm{DP}_-,\,\mathrm{DP}_+)\text{-CW center}} \\ \underbrace{(a_{j_1j_2s_1}, a_{j_2j_1}1)}_{(\mathrm{UP}_+,\,\mathrm{UP}_+)\text{-positive saddle}} \end{array} \right\}.$$

The (DP,LS)-down-parabola lower-saddle is a (DP,LS)-appearing bifurcation for center and saddle.

- For $a_{j_1j_2}0 < 0$ and $a_{j_2j_1}0 < 0$,

$$\underbrace{(a_{j_1j_2s_1}, a_{j_2j_1}1)}_{(\mathrm{UP},\,\mathrm{LS})\text{-appearing bifurcation}} \rightleftharpoons \left\{ \begin{array}{c} \underbrace{(a_{j_1j_2s_1}, a_{j_2j_1}2)}_{(\mathrm{UP}_-,\,\mathrm{UP}_-)\text{-negative saddle}} \\ \underbrace{(a_{j_1j_2s_1}, a_{j_2j_1}1)}_{(\mathrm{DP}_+,\,\mathrm{DP}_-)\text{-CCW center}} \end{array} \right\}.$$

The (UP,LS)-up-parabola lower-saddle is a (UP,LS)-appearing bifurcation for center and saddle.

(xvi$_2$) From cases (vi$_2$) and (xii), the corresponding bifurcations are obtained. The equilibrium of $(x^*_{j_2}, x^*_{j_1}) = (a_{j_1 j_2 s_2}, a_{j_2 j_1 1})$ has the following bifurcations as in Eqs. (2.130)–(2.133):

- For $a_{j_1 j_2 0}(a_{j_1 j_2 s_2} - a_{j_1 j_2 s_1}) > 0$ and $a_{j_2 j_1 0} > 0$,

$$\underbrace{(a_{j_1 j_2 s_2}, a_{j_2 j_1 1})}_{(\mathrm{II}_+,\mathrm{II}_+)\text{-appearing bifurcation}} \rightleftharpoons \left\{ \begin{array}{cc} \underbrace{(a_{j_1 j_2 (s_2)_2}, a_{j_2 j_1 1})}_{(\mathrm{DP}_-,\mathrm{DP}_+)\text{-CW center}} & \underbrace{(a_{j_1 j_2 (s_2)_2}, a_{j_2 j_1 2})}_{(\mathrm{UP}_+,\mathrm{UP}_+)\text{-positive saddle}} \\ \underbrace{(a_{j_1 j_2 (s_2)_1}, a_{j_2 j_1 1})}_{(\mathrm{UP}_-,\mathrm{UP}_-)\text{-negative saddle}} & \underbrace{(a_{j_1 j_2 (s_2)_1}, a_{j_2 j_1 2})}_{(\mathrm{DP}_+,\mathrm{DP}_-)\text{-CCW center}} \end{array} \right\}.$$

  The (II$_+$,II$_+$)-double increasing-inflection equilibrium is an (II$_+$,II$_+$)-appearing bifurcation for a network of saddles and centers.
- For $a_{j_1 j_2 0}(a_{j_1 j_2 s_2} - a_{j_1 j_2 s_1}) < 0$ and $a_{j_2 j_1 0} > 0$,

$$\underbrace{(a_{j_1 j_2 s_2}, a_{j_2 j_1 1})}_{(\mathrm{DI}_+,\mathrm{DI}_-)\text{-appearing bifurcation}} \rightleftharpoons \left\{ \begin{array}{cc} \underbrace{(a_{j_1 j_2 (s_2)_2}, a_{j_2 j_1 1})}_{(\mathrm{UP}_-,\mathrm{UP}_-)\text{-negative saddle}} & \underbrace{(a_{j_1 j_2 (s_2)_2}, a_{j_2 j_1 2})}_{(\mathrm{DP}_+,\mathrm{DP}_-)\text{-CCW center}} \\ \underbrace{(a_{j_1 j_2 (s_2)_1}, a_{j_2 j_1 1})}_{(\mathrm{DP}_-,\mathrm{DP}_+)\text{-CW center}} & \underbrace{(a_{j_1 j_2 (s_2)_1}, a_{j_2 j_1 2})}_{(\mathrm{UP}_+,\mathrm{UP}_+)\text{-positive saddle}} \end{array} \right\}.$$

  The (DI$_+$,DI$_-$)-double decreasing-inflection equilibrium is a (DI$_+$,DI$_-$)-appearing bifurcation for a network of saddles and centers.
- For $a_{j_1 j_2 0}(a_{j_1 j_2 s_2} - a_{j_1 j_2 s_1}) > 0$ and $a_{j_2 j_1 0} < 0$,

$$\underbrace{(a_{j_1 j_2 s_2}, a_{j_2 j_1 1})}_{(\mathrm{DI}_-,\mathrm{DI}_+)\text{-appearing bifurcation}} \rightleftharpoons \left\{ \begin{array}{cc} \underbrace{(a_{j_1 j_2 (s_2)_2}, a_{j_2 j_1 1})}_{(\mathrm{UP}_+,\mathrm{UP}_+)\text{-positive saddle}} & \underbrace{(a_{j_1 j_2 (s_2)_2}, a_{j_2 j_1 2})}_{(\mathrm{DP}_-,\mathrm{DP}_+)\text{-CW center}} \\ \underbrace{(a_{j_1 j_2 (s_2)_1}, a_{j_2 j_1 1})}_{(\mathrm{DP}_+,\mathrm{DP}_-)\text{-CCW center}} & \underbrace{(a_{j_1 j_2 (s_2)_1}, a_{j_2 j_1 2})}_{(\mathrm{UP}_-,\mathrm{UP}_-)\text{-negative saddle}} \end{array} \right\}.$$

  The (DI$_-$,DI$_+$)-double decreasing-inflection equilibrium is a (DI$_-$,DI$_+$)-appearing bifurcation for a network of saddles and centers.
- For $a_{j_1 j_2 0}(a_{j_1 j_2 s_2} - a_{j_1 j_2 s_1}) < 0$ and $a_{j_2 j_1 0} < 0$,

$$\underbrace{(a_{j_1 j_2 s_2}, a_{j_2 j_1 1})}_{(\mathrm{II}_-,\mathrm{II}_-)\text{-appearing bifurcation}} \rightleftharpoons \left\{ \begin{array}{cc} \underbrace{(a_{j_1 j_2 (s_2)_2}, a_{j_2 j_1 1})}_{(\mathrm{DP}_+,\mathrm{DP}_-)\text{-CCW center}} & \underbrace{(a_{j_1 j_2 (s_2)_2}, a_{j_2 j_1 2})}_{(\mathrm{UP}_-,\mathrm{UP}_-)\text{-negative saddle}} \\ \underbrace{(a_{j_1 j_2 (s_2)_1}, a_{j_2 j_1 1})}_{(\mathrm{UP}_+,\mathrm{UP}_+)\text{-positive saddle}} & \underbrace{(a_{j_1 j_2 (s_2)_1}, a_{j_2 j_1 2})}_{(\mathrm{DP}_-,\mathrm{DP}_+)\text{-CW center}} \end{array} \right\}.$$

  The (II$_-$,II$_-$)-double increasing-inflection equilibrium is an (II$_-$,II$_-$)-appearing bifurcation for a network of saddles and centers.

(xvii) If

$$\Delta_{j_1j_2}=0 \text{ and } \Delta_{j_2j_1}=0, a_{j_2j_1s_1}=a_{j_2j_1s_2}=a_{j_2j_1 1},$$

the standard form is

$$\begin{aligned}
&\dot{x}_{j_1}=a_{j_1j_2 0}(x_{j_2}-a_{j_1j_2 1})^3,\\
&\dot{x}_{j_2}=a_{j_2j_1 0}(x_{j_1}-a_{j_2j_1 1})^2,\\
&j_1,j_2\in\{1,2\};j_1\neq j_2.
\end{aligned}$$

From cases (vii) and (xii), the corresponding bifurcations are obtained. The equilibrium of $(x^*_{j_2},x^*_{j_1})=(a_{j_1j_2 1},a_{j_2j_1 1})$ has the following appearing bifurcations as in Eqs. (3.136)–(3.139):

- For $a_{j_1j_2 0}>0$ and $a_{j_2j_1 0}>0$,

$$\underbrace{(a_{j_1j_2 1},a_{j_2j_1 1})}_{(3^{\text{rd}}\text{UP, US})\text{-appearing bifurcation}} \rightleftharpoons \left\{\begin{array}{cc}
\underbrace{(a_{j_1j_2 3},a_{j_2j_1 1})}_{(\text{DP}_-,\text{DP}_+)\text{-CW center}} & \underbrace{(a_{j_1j_2 3},a_{j_2j_1 2})}_{(\text{UP}_+,\text{UP}_+)\text{-positive saddle}}\\
\underbrace{(a_{j_1j_2 2},a_{j_2j_1 1})}_{(\text{UP}_-,\text{UP}_-)\text{-negative saddle}} & \underbrace{(a_{j_1j_2 1},a_{j_2j_1 2})}_{(\text{DP}_+,\text{DP}_-)\text{-CCW center}}\\
\underbrace{(a_{j_1j_2 1},a_{j_2j_1 1})}_{(\text{DP}_-,\text{DP}_+)\text{-CW center}} & \underbrace{(a_{j_1j_2 1},a_{j_2j_1 2})}_{(\text{UP}_+,\text{UP}_+)\text{-positive saddle}}
\end{array}\right\}.$$

  The equilibrium of $(x^*_{j_2},x^*_{j_1})=(a_{j_1j_2s_1},a_{j_2j_1 1})$ is a $(3^{\text{rd}}$UP,US)-third-order-up-parabola upper-saddle appearing bifurcation for a network of saddles and centers.

- For $a_{j_1j_2 0}<0$ and $a_{j_2j_1 0}>0$,

$$\underbrace{(a_{j_1j_2 1},a_{j_2j_1 1})}_{(3^{\text{rd}}\text{DP, US})\text{-appearing bifucation}} \rightleftharpoons \left\{\begin{array}{cc}
\underbrace{(a_{j_1j_2 3},a_{j_2j_1 1})}_{(\text{UP}_-,\text{UP}_-)\text{-negative saddle}} & \underbrace{(a_{j_1j_2 3},a_{j_2j_1 2})}_{(\text{DP}_+,\text{DP}_-)\text{-CCW center}}\\
\underbrace{(a_{j_1j_2 2},a_{j_2j_1 1})}_{(\text{DP}_-,\text{DP}_+)\text{-CW center}} & \underbrace{(a_{j_1j_2 1},a_{j_2j_1 2})}_{(\text{UP}_+,\text{UP}_+)\text{-positive saddle}}\\
\underbrace{(a_{j_1j_2 1},a_{j_2j_1 1})}_{(\text{UP}_-,\text{UP}_-)\text{-negative saddle}} & \underbrace{(a_{j_1j_2 1},a_{j_2j_1 2})}_{(\text{DP}_+,\text{UP}_-)\text{-CCW center}}
\end{array}\right\}.$$

  The equilibrium of $(x^*_{j_1},x^*_{j_2})=(a_{j_2j_1 1},a_{j_1j_2 1})$ is a $(3^{\text{rd}}$DP,US)-third-order down-parabola upper-saddle appearing bifurcation for a network of saddles and centers.

- For $a_{j_1j_2 0}>0$ and $a_{j_2j_1 0}<0$,

$$\underbrace{(a_{j_1j_2}1, a_{j_2j_1}1)}_{(3^{\text{rd}}\text{DP},\text{LS})\text{-appearing bifucation}} \rightleftharpoons \left\{\begin{array}{cc} \underbrace{(a_{j_1j_2}3, a_{j_2j_1}1)}_{(\text{UP}_+,\text{UP}_+)\text{-positive saddle}} & \underbrace{(a_{j_1j_2}3, a_{j_2j_1}2)}_{(\text{DP}_-,\text{DP}_+)\text{-CW center}} \\ \underbrace{(a_{j_1j_2}2, a_{j_2j_1}1)}_{(\text{DP}_+,\text{DP}_-)\text{-CCW center}} & \underbrace{(a_{j_1j_2}1, a_{j_2j_1}2)}_{(\text{UP}_-,\text{UP}_-)\text{-negative saddle}} \\ \underbrace{(a_{j_1j_2}1, a_{j_2j_1}1)}_{(\text{UP}_+,\text{UP}_+)\text{-positive saddle}} & \underbrace{(a_{j_1j_2}1, a_{j_2j_1}2)}_{(\text{DP}_-,\text{DP}_+)\text{-CW center}} \end{array}\right\}.$$

The equilibrium of $(x^*_{j_1}, x^*_{j_2}) = (a_{j_2j_1}1, a_{j_1j_2}1)$ is a $(3^{\text{rd}}$DP,LS)-third-order down-parabola lower-saddle appearing bifurcation for a network of saddles and centers.

- For $a_{j_1j_2 0} < 0$ and $a_{j_2j_1 0} < 0$,

$$\underbrace{(a_{j_1j_2}1, a_{j_2j_1}1)}_{(3^{\text{rd}}\text{UP},\text{LS})\text{-appearing bifurcation}} \rightleftharpoons \left\{\begin{array}{cc} \underbrace{(a_{j_1j_2}3, a_{j_2j_1}1)}_{(\text{DP}_+,\text{DP}_-)\text{-CCW center}} & \underbrace{(a_{j_1j_2}3, a_{j_2j_1}2)}_{(\text{UP}_-,\text{UP}_-)\text{-negative saddle}} \\ \underbrace{(a_{j_1j_2}2, a_{j_2j_1}1)}_{(\text{UP}_+,\text{UP}_+)\text{-positive saddle}} & \underbrace{(a_{j_1j_2}1, a_{j_2j_1}2)}_{(\text{DP}_-,\text{DP}_+)\text{-CW center}} \\ \underbrace{(a_{j_1j_2}1, a_{j_2j_1}1)}_{(\text{DP}_+,\text{DP}_-)\text{-CCW center}} & \underbrace{(a_{j_1j_2}1, a_{j_2j_1}2)}_{(\text{UP}_-,\text{UP}_-)\text{-negative saddle}} \end{array}\right\}.$$

The equilibrium of $(x^*_{j_1}, x^*_{j_2}) = (a_{j_2j_1}1, a_{j_1j_2}1)$ is a $(3^{\text{rd}}$UP,LS)-third-order-up-parabola lower-saddle appearing bifurcation for a network of saddles and centers.

(xix) Under

$$\Delta_{j_1j_2} = 0 \text{ and } \Delta_{j_2j_1} > 0,$$

the standard form is

$$\begin{aligned} &\dot{x}_{j_2} = a_{j_2j_1 0}(x_{j_1} - a_{j_2j_1}1)(x_{j_1} - a_{j_2j_1}2), \\ &\dot{x}_{j_1} = a_{j_1j_2 0}(x_{j_2} - a_{j_1j_2 s_1})(x_{j_2} - a_{j_1j_2 s_2})^2, \\ &j_1, j_2 \in \{1,2\}; j_1 \neq j_2. \end{aligned}$$

From cases (viii$_2$) and (xii), the corresponding bifurcations are obtained. The equilibrium of $(x^*_{j_2}, x^*_{j_1}) = (a_{j_1j_2 s_2}, a_{j_2j_1 l_1})$ (s_1, s_2, l_1, $l_2 \in \{1,2\}$, $s_1 \neq s_2$; $l_1 \neq l_2$) has the following appearing bifurcation properties as in Eqs. (3.142)–(3.145):

- For $a_{j_1j_2 0}(a_{j_1j_2 s_2} - a_{j_1j_2 s_1}) > 0$ and $a_{j_2j_1 0}(a_{j_2j_1 l_1} - a_{j_2j_1 l_2}) > 0$,

$$\underbrace{(a_{j_1j_2s_2}, a_{j_2j_1l_1})}_{(\mathrm{US},\mathrm{UP})\text{-appearing bifurcation}} \rightleftharpoons \left\{\begin{array}{c} \underbrace{(a_{j_1j_2(s_2)_2}, a_{j_2j_1l_1})}_{(\mathrm{UP}_+,\mathrm{UP}_+)\text{-positive saddle}} \\ \underbrace{(a_{j_1j_2(s_2)_1}, a_{j_2j_1l_1})}_{(\mathrm{DP}_+,\mathrm{DP}_-)\text{-CCW center}} \end{array}\right\}.$$

The equilibrium of $(x_{j_2}^*, x_{j_1}^*) = (a_{j_1j_2s_2}, a_{j_2j_1l_1})$ is a (US,UP)-up-parabola upper-saddle appearing bifurcation for saddle and center.

- For $a_{j_1j_20}(a_{j_1j_2s_2} - a_{j_1j_2s_1}) < 0$ and $a_{j_2j_10}(a_{j_2j_1l_1} - a_{j_2j_1l_2}) > 0$,

$$\underbrace{(a_{j_1j_2s_2}, a_{j_2j_1l_1})}_{(\mathrm{LS},\mathrm{DP})\text{-appearing bifurcation}} \rightleftharpoons \left\{\begin{array}{c} \underbrace{(a_{j_1j_2(s_2)_2}, a_{j_2j_1l_1})}_{(\mathrm{DP}_+,\mathrm{DP}_-)\text{-CCW saddle}} \\ \underbrace{(a_{j_1j_2(s_2)_1}, a_{j_2j_1l_1})}_{(\mathrm{UP}_+,\mathrm{UP}_+)\text{-positive saddle}} \end{array}\right\}.$$

The equilibrium of $(x_{j_2}^*, x_{j_1}^*) = (a_{j_1j_2s_2}, a_{j_2j_1l_1})$ is an (LS,DP)-down-parabola lower-saddle appearing bifurcation for saddle and center.

- For $a_{j_1j_20}(a_{j_1j_2s_2} - a_{j_1j_2s_1}) > 0$ and $a_{j_2j_10}(a_{j_2j_1l_1} - a_{j_2j_1l_2}) < 0$,

$$\underbrace{(a_{j_1j_2s_2}, a_{j_2j_1l_1})}_{(\mathrm{US},\mathrm{DP})\text{-appearing bifurcation}} \rightleftharpoons \left\{\begin{array}{c} \underbrace{(a_{j_1j_2(s_2)_2}, a_{j_2j_1l_1})}_{(\mathrm{DP}_-,\mathrm{DP}_+)\text{-CW center}} \\ \underbrace{(a_{j_1j_2(s_2)_1}, a_{j_2j_1l_1})}_{(\mathrm{UP}_-,\mathrm{UP}_-)\text{-negative saddle}} \end{array}\right\}.$$

The equilibrium of $(x_{j_2}^*, x_{j_1}^*) = (a_{j_1j_2s_2}, a_{j_2j_1l_1})$ is a (US,DP)-down-parabola upper-saddle appearing bifurcation for saddle and center.

- For $a_{j_1j_20}(a_{j_1j_2s_2} - a_{j_1j_2s_1}) < 0$ and $a_{j_2j_10}(a_{j_2j_1l_1} - a_{j_2j_1l_2}) < 0$,

$$\underbrace{(a_{j_1j_2s_2}, a_{j_2j_1l_1})}_{(\mathrm{LS},\mathrm{UP})\text{-appearing bifurcation}} \rightleftharpoons \left\{\begin{array}{c} \underbrace{(a_{j_1j_2(s_2)_2}, a_{j_2j_1l_1})}_{(\mathrm{UP}_-,\mathrm{UP}_-)\text{-negative saddle}} \\ \underbrace{(a_{j_1j_2(s_2)_1}, a_{j_2j_1l_1})}_{(\mathrm{DP}_-,\mathrm{DP}_+)\text{-CW center}} \end{array}\right\}.$$

The equilibrium of $(x_{j_2}^*, x_{j_1}^*) = (a_{j_1j_2s_2}, a_{j_2j_1l_1})$ is an (LS,UP)-up-parabola lower-saddle appearing bifurcation for saddle and center.

(xx) If

$$\Delta_{j_1j_2}=0 \text{ and } \Delta_{j_2j_1}>0, a_{j_1j_2s_1}=a_{j_1j_2s_2}=a_{j_1j_21},$$

the standard form is

$$\begin{aligned}
&\dot{x}_{j_2}=a_{j_2j_10}(x_{j_1}-a_{j_2j_11})(x_{j_1}-a_{j_2j_12}),\\
&\dot{x}_{j_1}=a_{j_1j_20}(x_{j_2}-a_{j_1j_21})^3,\\
&j_1,j_2\in\{1,2\};j_1\neq j_2.
\end{aligned}$$

From cases (ix) and (xii), the appearing bifurcations are obtained. The equilibrium of $(x^*_{j_2},x^*_{j_1})=(a_{j_1j_21},a_{j_2j_1l_1})$ $(l_1,l_2\in\{1,2\},l_1\neq l_2)$ has the following bifurcations as in Eqs. (3.148)–(3.151):

- For $a_{j_1j_20}>0$ and $a_{j_2j_10}(a_{j_2j_1l_1}-a_{j_2j_1l_2})>0$,

$$\underbrace{(a_{j_1j_21},a_{j_2j_1l_1})}_{(\text{UP}_+,\text{UP}_+)\text{-positive saddle}} \rightleftharpoons \underbrace{(a_{j_1j_21},a_{j_2j_1l_1})}_{(3^{\text{rd}}\text{UP}_+,\text{UP}_+)\text{-appearing bifucation}} \rightleftharpoons \left\{\begin{array}{c}\underbrace{(a_{j_1j_23},a_{j_2j_1l_1})}_{(\text{UP}_+,\text{UP}_+)\text{-positive saddle}}\\ \underbrace{(a_{j_1j_22},a_{j_2j_1l_1})}_{(\text{DP}_+,\text{DP}_-)\text{-CCW center}}\\ \underbrace{(a_{j_1j_21},a_{j_2j_1l_1})}_{(\text{UP}_+,\text{UP}_+)\text{-positive saddle}}\end{array}\right\}.$$

  The $(3^{\text{rd}}\text{UP}_+,\text{UP}_+)$-third-order positive saddle is a $(3^{\text{rd}}\text{UP}_+,\text{UP}_+)$-appearing bifurcation from a positive saddle to positive saddle, counter-clockwise center, and positive saddle.

- For $a_{j_1j_20}<0$ and $a_{j_2j_10}(a_{j_2j_1l_1}-a_{j_2j_1l_2})>0$,

$$\underbrace{(a_{j_1j_21},a_{j_2j_1l_1})}_{(\text{DP}_+,\text{DP}_-)\text{-CCW center}} \rightleftharpoons \underbrace{(a_{j_1j_21},a_{j_2j_1l_1})}_{(3^{\text{rd}}\text{DP}_+,\text{DP}_-)\text{-appearing bifucation}} \rightleftharpoons \left\{\begin{array}{c}\underbrace{(a_{j_1j_23},a_{j_2j_1l_1})}_{(\text{DP}_+,\text{DP}_-)\text{-CCW center}}\\ \underbrace{(a_{j_1j_22},a_{j_2j_1l_1})}_{(\text{UP}_+,\text{UP}_+)\text{-positive saddle}}\\ \underbrace{(a_{j_1j_21},a_{j_2j_1l_1})}_{(\text{DP}_+,\text{DP}_-)\text{-CCW center}}\end{array}\right\}.$$

  The $(3^{\text{rd}}\text{DP}_-,\text{DP}_+)$-third-order-clockwise center is a$(3^{\text{rd}}\text{DP}_-,\text{DP}_+)$-appearing bifurcation from a counter-clockwise center to counter-clockwise center, positive saddle, and counter-clockwise center.

- For $a_{j_1j_20}>0$ and $a_{j_2j_10}(a_{j_2j_1l_1}-a_{j_2j_1l_2})<0$,

$$\underbrace{(a_{j_1j_2}1, a_{j_2j_1l_1})}_{(\mathrm{DP}_-,\mathrm{DP}_+)\text{-CW center}} \rightleftharpoons \underbrace{(a_{j_1j_2}1, a_{j_2j_1l_1})}_{(3^{\mathrm{rd}}\mathrm{DP}_+,\mathrm{DP}_-)\text{-appearing bifucation}} \rightleftharpoons \left\{ \begin{array}{c} \underbrace{(a_{j_1j_2}3, a_{j_2j_1l_1})}_{(\mathrm{DP}_-,\mathrm{DP}_+)\text{-CW center}} \\ \underbrace{(a_{j_1j_2}2, a_{j_2j_1l_1})}_{(\mathrm{UP}_-,\mathrm{UP}_-)\text{-negative saddle}} \\ \underbrace{(a_{j_1j_2}1, a_{j_2j_1l_1})}_{(\mathrm{DP}_-,\mathrm{DP}_+)\text{-CW center}} \end{array} \right\}.$$

The $(3^{\mathrm{rd}}\mathrm{DP}_+,\mathrm{DP}_-)$-third-order clockwise center is a $(3^{\mathrm{rd}}\mathrm{DP}_+,\mathrm{DP}_-)$-appearing bifurcation from a clockwise center to clockwise center, negative saddle, and clockwise center.

- For $a_{j_1j_2 0} < 0$ and $a_{j_2j_1 0}(a_{j_2j_1l_1} - a_{j_2j_1l_2}) < 0$,

$$\underbrace{(a_{j_1j_2}1, a_{j_2j_1l_1})}_{(\mathrm{UP}_-,\mathrm{UP}_-)\text{-negative saddle}} \rightleftharpoons \underbrace{(a_{j_1j_2}1, a_{j_2j_1l_1})}_{(3^{\mathrm{rd}}\mathrm{UP}_-,\mathrm{UP}_-)\text{-appearing bifucation}} \rightleftharpoons \left\{ \begin{array}{c} \underbrace{(a_{j_1j_2}3, a_{j_2j_1l_1})}_{(\mathrm{UP}_-,\mathrm{UP}_-)\text{-negative saddle}} \\ \underbrace{(a_{j_1j_2}2, a_{j_2j_1l_1})}_{(\mathrm{DP}_-,\mathrm{DP}_+)\text{-CW center}} \\ \underbrace{(a_{j_1j_2}1, a_{j_2j_1l_1})}_{(\mathrm{UP}_-,\mathrm{UP}_-)\text{-negative saddle}} \end{array} \right\}.$$

The $(3^{\mathrm{rd}}\mathrm{UP}_-,\mathrm{UP}_-)$-third-order negative saddle is a $(3^{\mathrm{rd}}\mathrm{UP}_-,\mathrm{UP}_-)$-appearing bifurcation from a negative saddle to negative saddle, clockwise center, and negative saddle.

In the end, from the above local analysis and the first integral of such a dynamical system, this theorem is easily proved. ■

3.3 Singular Equilibriums

From the previous theorem, the bifurcations and global dynamics of a cubic dynamical system possessing variable-crossing univariate quadratic and cubic vector fields. The singularity and bifurcations for saddle and centers of equilibriums will be presented.

3.3.1 Third-Order Parabola-Saddles

Consider a dynamical system as

$$\begin{aligned}\dot{x}_1 &= a_{120}(x_2 - a_{121})^3,\\ \dot{x}_2 &= a_{210}(x_1 - a_{211})^2,\end{aligned} \tag{3.152}$$

and the corresponding first integral manifold is determined by

$$\frac{1}{4}a_{120}\left[(x_2 - a_{121})^4 - (x_{20} - a_{121})^4\right] = \frac{1}{3}a_{210}\left[(x_1 - a_{211})^3 - (x_{10} - a_{211})^3\right]. \tag{3.153}$$

The corresponding phase portraits near the third-order parabola-upper-saddle and parabola-lower-saddle bifurcations are presented in Fig. 3.1. From the first integral manifold, the phase portraits for two-dimensional system for $x_1^* = a_{211}$ in the x_1-direction for the x_2-direction and $x_2^* = a_{121}$ in the x_2-direction for the x_1-direction are sketched. The equilibriums of $(x_2^*, x_1^*) = (a_{121}, a_{211})$ have the following properties: In Fig. 3.1a–d, a $(3^{\text{rd}}\text{UP}, \text{US})$-third-order up-parabola upper-saddle ($a_{120} > 0$, $a_{210} > 0$), a $(3^{\text{rd}}\text{DP}, \text{US})$-third-order down-parabola upper-saddle ($a_{120} < 0$, $a_{210} > 0$), a $(3^{\text{rd}}\text{DP}, \text{LS})$-third-order down-parabola lower-saddle ($a_{120} > 0$, $a_{210} < 0$), and a $(3^{\text{rd}}\text{UP}, \text{LS})$)-third-order up-parabola lower-saddle ($a_{120} < 0$, $a_{210} < 0$); respectively, are presented. The third-order up-parabola and down-parabola, upper-saddle, and lower-saddles are the parabola-like saddle globally, which are defined as follows:

$$(a_{121}, a_{211}) = \begin{cases} \underbrace{(3^{\text{rd}}\text{UP}, \text{US})}_{\text{third-order up-parabola upper-saddle}} & \text{for } a_{120} > 0, a_{210} > 0; \\ \underbrace{(3^{\text{rd}}\text{DP}, \text{US})}_{\text{third-order down-parabola upper-saddle}} & \text{for } a_{120} < 0, a_{210} > 0; \\ \underbrace{(3^{\text{rd}}\text{DP}, \text{LS})}_{\text{third-order down-parabola lower-saddle}} & \text{for } a_{120} > 0, a_{210} < 0; \\ \underbrace{(3^{\text{rd}}\text{UP}, \text{LS})}_{\text{third-order up-parabola lower-saddle}} & \text{for } a_{120} < 0, a_{210} < 0. \end{cases} \tag{3.154}$$

Consider a dynamical system as

$$\begin{aligned}\dot{x}_1 &= a_{120}(x_2 - a_{121})^2,\\ \dot{x}_2 &= a_{210}(x_1 - a_{211})^3,\end{aligned} \tag{3.155}$$

and the corresponding first integral manifold is determined by

$$\frac{1}{3}a_{120}\left[(x_2 - a_{121})^3 - (x_{20} - a_{121})^3\right] = \frac{1}{4}a_{210}\left[(x_1 - a_{211})^4 - (x_{10} - a_{211})^4\right]. \tag{3.156}$$

The corresponding phase portraits near the parabola-upper-saddle and parabola-lower-saddle bifurcation are presented in Fig. 3.2. From the first integral manifold,

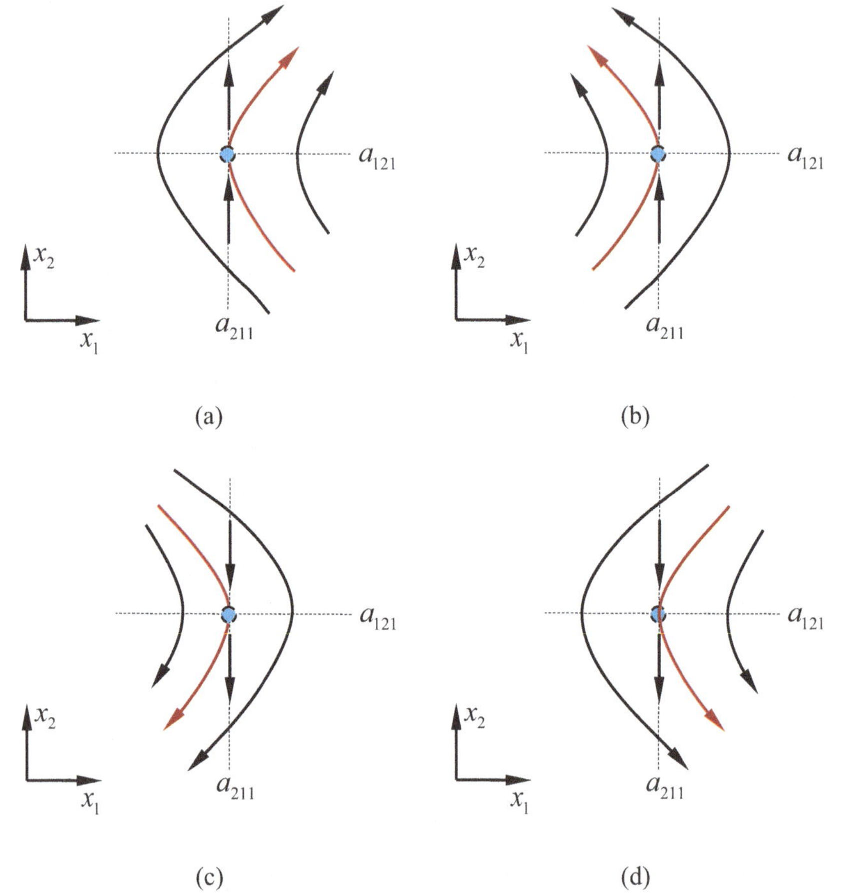

Fig. 3.1 Phase portraits for two-dimensional systems with $x_1^* = a_{211}$ in the x_1-direction for the x_2-direction and $x_2^* = a_{121}$ in the x_2-direction for the x_1-direction. The equilibriums of $(x_2^*, x_1^*) = (a_{121}, a_{211})$: **(a)** a (3rdUP,US)-third-order up-parabola upper-saddle ($a_{120} > 0$, $a_{210} > 0$), **(b)** a (3rdDP,US)-third-order down-parabola upper-saddle ($a_{120} < 0$, $a_{210} > 0$), **(c)** a (3rdDP,LS)-third-order down-parabola lower-saddle ($a_{120} > 0$, $a_{210} < 0$), **(d)** a (3rdDP,LS)-third-order down-parabola lower-saddle ($a_{120} < 0$, $a_{210} < 0$)

the phase portraits for the two-dimensional cubic system for $x_1^* = a_{211}$ in the x_1-direction for the x_2-direction and $x_2^* = a_{121}$ in the x_2-direction for the x_1-direction are sketched, and the equilibriums of $(x_2^*, x_1^*) = (a_{121}, a_{211})$ have the following properties: In Fig. 3.2a–d, a (US, 3rdUP)-third-order up-parabola upper-saddle ($a_{120} > 0$, $a_{210} > 0$), a (LS, 3rdDP)-third-order down-parabola lower-saddle ($a_{120} < 0$, $a_{210} > 0$), a (US, 3rdDP)-third-order down-parabola lower-saddle ($a_{120} > 0$, $a_{210} < 0$), and a (LS, 3rdUP))-third-order up-parabola lower-saddle ($a_{120} < 0$, $a_{210} < 0$) are presented. The third-order up-parabola and down-parabola, upper-saddle, and lower-saddle are the parabola-like saddle globally, which are given by four equilibriums as follows:

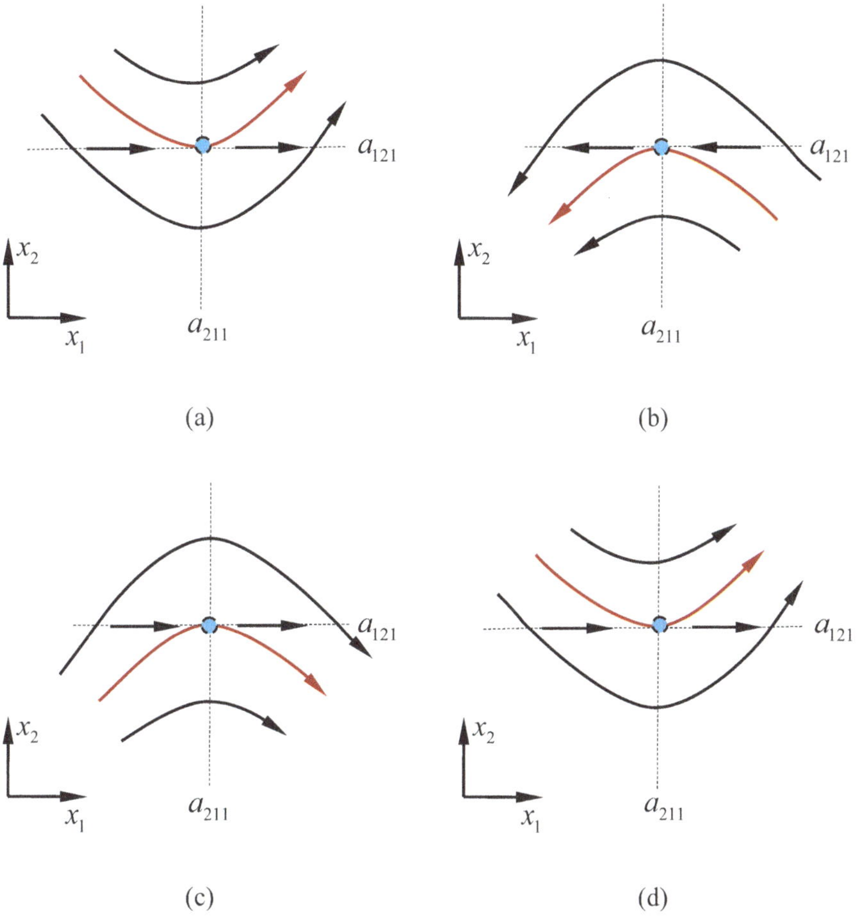

Fig. 3.2 Phase portraits for two-dimensional systems with $x_1^* = a_{211}$ in the x_1-direction for the x_2-direction and $x_2^* = a_{121}$ in the x_2-direction for the x_1-direction. The equilibriums of $(x_2^*, x_1^*) = (a_{121}, a_{211})$: (**a**) a (US, 3rdUP)-third-order up-parabola upper-saddle ($a_{120} > 0$, $a_{210} > 0$), (**b**) an (LS, 3rdDP)-third-order down-parabola lower-saddle ($a_{120} < 0$, $a_{210} > 0$), (**c**) a (US,3rdDP)-third-order down-parabola upper-saddle ($a_{120} > 0$, $a_{210} < 0$), (**d**) an (LS, 3rdDP)-third-order down-parabola lower-saddle ($a_{120} < 0$, $a_{210} < 0$)

$$(a_{121}, a_{211}) = \begin{cases} \underbrace{(\text{US}, 3^{\text{rd}}\text{UP})}_{\text{third-order up-parabola upper-saddle}} & \text{for } a_{120} > 0,\ a_{210} > 0; \\ \underbrace{(\text{LS}, 3^{\text{rd}}\text{DP})}_{\text{third-order down-parabola lower-saddle}} & \text{for } a_{120} < 0,\ a_{210} > 0; \\ \underbrace{(\text{US}, 3^{\text{rd}}\text{DP})}_{\text{third-order down-parabola upper-saddle}} & \text{for } a_{120} > 0,\ a_{210} < 0; \\ \underbrace{(\text{LS}, 3^{\text{rd}}\text{UP})}_{\text{third-order up-parabola lower-saddle}} & \text{for } a_{120} < 0,\ a_{210} < 0. \end{cases} \tag{3.157}$$

3.3.2 Third-Order Saddles and Centers

Consider a dynamical system as

$$\begin{aligned}\dot{x}_2 &= a_{210}(x_1 - a_{211})(x_1 - a_{212}),\\ \dot{x}_1 &= a_{120}(x_2 - a_{121})^3.\end{aligned} \tag{3.158}$$

and the first integral manifold is given by

$$\begin{aligned}&\frac{1}{4}a_{120}\left[(x_2 - a_{121})^4 - (x_{20} - a_{121})^4\right]\\ &= a_{210}\Big\{\frac{1}{3}\left[(x_1 - a_{211})^3 - (x_1 - a_{211})^3\right]\\ &+\frac{1}{2}(a_{211} - a_{212})\left[(x_1 - a_{211})^2 - (x_{10} - a_{211})^2\right]\Big\}.\end{aligned} \tag{3.159}$$

The phase portraits of third-order center and saddle are presented in Fig. 3.3. Phase portraits for two-dimensional cubic systems near the third-order center and third-order saddle bifurcations of $(x_2^*, x_1^*) = (a_{121}, a_{211})$, (a_{121}, a_{212}) are presented in Fig. 3.3a–d for four cases. The equilibriums of $(x_2^*, x_1^*) = (a_{121}, a_{211})$, (a_{121}, a_{212}) are (a) (3^{rd}DP$_-$,DP$_+$)-third-order clockwise center and (3^{rd}UP$_+$, UP$_+$)-third-order positive saddle ($a_{120} > 0$, $a_{210} > 0$), (b) (3^{rd}UP$_-$,UP$_-$)-third-order negative saddle and (3^{rd}DP$_+$,DP$_-$)-third-order clockwise center ($a_{120} < 0$, $a_{210} > 0$), (c) (3^{rd}UP$_+$, UP$_+$)-parabola-saddle and (3^{rd}DP$_+$,DP$_-$)-third-order counter-clockwise center ($a_{120} > 0$, $a_{210} < 0$) , and (d) (3^{rd}DP$_+$, DP$_-$)-third-order counter-clockwise center and (3^{rd}UP$_-$, UP$_-$)-third-order negative saddle ($a_{120} < 0$, $a_{210} < 0$). The four cases are summarized as follows:

(i) For $a_{120} > 0$ and $a_{210} > 0$,

$$\begin{Bmatrix} (a_{121}, a_{211}) \\ (a_{121}, a_{212}) \end{Bmatrix} = \begin{Bmatrix} \underbrace{(3^{rd}\text{DP}_-, \text{DP}_+)}_{\text{third-order CW center}} \\ \underbrace{(3^{rd}\text{UP}_+, \text{UP}_+)}_{\text{third-order positive saddle}} \end{Bmatrix}. \tag{3.160}$$

(ii) For $a_{120} < 0$ and $a_{210} > 0$,

$$\begin{Bmatrix} (a_{121}, a_{211}) \\ (a_{121}, a_{212}) \end{Bmatrix} = \begin{Bmatrix} \underbrace{(3^{rd}\text{UP}_-, \text{UP}_-)}_{\text{third-order negative saddle}} \\ \underbrace{(3^{rd}\text{DP}_+, \text{DP}_-)}_{\text{third-order CCW center}} \end{Bmatrix}. \tag{3.161}$$

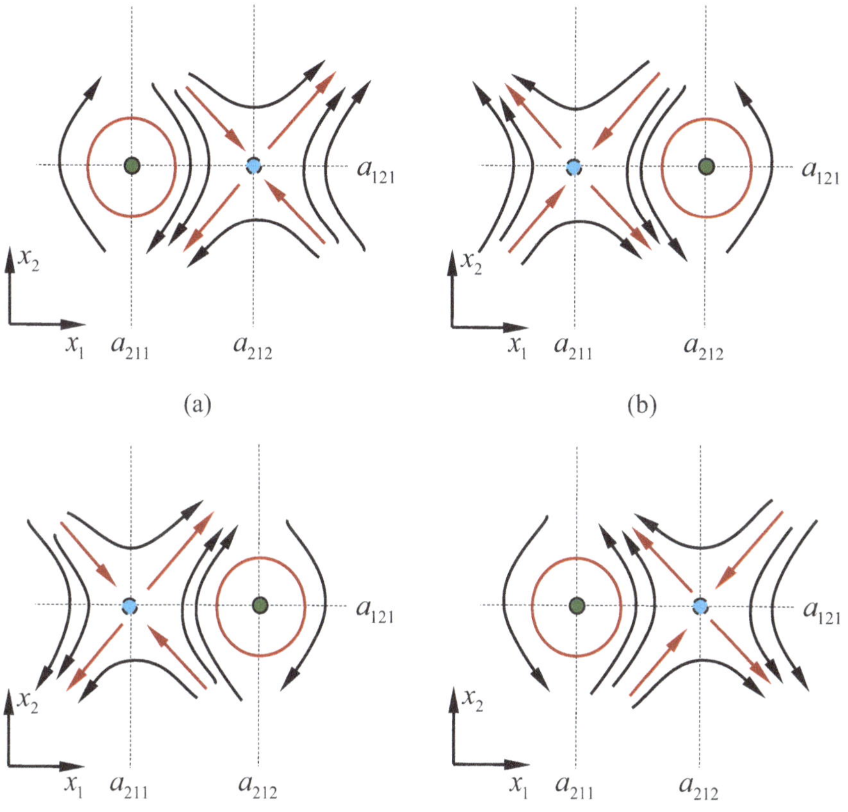

Fig. 3.3 Phase portraits for two-dimensional systems in the x_1-direction with $x_1^* = a_{211}, a_{212}$ for the x_2-direction and in the x_2-direction with $x_2^* = a_{121}$ for the x_1-direction. The equilibriums of $(x_2^*, x_1^*) = (a_{121}, a_{211})$, (a_{121}, a_{212}): (**a**) ($3^{\rm rd}$DP$_-$,DP$_+$)-third-order clockwise center and ($3^{\rm rd}$UP$_+$, UP$_+$)-third-order positive saddle ($a_{120} > 0, a_{210} > 0$), (**b**) ($3^{\rm rd}$UP$_-$,UP$_-$)-third-order negative saddle and ($3^{\rm rd}$DP$_+$,DP$_-$)-third-order counter-clockwise center ($a_{210} > 0$, $a_{120} < 0$), (**c**) ($3^{\rm rd}$UP$_+$,UP$_+$)-parabola-saddle and ($3^{\rm rd}$DP$_+$,DP$_-$)-third-order clockwise center ($a_{120} > 0$, $a_{210} < 0$), (**d**) ($3^{\rm rd}$DP$_+$, DP$_-$)-third-order counter-clockwise center and (LS,UP)-third-order negative saddle ($a_{210} < 0$, $a_{120} < 0$)

(iii) For $a_{120} > 0$ and $a_{210} < 0$,

$$\left\{ \begin{array}{c} (a_{121}, a_{211}) \\ (a_{121}, a_{212}) \end{array} \right\} = \left\{ \begin{array}{c} \underbrace{(3^{\rm rd}{\rm UP}_+, {\rm UP}_+)}_{\text{third-order positive saddle}} \\ \underbrace{(3^{\rm rd}{\rm DP}_-, {\rm DP}_+)}_{\text{third-order CW center}} \end{array} \right\}. \tag{3.162}$$

(iv) For $a_{120} < 0$ and $a_{210} < 0$,

$$\left\{\begin{array}{l}(a_{121}, a_{211}) \\ (a_{121}, a_{212})\end{array}\right\} = \left\{\begin{array}{c}\underbrace{(3^{\mathrm{rd}}\mathrm{DP}_{+}, \mathrm{DP}_{-})}_{\text{third-order CCW-center}} \\ \underbrace{(3^{\mathrm{rd}}\mathrm{UP}_{-}, \mathrm{UP}_{-})}_{\text{third-order negative saddle}}\end{array}\right\}. \tag{3.163}$$

The global flows for the third-order saddles and centers are determined by the third-order parabola-saddles, as shown in Fig. 3.1.

Similarly, consider a cubic dynamical system as

$$\begin{aligned}\dot{x}_2 &= a_{210}(x_1 - a_{211})^3, \\ \dot{x}_1 &= a_{120}(x_2 - a_{121})(x_2 - a_{122})\end{aligned} \tag{3.164}$$

and the first integral manifold is given by

$$\begin{aligned}&\frac{1}{4}a_{210}\left[(x_1 - a_{211})^4 - (x_{10} - a_{211})^4\right] \\ &= a_{120}\Big\{\frac{1}{3}\left[(x_2 - a_{121})^3 - (x_2 - a_{121})^3\right] \\ &+(a_{121} - a_{122})\left[(x_2 - a_{121})^2 - (x_{20} - a_{121})^2\right] \\ &+(a_{121} - a_{122})^2\left[(x_2 - a_{121}) - (x_{20} - a_{121})\right]\Big\}.\end{aligned} \tag{3.165}$$

The phase portraits of third-order canters and saddles are presented in Fig. 3.4. Phase portraits for two-dimensional cubic systems near the bifurcation with two equilibriums of $(x_1^*, x_2^*) = (a_{211}, a_{121})$, (a_{211}, a_{122}) are illustrated. In Fig. 3.4a–d, the four cases are the $(\mathrm{DP}_{+}, 3^{\mathrm{rd}}\mathrm{DP}_{-})$-third-order counter-clockwise center and $(\mathrm{UP}_{+}, 3^{\mathrm{rd}}\mathrm{UP}_{+})$-third-order positive saddle ($a_{120} > 0$, $a_{210} > 0$); the $(\mathrm{UP}_{+}, 3^{\mathrm{rd}}\mathrm{UP}_{+})$-third-order positive saddle and $(\mathrm{DP}_{+}, 3^{\mathrm{rd}}\mathrm{DP}_{-})$-third-order counter-clockwise center ($a_{120} < 0$, $a_{210} > 0$); the $(\mathrm{UP}_{-}, 3^{\mathrm{rd}}\mathrm{UP}_{-})$-third-order negative saddle and $(\mathrm{DP}_{-}, 3^{\mathrm{rd}}\mathrm{DP}_{+})$-third-order clockwise center $a_{120} > 0$, $a_{210} < 0$); and the $(\mathrm{DP}_{-}, 3^{\mathrm{rd}}\mathrm{DP}_{+})$-third-order clockwise center and $(\mathrm{UP}_{-}, 3^{\mathrm{rd}}\mathrm{UP}_{-})$-third-order negative saddle ($a_{120} < 0$, $a_{210} < 0$).

The four cases are summarized as follows:

(i) For $a_{120} > 0$ and $a_{210} > 0$,

$$\left\{\begin{array}{l}(a_{122}, a_{211}) \\ (a_{121}, a_{211})\end{array}\right\} = \left\{\begin{array}{c}\underbrace{(\mathrm{UP}_{+}, 3^{\mathrm{rd}}\mathrm{UP}_{+})}_{\text{third-order positive-saddle}} \\ \underbrace{(\mathrm{DP}_{+}, 3^{\mathrm{rd}}\mathrm{DP}_{-})}_{\text{third-order CCW-center}}\end{array}\right\}. \tag{3.166}$$

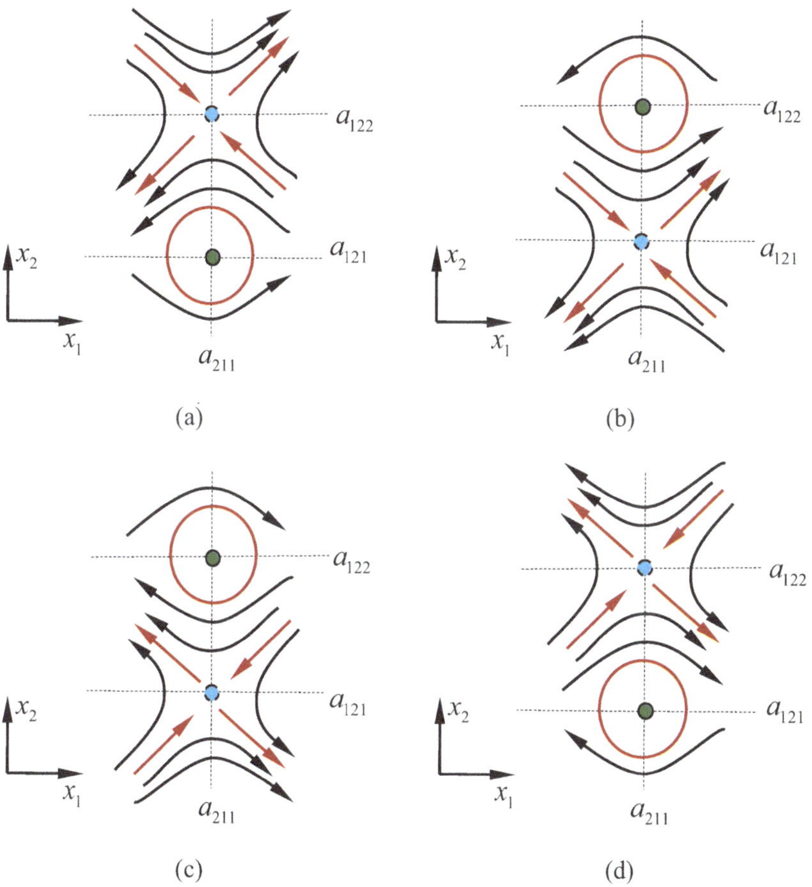

Fig. 3.4 Phase portraits for two-dimensional systems in the x_1-direction with $x_1^* = a_{211}$ and in the x_2-direction with $x_2^* = a_{121}, a_{122}$. The equilibriums of $(x_1^*, x_2^*) = (a_{211}, a_{121})$, (a_{211}, a_{122}): (**a**) (DP_+,$3^{rd}DP_-$)-third-order counter-clockwise center and (UP_+,$3^{rd}UP_+$)-third-order positive saddle ($a_{120} > 0$, $a_{210} > 0$), (**b**) (UP_+,$3^{rd}UP_+$)-third-order positive saddle and (DP_+,$3^{rd}DP_-$)-third-order counter-clockwise center ($a_{120} < 0$, $a_{210} > 0$), (**c**) ($3^{rd}UP_-$,UP_-)-third-order negative saddle and ($3^{rd}DP_-$,DP_+)-third-order clockwise center ($a_{120} > 0$, $a_{210} < 0$), (**d**) (DP_-,$3^{rd}DP_+$)-third-order clockwise center and (UP_-,$3^{rd}UP_-$)-third-order negative saddle ($a_{120} < 0$, $a_{210} < 0$)

(ii) For $a_{120} < 0$ and $a_{210} > 0$,

$$\begin{Bmatrix} (a_{122}, a_{211}) \\ (a_{121}, a_{211}) \end{Bmatrix} = \begin{Bmatrix} \underbrace{(\mathrm{DP}_+, 3^{\mathrm{rd}}\mathrm{DP}_-)}_{\text{third-order CCW center}} \\ \underbrace{(\mathrm{UP}_+, 3^{\mathrm{rd}}\mathrm{UP}_+)}_{\text{third-order positive-saddle}} \end{Bmatrix}. \tag{3.167}$$

(iii) For $a_{120} > 0$ and $a_{210} < 0$,

$$\left\{ \begin{array}{c} (a_{122}, a_{211}) \\ (a_{121}, a_{211}) \end{array} \right\} = \left\{ \begin{array}{c} \underbrace{(\mathrm{DP}_{-}, 3^{\mathrm{rd}}\mathrm{DP}_{+})}_{\text{third-order CW center}} \\ \underbrace{(\mathrm{UP}_{-}, 3^{\mathrm{rd}}\mathrm{UP}_{-})}_{\text{third-order negative saddle}} \end{array} \right\}. \tag{3.168}$$

(iv) For $a_{120} < 0$ and $a_{210} < 0$,

$$\left\{ \begin{array}{c} (a_{122}, a_{211}) \\ (a_{121}, a_{211}) \end{array} \right\} = \left\{ \begin{array}{c} \underbrace{(\mathrm{UP}_{-}, 3^{\mathrm{rd}}\mathrm{UP}_{-})}_{\text{third-order negative-saddle}} \\ \underbrace{(\mathrm{DP}_{-}, 3^{\mathrm{rd}}\mathrm{DP}_{+})}_{\text{third-order CW-center}} \end{array} \right\}. \tag{3.169}$$

The global flows for the third-order saddles and centers are also determined by the third-order parabola-saddle bifurcations, as shown in Fig. 3.2.

The third-order parabola upper-saddle and lower-saddle are the appearing bifurcations for the third-order centers and saddles. However, the third-order parabola upper-saddle and lower-saddle are also the appearing bifurcations for the third-order clockwise center with positive saddle and the third-order counter-clockwise center with negative saddle.

3.3.3 Parabola-Saddles Series

Consider a cubic dynamical system as

$$\begin{aligned} \dot{x}_2 &= a_{210}(x_1 - a_{211})(x_1 - a_{212})(x_1 - a_{213}), \\ \dot{x}_1 &= a_{120}(x_2 - a_{121})^2. \end{aligned} \tag{3.170}$$

and the first integral manifold is given by

$$\begin{aligned} &\frac{1}{3}a_{210}\left[(x_1 - a_{211})^3 - (x_{10} - a_{211})^3\right] \\ &= a_{120}\{\frac{1}{4}\left[(x_2 - a_{12s_1})^4 - (x_2 - a_{12s_1})^4\right] \\ &+ \frac{1}{3}\sum\nolimits_{s_2=1, s_2 \neq s_1}^{3} (a_{12s_1} - a_{12s_2})\left[(x_2 - a_{12s_1})^3 - (x_2 - a_{12s_1})^3\right] \\ &+ \frac{1}{2}\prod\nolimits_{s_2=1, s_2 \neq s_1}^{3} (a_{12s_1} - a_{12s_2})\left[(x_2 - a_{12s_1})^2 - (x_2 - a_{12s_1})^2\right]\}. \end{aligned} \tag{3.171}$$

Phase portraits for the two-dimensional system in the x_1-direction with $x_1^* = a_{211}$, a_{212}, a_{213} for the x_2-direction and in the x_2-direction with $x_2^* = a_{121}$ for the x_1-direction. The equilibriums of $(x_2^*, x_1^*) = (a_{121}, a_{211})$, (a_{121}, a_{212}), (a_{121}, a_{213}) are shown in Fig. 3.5a–d for (US,UP)-up-parabola upper-saddle, (US,DP)-down-parabola upper-saddle, and (US,UP)-up-parabola upper-saddle ($a_{120} > 0$, $a_{210} > 0$); (LS, DP)-down-parabola lower-saddle, (LS,UP)-up-parabola lower-saddle, and (LS, DP)-down-parabola lower-saddle ($a_{120} < 0$, $a_{210} > 0$); (US,DP)-down-parabola upper-saddle, (US,UP)-down-parabola upper-saddle, and (US,DP)-down-parabola upper-saddle ($a_{120} > 0$, $a_{210} < 0$); (LS,UP)-up-parabola lower-saddle, (LS,DP)-down-parabola lower-saddle, and (LS,UP)-up-parabola lower-saddle ($a_{210} < 0$, $a_{120} < 0$). The parabola-saddle series can be listed as follows:

(i) For $a_{210} > 0$ and $a_{120} > 0$,

$$\left\{\begin{array}{l}(a_{121}, a_{211})\\(a_{121}, a_{212})\\(a_{121}, a_{213})\end{array}\right\} = \left\{\begin{array}{c}\underbrace{(\text{US}, \text{UP})}_{\text{up-parabola upper-saddle}}\\\underbrace{(\text{US}, \text{DP})}_{\text{down-parabola upper-saddle}}\\\underbrace{(\text{US}, \text{UP})}_{\text{up-parabola upper-saddle}}\end{array}\right\}. \tag{3.172}$$

(ii) For $a_{120} < 0$ and $a_{210} > 0$,

$$\left\{\begin{array}{l}(a_{121}, a_{211})\\(a_{121}, a_{212})\\(a_{121}, a_{213})\end{array}\right\} = \left\{\begin{array}{c}\underbrace{(\text{LS}, \text{DP})}_{\text{down-parabola lower-saddle}}\\\underbrace{(\text{LS}, \text{UP})}_{\text{up-parabola lower-saddle}}\\\underbrace{(\text{LS}, \text{DP})}_{\text{down-parabola lower-saddle}}\end{array}\right\}. \tag{3.173}$$

(iii) For $a_{120} > 0$ and $a_{210} < 0$,

$$\left\{\begin{array}{l}(a_{121}, a_{211})\\(a_{121}, a_{212})\\(a_{121}, a_{213})\end{array}\right\} = \left\{\begin{array}{c}\underbrace{(\text{US}, \text{DP})}_{\text{down-parabola upper-saddle}}\\\underbrace{(\text{US}, \text{UP})}_{\text{up-parabola upper-saddle}}\\\underbrace{(\text{US}, \text{DP})}_{\text{down-parabola upper-saddle}}\end{array}\right\}. \tag{3.174}$$

(iv) For $a_{120} < 0$ and $a_{210} < 0$,

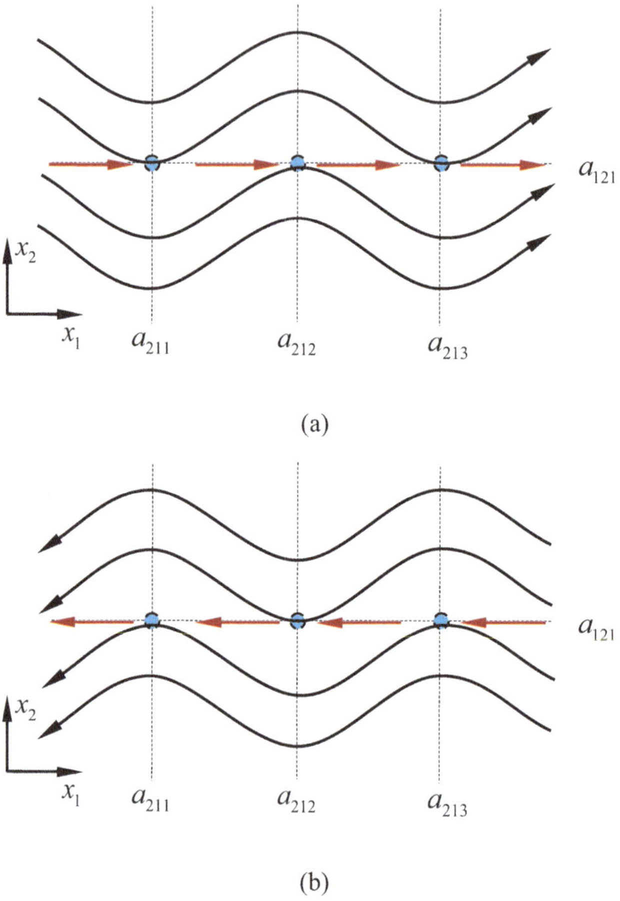

Fig. 3.5 Phase portraits for two-dimensional systems in the x_1-direction with $x_1^* = a_{211}, a_{212}, a_{213}$ for the x_2-direction and in the x_2-direction with $x_2^* = a_{121}$ for the x_1-direction. The equilibriums of $(x_2^*, x_1^*) = (a_{121}, a_{211})$, (a_{121}, a_{212}), (a_{121}, a_{213}): (**a**) (US,UP)-up-parabola upper-saddle, (US,DP)-down-parabola upper-saddle; and (US,UP)-up-parabola upper-saddle ($a_{120} > 0$, $a_{210} > 0$); (**b**) (LS, DP)-down-parabola lower-saddle, (LS,UP)-up-parabola lower-saddle, and (LS,DP)-down-parabola lower-saddle ($a_{120} < 0$, $a_{210} > 0$); (**c**) (US,DP)-down-parabola upper-saddle, (US,UP)-down-parabola upper-saddle, and (US,DP)-down-parabola upper-saddle ($a_{120} > 0$, $a_{210} < 0$); (**d**) (LS, UP)-up-parabola lower-saddle, (LS,DP)-down-parabola lower-saddle, and (LS,UP)-up-parabola lower-saddle ($a_{210} < 0$, $a_{120} < 0$)

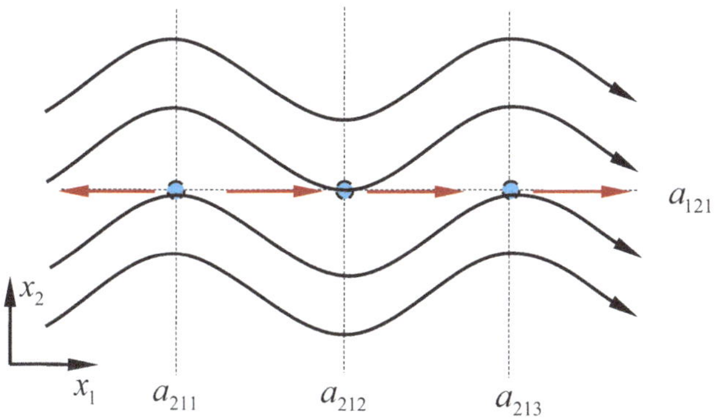

(c)

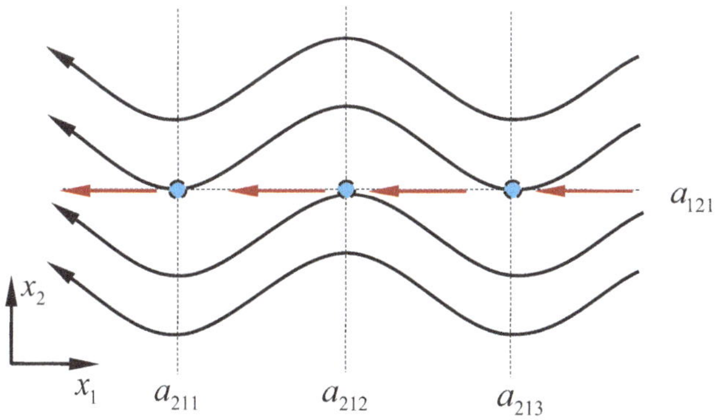

(d)

Fig. 3.5 (continued)

$$\left\{\begin{array}{c}(a_{121}, a_{211})\\(a_{121}, a_{212})\\(a_{121}, a_{213})\end{array}\right\} = \left\{\begin{array}{c}\underbrace{(\mathrm{LS}, \mathrm{UP})}_{\text{up-parabola lower-saddle}}\\\underbrace{(\mathrm{LS}, \mathrm{DP})}_{\text{down-parabola lower-saddle}}\\\underbrace{(\mathrm{LS}, \mathrm{UP})}_{\text{up-parabola lower-saddle}}\end{array}\right\}. \tag{3.175}$$

The global flows for three parabola-saddles are also the appearing bifurcations of third-order parabola-saddles.

Consider a dynamical system as

$$\begin{aligned} \dot{x}_2 &= a_{210}(x_1 - a_{211})^2, \\ \dot{x}_1 &= a_{120}(x_2 - a_{121})(x_2 - a_{122})(x_2 - a_{123}). \end{aligned} \tag{3.176}$$

and the first integral manifold is given by

$$\begin{aligned} &\frac{1}{3}a_{120}\left[(x_2 - a_{121})^3 - (x_{20} - a_{121})^3\right] \\ &= a_{210}\{\frac{1}{4}\left[(x_1 - a_{21s_1})^4 - (x_1 - a_{21s_1})^4\right] \\ &+ \frac{1}{3}\sum\nolimits_{s_2=1, s_2 \neq s_1}^{3}(a_{21s_1} - a_{21s_2})[(x_1 - a_{21s_1})^3 - (x_1 - a_{21s_1})^3 \\ &+ \frac{1}{2}\prod\nolimits_{s_2=1, s_2 \neq s_1}^{3}(a_{21s_1} - a_{21s_2})\left[(x_1 - a_{211})^2 - (x_{10} - a_{211})^2\right]\}. \end{aligned} \tag{3.177}$$

From the first integral manifold, the phase portraits for the two-dimensional system in the x_1-direction with $x_1^* = a_{211}, a_{212}, a_{213}$ for the x_2-direction and in the x_2-direction with $x_2^* = a_{121}$ for the x_1-direction are presented in Fig. 3.6. The equilibriums of $(x_2^*, x_1^*) = (a_{121}, a_{211})$, (a_{122}, a_{211}), (a_{123}, a_{211}) are (UP,US)-up-parabola upper-saddle, (DP,US)-down-parabola upper-saddle, and (UP,US)-up-parabola upper-saddle ($a_{120} > 0$, $a_{210} > 0$); (DP,US)-down-parabola upper-saddle, (UP,US)-up-parabola upper-saddle, and (DP,US)-down-parabola upper-saddle ($a_{210} > 0$, $a_{120} < 0$); and (DP,US)-down-parabola lower-saddle, (UP,US)-down-parabola lower-saddle, and (DP,US)-down-parabola lower-saddle ($a_{120} > 0$, $a_{210} < 0$); (UP, LS)-up-parabola lower-saddle, (DP,LS)-down-parabola lower-saddle, and (UP,LS)-up-parabola lower-saddle ($a_{210} < 0$, $a_{120} < 0$). The properties are listed as follows:

(i) For $a_{210} > 0$ and $a_{120} > 0$,

$$\begin{Bmatrix} (a_{123}, a_{211}) \\ (a_{122}, a_{211}) \\ (a_{121}, a_{211}) \end{Bmatrix} = \begin{Bmatrix} \underbrace{(\text{UP, US})}_{\text{up-parabola upper-saddle}} \\ \underbrace{(\text{DP, US})}_{\text{down-parabola upper-saddle}} \\ \underbrace{(\text{UP, US})}_{\text{up-parabola upper-saddle}} \end{Bmatrix}. \tag{3.178}$$

(ii) For $a_{120} < 0$ and $a_{210} > 0$,

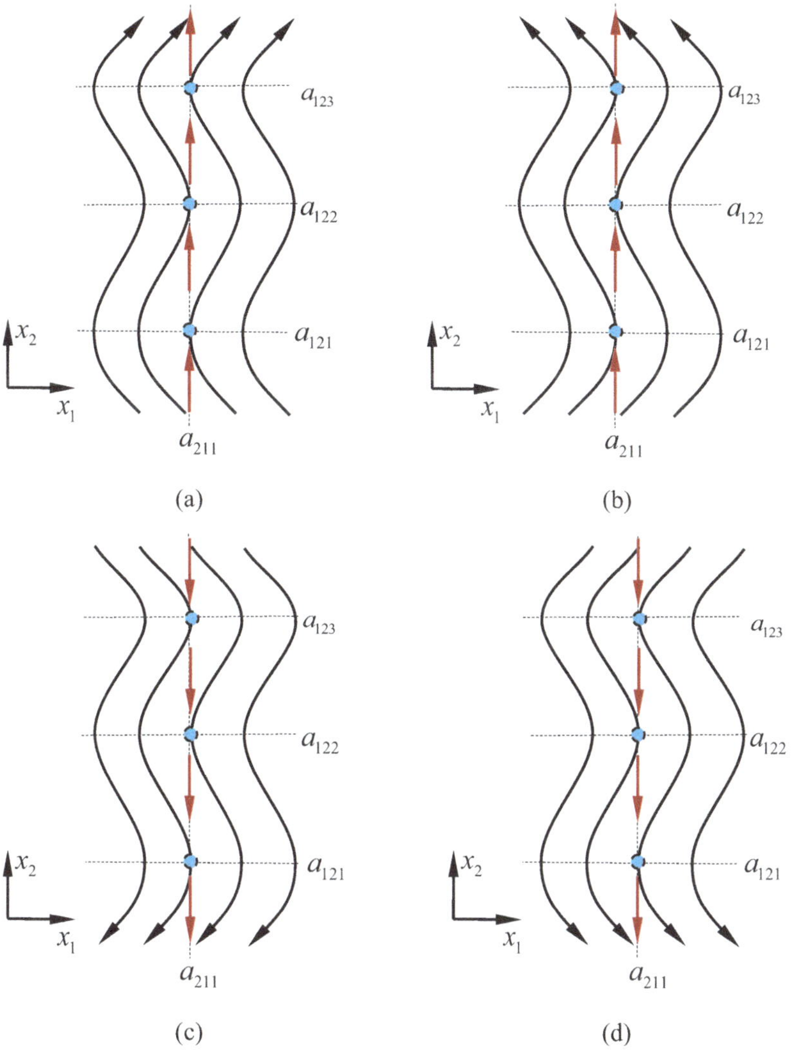

Fig. 3.6 Phase portraits for two-dimensional systems in the x_1-direction with $x_1^* = a_{211}$ and in the x_2-direction with $x_2^* = a_{121}, a_{122}, a_{123}$. The equilibriums of $(x_1^*, x_2^*) = (a_{211}, a_{121})$, (a_{211}, a_{122}), (a_{211}, a_{123}): (**a**) (UP,US)-up-parabola upper-saddle, (DP,US)-down-parabola upper-saddle, (UP, US)-up-parabola upper-saddle ($a_{120} > 0, a_{210} > 0$); (**b**) (DP,US)-down-parabola upper-saddle, (UP,US)-up-parabola upper-saddle, (DP,US)-down-parabola upper-saddle ($a_{120} < 0$, $a_{210} > 0$); (**c**) (DP,LS)-down-parabola lower-saddle, (UP,LS)-up-parabola lower-saddle, (DP,LS)-down-parabola lower-saddle($a_{120} > 0$, $a_{210} < 0$); (**d**) (UP,LS)-up-parabola lower-saddle, (DP,LS)-down-parabola lower-saddle, (UP,LS)-up-parabola lower-saddle ($a_{120} < 0$, $a_{210} < 0$)

$$\left\{\begin{matrix}(a_{123}, a_{211})\\(a_{122}, a_{211})\\(a_{121}, a_{211})\end{matrix}\right\} = \left\{\begin{matrix}\underbrace{(\text{DP}, \text{US})}_{\text{down-parabola upper-saddle}}\\\underbrace{(\text{UP}, \text{US})}_{\text{up-parabola upper-saddle}}\\\underbrace{(\text{DP}, \text{US})}_{\text{down-parabola upper-saddle}}\end{matrix}\right\}. \tag{3.179}$$

(iii) For $a_{120} > 0$ and $a_{210} < 0$,

$$\left\{\begin{matrix}(a_{123}, a_{211})\\(a_{122}, a_{211})\\(a_{121}, a_{211})\end{matrix}\right\} = \left\{\begin{matrix}\underbrace{(\text{DP}, \text{LS})}_{\text{down-parabola lower-saddle}}\\\underbrace{(\text{UP}, \text{LS})}_{\text{up-parabola lower-saddle}}\\\underbrace{(\text{DP}, \text{LS})}_{\text{down-parabola lower-saddle}}\end{matrix}\right\}. \tag{3.180}$$

(iv) For $a_{120} < 0$ and $a_{210} < 0$,

$$\left\{\begin{matrix}(a_{123}, a_{211})\\(a_{122}, a_{211})\\(a_{121}, a_{213})\end{matrix}\right\} = \left\{\begin{matrix}\underbrace{(\text{UP}, \text{LS})}_{\text{up-parabola lower-saddle}}\\\underbrace{(\text{DP}, \text{LS})}_{\text{down-parabola lower-saddle}}\\\underbrace{(\text{UP}, \text{LS})}_{\text{up-parabola lower-saddle}}\end{matrix}\right\}. \tag{3.181}$$

The global flows for three parabola-saddles are appearing bifurcations of the third-order parabola-saddles.

3.4 Limit Cycles and Homoclinic Networks

Consider a quadratic dynamical system as

$$\begin{aligned}\dot{x}_1 &= a_{120}(x_2 - a_{121})(x_2 - a_{122}),\\\dot{x}_2 &= a_{210}(x_1 - a_{211})(x_1 - a_{212})(x_1 - a_{213}).\end{aligned} \tag{3.182}$$

and the first integral manifold is determined by

$$
\begin{aligned}
& a_{120}\{\frac{1}{3}\left[(x_2-a_{121})^3-(x_{20}-a_{121})^3\right] \\
& +\frac{1}{2}(a_{211}-a_{212})\left[(x_2-a_{211})^2-(x_{20}-a_{211})^2\right]\} \\
& =a_{210}\{\frac{1}{4}\left[(x_1-a_{211})^4-(x_{10}-a_{211})^4\right] \\
& +\frac{1}{3}(2a_{211}-a_{212}-a_{213})\left[(x_1-a_{211})^3-(x_{10}-a_{211})^3\right] \\
& +\frac{1}{2}(a_{211}-a_{212})(a_{211}-a_{213})\left[(x_1-a_{211})^2-(x_{10}-a_{211})^2\right]\}.
\end{aligned}
\tag{3.183}
$$

Phase portraits for two-dimensional systems for the saddle and center are presented in Fig. 3.7a–d for ($a_{120}>0$, $a_{210}>0$), ($a_{120}<0$, $a_{210}>0$), ($a_{120}>0$, $a_{210}<0$) and ($a_{120}<0$, $a_{210}<0$). The six equilibriums of $(x_2^*, x_1^*)=(a_{121}, a_{211})$, (a_{121}, a_{212}), (a_{121}, a_{213}), (a_{122}, a_{211}), (a_{122}, a_{212}), (a_{122}, a_{213}) are based on third-order parabola-saddle appearing bifurcations or the third-order center and third-order saddle switching bifurcations or the three parabola-saddle appearing bifurcations. The six equilibriums are three saddles and three centers. There are three limit cycles around the centers and three closed separatrices connecting three saddles. This is a global (2,3)-parabola-saddle enclosing the three limit cycles and three heteroclinic orbits, plus three saddles and three centers. The properties of the six simple equilibriums are listed in Table 3.1. The blue curves are for global third-order parabola-saddle flows and heteroclinic orbits.

(i) For $a_{120}>0$ and $a_{210}>0$,

$$
\begin{aligned}
& \left\{\begin{array}{ccc}
(a_{122}, a_{211}) & (a_{122}, a_{212}) & (a_{122}, a_{213}) \\
(a_{121}, a_{211}) & (a_{121}, a_{212}) & (a_{121}, a_{213})
\end{array}\right\} \\
& =\left\{\begin{array}{ccc}
\underbrace{(\mathrm{UP}_{+}, \mathrm{UP}_{+})}_{\text{positive saddle}} & \underbrace{(\mathrm{DP}_{-}, \mathrm{DP}_{+})}_{\text{CW center}} & \underbrace{(\mathrm{UP}_{+}, \mathrm{UP}_{+})}_{\text{positive saddle}} \\
\underbrace{(\mathrm{DP}_{+}, \mathrm{DP}_{-})}_{\text{CCW center}} & \underbrace{(\mathrm{UP}_{-}, \mathrm{UP}_{-})}_{\text{negative saddle}} & \underbrace{(\mathrm{DP}_{+}, \mathrm{DP}_{-})}_{\text{CCW center}}
\end{array}\right\}.
\end{aligned}
\tag{3.184}
$$

(ii) For $a_{120}<0$ and $a_{210}>0$,

$$
\begin{aligned}
& \left\{\begin{array}{ccc}
(a_{122}, a_{211}) & (a_{122}, a_{212}) & (a_{122}, a_{213}) \\
(a_{121}, a_{211}) & (a_{121}, a_{212}) & (a_{121}, a_{213})
\end{array}\right\} \\
& =\left\{\begin{array}{ccc}
\underbrace{(\mathrm{DP}_{+}, \mathrm{DP}_{-})}_{\text{CCW center}} & \underbrace{(\mathrm{UP}_{-}, \mathrm{UP}_{-})}_{\text{negative saddle}} & \underbrace{(\mathrm{DP}_{+}, \mathrm{DP}_{-})}_{\text{CCW center}} \\
\underbrace{(\mathrm{UP}_{+}, \mathrm{UP}_{+})}_{\text{positive saddle}} & \underbrace{(\mathrm{DP}_{-}, \mathrm{DP}_{+})}_{\text{CW center}} & \underbrace{(\mathrm{UP}_{+}, \mathrm{UP}_{+})}_{\text{positive saddle}}
\end{array}\right\}.
\end{aligned}
\tag{3.185}
$$

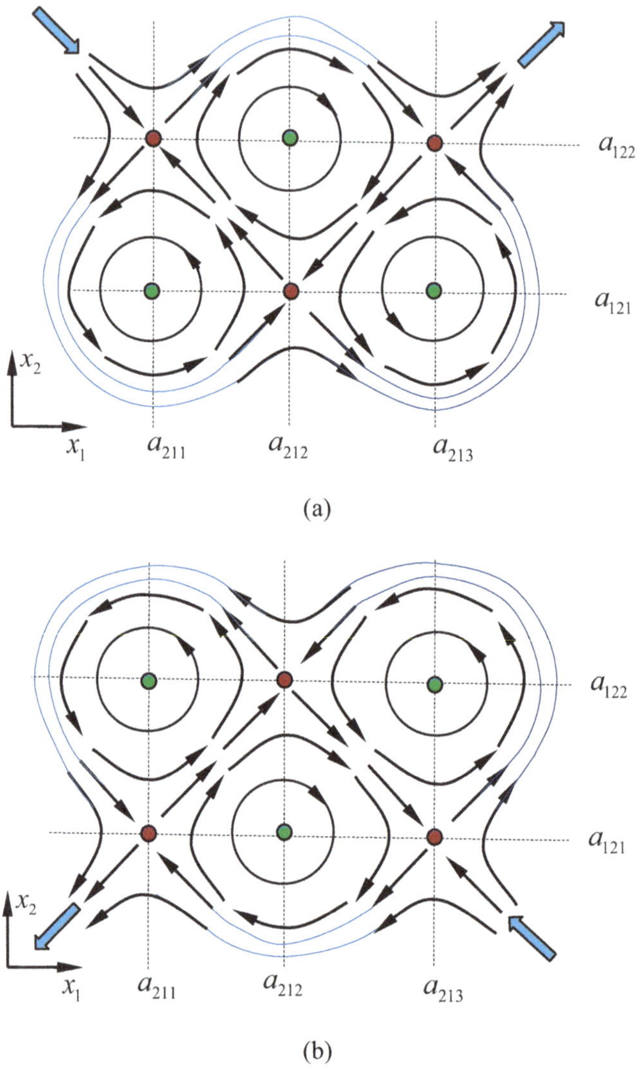

Fig. 3.7 Phase portraits for two-dimensional systems in the x_1-direction with $x_1^* = a_{211}, a_{212}, a_{213}$ and in the x_2-direction with $x_1^* = a_{121}, a_{122}$. The four sets of six simple saddle and center: (**a**) ($a_{120} > 0$, $a_{210} > 0$), (**b**) ($a_{120} < 0$, $a_{210} > 0$), (**c**) ($a_{120} > 0$, $a_{210} < 0$), (**d**) ($a_{120} < 0$, $a_{210} < 0$)

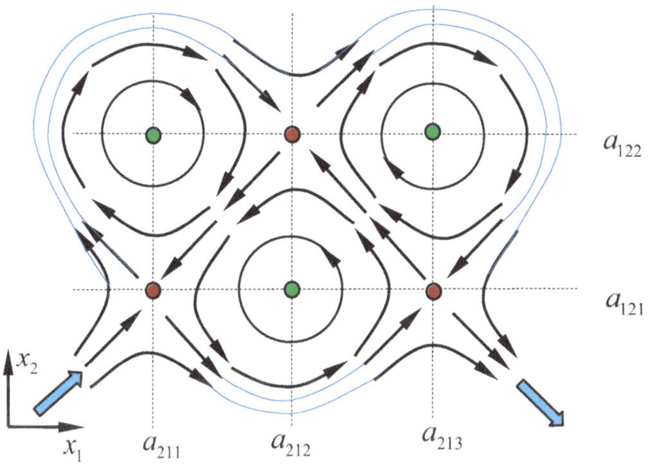

(c)

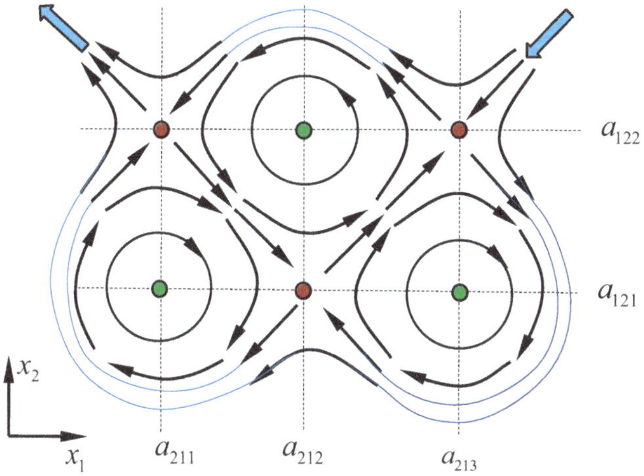

(d)

Fig. 3.7 (continued)

Table 3.1 Local saddles and centers for two-dimensional cubic nonlinear systems

Fig. 3.7	(a)	(b)	(c)	(d)
(a_{120}, a_{210})	(+,+)	(-,+)	(+,-)	(-,-)
(a_{122}, a_{211})	positive saddle	CCW center	CW center	negative saddle
(a_{122}, a_{212})	CW center	negative saddle	positive saddle	CCW center
(a_{122}, a_{213})	positive saddle	CCW center	CW center	negative saddle
(a_{121}, a_{211})	CCW center	positive saddle	negative saddle	CW center
(a_{121}, a_{212})	negative saddle	CW center	CCW center	positive saddle
(a_{121}, a_{213})	CCW center	positive saddle	negative saddle	CW center

(iii) For $a_{120} > 0$ and $a_{210} < 0$,

$$\begin{Bmatrix} (a_{122}, a_{211}) & (a_{122}, a_{212}) & (a_{122}, a_{213}) \\ (a_{121}, a_{211}) & (a_{121}, a_{212}) & (a_{121}, a_{213}) \end{Bmatrix} = \begin{Bmatrix} \underbrace{(\mathrm{DP}_-, \mathrm{DP}_+)}_{\text{CW center}} & \underbrace{(\mathrm{UP}_+, \mathrm{UP}_+)}_{\text{positive saddle}} & \underbrace{(\mathrm{DP}_-, \mathrm{DP}_+)}_{\text{CW center}} \\ \underbrace{(\mathrm{UP}_-, \mathrm{UP}_-)}_{\text{negative saddle}} & \underbrace{(\mathrm{DP}_+, \mathrm{DP}_-)}_{\text{CCW center}} & \underbrace{(\mathrm{UP}_-, \mathrm{UP}_-)}_{\text{negative saddle}} \end{Bmatrix}. \tag{3.186}$$

(iv) For $a_{120} < 0$ and $a_{210} < 0$,

$$\begin{Bmatrix} (a_{122}, a_{211}) & (a_{122}, a_{212}) & (a_{122}, a_{213}) \\ (a_{121}, a_{211}) & (a_{121}, a_{212}) & (a_{121}, a_{213}) \end{Bmatrix} = \begin{Bmatrix} \underbrace{(\mathrm{UP}_-, \mathrm{UP}_-)}_{\text{negative saddle}} & \underbrace{(\mathrm{DP}_+, \mathrm{DP}_-)}_{\text{CCW center}} & \underbrace{(\mathrm{UP}_-, \mathrm{UP}_-)}_{\text{negative saddle}} \\ \underbrace{(\mathrm{DP}_-, \mathrm{DP}_+)}_{\text{CW center}} & \underbrace{(\mathrm{UP}_+, \mathrm{UP}_+)}_{\text{positive saddle}} & \underbrace{(\mathrm{DP}_-, \mathrm{DP}_+)}_{\text{CW center}} \end{Bmatrix}. \tag{3.187}$$

For the above four cases, without the local centers and saddles, the global characteristics of such a dynamical system with the four equilibriums are given by the double-inflections flows, as shown in Fig. 3.7. The global flows are given by the (US,3^{rd}UP)-third-order up-parabola upper-saddle ($a_{120} > 0$, $a_{210} > 0$), the (LS, 3^{rd}DP)-third-order down-parabola lower-saddle ($a_{120} < 0$, $a_{210} > 0$), the (US, 3^{rd}DP)-third-order down-parabola upper-saddle ($a_{120} > 0$, $a_{210} < 0$), and the (LS, 3^{rd}UP)-third-order up-parabola lower-saddle ($a_{120} < 0$, $a_{210} < 0$).

Consider a cubic dynamical system as

$$\begin{aligned} \dot{x}_2 &= a_{210}(x_1 - a_{211})(x_1 - a_{212}), \\ \dot{x}_1 &= a_{120}(x_2 - a_{121})(x_1 - a_{122})(x_1 - a_{123}). \end{aligned} \tag{3.188}$$

and the first integral manifold is determined by

$$\begin{aligned} & a_{210}\{\frac{1}{3}\left[(x_1 - a_{211})^3 - (x_{10} - a_{211})^3\right] \\ & + \frac{1}{2}(a_{211} - a_{212})\left[(x_1 - a_{211})^2 - (x_{10} - a_{211})^2\right]\} \\ & = a_{120}\{\frac{1}{4}\left[(x_2 - a_{121})^4 - (x_{20} - a_{121})^4\right] \\ & + \frac{1}{3}(2a_{121} - a_{122} - a_{123})\left[(x_2 - a_{121})^3 - (x_{20} - a_{121})^3\right] \\ & + \frac{1}{2}(a_{121} - a_{122})(a_{121} - a_{123})\left[(x_2 - a_{121})^2 - (x_{20} - a_{121})^2\right\}. \end{aligned} \tag{3.189}$$

From the first integral manifold, the corresponding center and saddle networks are given as follows:

(i) For $a_{120} > 0$ and $a_{210} > 0$,

$$\left\{ \begin{matrix} (a_{123}, a_{211}) & (a_{122}, a_{212}) \\ (a_{122}, a_{211}) & (a_{122}, a_{212}) \\ (a_{121}, a_{211}) & (a_{121}, a_{212}) \end{matrix} \right\} = \left\{ \begin{matrix} \underbrace{(\mathrm{DP}_-, \mathrm{DP}_+)}_{\text{CW center}} & \underbrace{(\mathrm{UP}_+, \mathrm{UP}_+)}_{\text{positive saddle}} \\ \underbrace{(\mathrm{UP}_-, \mathrm{UP}_-)}_{\text{negative saddle}} & \underbrace{(\mathrm{DP}_+, \mathrm{DP}_-)}_{\text{CCW center}} \\ \underbrace{(\mathrm{DP}_-, \mathrm{DP}_+)}_{\text{CW center}} & \underbrace{(\mathrm{UP}_+, \mathrm{UP}_+)}_{\text{positive saddle}} \end{matrix} \right\}. \tag{3.190}$$

(ii) For $a_{120} < 0$ and $a_{210} > 0$,

$$\left\{ \begin{matrix} (a_{123}, a_{211}) & (a_{122}, a_{212}) \\ (a_{122}, a_{211}) & (a_{122}, a_{212}) \\ (a_{121}, a_{211}) & (a_{121}, a_{212}) \end{matrix} \right\} = \left\{ \begin{matrix} \underbrace{(\mathrm{UP}_-, \mathrm{UP}_-)}_{\text{negative saddle}} & \underbrace{(\mathrm{DP}_+, \mathrm{DP}_-)}_{\text{CCW center}} \\ \underbrace{(\mathrm{DP}_-, \mathrm{DP}_+)}_{\text{CW center}} & \underbrace{(\mathrm{UP}_+, \mathrm{UP}_+)}_{\text{positive saddle}} \\ \underbrace{(\mathrm{UP}_-, \mathrm{UP}_-)}_{\text{negative saddle}} & \underbrace{(\mathrm{DP}_+, \mathrm{DP}_-)}_{\text{CCW center}} \end{matrix} \right\}. \tag{3.191}$$

(iii) For $a_{120} > 0$ and $a_{210} < 0$,

$$\left\{\begin{matrix}(a_{123},a_{211}) & (a_{122},a_{212})\\ (a_{122},a_{211}) & (a_{122},a_{212})\\ (a_{121},a_{211}) & (a_{121},a_{212})\end{matrix}\right\}=\left\{\begin{matrix}\underbrace{(\mathrm{UP}_{+},\mathrm{UP}_{+})}_{\text{positive saddle}} & \underbrace{(\mathrm{DP}_{-},\mathrm{DP}_{+})}_{\text{CW center}}\\ \underbrace{(\mathrm{DP}_{+},\mathrm{DP}_{-})}_{\text{CCW center}} & \underbrace{(\mathrm{UP}_{-},\mathrm{UP}_{-})}_{\text{negative saddle}}\\ \underbrace{(\mathrm{UP}_{+},\mathrm{UP}_{+})}_{\text{positive saddle}} & \underbrace{(\mathrm{DP}_{-},\mathrm{DP}_{+})}_{\text{CW center}}\end{matrix}\right\}. \tag{3.192}$$

(iv) For $a_{120}<0$ and $a_{210}<0$,

$$\left\{\begin{matrix}(a_{123},a_{211}) & (a_{122},a_{212})\\ (a_{122},a_{211}) & (a_{122},a_{212})\\ (a_{121},a_{211}) & (a_{121},a_{212})\end{matrix}\right\}=\left\{\begin{matrix}\underbrace{(\mathrm{DP}_{+},\mathrm{DP}_{-})}_{\text{CCW center}} & \underbrace{(\mathrm{UP}_{-},\mathrm{UP}_{-})}_{\text{negative saddle}}\\ \underbrace{(\mathrm{UP}_{+},\mathrm{UP}_{+})}_{\text{positive saddle}} & \underbrace{(\mathrm{DP}_{-},\mathrm{DP}_{+})}_{\text{CW center}}\\ \underbrace{(\mathrm{DP}_{+},\mathrm{DP}_{-})}_{\text{CCW center}} & \underbrace{(\mathrm{UP}_{-},\mathrm{UP}_{-})}_{\text{negative saddle}}\end{matrix}\right\}. \tag{3.193}$$

For the above four cases, without the local centers and saddles, the global characteristics of such a dynamical system with the four equilibriums are shown in Fig. 3.8a–d. The global flows are given by the (3^{rd}UP,US)-third-order up-parabola upper-saddle ($a_{120}>0$, $a_{210}>0$), the (3^{rd}DP, US)-third-order down-parabola upper-saddle ($a_{120}<0$, $a_{210}>0$), the (3^{rd}DP, LS)-third-order down-parabola lower-saddle ($a_{120}>0$, $a_{210}<0$), and the (3^{rd}UP, LS)-third-order up-parabola lower-saddle ($a_{120}<0$, $a_{210}<0$).

The four local equilibriums inside of the homoclinic orbits are treated as a local point. The flows near the six equilibriums are called a local network of the center and saddle with limit cycle and homoclinic orbits. The local network is formed by the homoclinic orbits, and the saddles are nodes for such a local network, which are different from the network of sinks, sources, and saddles.

3.5 Bifurcation Routes

Consider a standard form of the quadratic dynamical system as

$$\begin{aligned}&\dot{x}_{j_1}=a_{j_1j_20}(x_{j_2}-b_{j_1j_21})\big[(x_{j_2}-\tfrac{1}{2}B_{j_1j_2})^2+\tfrac{1}{4}(-\Delta_{j_1j_2})\big],\\&\dot{x}_{j_2}=a_{j_2j_10}\big[(x_{j_1}-\tfrac{1}{2}B_{j_2j_1})^2+\tfrac{1}{4}(-\Delta_{j_2j_1})\big],\\&j_1,j_2\in\{1,2\};j_1\neq j_2.\end{aligned} \tag{3.194}$$

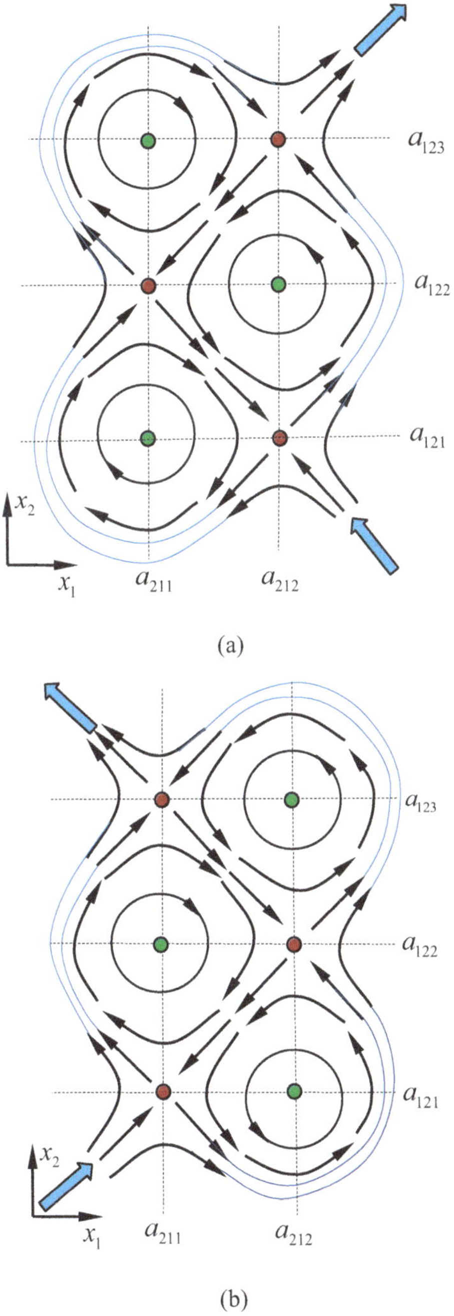

Fig. 3.8 Phase portraits for two-dimensional systems in the x_1-direction with $x_1^* = a_{211}, a_{212}, a_{213}$ and in the x_2-direction with $x_1^* = a_{121}, a_{122}$. The four sets of six simple saddle and center: (**a**) ($a_{120} > 0$, $a_{210} > 0$), (**b**) ($a_{120} < 0$, $a_{210} > 0$), (**c**) ($a_{120} > 0$, $a_{210} < 0$), (**d**) ($a_{120} < 0$, $a_{210} < 0$)

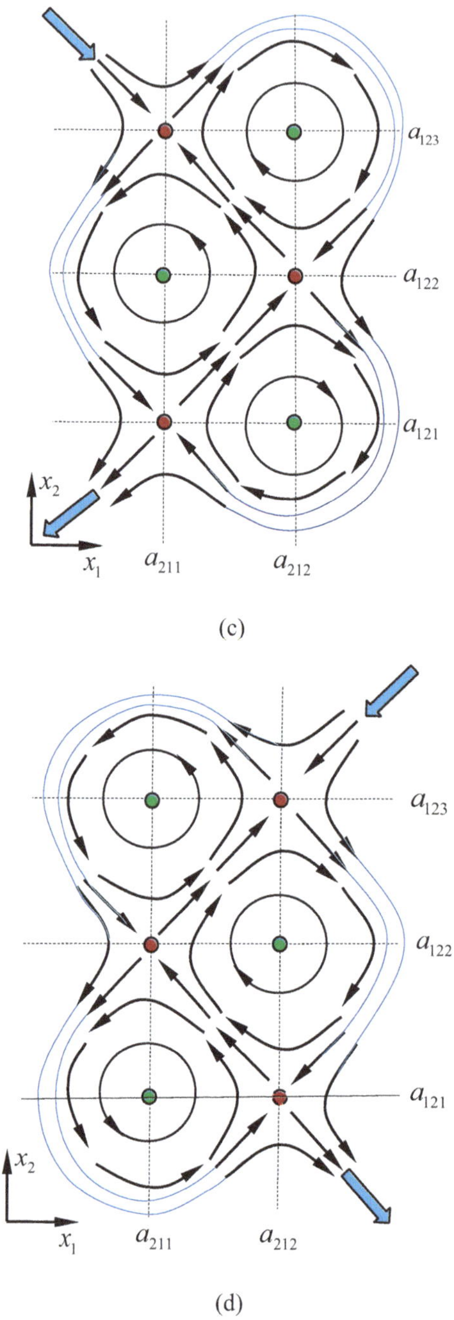

Fig. 3.8 (continued)

(**A**) With $\Delta_{j_1j_2}=0, a_{j_1j_2 1}=b_{j_1j_2 1}=\frac{1}{2}B_{j_1j_2 1}$, the flow appearing bifurcation system is

$$\begin{aligned}
&\dot{x}_{j_1}=a_{j_1j_2 0}(x_{j_2}-a_{j_1j_2 1})^3,\\
&\dot{x}_{j_2}=a_{j_2j_1 0}\left[(x_{j_1}-\frac{1}{2}B_{j_2j_1})^2+\frac{1}{4}(-\Delta_{j_2j_1})\right],\\
&j_1,j_2\in\{1,2\};j_1\neq j_2.
\end{aligned}\tag{3.195}$$

(**B**) With $\Delta_{j_1j_2}=0, a_{j_1j_2 1}=b_{j_1j_2 1}<\frac{1}{2}B_{j_1j_2}=a_{j_1j_2 2}$, the flow appearing bifurcation system is

$$\begin{aligned}
&\dot{x}_{j_1}=a_{j_1j_2 0}(x_{j_2}-a_{j_1j_2 1})(x_{j_2}-a_{j_1j_2 2})^2,\\
&\dot{x}_{j_2}=a_{j_2j_1 0}\left[(x_{j_1}-\frac{1}{2}B_{j_2j_1})^2+\frac{1}{4}(-\Delta_{j_2j_1})\right],\\
&j_1,j_2\in\{1,2\};j_1\neq j_2.
\end{aligned}\tag{3.196}$$

(**C**) With $\Delta_{j_2j_1}=0, a_{j_1j_2 2}=b_{j_1j_2 1}>\frac{1}{2}B_{j_2j_1 1}=a_{j_1j_2 1}$, the flow appearing bifurcation system is

$$\begin{aligned}
&\dot{x}_{j_1}=a_{j_1j_2 0}(x_{j_2}-a_{j_1j_2 1})^2(x_{j_2}-a_{j_1j_2 2}),\\
&\dot{x}_{j_2}=a_{j_2j_1 0}\left[(x_{j_1}-\frac{1}{2}B_{j_2j_1})^2+\frac{1}{4}(-\Delta_{j_2j_1})\right],\\
&j_1,j_2\in\{1,2\};j_1\neq j_2.
\end{aligned}\tag{3.197}$$

After the appearing bifurcation, the post-bifurcation system is

$$\begin{aligned}
&\dot{x}_{j_1}=a_{j_1j_2 0}(x_{j_2}-a_{j_1j_2 1})(x_{j_2}-a_{j_1j_2 2})(x_{j_2}-a_{j_1j_2 3}),\\
&\dot{x}_{j_2}=a_{j_2j_1 0}\left[(x_{j_1}-\frac{1}{2}B_{j_2j_1})^2+\frac{1}{4}(-\Delta_{j_2j_1})\right],\\
&j_1,j_2\in\{1,2\};j_1\neq j_2.
\end{aligned}\tag{3.198}$$

With $\Delta_{j_2j_1}=0, a_{j_2j_1 1}=\frac{1}{2}B_{j_2j_1}$, the equilibrium bifurcation system is

$$\begin{aligned}
&\dot{x}_{j_1}=a_{j_1j_2 0}(x_{j_2}-a_{j_1j_2 1})(x_{j_2}-a_{j_1j_2 2})(x_{j_2}-a_{j_1j_2 3}),\\
&\dot{x}_{j_2}=a_{j_2j_1 0}(x_{j_1}-a_{j_2j_1 1})^2,\\
&j_1,j_2\in\{1,2\};j_1\neq j_2.
\end{aligned}\tag{3.199}$$

After the appearing bifurcation, the post-bifurcation system is

$$\begin{aligned}
&\dot{x}_{j_1}=a_{j_1j_2 0}(x_{j_2}-a_{j_1j_2 1})(x_{j_2}-a_{j_1j_2 2})(x_{j_2}-a_{j_1j_2 3}),\\
&\dot{x}_{j_2}=a_{j_2j_1 0}(x_{j_1}-a_{j_2j_1 1})(x_{j_1}-a_{j_2j_1 2}),\\
&j_1,j_2\in\{1,2\};j_1\neq j_2.
\end{aligned}\tag{3.200}$$

The bifurcation routes from a parabola flow to a saddle and center network are presented as follows:

(i) For $a_{j_1 j_2 0} > 0$ and $a_{j_2 j_1 0} > 0$, the (3rdUP,pF)-flow appearing bifurcation and the up-parabola and down-parabola upper-saddle bifurcations are

$$\underbrace{(\text{UP}, \text{pF})}_{\text{up-parabola flow } (+)} \rightleftharpoons \underbrace{(3^{\text{rd}}\text{UP}, \text{pF})}_{\text{appearing bifurcation}} \rightleftharpoons \left\{ \begin{array}{c} \underbrace{(\text{UP}, \text{pF})}_{\text{up-parabola flow } (+)} \\ \underbrace{(\text{DP}, \text{pF})}_{\text{down-parabola flow } (+)} \\ \underbrace{(\text{UP}, \text{pF})}_{\text{up-parabola flow } (+)} \end{array} \right\}, \tag{3.201}$$

$$\underbrace{(\text{UP}, \text{pF})}_{\text{up-parabola flow } (+)} \rightleftharpoons \left\{ \begin{array}{c} \underbrace{(\text{II}, \text{pF})}_{\text{appearing bifurcation}} \\ \underbrace{(\text{UP}, \text{pF})}_{\text{up-parabola flow } (+)} \end{array} \right\} \rightleftharpoons \left\{ \begin{array}{c} \underbrace{(\text{UP}, \text{pF})}_{\text{up-parabola flow } (+)} \\ \underbrace{(\text{DP}, \text{pF})}_{\text{down-parabola flow } (+)} \\ \underbrace{(\text{UP}, \text{pF})}_{\text{up-parabola flow } (+)} \end{array} \right\}, \tag{3.202}$$

$$\underbrace{(\text{UP}, \text{pF})}_{\text{up-parabola flow } (+)} \rightleftharpoons \left\{ \begin{array}{c} \underbrace{(\text{UP}, \text{pF})}_{\text{up-parabola flow } (+)} \\ \underbrace{(\text{DI}, \text{pF})}_{\text{appearing bifurcation}} \end{array} \right\} \rightleftharpoons \left\{ \begin{array}{c} \underbrace{(\text{UP}, \text{pF})}_{\text{up-parabola flow } (+)} \\ \underbrace{(\text{DP}, \text{pF})}_{\text{down-parabola flow } (+)} \\ \underbrace{(\text{UP}, \text{pF})}_{\text{up-parabola flow } (+)} \end{array} \right\}, \tag{3.203}$$

with

$$\left\{ \begin{array}{c} \underbrace{(\text{UP}, \text{pF})}_{\text{up-parabola flow } (+)} \\ \underbrace{(\text{DP}, \text{pF})}_{\text{down-parabola flow } (+)} \\ \underbrace{(\text{UP}, \text{pF})}_{\text{up-parabola flow } (+)} \end{array} \right\} \rightleftharpoons \left\{ \begin{array}{c} \underbrace{(\text{UP}, \text{US})}_{\text{appearing bifurcation}} \\ \underbrace{(\text{DP}, \text{US})}_{\text{appearing bifurcation}} \\ \underbrace{(\text{UP}, \text{US})}_{\text{appearing bifurcation}} \end{array} \right\} \rightleftharpoons \left\{ \begin{array}{cc} \underbrace{(\text{DP}_-, \text{DP}_+)}_{\text{CW center}} & \underbrace{(\text{UP}_+, \text{UP}_+)}_{\text{positive saddle}} \\ \underbrace{(\text{UP}_-, \text{UP}_-)}_{\text{negative saddle}} & \underbrace{(\text{DP}_+, \text{DP}_-)}_{\text{CCW center}} \\ \underbrace{(\text{DP}_-, \text{DP}_+)}_{\text{CW center}} & \underbrace{(\text{UP}_+, \text{UP}_+)}_{\text{positive saddle}} \end{array} \right\}. \tag{3.204}$$

(ii) For $a_{j_1 j_2 0} < 0$ and $a_{j_2 j_1 0} > 0$, the (3rdDP,pF)-flow appearing bifurcation and the up-parabola and down-parabola upper-saddle appearing bifurcations are

$$\underbrace{(\mathrm{DP},\mathrm{pF})}_{\text{down-parabola flow }(+)} \rightleftharpoons \underbrace{(3^{\mathrm{rd}}\mathrm{DP},\mathrm{pF})}_{\text{appearing bifurcation}} \rightleftharpoons \left\{\begin{array}{c} \underbrace{(\mathrm{DP},\mathrm{pF})}_{\text{down-parabola flow }(+)} \\ \underbrace{(\mathrm{UP},\mathrm{pF})}_{\text{up-parabola flow }(+)} \\ \underbrace{(\mathrm{DP},\mathrm{pF})}_{\text{down-parabola flow }(+)} \end{array}\right\}, \tag{3.205}$$

$$\underbrace{(\mathrm{DP},\mathrm{pF})}_{\text{down-parabola flow }(+)} \rightleftharpoons \left\{\begin{array}{c} \underbrace{(\mathrm{DI},\mathrm{pF})}_{\text{appearing bifurcation}} \\ \underbrace{(\mathrm{DP},\mathrm{pF})}_{\text{down-parabola flow }(+)} \end{array}\right\} \rightleftharpoons \left\{\begin{array}{c} \underbrace{(\mathrm{DP},\mathrm{pF})}_{\text{down-parabola flow }(+)} \\ \underbrace{(\mathrm{UP},\mathrm{pF})}_{\text{up-parabola flow }(+)} \\ \underbrace{(\mathrm{DP},\mathrm{pF})}_{\text{down-parabola flow }(+)} \end{array}\right\}, \tag{3.206}$$

$$\underbrace{(\mathrm{DP},\mathrm{pF})}_{\text{down-parabola flow }(+)} \rightleftharpoons \left\{\begin{array}{c} \underbrace{(\mathrm{DP},\mathrm{pF})}_{\text{down-parabola flow }(+)} \\ \underbrace{(\mathrm{II},\mathrm{pF})}_{\text{appearing bifurcation}} \end{array}\right\} \rightleftharpoons \left\{\begin{array}{c} \underbrace{(\mathrm{DP},\mathrm{pF})}_{\text{down-parabola flow }(+)} \\ \underbrace{(\mathrm{UP},\mathrm{pF})}_{\text{up-parabola flow }(+)} \\ \underbrace{(\mathrm{DP},\mathrm{pF})}_{\text{down-parabola flow }(+)} \end{array}\right\}, \tag{3.207}$$

with

$$\left\{\begin{array}{c} \underbrace{(\mathrm{DP},\mathrm{pF})}_{\text{down-parabola flow }(+)} \\ \underbrace{(\mathrm{UP},\mathrm{pF})}_{\text{up-parabola flow }(+)} \\ \underbrace{(\mathrm{DP},\mathrm{pF})}_{\text{down-parabola flow }(+)} \end{array}\right\} \rightleftharpoons \left\{\begin{array}{c} \underbrace{(\mathrm{DP},\mathrm{US})}_{\text{appearing bifurcation}} \\ \underbrace{(\mathrm{UP},\mathrm{US})}_{\text{appearing bifurcation}} \\ \underbrace{(\mathrm{DP},\mathrm{US})}_{\text{appearing bifurcation}} \end{array}\right\} \rightleftharpoons \left\{\begin{array}{cc} \underbrace{(\mathrm{UP}_-,\mathrm{UP}_-)}_{\text{negative saddle}} & \underbrace{(\mathrm{DP}_+,\mathrm{DP}_-)}_{\text{CCW center}} \\ \underbrace{(\mathrm{DP}_-,\mathrm{DP}_+)}_{\text{CW center}} & \underbrace{(\mathrm{UP}_+,\mathrm{UP}_+)}_{\text{positive saddle}} \\ \underbrace{(\mathrm{UP}_-,\mathrm{UP}_-)}_{\text{negative saddle}} & \underbrace{(\mathrm{DP}_+,\mathrm{DP}_-)}_{\text{CCW center}} \end{array}\right\}. \tag{3.208}$$

(iii) For $a_{j_1j_20} > 0$ and $a_{j_2j_10} < 0$, the (3^{rd}DP,nF)-flow appearing bifurcation and the up-parabola and down-parabola lower-saddle appearing bifurcations are

$$\underbrace{(\mathrm{DP},\mathrm{nF})}_{\text{down-parabola flow }(-)} \rightleftharpoons \underbrace{(3^{\mathrm{rd}}\mathrm{DP},\mathrm{nF})}_{\text{appearing bifurcation}} \rightleftharpoons \left\{\begin{array}{c} \underbrace{(\mathrm{DP},\mathrm{nF})}_{\text{down-parabola flow }(-)} \\ \underbrace{(\mathrm{UP},\mathrm{nF})}_{\text{up-parabola flow }(-)} \\ \underbrace{(\mathrm{DP},\mathrm{nF})}_{\text{down-parabola flow }(-)} \end{array}\right\}, \tag{3.209}$$

$$\underbrace{(\mathrm{DP,nF})}_{\text{down-parabola flow }(-)} \rightleftharpoons \left\{ \begin{array}{c} \underbrace{(\mathrm{DI,nF})}_{\text{appearing bifurcation}} \\ \underbrace{(\mathrm{DP,nF})}_{\text{down-parabola flow }(-)} \end{array} \right\} \rightleftharpoons \left\{ \begin{array}{c} \underbrace{(\mathrm{DP,nF})}_{\text{down-parabola flow }(-)} \\ \underbrace{(\mathrm{UP,nF})}_{\text{up-parabola flow }(-)} \\ \underbrace{(\mathrm{DP,nF})}_{\text{down-parabola flow }(-)} \end{array} \right\}, \tag{3.210}$$

$$\underbrace{(\mathrm{DP,nF})}_{\text{down-parabola flow }(-)} \rightleftharpoons \left\{ \begin{array}{c} \underbrace{(\mathrm{DP,nF})}_{\text{down-parabola flow }(-)} \\ \underbrace{(\mathrm{II,nF})}_{\text{appearing bifurcation}} \end{array} \right\} \rightleftharpoons \left\{ \begin{array}{c} \underbrace{(\mathrm{DP,nF})}_{\text{down-parabola flow }(-)} \\ \underbrace{(\mathrm{UP,nF})}_{\text{up-parabola flow }(-)} \\ \underbrace{(\mathrm{DP,nF})}_{\text{down-parabola flow }(-)} \end{array} \right\}, \tag{3.211}$$

with

$$\left\{ \begin{array}{c} \underbrace{(\mathrm{DP,nF})}_{\text{down-parabola flow }(-)} \\ \underbrace{(\mathrm{UP,nF})}_{\text{up-parabola flow }(-)} \\ \underbrace{(\mathrm{DP,nF})}_{\text{down-parabola flow }(-)} \end{array} \right\} \rightleftharpoons \left\{ \begin{array}{c} \underbrace{(\mathrm{DP,LS})}_{\text{appearing bifurcation}} \\ \underbrace{(\mathrm{UP,LS})}_{\text{appearing bifurcation}} \\ \underbrace{(\mathrm{DP,LS})}_{\text{appearing bifurcation}} \end{array} \right\} \rightleftharpoons \left\{ \begin{array}{cc} \underbrace{(\mathrm{UP}_+,\mathrm{UP}_+)}_{\text{positive saddle}} & \underbrace{(\mathrm{DP}_-,\mathrm{DP}_+)}_{\text{CW center}} \\ \underbrace{(\mathrm{DP}_+,\mathrm{DP}_-)}_{\text{CCW center}} & \underbrace{(\mathrm{UP}_-,\mathrm{UP}_-)}_{\text{negative saddle}} \\ \underbrace{(\mathrm{UP}_+,\mathrm{UP}_+)}_{\text{positive saddle}} & \underbrace{(\mathrm{DP}_-,\mathrm{DP}_+)}_{\text{CW center}} \end{array} \right\}. \tag{3.212}$$

(iv) For $a_{j_1j_20} < 0$ and $a_{j_2j_10} < 0$, the (3rdUP,nF)-flow appearing bifurcation and the up-parabola and down-parabola lower-saddle appearing bifurcations are

$$\underbrace{(\mathrm{UP,nF})}_{\text{up-parabola flow }(-)} \rightleftharpoons \underbrace{(3^{\text{rd}}\mathrm{UP,nF})}_{\text{appearing bifurcation}} \rightleftharpoons \left\{ \begin{array}{c} \underbrace{(\mathrm{UP,nF})}_{\text{up-parabola flow }(-)} \\ \underbrace{(\mathrm{DP,nF})}_{\text{down-parabola flow }(-)} \\ \underbrace{(\mathrm{UP,nF})}_{\text{up-parabola flow }(-)} \end{array} \right\}, \tag{3.213}$$

$$\underbrace{(\mathrm{UP,nF})}_{\text{up-parabola flow }(-)} \rightleftharpoons \left\{ \begin{array}{c} \underbrace{(\mathrm{II,nF})}_{\text{appearing bifurcation}} \\ \underbrace{(\mathrm{UP,nF})}_{\text{up-parabola flow }(-)} \end{array} \right\} \rightleftharpoons \left\{ \begin{array}{c} \underbrace{(\mathrm{UP,nF})}_{\text{up-parabola flow }(-)} \\ \underbrace{(\mathrm{DP,nF})}_{\text{down-parabola flow }(-)} \\ \underbrace{(\mathrm{UP,nF})}_{\text{up-parabola flow }(-)} \end{array} \right\}, \tag{3.214}$$

$$\underbrace{(\mathrm{UP,nF})}_{\text{up-parabola flow }(-)} \rightleftharpoons \left\{ \begin{array}{c} \underbrace{(\mathrm{UP,nF})}_{\text{up-parabola flow }(-)} \\ \underbrace{(\mathrm{DI,nF})}_{\text{appearing bifurcation}} \end{array} \right\} \rightleftharpoons \left\{ \begin{array}{c} \underbrace{(\mathrm{UP,nF})}_{\text{up-parabola flow }(-)} \\ \underbrace{(\mathrm{DP,nF})}_{\text{down-parabola flow }(-)} \\ \underbrace{(\mathrm{UP,nF})}_{\text{up-parabola flow }(-)} \end{array} \right\}, \tag{3.215}$$

with

$$\left\{ \begin{array}{c} \underbrace{(\mathrm{UP,nF})}_{\text{up-parabola flow }(-)} \\ \underbrace{(\mathrm{DP,nF})}_{\text{down-parabola flow }(-)} \\ \underbrace{(\mathrm{UP,nF})}_{\text{up-parabola flow }(-)} \end{array} \right\} \rightleftharpoons \left\{ \begin{array}{c} \underbrace{(\mathrm{UP,LS})}_{\text{appearing bifurcation}} \\ \underbrace{(\mathrm{DP,LS})}_{\text{appearing bifurcation}} \\ \underbrace{(\mathrm{UP,LS})}_{\text{appearing bifurcation}} \end{array} \right\} \rightleftharpoons \left\{ \begin{array}{cc} \underbrace{(\mathrm{DP}_+,\mathrm{DP}_-)}_{\text{CCW center}} & \underbrace{(\mathrm{UP}_-,\mathrm{UP}_-)}_{\text{negative saddle}} \\ \underbrace{(\mathrm{UP}_+,\mathrm{UP}_+)}_{\text{positive saddle}} & \underbrace{(\mathrm{DP}_-,\mathrm{DP}_+)}_{\text{CW center}} \\ \underbrace{(\mathrm{DP}_+,\mathrm{DP}_-)}_{\text{CCW center}} & \underbrace{(\mathrm{UP}_-,\mathrm{UP}_-)}_{\text{negative saddle}} \end{array} \right\}. \tag{3.216}$$

In Figs. 3.9 and 3.10, the bifurcation routes are from parabola flows to saddle and center networks.

In Fig. 3.9, the bifurcation routes from parabola flows to saddle and center networks for $x_1^* = a_{j_2j_1 1}, a_{j_2j_1 2}$ and $x_{j_2}^* = a_{j_1j_2 1}, a_{j_1j_2 2}, a_{j_1j_2 3}$ are presented. The bifurcations are the (3^{rd}DP,pF)-positive third-order up-parabola flow appearing bifurcation with parabola upper-saddle and lower-saddle appearing bifurcations ($a_{j_1j_2 0} > 0$, $a_{j_2j_1 0} > 0$); the (3^{rd}DP,pF)-positive third-order down-parabola flow appearing bifurcation with parabola upper-saddle and lower-saddle appearing bifurcations ($a_{j_1j_2 0} < 0$, $a_{j_2j_1 0} > 0$); the (3^{rd}DP,nF)-negative third-order down-parabola flow appearing bifurcation with parabola upper-saddle and lower-saddle appearing bifurcations ($a_{j_1j_2 0} > 0$, $a_{j_2j_1 0} < 0$); and the (3^{rd}UP, nF)-negative third-order up-parabola flow appearing bifurcation with parabola upper-saddle and lower-saddle appearing bifurcations ($a_{j_1j_2 0} < 0$, $a_{j_2j_1 0} < 0$).

In Fig. 3.10a–d, the bifurcation routes from parabola flows to saddle and center networks for $x_1^* = a_{j_2j_1 1}, a_{j_2j_1 2}$ and $x_{j_2}^* = a_{j_1j_2 1}, a_{j_1j_2 2}, a_{j_1j_2 3}$ are presented through inflection flow appearing bifurcation and parabola upper-saddle and lower-saddle bifurcations. They are the (II,pF)-positive increasing-inflection flow appearing bifurcation with parabola upper-saddle and lower-saddle appearing bifurcations ($a_{j_1j_2 0} > 0$, $a_{j_2j_1 0} > 0$); the (DI,pF)-positive decreasing-inflection flow appearing bifurcation with parabola upper-saddle and lower-saddle appearing bifurcations ($a_{j_1j_2 0} > 0$, $a_{j_2j_1 0} < 0$); the (DI,nF)-negative decreasing-inflection flow appearing bifurcation with parabola upper-saddle and lower-saddle appearing bifurcations ($a_{j_1j_2 0} > 0$, $a_{j_2j_1 0} < 0$); and the (II,nF)-negative increasing-inflection flow appearing

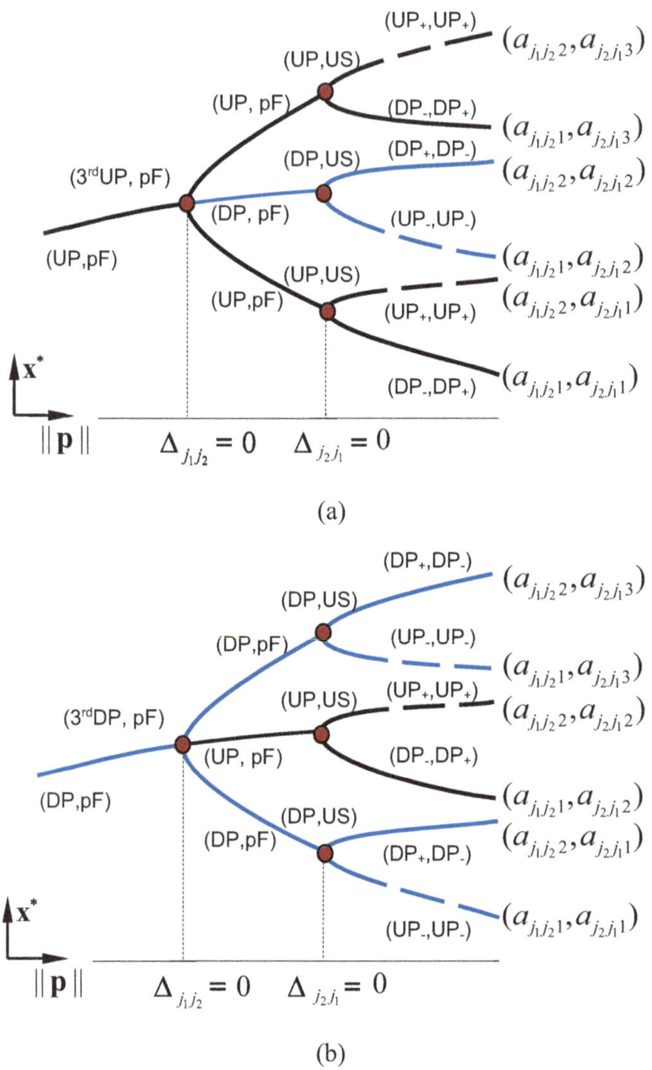

Fig. 3.9 The bifurcation routes from parabola flows to saddle and center networks for $x_1^* = a_{j_2j_1 1}$, $a_{j_2j_1 2}$ and $x_{j_2}^* = a_{j_1j_2 1}, a_{j_1j_2 2}, a_{j_1j_2 3}$: (**a**) (3rdUP,pF)-appearing bifurcation ($a_{j_1j_2 0} > 0,\ \ a_{j_2j_1 0} > 0$), (**b**) (3rdDP,pF)-appearing bifurcation ($a_{j_1j_2 0} > 0,\ \ a_{j_2j_1 0} < 0$), (**c**) (3rdDP,nF)-appearing bifurcation ($a_{j_1j_2 0} > 0,\ \ a_{j_2j_1 0} < 0$), (**d**) (3rdUP, nF)-appearing bifurcation ($a_{j_1j_2 0} < 0,\ \ a_{j_2j_1 0} < 0$). 3rdUP: third-order up-parabola flow, 3rdDP: third-order down-parabola flow. *pF* positive flow, *nF* negative flow. *US* upper-saddle, *LS* lower-saddle, *DP* down-parabola, *UP* up-parabola

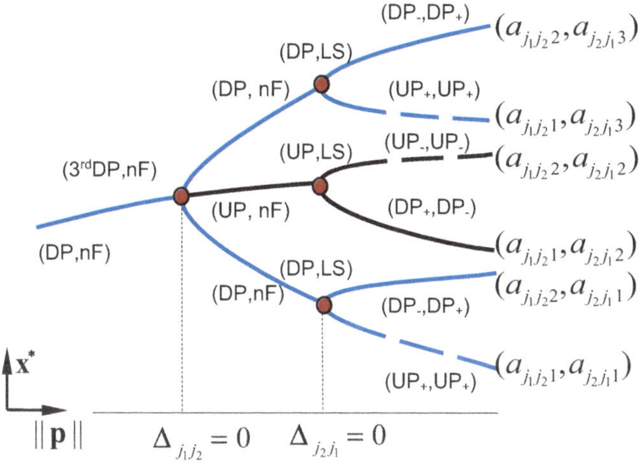

(c)

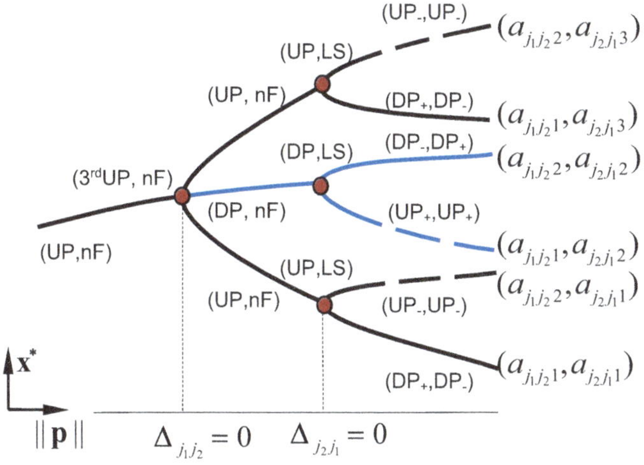

(d)

Fig. 3.9 (continued)

bifurcation with parabola upper-saddle and lower-saddle appearing bifurcations $(a_{j_1j_2 0} < 0,\ a_{j_2j_1 0} < 0)$.

Similar to Fig. 3.10, the bifurcation routes from parabola flows to saddle and center networks for $x_1^* = a_{j_2j_1 1}, a_{j_2j_1 2}$ and $x_{j_2}^* = a_{j_1j_2 1}, a_{j_1j_2 2}, a_{j_1j_2 3}$ can be presented through another set of inflection flow appearing bifurcation and parabola upper-saddle and lower-saddle bifurcations. The readers can be done as an exercise.

Consider a standard form of the quadratic dynamical system as

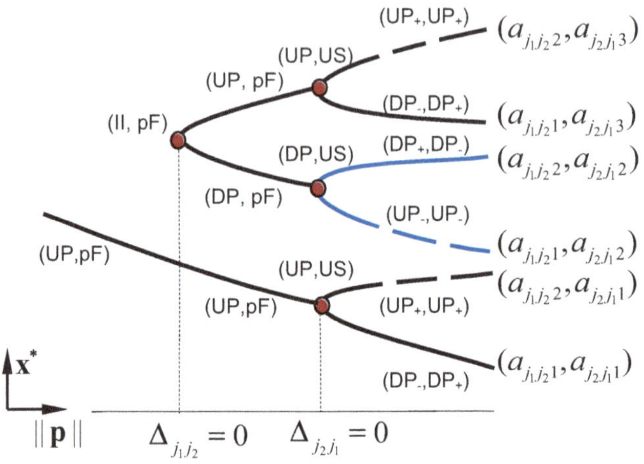

(a)

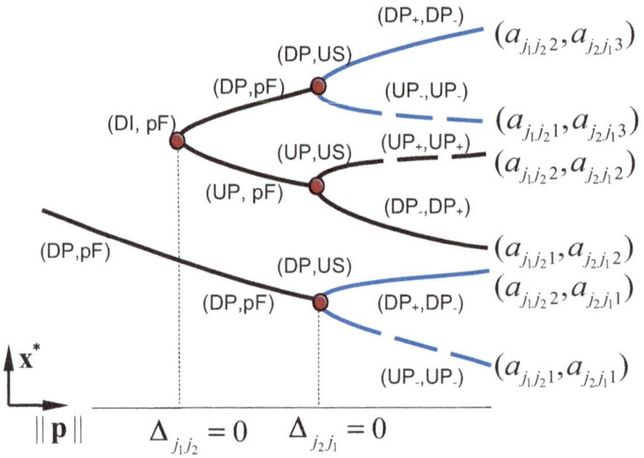

(b)

Fig. 3.10 The bifurcation routes from parabola flows to saddles and centers networks for $x_1^* = a_{j_2j_1 1}$, $a_{j_2j_1 2}, a_{j_2j_1 3}$ and $x_{j_2}^* = a_{j_1j_2 1}, a_{j_1j_2 2}$: (**a**) (II,pF)-appearing bifurcation ($a_{j_1j_2 0} > 0,\ a_{j_2j_1 0} > 0$), (**b**) (DI,pF)-appearing bifurcation ($a_{j_1j_2 0} > 0,\ a_{j_2j_1 0} < 0$), (**c**) (DI,nF)-appearing bifurcation ($a_{j_1j_2 0} > 0,\ a_{j_2j_1 0} < 0$), (**d**) (II,nF)-appearing bifurcation ($a_{j_1j_2 0} < 0,\ \ a_{j_2j_1 0} < 0$). DI: decreasing inflection, II: increasing inflection. pF: positive flow, nF: negative flow, US: upper-saddle, LS: lower-saddle, DP: down-parabola, UP: up-parabola

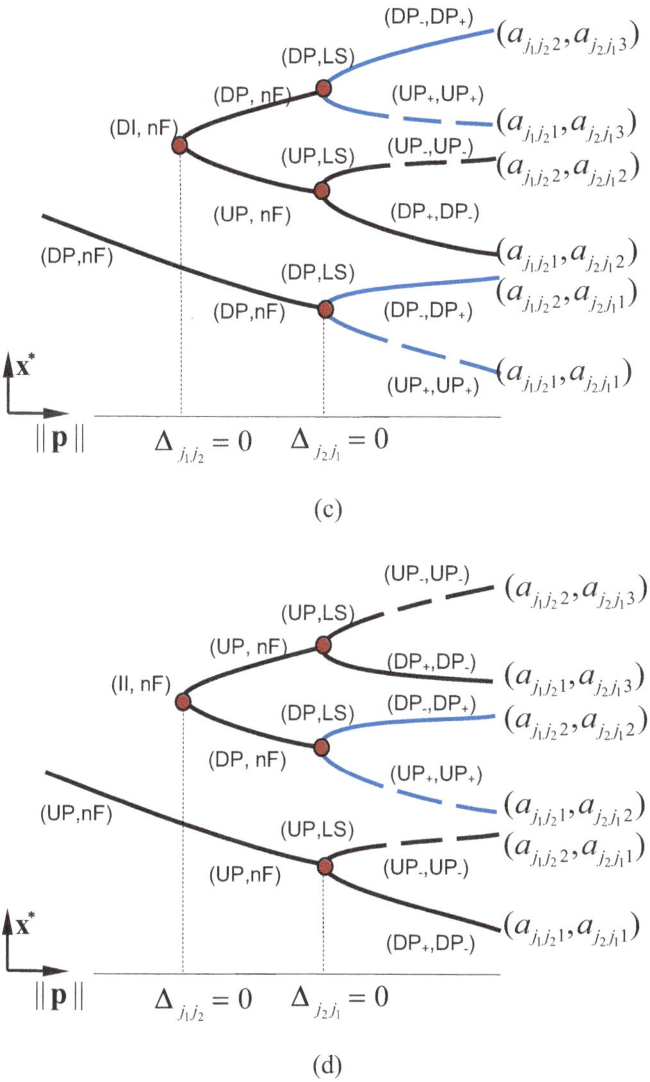

Fig. 3.10 (continued)

$$
\begin{aligned}
&\dot{x}_{j_1} = a_{j_1j_2 0}(x_{j_2} - b_{j_1j_2 1})\left[(x_{j_1} - \frac{1}{2}B_{j_1j_2})^2 + \frac{1}{4}(-\Delta_{j_1j_2})\right],\\
&\dot{x}_{j_2} = a_{j_2j_1 0}\left[(x_{j_1} - \frac{1}{2}B_{j_2j_1})^2 + \frac{1}{4}(-\Delta_{j_2j_1})\right],\\
&j_1, j_2 \in \{1,2\}; j_1 \neq j_2.
\end{aligned}
\tag{3.217}
$$

With $\Delta_{j_1j_2} = 0, a_{j_2j_1 1} = \frac{1}{2}B_{j_2j_1}, a_{j_1j_2 1} = b_{j_1j_2 1}$, the flow appearing bifurcation system is

$$\begin{aligned}
&\dot{x}_{j_1} = a_{j_1 j_2 0}(x_{j_2} - a_{j_1 j_2 1})\big[(x_{j_2} - \frac{1}{2}B_{j_1 j_2})^2 + \frac{1}{4}(-\Delta_{j_1 j_2})\big],\\
&\dot{x}_{j_2} = a_{j_2 j_1 0}(x_{j_1} - a_{j_2 j_1 1})^2,\\
&j_1, j_2 \in \{1,2\}; j_1 \neq j_2.
\end{aligned} \tag{3.218}$$

After the appearing bifurcation, the post-bifurcation system is

$$\begin{aligned}
&\dot{x}_{j_1} = a_{j_1 j_2 0}(x_{j_2} - a_{j_1 j_2 1})\big[(x_{j_2} - \frac{1}{2}B_{j_1 j_2})^2 + \frac{1}{4}(-\Delta_{j_1 j_2})\big],\\
&\dot{x}_{j_2} = a_{j_2 j_1 0}(x_{j_1} - a_{j_2 j_1 1})(x_{j_2} - a_{j_2 j_1 2}),\\
&j_1, j_2 \in \{1,2\}; j_1 \neq j_2.
\end{aligned} \tag{3.219}$$

(A) With $\Delta_{j_2 j_1} = 0$ and $a_{j_1 j_2 1} = b_{j_1 j_2 1} = \frac{1}{2}B_{j_1 j_2}$, the equilibrium bifurcation system is

$$\begin{aligned}
&\dot{x}_{j_1} = a_{j_1 j_2 0}(x_{j_2} - a_{j_1 j_2 1})^3,\\
&\dot{x}_{j_2} = a_{j_2 j_1 0}(x_{j_1} - a_{j_2 j_1 1})(x_{j_1} - a_{j_2 j_1 2}),\\
&j_1, j_2 \in \{1,2\}; j_1 \neq j_2.
\end{aligned} \tag{3.220}$$

(B) With $\Delta_{j_2 j_1} = 0$ and $a_{j_1 j_2 1} = b_{j_1 j_2 1} < \frac{1}{2}B_{j_1 j_2} < a_{j_1 j_2 2}$, the equilibrium bifurcation system is

$$\begin{aligned}
&\dot{x}_{j_1} = a_{j_1 j_2 0}(x_{j_2} - a_{j_1 j_2 1})(x_{j_2} - a_{j_1 j_2 2})^2,\\
&\dot{x}_{j_2} = a_{j_2 j_1 0}(x_{j_1} - a_{j_2 j_1 1})(x_{j_1} - a_{j_2 j_1 2}),\\
&j_1, j_2 \in \{1,2\}; j_1 \neq j_2.
\end{aligned} \tag{3.221}$$

(C) With $\Delta_{j_2 j_1} = 0$ and $a_{j_1 j_2 1} = b_{j_1 j_2 2} > \frac{1}{2}B_{j_1 j_2} = b_{j_1 j_2 1}$, the equilibrium bifurcation system is

$$\begin{aligned}
&\dot{x}_{j_1} = a_{j_1 j_2 0}(x_{j_2} - a_{j_1 j_2 1})^2(x_{j_2} - a_{j_1 j_2 2}),\\
&\dot{x}_{j_2} = a_{j_2 j_1 0}(x_{j_1} - a_{j_2 j_1 1})(x_{j_1} - a_{j_2 j_1 2}),\\
&j_1, j_2 \in \{1,2\}; j_1 \neq j_2.
\end{aligned} \tag{3.222}$$

After the appearing bifurcation, the post-bifurcation system is

$$\begin{aligned}
&\dot{x}_{j_1} = a_{j_1 j_2 0}(x_{j_2} - a_{j_1 j_2 1})(x_{j_2} - a_{j_1 j_2 2})(x_{j_2} - a_{j_1 j_2 3}),\\
&\dot{x}_{j_2} = a_{j_2 j_1 0}(x_{j_1} - a_{j_2 j_1 1})(x_{j_1} - a_{j_2 j_1 2}),\\
&j_1, j_2 \in \{1,2\}; j_1 \neq j_2.
\end{aligned} \tag{3.223}$$

The bifurcation routes from a parabola flow to a saddle and center network are presented as follows:

(i) For $a_{j_1 j_2 0} > 0$ and $a_{j_2 j_1 0} > 0$, the (UP,US)-parabola-saddle bifurcations is

$$(\mathrm{UP},\mathrm{pF}) \rightleftharpoons \underbrace{(\mathrm{UP},\mathrm{US})}_{\text{appearing bifurcation}} \rightleftharpoons \{\underbrace{(\mathrm{DP}_-,\mathrm{DP}_+)}_{\text{CW center}},\underbrace{(\mathrm{UP}_+,\mathrm{UP}_+)}_{\text{positive saddle}}\}, \tag{3.224}$$

with three cases for the further bifurcations as follows:

$$\begin{aligned}
\{\underbrace{(\mathrm{DP}_-,\mathrm{DP}_+)}_{\text{CW center}},\underbrace{(\mathrm{UP}_+,\mathrm{UP}_+)}_{\text{positive saddle}}\} &\rightleftharpoons \{\underbrace{(3^{\mathrm{rd}}\mathrm{DP}_-,\mathrm{DP}_+)}_{\text{third-order CW center}},\underbrace{(3^{\mathrm{rd}}\mathrm{UP}_+,\mathrm{UP}_+)}_{\text{third-order positive saddle}}\}\\
&\rightleftharpoons \left\{\begin{array}{cc}
\underbrace{(\mathrm{DP}_-,\mathrm{DP}_+)}_{\text{CW center}} & \underbrace{(\mathrm{UP}_+,\mathrm{UP}_+)}_{\text{positive saddle}}\\
\underbrace{(\mathrm{UP}_-,\mathrm{UP}_-)}_{\text{negative saddle}} & \underbrace{(\mathrm{DP}_+,\mathrm{DP}_-)}_{\text{CCW center}}\\
\underbrace{(\mathrm{DP}_-,\mathrm{DP}_+)}_{\text{CW center}} & \underbrace{(\mathrm{UP}_+,\mathrm{UP}_+)}_{\text{positive saddle}}
\end{array}\right\}.
\end{aligned} \tag{3.225}$$

$$\begin{aligned}
\{\underbrace{(\mathrm{DP}_-,\mathrm{DP}_+)}_{\text{CW center}},\underbrace{(\mathrm{UP}_+,\mathrm{UP}_+)}_{\text{positive saddle}}\} &\rightleftharpoons \left\{\begin{array}{cc}
\underbrace{(\mathrm{US},\mathrm{DP})}_{\text{appearing bifurcation}} & \underbrace{(\mathrm{US},\mathrm{UP})}_{\text{appearing bifurcation}}\\
\underbrace{(\mathrm{DP}_-,\mathrm{DP}_+)}_{\text{CW center}} & \underbrace{(\mathrm{UP}_+,\mathrm{UP}_+)}_{\text{positive saddle}}
\end{array}\right\}\\
&\rightleftharpoons \left\{\begin{array}{cc}
\underbrace{(\mathrm{DP}_-,\mathrm{DP}_+)}_{\text{CW center}} & \underbrace{(\mathrm{UP}_+,\mathrm{UP}_+)}_{\text{positive saddle}}\\
\underbrace{(\mathrm{UP}_-,\mathrm{UP}_-)}_{\text{negative saddle}} & \underbrace{(\mathrm{DP}_+,\mathrm{DP}_-)}_{\text{CCW center}}\\
\underbrace{(\mathrm{DP}_-,\mathrm{DP}_+)}_{\text{CW center}} & \underbrace{(\mathrm{UP}_+,\mathrm{UP}_+)}_{\text{positive saddle}}
\end{array}\right\}.
\end{aligned} \tag{3.226}$$

$$\begin{aligned}
\{\underbrace{(\mathrm{DP}_-,\mathrm{DP}_+)}_{\text{CW center}},\underbrace{(\mathrm{UP}_+,\mathrm{UP}_+)}_{\text{positive saddle}}\} &\rightleftharpoons \left\{\begin{array}{cc}
\underbrace{(\mathrm{DP}_-,\mathrm{DP}_+)}_{\text{CW center}} & \underbrace{(\mathrm{UP}_+,\mathrm{UP}_+)}_{\text{positive saddle}}\\
\underbrace{(\mathrm{LS},\mathrm{UP})}_{\text{appearing bifurcation}} & \underbrace{(\mathrm{LS},\mathrm{DP})}_{\text{appearing bifurcation}}
\end{array}\right\}\\
&\rightleftharpoons \left\{\begin{array}{cc}
\underbrace{(\mathrm{DP}_-,\mathrm{DP}_+)}_{\text{CW center}} & \underbrace{(\mathrm{UP}_+,\mathrm{UP}_+)}_{\text{positive saddle}}\\
\underbrace{(\mathrm{UP}_-,\mathrm{UP}_-)}_{\text{negative saddle}} & \underbrace{(\mathrm{DP}_+,\mathrm{DP}_-)}_{\text{CCW center}}\\
\underbrace{(\mathrm{DP}_-,\mathrm{DP}_+)}_{\text{CW center}} & \underbrace{(\mathrm{UP}_+,\mathrm{UP}_+)}_{\text{positive saddle}}
\end{array}\right\}.
\end{aligned} \tag{3.227}$$

(ii) For $a_{j_1j_20}<0$ and $a_{j_2j_10}>0$, the (DP,US)-parabola-saddle bifurcation is

$$(\mathrm{DP},\mathrm{pF}) \rightleftharpoons \underbrace{(\mathrm{DP},\mathrm{US})}_{\text{appearing bifurcation}} \rightleftharpoons \{\underbrace{(\mathrm{UP}_-,\mathrm{UP}_-)}_{\text{negative saddle}},\underbrace{(\mathrm{DP}_+,\mathrm{DP}_-)}_{\text{CCW center}}\}, \tag{3.228}$$

with three cases for the further bifurcations as follows:

$$\begin{aligned}
\{\underbrace{(\mathrm{UP}_-,\mathrm{UP}_-)}_{\text{negative saddle}},\underbrace{(\mathrm{DP}_+,\mathrm{DP}_-)}_{\text{CCW center}}\} &\rightleftharpoons \{\underbrace{(3^{\mathrm{rd}}\mathrm{UP}_-,\mathrm{UP}_-)}_{\text{third-order negative saddle}},\underbrace{(3^{\mathrm{rd}}\mathrm{DP}_+,\mathrm{DP}_-)}_{\text{CCW center}}\}\\
&\rightleftharpoons \left\{\begin{array}{cc}
\underbrace{(\mathrm{UP}_-,\mathrm{UP}_-)}_{\text{negative saddle}} & \underbrace{(\mathrm{DP}_+,\mathrm{DP}_-)}_{\text{CCW center}}\\
\underbrace{(\mathrm{DP}_-,\mathrm{DP}_+)}_{\text{CW center}} & \underbrace{(\mathrm{UP}_+,\mathrm{UP}_+)}_{\text{positive saddle}}\\
\underbrace{(\mathrm{UP}_-,\mathrm{UP}_-)}_{\text{negative saddle}} & \underbrace{(\mathrm{DP}_+,\mathrm{DP}_-)}_{\text{CCW center}}
\end{array}\right\}.
\end{aligned} \tag{3.229}$$

$$\begin{aligned}
\{\underbrace{(\mathrm{UP}_-,\mathrm{UP}_-)}_{\text{negative saddle}},\underbrace{(\mathrm{DP}_+,\mathrm{DP}_-)}_{\text{CCW center}}\} &\rightleftharpoons \left\{\begin{array}{cc}
\underbrace{(\mathrm{LS},\mathrm{UP})}_{\text{appearing bifurcation}} & \underbrace{(\mathrm{LS},\mathrm{DP})}_{\text{appearing bifurcation}}\\
\underbrace{(\mathrm{UP}_-,\mathrm{UP}_-)}_{\text{negative saddle}} & \underbrace{(\mathrm{DP}_+,\mathrm{DP}_-)}_{\text{CCW center}}
\end{array}\right\}\\
&\rightleftharpoons \left\{\begin{array}{cc}
\underbrace{(\mathrm{UP}_-,\mathrm{UP}_-)}_{\text{negative saddle}} & \underbrace{(\mathrm{DP}_+,\mathrm{DP}_-)}_{\text{CCW center}}\\
\underbrace{(\mathrm{DP}_-,\mathrm{DP}_+)}_{\text{CW center}} & \underbrace{(\mathrm{UP}_+,\mathrm{UP}_+)}_{\text{positive saddle}}\\
\underbrace{(\mathrm{UP}_-,\mathrm{UP}_-)}_{\text{negative saddle}} & \underbrace{(\mathrm{DP}_+,\mathrm{DP}_-)}_{\text{CCW center}}
\end{array}\right\}.
\end{aligned} \tag{3.230}$$

$$\begin{aligned}
\{\underbrace{(\mathrm{UP}_-,\mathrm{UP}_-)}_{\text{negative saddle}},\underbrace{(\mathrm{DP}_+,\mathrm{DP}_-)}_{\text{CCW center}}\} &\rightleftharpoons \left\{\begin{array}{cc}
\underbrace{(\mathrm{UP}_-,\mathrm{UP}_-)}_{\text{negative saddle}} & \underbrace{(\mathrm{DP}_+,\mathrm{DP}_-)}_{\text{CCW center}}\\
\underbrace{(\mathrm{US},\mathrm{DP})}_{\text{appearing bifurcation}} & \underbrace{(\mathrm{US},\mathrm{UP})}_{\text{appearing bifurcation}}
\end{array}\right\}\\
&\rightleftharpoons \left\{\begin{array}{cc}
\underbrace{(\mathrm{UP}_-,\mathrm{UP}_-)}_{\text{negative saddle}} & \underbrace{(\mathrm{DP}_+,\mathrm{DP}_-)}_{\text{CCW center}}\\
\underbrace{(\mathrm{DP}_-,\mathrm{DP}_+)}_{\text{CW center}} & \underbrace{(\mathrm{UP}_+,\mathrm{UP}_+)}_{\text{positive saddle}}\\
\underbrace{(\mathrm{UP}_-,\mathrm{UP}_-)}_{\text{negative saddle}} & \underbrace{(\mathrm{DP}_+,\mathrm{DP}_-)}_{\text{CCW center}}
\end{array}\right\}.
\end{aligned} \tag{3.231}$$

(iii) For $a_{j_1 j_2 0} > 0$ and $a_{j_2 j_1 0} < 0$, the (DP,LS)-parabola-saddle bifurcation is

$$(\mathrm{DP},\mathrm{nF}) \rightleftharpoons \underbrace{(\mathrm{DP},\mathrm{LS})}_{\text{appearing bifurcation}} \rightleftharpoons \{\underbrace{(\mathrm{UP}_+,\mathrm{UP}_+)}_{\text{positive saddle}},\underbrace{(\mathrm{DP}_-,\mathrm{DP}_+)}_{\text{CW center}}\}, \tag{3.232}$$

with three cases for further bifurcation as follows:

$$\begin{aligned}\{\underbrace{(\mathrm{UP}_+,\mathrm{UP}_+)}_{\text{positive saddle}},\underbrace{(\mathrm{DP}_-,\mathrm{DP}_+)}_{\text{CW center}}\} &\rightleftharpoons \{\underbrace{(3^{\mathrm{rd}}\mathrm{UP}_+,\mathrm{UP}_+)}_{\text{third-order positive saddle}},\underbrace{(3^{\mathrm{rd}}\mathrm{DP}_-,\mathrm{DP}_+)}_{\text{third-order CW center}}\}\\ &\rightleftharpoons \left\{\begin{matrix}\underbrace{(\mathrm{UP}_+,\mathrm{UP}_+)}_{\text{positive saddle}} & \underbrace{(\mathrm{DP}_-,\mathrm{DP}_+)}_{\text{CW center}}\\ \underbrace{(\mathrm{DP}_+,\mathrm{DP}_-)}_{\text{CCW center}} & \underbrace{(\mathrm{UP}_-,\mathrm{UP}_-)}_{\text{negative saddle}}\\ \underbrace{(\mathrm{UP}_+,\mathrm{UP}_+)}_{\text{positive saddle}} & \underbrace{(\mathrm{DP}_-,\mathrm{DP}_+)}_{\text{CW center}}\end{matrix}\right\}.\end{aligned} \tag{3.233}$$

$$\begin{aligned}\{\underbrace{(\mathrm{UP}_+,\mathrm{UP}_+)}_{\text{positive saddle}},\underbrace{(\mathrm{DP}_-,\mathrm{DP}_+)}_{\text{CW center}}\} &\rightleftharpoons \left\{\begin{matrix}\underbrace{(\mathrm{US},\mathrm{UP})}_{\text{appearing bifurcation}} & \underbrace{(\mathrm{US},\mathrm{DP})}_{\text{appearing bifurcation}}\\ \underbrace{(\mathrm{UP}_+,\mathrm{UP}_+)}_{\text{positive saddle}} & \underbrace{(\mathrm{DP}_-,\mathrm{DP}_+)}_{\text{CW center}}\end{matrix}\right\}\\ &\rightleftharpoons \left\{\begin{matrix}\underbrace{(\mathrm{UP}_+,\mathrm{UP}_+)}_{\text{positive saddle}} & \underbrace{(\mathrm{DP}_-,\mathrm{DP}_+)}_{\text{CW center}}\\ \underbrace{(\mathrm{DP}_+,\mathrm{DP}_-)}_{\text{CCW center}} & \underbrace{(\mathrm{UP}_-,\mathrm{UP}_-)}_{\text{negative saddle}}\\ \underbrace{(\mathrm{UP}_+,\mathrm{UP}_+)}_{\text{positive saddle}} & \underbrace{(\mathrm{DP}_-,\mathrm{DP}_+)}_{\text{CW center}}\end{matrix}\right\}.\end{aligned} \tag{3.234}$$

$$\begin{aligned}\{\underbrace{(\mathrm{UP}_+,\mathrm{UP}_+)}_{\text{positive saddle}},\underbrace{(\mathrm{DP}_-,\mathrm{DP}_+)}_{\text{CW center}}\} &\rightleftharpoons \left\{\begin{matrix}\underbrace{(\mathrm{UP}_+,\mathrm{UP}_+)}_{\text{positive saddle}} & \underbrace{(\mathrm{DP}_-,\mathrm{DP}_+)}_{\text{CW center}}\\ \underbrace{(\mathrm{LS},\mathrm{DP})}_{\text{appearing bifurcation}} & \underbrace{(\mathrm{LS},\mathrm{UP})}_{\text{appearing bifurcation}}\end{matrix}\right\}\\ &\rightleftharpoons \left\{\begin{matrix}\underbrace{(\mathrm{UP}_+,\mathrm{UP}_+)}_{\text{positive saddle}} & \underbrace{(\mathrm{DP}_-,\mathrm{DP}_+)}_{\text{CW center}}\\ \underbrace{(\mathrm{DP}_+,\mathrm{DP}_-)}_{\text{CCW center}} & \underbrace{(\mathrm{UP}_-,\mathrm{UP}_-)}_{\text{negative saddle}}\\ \underbrace{(\mathrm{UP}_+,\mathrm{UP}_+)}_{\text{positive saddle}} & \underbrace{(\mathrm{DP}_-,\mathrm{DP}_+)}_{\text{CW center}}\end{matrix}\right\}.\end{aligned} \tag{3.235}$$

(vi) For $a_{j_1 j_2 0} < 0$ and $a_{j_2 j_1 0} < 0$, the (UP,LS)-parabola-saddle bifurcation is

$$(\mathrm{UP},\mathrm{nF}) \rightleftharpoons \underbrace{(\mathrm{UP},\mathrm{LS})}_{\text{appearing bifurcation}} \rightleftharpoons \{\underbrace{(\mathrm{DP}_+,\mathrm{DP}_-)}_{\text{CCW center}},\underbrace{(\mathrm{UP}_-,\mathrm{UP}_-)}_{\text{negative saddle}}\}, \tag{3.236}$$

with three cases for the further bifurcations as follows:

$$\begin{aligned}
&\{\underbrace{(\mathrm{DP}_+,\mathrm{DP}_-)}_{\text{CCW center}},\underbrace{(\mathrm{UP}_-,\mathrm{UP}_-)}_{\text{negative saddle}}\} \rightleftharpoons \{\underbrace{(3^{\mathrm{rd}}\mathrm{DP}_+,\mathrm{DP}_-)}_{\text{third-order CCW center}},\underbrace{(3^{\mathrm{rd}}\mathrm{UP}_-,\mathrm{UP}_-)}_{\text{third-order positive saddle}}\}\\
&\rightleftharpoons \left\{\begin{array}{cc}
\underbrace{(\mathrm{DP}_+,\mathrm{DP}_-)}_{\text{CCW center}} & \underbrace{(\mathrm{UP}_-,\mathrm{UP}_-)}_{\text{negative saddle}}\\
\underbrace{(\mathrm{UP}_+,\mathrm{UP}_+)}_{\text{positive saddle}} & \underbrace{(\mathrm{DP}_-,\mathrm{DP}_+)}_{\text{CW center}}\\
\underbrace{(\mathrm{DP}_+,\mathrm{DP}_-)}_{\text{CCW center}} & \underbrace{(\mathrm{UP}_-,\mathrm{UP}_-)}_{\text{negative saddle}}
\end{array}\right\}.
\end{aligned} \tag{3.237}$$

$$\begin{aligned}
&\{\underbrace{(\mathrm{DP}_+,\mathrm{DP}_-)}_{\text{CCW center}},\underbrace{(\mathrm{UP}_-,\mathrm{UP}_-)}_{\text{negative saddle}}\} \rightleftharpoons \left\{\begin{array}{cc}
\underbrace{(\mathrm{LS},\mathrm{DP})}_{\text{appearing bifurcation}} & \underbrace{(\mathrm{LS},\mathrm{UP})}_{\text{appearing bifurcation}}\\
\underbrace{(\mathrm{DP}_+,\mathrm{DP}_-)}_{\text{CCW center}} & \underbrace{(\mathrm{UP}_-,\mathrm{UP}_-)}_{\text{negative saddle}}
\end{array}\right\}\\
&\rightleftharpoons \left\{\begin{array}{cc}
\underbrace{(\mathrm{DP}_+,\mathrm{DP}_-)}_{\text{CCW center}} & \underbrace{(\mathrm{UP}_-,\mathrm{UP}_-)}_{\text{negative saddle}}\\
\underbrace{(\mathrm{UP}_+,\mathrm{UP}_+)}_{\text{positive saddle}} & \underbrace{(\mathrm{DP}_-,\mathrm{DP}_+)}_{\text{CW center}}\\
\underbrace{(\mathrm{DP}_+,\mathrm{DP}_-)}_{\text{CCW center}} & \underbrace{(\mathrm{UP}_-,\mathrm{UP}_-)}_{\text{negative saddle}}
\end{array}\right\}.
\end{aligned} \tag{3.238}$$

$$\begin{aligned}
&\{\underbrace{(\mathrm{DP}_-,\mathrm{DP}_+)}_{\text{CW center}},\underbrace{(\mathrm{UP}_-,\mathrm{UP}_-)}_{\text{negative saddle}}\} \rightleftharpoons \left\{\begin{array}{cc}
\underbrace{(\mathrm{DP}_+,\mathrm{DP}_-)}_{\text{CCW center}} & \underbrace{(\mathrm{UP}_-,\mathrm{UP}_-)}_{\text{negative saddle}}\\
\underbrace{(\mathrm{US},\mathrm{UP})}_{\text{appearing bifurcation}} & \underbrace{(\mathrm{US},\mathrm{DP})}_{\text{appearing bifurcation}}
\end{array}\right\}\\
&\rightleftharpoons \left\{\begin{array}{cc}
\underbrace{(\mathrm{DP}_+,\mathrm{DP}_-)}_{\text{CCW center}} & \underbrace{(\mathrm{UP}_-,\mathrm{UP}_-)}_{\text{negative saddle}}\\
\underbrace{(\mathrm{UP}_+,\mathrm{UP}_+)}_{\text{positive saddle}} & \underbrace{(\mathrm{DP}_-,\mathrm{DP}_+)}_{\text{CW center}}\\
\underbrace{(\mathrm{DP}_+,\mathrm{DP}_-)}_{\text{CCW center}} & \underbrace{(\mathrm{UP}_-,\mathrm{UP}_-)}_{\text{negative saddle}}
\end{array}\right\}.
\end{aligned} \tag{3.239}$$

The corresponding bifurcation routes to the saddle and center network are sketched in Figs. 3.11 and 3.12.

In Fig. 3.11a–d, the bifurcation routes from parabola flows to saddle and center networks for $x_1^* = a_{j_2 j_1 1}, a_{j_2 j_1 2}$ and $x_{j_2}^* = a_{j_1 j_2 1}, a_{j_1 j_2 2}, a_{j_1 j_2 3}$ are presented. The two bifurcations are the (UP,US)-up-parabola-upper-saddle appearing bifurcation and third-order saddle and center appearing bifurcations ($a_{j_1 j_2 0} > 0, a_{j_2 j_1 0} > 0$); the (DP, US)-down-parabola upper-saddle appearing bifurcation and third-order saddle and center appearing bifurcations ($a_{j_1 j_2 0} < 0, a_{j_2 j_1 0} > 0$); the (DP,LS)-down-parabola lower-saddle appearing bifurcation and third-order saddle and center appearing bifurcations ($a_{j_1 j_2 0} > 0, a_{j_2 j_1 0} < 0$); and the (UP,LS)-up-parabola lower-saddle appearing bifurcation and third-order saddle and center appearing bifurcations ($a_{j_1 j_2 0} < 0, a_{j_2 j_1 0} < 0$).

In Fig. 3.12a-d, the bifurcation routes from parabola flows to saddle and center networks for $x_1^* = a_{j_2 j_1 1}, a_{j_2 j_1 2}$ and $x_{j_2}^* = a_{j_1 j_2 1}, a_{j_1 j_2 2}, a_{j_1 j_2 3}$ are presented. The two bifurcations are (i) (UP,US)-up-parabola-upper-saddle appearing bifurcation with (US,UP)-up-parabola upper-saddle and (US,DP)-down-parabola upper-saddle appearing bifurcations ($a_{j_1 j_2 0} > 0, a_{j_2 j_1 0} > 0$); the (DP,US)-down-parabola upper-saddle appearing bifurcation with (LS,DP)-down-parabola lower-saddle and (LS,UP)-up-parabola lower-saddle appearing bifurcations ($a_{j_1 j_2 0} < 0, a_{j_2 j_1 0} > 0$); the (DP,LS)-down-parabola lower-saddle appearing bifurcation with (US,DP)-down-parabola upper-saddle and (US,DP)-down-parabola upper-saddle appearing bifurcations ($a_{j_1 j_2 0} > 0, a_{j_2 j_1 0} < 0$); and the (UP, LS)-up-parabola lower-saddle appearing bifurcation with (LS,UP)-up-parabola upper-saddle and (LS,DP)-down-parabola upper-saddle appearing bifurcations ($a_{j_1 j_2 0} < 0, a_{j_2 j_1 0} < 0$).

Similar to Fig. 3.12, the bifurcation routes from parabola flows to saddle and center networks for $x_1^* = a_{j_2 j_1 1}, a_{j_2 j_1 2}$ and $x_{j_2}^* = a_{j_1 j_2 1}, a_{j_1 j_2 2}, a_{j_1 j_2 3}$ can be presented through another set of inflection flow appearing bifurcation and up-parabola and down-parabola upper-saddle or up-parabola and down-parabola lower-saddle appearing bifurcations. Such bifurcation routes can be done by readers as an exercise.

Consider a standard form of the crossing-cubic dynamical system as

$$\begin{aligned} \dot{x}_{j_1} &= a_{j_1 j_2 0}(x_{j_2} - a_{j_1 j_2 1})^3, \\ \dot{x}_{j_2} &= a_{j_2 j_1 0}(x_{j_1} - a_{j_2 j_1 1})^2, \\ j_1, j_2 &\in \{1,2\}; j_1 \neq j_2. \end{aligned} \tag{3.240}$$

(i) For $a_{j_1 j_2 0} > 0$ and $a_{j_2 j_1 0} > 0$, the (3rd UP,US)-third-order up-parabola upper-saddle appearing bifurcation of $(x_{j_1}^*, x_{j_2}^*) = (a_{j_1 j_1 1}, a_{j_2 j_2 1})$ is

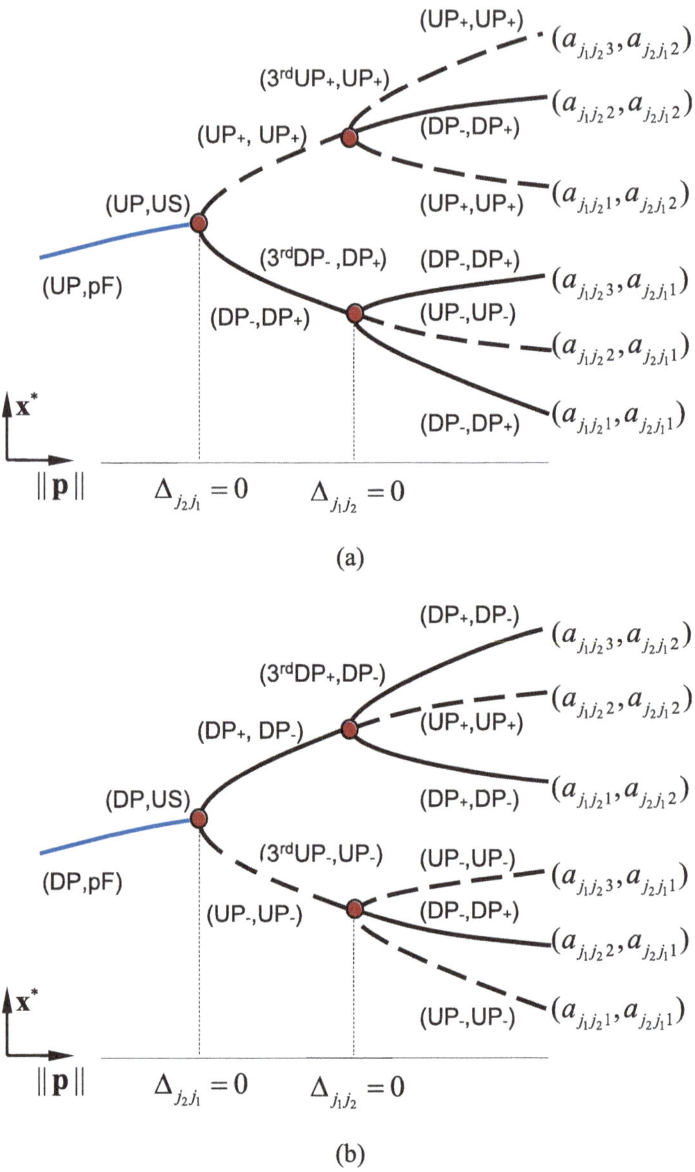

Fig. 3.11 The bifurcation routes from parabola flows to saddle and center networks for $x_1^* = a_{j_2j_1 1}, a_{j_2j_1 2}$ and $x_{j_2}^* = a_{j_1j_2 1}, a_{j_1j_2 2}, a_{j_1j_2 3}$: **(a)** (UP,US)-appearing bifurcation ($a_{j_1j_2 0} > 0, a_{j_2j_1 0} > 0$), **(b)** (DP,US)-appearing bifurcation ($a_{j_1j_2 0} > 0, a_{j_2j_1 0} < 0$), **(c)** (DP,LS)-appearing bifurcation ($a_{j_1j_2 0} > 0, a_{j_2j_1 0} < 0$), **(d)** (UP, LS)-appearing bifurcation ($a_{j_1j_2 0} < 0, a_{j_2j_1 0} < 0$). *pF* positive flow, *nF* negative flow. *US* upper-saddle, *LS* lower-saddle, *DP* down-parabola, *UP* up-parabola

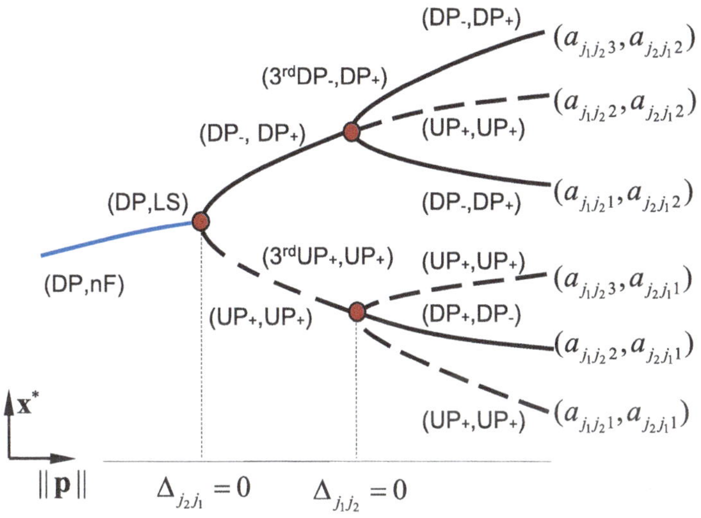

(c)

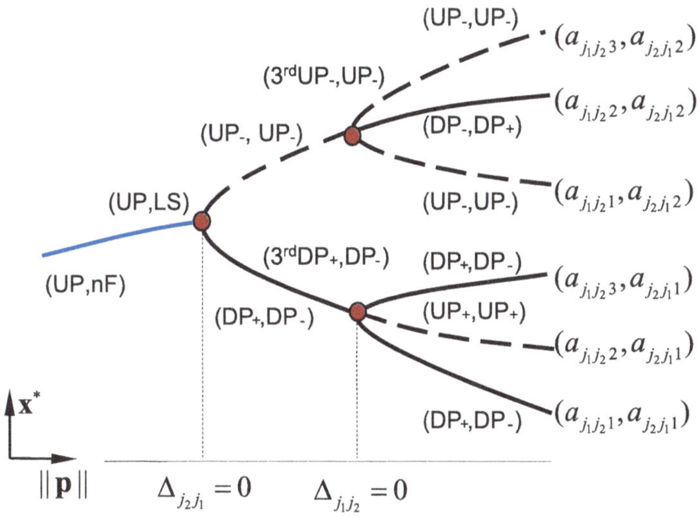

(d)

Fig. 3.11 (continued)

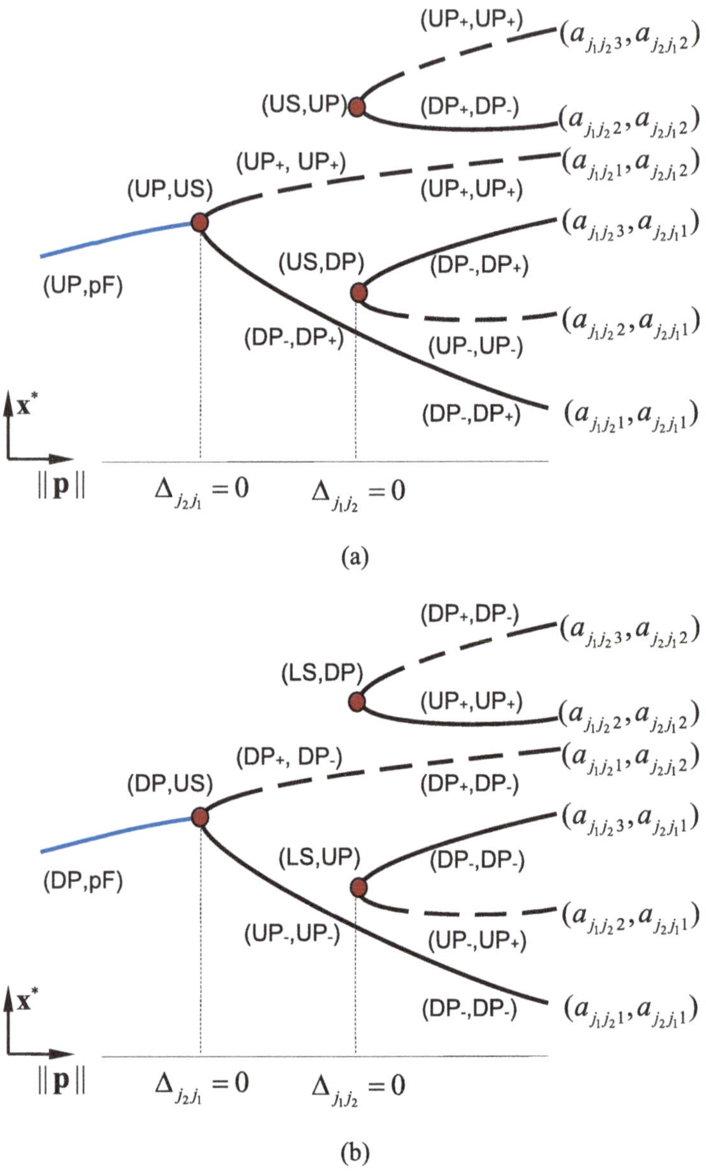

Fig. 3.12 The bifurcation routes from parabola flows to saddle and center networks for $x_1^* = a_{j_2j_1 1}, a_{j_2j_1 2}$ and $x_{j_2}^* = a_{j_1j_2 1}, a_{j_1j_2 2}, a_{j_1j_2 3}$: **(a)** (UP,US)-appearing bifurcation ($a_{j_1j_2 0} > 0, a_{j_2j_1 0} > 0$), **(b)** (DP,US)-appearing bifurcation ($a_{j_1j_2 0} > 0, a_{j_2j_1 0} < 0$), **(c)** (DP,LS)-appearing bifurcation ($a_{j_1j_2 0} > 0, a_{j_2j_1 0} < 0$), **(d)** (UP, LS)-appearing bifurcation ($a_{j_1j_2 0} < 0, a_{j_2j_1 0} < 0$). *pF* positive flow, *nF* negative flow. *US* upper-saddle, *LS* lower-saddle, *DP* down-parabola, *UP* up-parabola

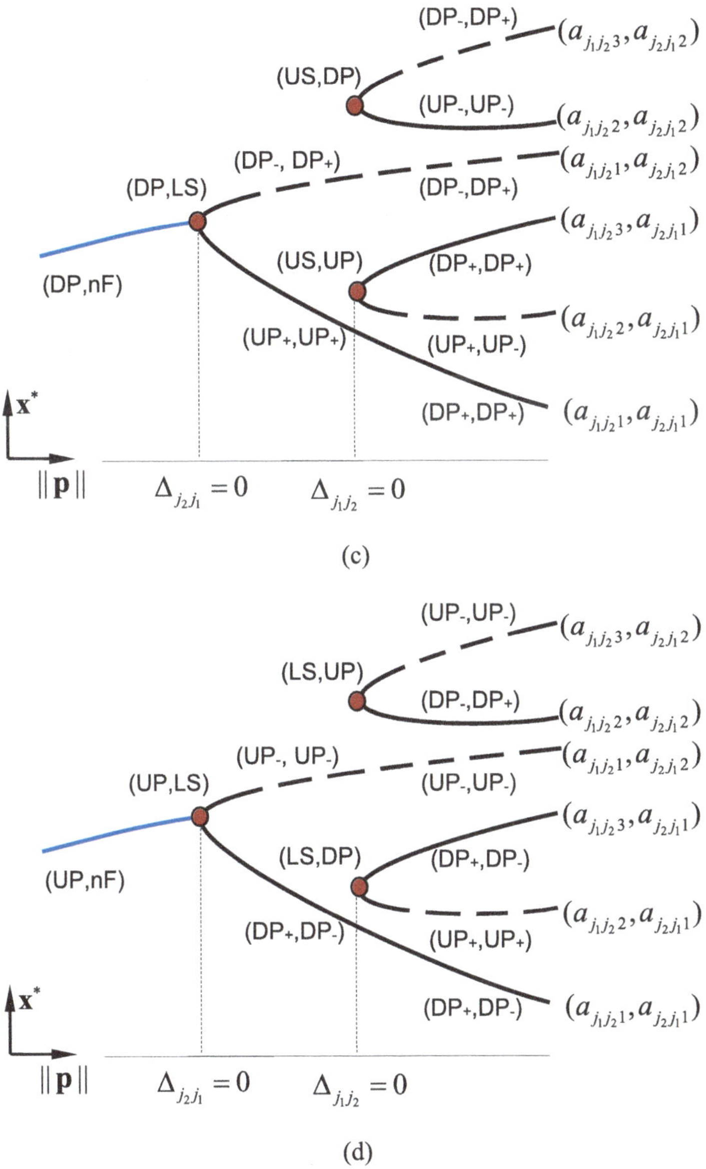

Fig. 3.12 (continued)

$$\underbrace{(\mathrm{UP},\mathrm{pF})}_{\text{up-parabola flow }(+)} \rightleftharpoons \underbrace{(3^{\mathrm{rd}}\mathrm{UP},\mathrm{US})}_{\text{appearing bifurcation}} \rightleftharpoons \left\{\begin{array}{cc} \underbrace{(\mathrm{DP}_-,\mathrm{DP}_+)}_{\text{CW center}} & \underbrace{(\mathrm{UP}_+,\mathrm{UP}_+)}_{\text{positive saddle}} \\ \underbrace{(\mathrm{UP}_-,\mathrm{UP}_-)}_{\text{negative saddle}} & \underbrace{(\mathrm{DP}_+,\mathrm{DP}_-)}_{\text{CCW center}} \\ \underbrace{(\mathrm{DP}_-,\mathrm{DP}_+)}_{\text{CW center}} & \underbrace{(\mathrm{UP}_+,\mathrm{UP}_+)}_{\text{positive saddle}} \end{array}\right\} \tag{3.241}$$

(ii) For $a_{j_1j_20}<0$ and $a_{j_2j_10}>0$, the (3rdDP,US)-third-order down-parabola upper-saddle appearing bifurcation of $(x_{j_1}^*,x_{j_2}^*)=(a_{j_2j_11},a_{j_1j_21})$ is

$$\underbrace{(\mathrm{DP},\mathrm{pF})}_{\text{down-parabola flow }(+)} \rightleftharpoons \underbrace{(3^{\mathrm{rd}}\mathrm{DP},\mathrm{US})}_{\text{appearing bifurcation}} \rightleftharpoons \left\{\begin{array}{cc} \underbrace{(\mathrm{UP}_-,\mathrm{UP}_-)}_{\text{negative saddle}} & \underbrace{(\mathrm{DP}_+,\mathrm{DP}_-)}_{\text{CCW center}} \\ \underbrace{(\mathrm{DP}_-,\mathrm{DP}_+)}_{\text{CW center}} & \underbrace{(\mathrm{UP}_+,\mathrm{UP}_+)}_{\text{positive saddle}} \\ \underbrace{(\mathrm{UP}_-,\mathrm{UP}_-)}_{\text{negative saddle}} & \underbrace{(\mathrm{DP}_+,\mathrm{DP}_-)}_{\text{CCW center}} \end{array}\right\}. \tag{3.242}$$

(iii) For $a_{j_1j_20}>0$ and $a_{j_2j_10}<0$, the (3rdDP,LS)-third-order down-parabola lower-saddle appearing bifurcation of $(x_{j_1}^*,x_{j_2}^*)=(a_{j_1j_11},a_{j_2j_21})$ is

$$\underbrace{(\mathrm{DP},\mathrm{nF})}_{\text{down-parabola flow }(-)} \rightleftharpoons \underbrace{(3^{\mathrm{rd}}\mathrm{DP},\mathrm{LS})}_{\text{appearing bifurcation}} \rightleftharpoons \left\{\begin{array}{cc} \underbrace{(\mathrm{UP}_+,\mathrm{UP}_+)}_{\text{positive saddle}} & \underbrace{(\mathrm{DP}_-,\mathrm{DP}_+)}_{\text{CW center}} \\ \underbrace{(\mathrm{DP}_+,\mathrm{DP}_-)}_{\text{CCW center}} & \underbrace{(\mathrm{UP}_-,\mathrm{UP}_-)}_{\text{negative saddle}} \\ \underbrace{(\mathrm{UP}_+,\mathrm{UP}_+)}_{\text{positive saddle}} & \underbrace{(\mathrm{DP}_-,\mathrm{DP}_+)}_{\text{CW center}} \end{array}\right\}. \tag{3.243}$$

(iv) For $a_{j_1j_20}<0$ and $a_{j_2j_10}<0$, the (3rdUP,LS)-third-order up-parabola lower-saddle appearing bifurcation of $(x_{j_2}^*,x_{j_1}^*)=(a_{j_1j_21},a_{j_2j_11})$ is

$$\underbrace{(\mathrm{UP},\mathrm{nF})}_{\text{up-parabola flow }(-)} \rightleftharpoons \underbrace{(3^{\mathrm{rd}}\mathrm{UP},\mathrm{LS})}_{\text{appearing bifurcation}} \rightleftharpoons \left\{ \begin{array}{cc} \underbrace{(\mathrm{DP}_+,\mathrm{DP}_-)}_{\text{CCW center}} & \underbrace{(\mathrm{UP}_-,\mathrm{UP}_-)}_{\text{negative saddle}} \\ \underbrace{(\mathrm{UP}_+,\mathrm{UP}_+)}_{\text{positive saddle}} & \underbrace{(\mathrm{DP}_-,\mathrm{DP}_+)}_{\text{CW center}} \\ \underbrace{(\mathrm{DP}_+,\mathrm{DP}_-)}_{\text{CCW center}} & \underbrace{(\mathrm{UP}_-,\mathrm{UP}_-)}_{\text{negative saddle}} \end{array} \right\}. \tag{3.244}$$

In Fig. 3.13, the appearing bifurcations of third-order up-parabola upper-saddle and lower-saddle for $x_1^* = a_{j_2j_1 1}, a_{j_2j_1 2}$ and $x_{j_2}^* = a_{j_1j_2 1}, a_{j_1j_2 2}, a_{j_1j_2 3}$ are sketched. The corresponding bifurcations are the (3rdUP,US)-appearing bifurcation ($a_{j_1j_2 0} > 0$, $a_{j_2j_1 0} > 0$), the (3rdDP,US)-appearing bifurcation ($a_{j_1j_2 0} > 0$, $a_{j_2j_1 0} < 0$), the (3rdDP, LS)-appearing bifurcation ($a_{j_1j_2 0} < 0$, $a_{j_2j_1 0} > 0$), and the (3rdUP,LS)-appearing bifurcation ($a_{j_1j_2 0} < 0$, $a_{j_2j_1 0} < 0$), as shown in Fig. 3.13a–d, respectively.

Consider a standard form of the quadratic dynamical system as

$$\begin{aligned} \dot{x}_{j_1} &= a_{j_1j_2 0}(x_{j_2} - a_{j_1j_2 1})^2, \\ \dot{x}_{j_2} &= a_{j_2j_1 0}(x_{j_1} - a_{j_2j_1 1})^3, \\ j_1, j_2 &\in \{1,2\}; j_1 \neq j_2. \end{aligned} \tag{3.345}$$

(i) For $a_{j_1j_2 0} > 0$ and $a_{j_2j_1 0} > 0$, the (US,3rdUP)-third-order up-parabola upper-saddle appearing bifurcation of $(x_{j_2}^*, x_{j_1}^*) = (a_{j_1j_2 1}, a_{j_2j_1 1})$ is

$$\begin{aligned} &\underbrace{(\mathrm{pF},\mathrm{UP})}_{\text{up-parabola flow }(+)} \rightleftharpoons \underbrace{(\mathrm{US},3^{\mathrm{rd}}\mathrm{UP})}_{\text{appearing bifurcation}} \\ &\rightleftharpoons \left\{ \begin{array}{ccc} \underbrace{(\mathrm{UP}_+,\mathrm{UP}_+)}_{\text{positive saddle}} & \underbrace{(\mathrm{DP}_-,\mathrm{DP}_+)}_{\text{CW center}} & \underbrace{(\mathrm{UP}_+,\mathrm{UP}_+)}_{\text{positive saddle}} \\ \underbrace{(\mathrm{DP}_+,\mathrm{DP}_-)}_{\text{CCW center}} & \underbrace{(\mathrm{UP}_-,\mathrm{UP}_-)}_{\text{negative saddle}} & \underbrace{(\mathrm{DP}_+,\mathrm{DP}_-)}_{\text{CCW center}} \end{array} \right\}. \end{aligned} \tag{3.246}$$

(ii) For $a_{j_1j_2 0} < 0$ and $a_{j_2j_1 0} > 0$, the (US,3rdDP)-third-order down-parabola upper-saddle appearing bifurcation of $(x_{j_2}^*, x_{j_1}^*) = (a_{j_1j_2 1}, a_{j_2j_1 1})$ is

$$\begin{aligned} &\underbrace{(\mathrm{nF},\mathrm{DP})}_{\text{down-parabola flow }(-)} \rightleftharpoons \underbrace{(\mathrm{LS},3^{\mathrm{rd}}\mathrm{DP})}_{\text{switching bifurcation}} \\ &\rightleftharpoons \left\{ \begin{array}{ccc} \underbrace{(\mathrm{DP}_+,\mathrm{DP}_-)}_{\text{CCW center}} & \underbrace{(\mathrm{UP}_-,\mathrm{UP}_-)}_{\text{negative saddle}} & \underbrace{(\mathrm{DP}_+,\mathrm{DP}_-)}_{\text{CCW center}} \\ \underbrace{(\mathrm{UP}_+,\mathrm{UP}_+)}_{\text{positive saddle}} & \underbrace{(\mathrm{DP}_-,\mathrm{DP}_+)}_{\text{CW center}} & \underbrace{(\mathrm{UP}_+,\mathrm{UP}_+)}_{\text{positive saddle}} \end{array} \right\}. \end{aligned} \tag{3.247}$$

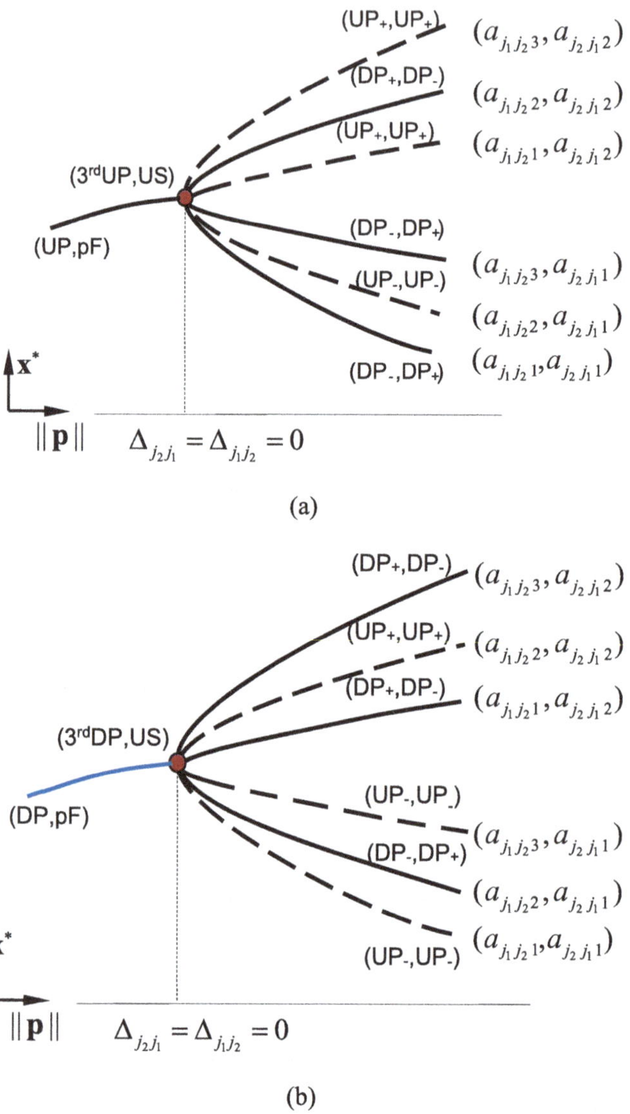

Fig. 3.13 The appearing bifurcations of (3,2)-parabola-saddles for $x_1^* = a_{j_2j_11}, a_{j_2j_12}$ and $x_{j_2}^* = a_{j_1j_21}, a_{j_1j_22}$: (**a**) ($3^{rd}$UP,US)-appearing bifurcation ($a_{j_1j_20} > 0$, $a_{j_2j_10} > 0$), (**b**) (3^{rd}DP,US)-appearing bifurcation ($a_{j_1j_20} > 0$, $a_{j_2j_10} < 0$), (**c**) (3^{rd}DP,LS)-appearing bifurcation ($a_{j_1j_20} < 0$, $a_{j_2j_10} > 0$), (**d**) (3^{rd}UP,LS)-appearing bifurcation ($a_{j_1j_20} < 0$, $a_{j_2j_10} < 0$). US: upper-saddle, LS: lower-saddle, DP: down-parabola, UP: up-parabola

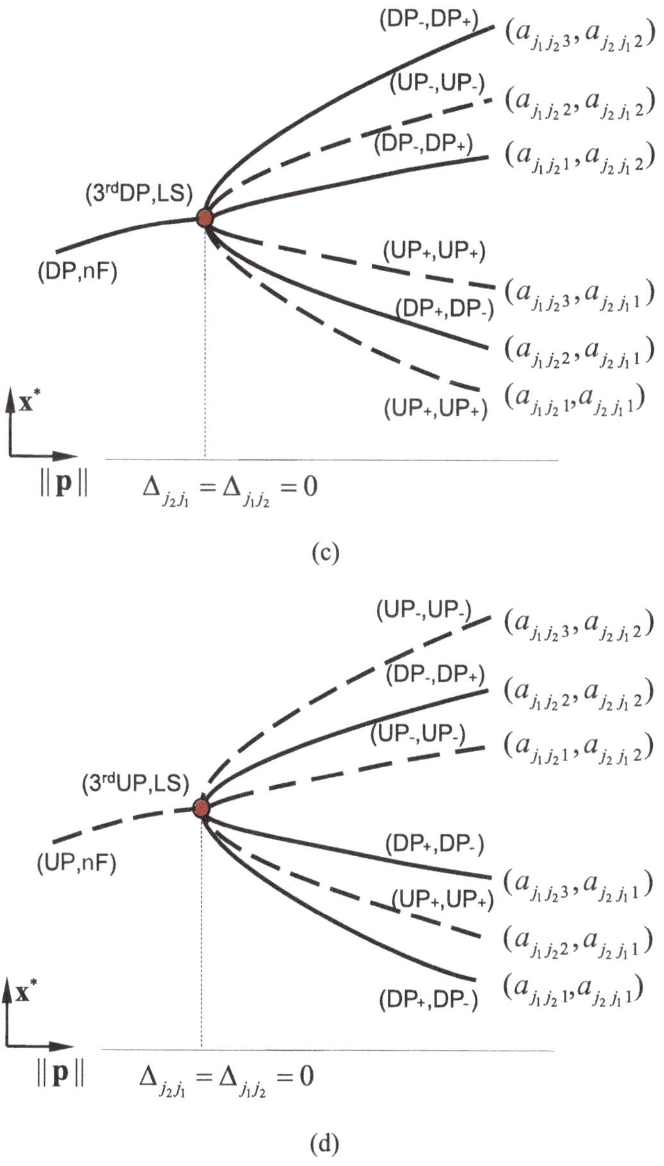

Fig. 3.13 (continued)

(iii) For $a_{j_1 j_2 0} > 0$ and $a_{j_2 j_1 0} < 0$, the (LS,3[rd]DP)-third-order down-parabola lower-saddle appearing bifurcation of $(x_{j_2}^*, x_{j_1}^*) = (a_{j_1 j_2 1}, a_{j_2 j_1 1})$ is

$$\begin{aligned}
&\underbrace{(\mathrm{pF},\mathrm{DP})}_{\text{down-parabola flow }(+)} \rightleftharpoons \underbrace{(\mathrm{US},3^{\mathrm{rd}}\mathrm{DP})}_{\text{appearing bifurcation}}\\
&\rightleftharpoons \left\{\begin{array}{ccc}
\underbrace{(\mathrm{DP}_-,\mathrm{DP}_+)}_{\text{CW center}} & \underbrace{(\mathrm{UP}_+,\mathrm{UP}_+)}_{\text{positive sadde}} & \underbrace{(\mathrm{DP}_-,\mathrm{DP}_+)}_{\text{CW center}}\\
\underbrace{(\mathrm{UP}_-,\mathrm{UP}_-)}_{\text{negative sadde}} & \underbrace{(\mathrm{DP}_+,\mathrm{DP}_-)}_{\text{CCW center}} & \underbrace{(\mathrm{UP}_-,\mathrm{UP}_-)}_{\text{negative sadde}}
\end{array}\right\}.
\end{aligned} \tag{3.248}$$

(iv) For $a_{j_1 j_2 0} < 0$ and $a_{j_2 j_1 0} < 0$, the (LS,3[rd]UP)-third-order up-parabola lower-saddle appearing bifurcation of $(x_{j_2}^*, x_{j_1}^*) = (a_{j_1 j_2 1}, a_{j_2 j_1 1})$ is

$$\begin{aligned}
&\underbrace{(\mathrm{nF},\mathrm{UP})}_{\text{up-parabola flow }(-)} \rightleftharpoons \underbrace{(\mathrm{LS},3^{\mathrm{nd}}\mathrm{UP})}_{\text{appearing bifurcation}}\\
&\rightleftharpoons \left\{\begin{array}{ccc}
\underbrace{(\mathrm{UP}_-,\mathrm{UP}_-)}_{\text{negative sadde}} & \underbrace{(\mathrm{DP}_+,\mathrm{DP}_-)}_{\text{CCW center}} & \underbrace{(\mathrm{UP}_-,\mathrm{UP}_-)}_{\text{negative saddle}}\\
\underbrace{(\mathrm{DP}_-,\mathrm{DP}_+)}_{\text{CW center}} & \underbrace{(\mathrm{UP}_+,\mathrm{UP}_+)}_{\text{positive saddle}} & \underbrace{(\mathrm{DP}_-,\mathrm{DP}_+)}_{\text{CW center}}
\end{array}\right\}.
\end{aligned} \tag{3.249}$$

Chapter 4
Two Crossing-Cubic Vector Fields

In this chapter, a cubic dynamical system possessing two crossing-variable cubic vector fields are discussed. The appearing and switching bifurcations for centers and saddles of equilibriums are discussed, and the limit cycle and homoclinic networks are presented through the first integral manifolds. The corresponding equilibriums include simple center and saddles; parabola-saddle; and third-order saddles and centers, (2,2)-double-diagonal inflection saddles, (3,2)-parabola-saddle, and (3,3)-double-third-order saddles and centers. The double third-order saddle and centers are the appearing bifurcations of the 3×3 network of saddles and centers with homoclinic orbits. The double-third-order saddles and centers are also the switching bifurcations of parabola-saddle with saddles and centers. The double-third-order saddles and centers are also the switching bifurcation for the networks of center (saddle), parabola-saddles, and inflection saddles. The networks of saddle and center for such nonlinear systems with two cubic variable-crossing vector fields are presented.

4.1 Two-Crossing-Cubic Systems

Consider a dynamical system possessing two crossing-cubic vector fields. The corresponding dynamical behaviors will be presented through the following theorem.

Theorem 4.1 *Consider a two-dimensional, dynamical system with two crossing-cubic vector fields as*

$$\begin{aligned}\dot{x}_{j_1} &= f_{j_1}(\mathbf{x},\mathbf{p}) = A_{j_1j_2 3} + A_{j_1j_2 2}x_{j_2} + A_{j_1j_2 1}x_{j_2}^2 + A_{j_1j_2 0}x_{j_2}^3,\\ \dot{x}_{j_2} &= f_{j_2}(\mathbf{x},\mathbf{p}) = A_{j_2j_1 3} + A_{j_2j_1 2}x_{j_1} + A_{j_2j_1 1}x_{j_1}^2 + A_{j_2j_1 0}x_{j_1}^3.\end{aligned} \tag{4.1}$$

A. C. J. Luo, *Two-dimensional Crossing-Variable Cubic Nonlinear Systems*,
https://doi.org/10.1007/978-3-031-62810-8_4

For $A_{j_1j_20} \neq 0$ and $A_{j_2j_10} \neq 0$, the two-dimensional dynamical system possesses two variable-crossing univariate cubic vector fields as

$$\begin{aligned}\dot{x}_{j_2} &= a_{j_2j_10}(x_{j_1} - b_{j_2j_11})(x_{j_1}^2 + B_{j_2j_1}x_{j_1} + C_{j_2j_1}),\\ \dot{x}_{j_1} &= a_{j_1j_20}(x_{j_2} - b_{j_1j_21})(x_{j_2}^2 + B_{j_1j_2}x_{j_2} + C_{j_1j_2})\end{aligned} \tag{4.2}$$

with

$$\begin{aligned}&\Delta_{j_2j_1} = (B_{j_2j_1})^2 - 4C_{j_2j_1}, \Delta_{j_1j_2} = (B_{j_1j_2})^2 - 4C_{j_1j_2}\\ &\text{for } j_1, j_2 \in \{1,2\}, j_1 \neq j_2.\end{aligned} \tag{4.3}$$

(i) If

$$\Delta_{j_1j_2} < 0 \text{ and } \Delta_{j_2j_1} < 0, \tag{4.4}$$

then

$$\begin{aligned}\dot{x}_{j_1} &= a_{j_1j_20}(x_{j_2} - a_{j_1j_21})\left[(x_{j_2} - a_{j_1j_2})^2 + b_{j_1j_2}\right],\\ \dot{x}_{j_2} &= a_{j_2j_10}(x_{j_1} - a_{j_2j_11})\left[(x_{j_1} - a_{j_2j_1})^2 + b_{j_2j_1}\right],\\ &j_1, j_2 \in \{1,2\}, j_2 \neq j_1.\end{aligned} \tag{4.5}$$

The first integral manifold is

$$\begin{aligned}&a_{j_2j_10}\Big\{\frac{1}{4}\left[(x_{j_1} - a_{j_2j_11})^4 - (x_{j_10} - a_{j_2j_11})^4\right]\\ &+\frac{2}{3}(a_{j_2j_11} - a_{j_2j_1})\left[(x_{j_1} - a_{j_2j_11})^3 - (x_{j_10} - a_{j_2j_11})^3\right]\\ &+\frac{1}{2}\left[(a_{j_2j_11} - a_{j_2j_1})^2 + b_{j_2j_1}\right]\left[(x_{j_1} - a_{j_2j_11})^2 - (x_{j_10} - a_{j_2j_11})^2\right]\Big\}\\ &= a_{j_1j_20}\Big\{\frac{1}{4}\left[(x_{j_2} - a_{j_1j_21})^4 - (x_{j_20} - a_{j_1j_21})^4\right]\\ &+\frac{2}{3}(a_{j_1j_21} - a_{j_1j_2})\left[(x_{j_2} - a_{j_1j_21})^3 - (x_{j_20} - a_{j_1j_21})^3\right]\\ &+\frac{1}{2}\left[(a_{j_1j_21} - a_{j_1j_2})^2 + b_{j_1j_2}\right]\left[(x_{j_2} - a_{j_1j_21})^2 - (x_{j_20} - a_{j_1j_21})^2\right]\Big\}.\end{aligned} \tag{4.6}$$

The equilibrium of $(x_{j_2}^*, x_{j_1}^*) = (a_{j_1j_21}, a_{j_2j_11})$ has the following properties.

- *For $a_{j_1j_20} > 0$ and $a_{j_2j_10} > 0$,*

$$(a_{j_1j_21}, a_{j_2j_11}) = \underbrace{(\text{UP}_+, \text{UP}_+)}_{\text{positive saddle}}. \tag{4.7}$$

The equilibrium of $(x_{j_2}^*, x_{j_1}^*) = (a_{j_1j_21}, a_{j_2j_11})$ is a $(\text{UP}_+,\text{UP}_+)$*-positive saddle.*

- *For* $a_{j_1j_20} < 0$ *and* $a_{j_2j_10} > 0$,

$$(a_{j_1j_21}, a_{j_2j_11}) = \underbrace{(\mathrm{DP}_+, \mathrm{DP}_-)}_{\text{CCW center}}. \tag{4.8}$$

The equilibrium of $(x^*_{j_2}, x^*_{j_1}) = (a_{j_1j_21}, a_{j_2j_11})$ *is a* (DP$_+$,DP$_-$)*-counter-clockwise center.*

- *For* $a_{j_1j_20} > 0$ *and* $a_{j_2j_10} < 0$,

$$(a_{j_1j_21}, a_{j_2j_11}) = \underbrace{(\mathrm{DP}_-, \mathrm{DP}_+)}_{\text{CW center}}. \tag{4.9}$$

The equilibrium of $(x^*_{j_2}, x^*_{j_1}) = (a_{j_1j_21}, a_{j_2j_11})$ *is a* (DP$_-$,DP$_+$)*-clockwise center.*

- *For* $a_{j_1j_20} < 0$ *and* $a_{j_2j_10} < 0$,

$$(a_{j_1j_21}, a_{j_2j_11}) = \underbrace{(\mathrm{UP}_-, \mathrm{UP}_-)}_{\text{negative saddle}}. \tag{4.10}$$

The equilibrium of $(x^*_{j_2}, x^*_{j_1}) = (a_{j_1j_21}, a_{j_2j_11})$ *is a* (UP$_-$,UP$_-$)*-negative saddle.*

(ii) If

$$\Delta_{j_1j_2} < 0 \,\text{and}\, \Delta_{j_2j_1} = 0, a_{j_2j_1l_1} \neq a_{j_2j_1l_2} \tag{4.11}$$

then

$$\begin{aligned} &\dot{x}_{j_2} = a_{j_2j_10}(x_{j_1} - a_{j_2j_1l_1})(x_{j_1} - a_{j_2j_1l_2})^2, \\ &\dot{x}_{j_1} = a_{j_1j_20}(x_{j_2} - a_{j_1j_21})\left[(x_{j_2} - a_{j_1j_2})^2 + b_{j_ij_2}\right], \\ &j_1, j_2, l_1, l_2 \in \{1,2\}; j_1 \neq j_2; l_1 \neq l_2. \end{aligned} \tag{4.12}$$

The first integral manifold is

$$\begin{aligned} &a_{j_2j_10}\Big\{\frac{1}{4}\left[(x_{j_1} - a_{j_2j_1l_1})^4 - (x_{j_10} - a_{j_2j_1l_1})^4\right] \\ &+\frac{2}{3}(a_{j_2j_1l_1} - a_{j_2j_1l_2})\left[(x_{j_1} - a_{j_2j_1l_1})^3 - (x_{j_10} - a_{j_2j_1l_1})^3\right] \\ &+\frac{1}{2}(a_{j_2j_1l_1} - a_{j_2j_2l_2})^2\left[(x_{j_1} - a_{j_2j_1l_1})^2 - (x_{j_10} - a_{j_2j_1l_1})^2\right]\Big\} \\ &= a_{j_1j_20}\Big\{\frac{1}{4}\left[(x_{j_1} - a_{j_1j_21})^4 - (x_{j_10} - a_{j_1j_21})^4\right] \\ &+\frac{2}{3}(a_{j_1j_21} - a_{j_1j_2})\left[(x_{j_2} - a_{j_1j_21})^3 - (x_{j_20} - a_{j_1j_21})^3\right] \\ &+\frac{1}{2}\left[(a_{j_1j_21} - a_{j_1j_2})^2 + b_{j_1j_2}\right]\left[(x_{j_2} - a_{j_1j_21})^2 - (x_{j_20} - a_{j_1j_21})^2\right]\Big\}. \end{aligned} \tag{4.13}$$

(ii$_1$) The equilibrium of $(x^*_{j_2}, x^*_{j_1}) = (a_{j_1 j_2 1}, a_{j_2 j_1 l_1})$ $(l_1, l_2 \in \{1,2\}, l_1 \neq l_2)$ *has the following properties:*

- *For* $a_{j_1 j_2 0} > 0$ *and* $a_{j_2 j_1 0}(a_{j_2 j_1 l_1} - a_{j_2 j_1 l_2})^2 > 0$,

$$(a_{j_1 j_2 1}, a_{j_2 j_1 l_1}) = \underbrace{(\mathrm{UP}_+, \mathrm{UP}_+)}_{\text{positive saddle}}. \tag{4.14}$$

  The equilibrium of $(x^*_{j_2}, x^*_{j_1}) = (a_{j_1 j_2 1}, a_{j_2 j_1 l_1})$ *is a* (UP$_+$,UP$_+$)*-positive saddle.*

- *For* $a_{j_1 j_2 0} < 0$ *and* $a_{j_2 j_1 0}(a_{j_2 j_1 l_1} - a_{j_2 j_1 l_2})^2 > 0$,

$$(a_{j_1 j_2 1}, a_{j_2 j_1 l_1}) = \underbrace{(\mathrm{DP}_+, \mathrm{DP}_-)}_{\text{CCW center}}. \tag{4.15}$$

  The equilibrium of $(x^*_{j_2}, x^*_{j_1}) = (a_{j_1 j_2 1}, a_{j_2 j_1 l_1})$ *is a* (DP$_+$,DP$_-$)*-counter-clockwise center.*

- *For* $a_{j_1 j_2 0} > 0$ *and* $a_{j_2 j_1 0}(a_{j_2 j_1 l_1} - a_{j_2 j_1 l_2})^2 < 0$,

$$(a_{j_1 j_2 1}, a_{j_2 j_1 l_1}) = \underbrace{(\mathrm{DP}_-, \mathrm{DP}_+)}_{\text{CW center}}. \tag{4.16}$$

  The equilibrium of $(x^*_{j_2}, x^*_{j_1}) = (a_{j_1 j_2 1}, a_{j_2 j_1 l_1})$ *is a* (DP$_-$,DP$_+$)*-clockwise center.*

- *For* $a_{j_1 j_2 0} < 0$ *and* $a_{j_2 j_1 0}(a_{j_2 j_1 l_1} - a_{j_2 j_1 l_2})^2 < 0$,

$$(a_{j_1 j_2 1}, a_{j_2 j_1 l_1}) = \underbrace{(\mathrm{UP}_-, \mathrm{UP}_-)}_{\text{negative saddle}}. \tag{4.17}$$

  The equilibrium of $(x^*_{j_2}, x^*_{j_1}) = (a_{j_1 j_2 1}, a_{j_2 j_1 l_1})$ *is a* (UP$_-$,UP$_-$)*-negative saddle.*

(ii$_2$) The equilibrium of $(x^*_{j_2}, x^*_{j_1}) = (a_{j_1 j_2 1}, a_{j_2 j_1 l_2})$ $(l_1, l_2 \in \{1,2\}, l_1 \neq l_2)$ *has the following properties.*

- *For* $a_{j_1 j_2 0} > 0$ *and* $a_{j_2 j_1 0}(a_{j_2 j_1 l_2} - a_{j_2 j_1 l_1}) > 0$,

$$(a_{j_1 j_2 1}, a_{j_2 j_1 l_2}) = \underbrace{(\mathrm{UP}, \mathrm{US})}_{\text{up-parabola upper-saddle}}. \tag{4.18}$$

  The equilibrium of $(x^*_{j_2}, x^*_{j_1}) = (a_{j_1 j_2 1}, a_{j_2 j_1 l_2})$ *is a* (UP,US)*-up-parabola upper-saddle.*

- *For* $a_{j_1 j_2 0} < 0$ *and* $a_{j_2 j_1 0}(a_{j_2 j_1 l_2} - a_{j_2 j_1 l_1}) > 0$,

$$(a_{j_1 j_2 1}, a_{j_2 j_1 l_2}) = \underbrace{(\mathrm{DP}, \mathrm{US})}_{\text{down-parabola upper-saddle}}. \tag{4.19}$$

  The equilibrium of $(x^*_{j_2}, x^*_{j_1}) = (a_{j_1 j_2 1}, a_{j_2 j_1 l_2})$ *is a* (DP,US)*-down-parabola upper-saddle.*

- *For $a_{j_1j_2 0} > 0$ and $a_{j_2j_1 0}(a_{j_2j_1l_2} - a_{j_2j_1l_1}) < 0$,*

$$(a_{j_1j_2 1}, a_{j_2j_1l_2}) = \underbrace{(\text{DP}, \text{LS})}_{\text{down-parabola lower-saddlle}}. \tag{4.20}$$

The equilibrium of $(x_{j_2}^*, x_{j_1}^*) = (a_{j_1j_2 1}, a_{j_2j_1l_2})$ is a (DP,LS)*-down-parabola lower-saddle.*

- *For $a_{j_1j_2 0} < 0$ and $a_{j_2j_1 0}(a_{j_2j_1l_2} - a_{j_2j_1l_1}) < 0$,*

$$(a_{j_1j_2 1}, a_{j_2j_1l_2}) = \underbrace{(\text{UP}, \text{LS})}_{\text{up-parabola lower-saddle}}. \tag{4.21}$$

The equilibrium of $(x_{j_2}^*, x_{j_1}^*) = (a_{j_1j_2 1}, a_{j_2j_1l_2})$ is a (UP,LS)*-up-parabola lower-saddle.*

(iii) If

$$\Delta_{j_1j_2} < 0 \,\text{and}\, \Delta_{j_2j_1} = 0, a_{j_2j_1l_1} = a_{j_2j_2l_2} = a_{j_2j_1 1} \tag{4.22}$$

then

$$\begin{aligned}
&\dot{x}_{j_2} = a_{j_2j_1 0}(x_{j_1} - a_{j_2j_1 1})^3, \\
&\dot{x}_{j_1} = a_{j_1j_2 0}(x_{j_2} - a_{j_1j_2 1})\left[(x_{j_2} - a_{j_1j_2})^2 + b_{j_1j_2}\right], \\
&j_1, j_2, l_1, l_2 \in \{1, 2\}; j_1 \neq j_2; l_1 \neq l_2.
\end{aligned} \tag{4.23}$$

The first integral manifold is

$$\begin{aligned}
&\frac{1}{4} a_{j_2j_1 0}\left[(x_{j_1} - a_{j_2j_1 1})^4 - (x_{j_1 0} - a_{j_2j_1 1})^4\right] \\
&= a_{j_1j_2 0}\Big\{\frac{1}{4}\left[(x_{j_1} - a_{j_1j_2 1})^4 - (x_{j_1 0} - a_{j_1j_2 1})^4\right] \\
&+ \frac{2}{3}(a_{j_1j_2 1} - a_{j_1j_2})\left[(x_{j_2} - a_{j_1j_2 1})^3 - (x_{j_2 0} - a_{j_1j_2 1})^3\right] \\
&+ \frac{1}{2}\left[(a_{j_1j_2 1} - a_{j_1j_2})^2 + b_{j_1j_2}\right]\left[(x_{j_2} - a_{j_1j_2 1})^2 - (x_{j_2 0} - a_{j_1j_2 1})^2\right]\Big\}.
\end{aligned} \tag{4.24}$$

The equilibrium of $(x_{j_2}^*, x_{j_1}^*) = (a_{j_1j_2 1}, a_{j_2j_1 1})$ has the flowing properties:

- *For $a_{j_1j_2 0} > 0$ and $a_{j_2j_1 0} > 0$,*

$$(a_{j_1j_2 1}, a_{j_2j_1 1}) = \underbrace{(\text{UP}_+, 3^{\text{rd}}\text{UP}_+)}_{\text{third-order positive saddle}}. \tag{4.25}$$

The equilibrium of $(x_{j_2}^*, x_{j_1}^*) = (a_{j_1j_2 1}, a_{j_2j_1 1})$ is a (UP_+,3^{rd}UP_+)*-third-order positive saddle.*

- *For* $a_{j_1j_2 0} < 0$ *and* $a_{j_2j_1 0} > 0$,

$$(a_{j_1j_2 1}, a_{j_2j_1 1}) = \underbrace{(\mathrm{DP}_+, 3^{\mathrm{rd}}\mathrm{DP}_-)}_{\text{third-order CCW center}}. \tag{4.26}$$

The equilibrium of $(x^*_{j_2}, x^*_{j_1}) = (a_{j_1j_2 1}, a_{j_2j_1 1})$ *is a* $(\mathrm{DP}_+, 3^{\mathrm{rd}}\mathrm{DP}_-)$*-third-order counter-clockwise center.*

- *For* $a_{j_1j_2 0} > 0$ *and* $a_{j_2j_1 0} < 0$,

$$(a_{j_1j_2 1}, a_{j_2j_1 1}) = \underbrace{(\mathrm{DP}_-, 3^{\mathrm{rd}}\mathrm{DP}_+)}_{\text{third-order CW center}}. \tag{4.27}$$

The equilibrium of $(x^*_{j_2}, x^*_{j_1}) = (a_{j_1j_2 1}, a_{j_2j_1 1})$ *is a* $(\mathrm{DP}_-, \mathrm{DP}_+)$*-third-order clockwise center.*

- *For* $a_{j_1j_2 0} < 0$ *and* $a_{j_2j_1 0} < 0$,

$$(a_{j_1j_2 1}, a_{j_2j_1 1}) = \underbrace{(\mathrm{UP}_-, 3^{\mathrm{rd}}\mathrm{UP}_-)}_{\text{third-order negative saddle}}. \tag{4.28}$$

The equilibrium of $(x^*_{j_2}, x^*_{j_1}) = (a_{j_1j_2 1}, a_{j_2j_1 1})$ *is a* $(\mathrm{UP}_-, 3^{\mathrm{rd}}\mathrm{UP}_-)$*-third-order negative saddle.*

(iv) If

$$\Delta_{j_1j_2} < 0 \,\text{and}\, \Delta_{j_2j_1} > 0, \tag{4.29}$$

then

$$\begin{aligned} \dot{x}_{j_2} &= a_{j_2j_1 0}(x_{j_1} - a_{j_2j_1 1})(x_{j_1} - a_{j_2j_1 2})(x_{j_1} - a_{j_2j_1 3}), \\ \dot{x}_{j_1} &= a_{j_1j_2 0}(x_{j_2} - a_{j_1j_2 1})\left[(x_{j_2} - a_{j_1j_2})^2 + b_{j_1j_2}\right], \\ j_1, j_2 &\in \{1,2\}; j_1 \neq j_2 \end{aligned} \tag{4.30}$$

where

$$\begin{aligned} &\{a_{j_2j_1 1}, a_{j_2j_1 2}, a_{j_2j_1 3}\} = \mathrm{sort}\{b_{j_2j_1 1}, b_{j_2j_1 2}, b_{j_2j_1 3}\}, a_{j_2j_1 1} < a_{j_2j_1 2} < a_{j_2j_1 3}; \\ &b_{j_2j_1 2}, b_{j_2j_1 3} = -\frac{1}{2}(B_{j_2j_1} \pm \sqrt{\Delta_{j_2j_1}}), a_{j_2j_1 0} = A_{j_2j_1 0}; \\ &a_{j_1j_2} = -\frac{1}{2}B_{j_1j_2}, b_{j_1j_2} = \frac{1}{4}(-\Delta_{j_1j_2}). \end{aligned} \tag{4.31}$$

The first integral manifold is

$$\begin{aligned}
&a_{j_2j_10}\Big\{\frac{1}{4}\left[(x_{j_1}-a_{j_2j_1l_1})^4-(x_{j_10}-a_{j_2j_1l_1})^4\right]\\
&+\frac{1}{3}\sum\nolimits_{l_2=1,l_2\neq l_1}^{3}(a_{j_2j_1l_1}-a_{j_2j_1l_2})\left[(x_{j_1}-a_{j_2j_1l_1})^3-(x_{j_10}-a_{j_2j_1l_1})^3\right]\\
&+\frac{1}{2}\prod\nolimits_{2=1,l_2\neq l_1}^{3}(a_{j_2j_1l_1}-a_{j_2j_1l_2})\left[(x_{j_1}-a_{j_2j_1l_1})^2-(x_{j_10}-a_{j_2j_1l_1})^2\right]\Big\}\\
&=a_{j_1j_20}\Big\{\frac{1}{4}\left[(x_{j_2}-a_{j_1j_21})^4-(x_{j_20}-a_{j_1j_21})^4\right]\\
&+\frac{2}{3}(a_{j_1j_21}-a_{j_1j_2})\left[(x_{j_2}-a_{j_1j_21})^3-(x_{j_20}-a_{j_1j_21})^3\right]\\
&+\frac{1}{2}\left[(a_{j_1j_21}-a_{j_1j_2})^2+b_{j_1j_2}\right]\left[(x_{j_2}-a_{j_1j_21})^2-(x_{j_20}-a_{j_1j_21})^2\right]\Big\}.
\end{aligned}\tag{4.32}$$

The equilibrium of $(x_{j_2}^*,x_{j_1}^*)=(a_{j_1j_21},a_{j_2j_1l_1})(l_1,l_2\in\{1,2,3\},l_1\neq l_2)$ *has the following properties.*

- *For* $a_{j_1j_20}>0$ *and* $a_{j_2j_10}\prod_{l_2=1,l_2\neq l_1}^{3}(a_{j_2j_1l_1}-a_{j_2j_1l_2})>0$,

$$(a_{j_1j_21},a_{j_2j_1l_1})=\underbrace{(\mathrm{UP}_+,\mathrm{UP}_+)}_{\text{positive saddle}}.\tag{4.33}$$

The equilibrium of $(x_{j_2}^*,x_{j_1}^*)=(a_{j_1j_21},a_{j_2j_1l_1})$ *is a* $(\mathrm{UP}_+,\mathrm{UP}_+)$*-positive saddle.*

- *For* $a_{j_1j_20}<0$ *and* $a_{j_2j_10}\prod_{l_2=1,l_2\neq l_1}^{3}(a_{j_2j_1l_1}-a_{j_2j_1l_2})>0$,

$$(a_{j_1j_21},a_{j_2j_1l_1})=\underbrace{(\mathrm{DP}_+,\mathrm{DP}_-)}_{\text{CCW center}}.\tag{4.34}$$

The equilibrium of $(x_{j_2}^*,x_{j_1}^*)=(a_{j_1j_21},a_{j_2j_1l_1})$ *is a* $(\mathrm{DP}_+,\mathrm{DP}_-)$*-counter-clockwise center.*

- *For* $a_{j_1j_20}>0$ *and* $a_{j_2j_10}\prod_{l_2=1,l_2\neq l_1}^{3}(a_{j_2j_1l_1}-a_{j_2j_1l_2})<0$,

$$(a_{j_1j_21},a_{j_2j_1l_1})=\underbrace{(\mathrm{DP}_-,\mathrm{DP}_+)}_{\text{CW center}}.\tag{4.35}$$

The equilibrium of $(x_{j_2}^*,x_{j_1}^*)=(a_{j_1j_21},a_{j_2j_1l_1})$ *is a* $(\mathrm{DP}_-,\mathrm{DP}_+)$*-clockwise center.*

- *For* $a_{j_1j_20}<0$ *and* $a_{j_2j_10}\prod_{l_2=1,l_2\neq l_1}^{3}(a_{j_2j_1l_1}-a_{j_2j_1l_2})<0$,

$$(a_{j_1j_21},a_{j_2j_1l_1})=\underbrace{(\mathrm{UP}_-,\mathrm{UP}_-)}_{\text{negative saddle}}.\tag{4.36}$$

The equilibrium of $(x_{j_2}^*,x_{j_1}^*)=(a_{j_1j_21},a_{j_2j_1l_1})$ *is a* $(\mathrm{UP}_-,\mathrm{UP}_+)$*-negative saddle.*

(v) If

$$\Delta_{j_1j_2} = 0 \text{ and } \Delta_{j_2j_1} < 0, a_{j_1j_2s_1} \neq a_{j_1j_2s_2}, \tag{4.37}$$

then

$$\begin{aligned}
&\dot{x}_{j_1} = a_{j_1j_20}(x_{j_2} - a_{j_1j_2s_1})(x_{j_2} - a_{j_1j_2s_2})^2,\\
&\dot{x}_{j_2} = a_{j_2j_10}(x_{j_1} - a_{j_2j_11})\left[(x_{j_1} - a_{j_2j_1})^2 + b_{j_2j_1}\right]\\
&j_1, j_2 \in \{1,2\}, j_2 \neq j_1.
\end{aligned} \tag{4.38}$$

The first integral manifold is

$$\begin{aligned}
&a_{j_2j_10}\Big\{\frac{1}{4}\left[(x_{j_1} - a_{j_2j_11})^4 - (x_{j_10} - a_{j_2j_11})^4\right]\\
&+\frac{2}{3}(a_{j_2j_11} - a_{j_2j_1})\left[(x_{j_1} - a_{j_2j_11})^3 - (x_{j_10} - a_{j_2j_11})^3\right]\\
&+\frac{1}{2}\left[(a_{j_2j_11} - a_{j_2j_1})^2 + b_{j_2j_1}\right]\left[(x_{j_1} - a_{j_2j_11})^2 - (x_{j_10} - a_{j_2j_11})^2\right]\Big\}\\
&= a_{j_1j_20}\Big\{\frac{1}{4}\left[(x_{j_2} - a_{j_1j_2s_1})^4 - (x_{j_20} - a_{j_1j_2s_1})^4\right]\\
&+\frac{2}{3}(a_{j_1j_2s_1} - a_{j_1j_2s_2})\left[(x_{j_2} - a_{j_1j_2s_1})^3 - (x_{j_20} - a_{j_1j_2s_1})^3\right]\\
&+\frac{1}{2}(a_{j_1j_2s_1} - a_{j_1j_2s_2})^2\left[(x_{j_2} - a_{j_1j_2s_1})^2 - (x_{j_20} - a_{j_1j_2s_1})^2\right]\Big\}.
\end{aligned} \tag{4.39}$$

(v_1) The equilibrium of $(x^*_{j_2}, x^*_{j_1}) = (a_{j_1j_2s_1}, a_{j_1j_21})$ $(s_1, s_2 \in \{1,2\}, s_1 \neq s_2)$ *has the following properties.*

- *For* $a_{j_1j_20}(a_{j_1j_2s_1} - a_{j_1j_2s_2})^2 > 0$ *and* $a_{j_2j_10} > 0$,

$$(a_{j_1j_2s_1}, a_{j_1j_21}) = \underbrace{(\mathrm{UP}_+, \mathrm{UP}_+)}_{\text{positive saddle}}. \tag{4.40}$$

  The equilibrium of $(x^*_{j_2}, x^*_{j_1}) = (a_{j_1j_2s_1}, a_{j_1j_21})$ *is a* $(\mathrm{UP}_+,\mathrm{UP}_+)$*-positive saddle.*

- *For* $a_{j_1j_20}(a_{j_1j_2s_1} - a_{j_1j_2s_2})^2 < 0$ *and* $a_{j_2j_10} > 0$,

$$(a_{j_1j_2s_1}, a_{j_1j_21}) = \underbrace{(\mathrm{DP}_+, \mathrm{DP}_-)}_{\text{CCW center}}. \tag{4.41}$$

  The equilibrium of $(x^*_{j_2}, x^*_{j_1}) = (a_{j_1j_2s_1}, a_{j_1j_21})$ *is a* $(\mathrm{DP}_+,\mathrm{DP}_-)$*-counter-clockwise center.*

- *For* $a_{j_1j_20}(a_{j_1j_2s_1} - a_{j_1j_2s_2})^2 > 0$ *and* $a_{j_2j_10} < 0$,

$$(a_{j_1j_2s_1}, a_{j_1j_21}) = \underbrace{(\mathrm{DP}_-, \mathrm{DP}_+)}_{\text{CW center}}. \tag{4.42}$$

  The equilibrium of $(x^*_{j_2}, x^*_{j_1}) = (a_{j_1j_2s_1}, a_{j_1j_21})$ *is a* $(\mathrm{DP}_-,\mathrm{DP}_+)$*-clockwise center.*

- *For* $a_{j_1j_20}(a_{j_1j_2s_1}-a_{j_1j_2s_2})^2<0$ *and* $a_{j_2j_10}<0$,

$$(a_{j_1j_2s_1},a_{j_1j_21})=\underbrace{(\mathrm{UP}_-,\mathrm{UP}_-)}_{\text{negative saddle}}. \tag{4.43}$$

The equilibrium of $(x^*_{j_2},x^*_{j_1})=(a_{j_1j_2s_1},a_{j_1j_21})$ *is a* (UP$_-$,UP$_-$)*-negative saddle.*

(v₂) The equilibrium of $(x^*_{j_2},x^*_{j_1})=(a_{j_1j_2s_2},a_{j_2j_11})$ $(s_1,s_2\in\{1,2\},s_1\neq s_2)$ on the x_{j_2}-direction *has the following properties.*

- *For* $a_{j_1j_20}(a_{j_1j_2s_2}-a_{j_1j_2s_1})>0$ *and* $a_{j_2j_10}>0$,

$$(a_{j_1j_2s_2},a_{j_2j_11})=\underbrace{(\mathrm{US},\mathrm{UP})}_{\text{up-parabola upper-saddle}}. \tag{4.44}$$

The equilibrium $(x^*_{j_2},x^*_{j_1})=(a_{j_1j_2s_2},a_{j_2j_11})$ *is a* (US,UP)*-up-parabola upper-saddle.*

- *For* $a_{j_1j_20}(a_{j_1j_2s_2}-a_{j_1j_2s_1})<0$ *and* $a_{j_2j_10}>0$,

$$(a_{j_1j_2s_2},a_{j_2j_11})=\underbrace{(\mathrm{LS},\mathrm{DP})}_{\text{down-parabola lower-saddle}}. \tag{4.45}$$

The equilibrium $(x^*_{j_2},x^*_{j_1})=(a_{j_1j_2s_2},a_{j_2j_11})$ *is an* (LS,DP)*-down-parabola lower-saddle.*

- *For* $a_{j_1j_20}(a_{j_1j_2s_2}-a_{j_1j_2s_1})>0$ *and* $a_{j_2j_10}<0$,

$$(a_{j_1j_2s_2},a_{j_2j_11})=\underbrace{(\mathrm{US},\mathrm{DP})}_{\text{down-parabola upper-saddle}}. \tag{4.46}$$

The equilibrium $(x^*_{j_2},x^*_{j_1})=(a_{j_1j_2s_2},a_{j_2j_11})$ *is a* (US,DP)*-down-parabola upper-saddle.*

- *For* $a_{j_1j_20}(a_{j_1j_2s_2}-a_{j_1j_2s_1})<0$ *and* $a_{j_2j_10}<0$,

$$(a_{j_1j_2s_2},a_{j_2j_11})=\underbrace{(\mathrm{LS},\mathrm{UP})}_{\text{up-parabola lower-saddle}}. \tag{4.47}$$

The equilibrium $(x^*_{j_2},x^*_{j_1})=(a_{j_1j_2s_2},a_{j_2j_11})$ *is an* (LS,UP)*-up-parabola lower-saddle.*

(vi) If

$$\Delta_{j_1j_2}=0 \text{ and } \Delta_{j_2j_1}<0, a_{j_1j_2s_1}=a_{j_1j_2s_2}=a_{j_1j_21}, \tag{4.48}$$

then

$$\begin{aligned}
&\dot{x}_{j_1} = a_{j_1j_20}(x_{j_2} - a_{j_1j_21})^3,\\
&\dot{x}_{j_2} = a_{j_2j_10}(x_{j_1} - a_{j_2j_11})\left[(x_{j_1} - a_{j_2j_1})^2 + b_{j_2j_1}\right]\\
&j_1, j_2 \in \{1,2\}, j_2 \neq j_1.
\end{aligned} \tag{4.49}$$

The first integral manifold is

$$\begin{aligned}
&a_{j_2j_10}\Big\{\frac{1}{4}\left[(x_{j_1} - a_{j_2j_11})^4 - (x_{j_10} - a_{j_2j_11})^4\right]\\
&+\frac{2}{3}(a_{j_2j_11} - a_{j_2j_1})\left[(x_{j_1} - a_{j_2j_11})^3 - (x_{j_10} - a_{j_2j_11})^3\right]\\
&+\frac{1}{2}\left[(a_{j_2j_11} - a_{j_2j_1})^2 + b_{j_2j_1}\right]\left[(x_{j_1} - a_{j_2j_11})^2 - (x_{j_10} - a_{j_2j_11})^2\right]\Big\}\\
&= \frac{1}{4}a_{j_1j_20}\left[(x_{j_2} - a_{j_1j_21})^4 - (x_{j_20} - a_{j_1j_21})^4\right].
\end{aligned} \tag{4.50}$$

The equilibrium of $(x^*_{j_2}, x^*_{j_1}) = (a_{j_1j_21}, a_{j_2j_11})$ *has the following properties.*

- *For* $a_{j_1j_20} > 0$ *and* $a_{j_2j_10} > 0$,

$$(a_{j_1j_21}, a_{j_2j_11}) = \underbrace{(3^{\text{rd}}\text{UP}_+, \text{UP}_+)}_{\text{third-order positive saddle}}. \tag{4.51}$$

The equilibrium of $(x^*_{j_2}, x^*_{j_1}) = (a_{j_1j_21}, a_{j_2j_11})$ *is a* $(3^{\text{rd}}\text{UP}_+,\text{UP}_+)$*-third-order positive saddle.*

- *For* $a_{j_1j_20} < 0$ *and* $a_{j_2j_10} > 0$,

$$(a_{j_1j_21}, a_{j_2j_11}) = \underbrace{(3^{\text{rd}}\text{DP}_+, \text{DP}_-)}_{\text{third-order CCW center}}. \tag{4.52}$$

The equilibrium of $(x^*_{j_2}, x^*_{j_1}) = (a_{j_1j_21}, a_{j_2j_11})$ *is a* $(3^{\text{rd}}\text{DP}_+,\text{DP}_-)$*-third-order counter-clockwise center.*

- *For* $a_{j_1j_20} > 0$ *and* $a_{j_2j_10} < 0$,

$$(a_{j_1j_21}, a_{j_2j_11}) = \underbrace{(3^{\text{rd}}\text{DP}_-, \text{DP}_+)}_{\text{third-order CW center}}. \tag{4.53}$$

The equilibrium of $(x^*_{j_2}, x^*_{j_1}) = (a_{j_1j_21}, a_{j_2j_11})$ *is a* $(3^{\text{rd}}\text{DP}_-,\text{DP}_+)$*-third-order clockwise center.*

- *For* $a_{j_1j_20} < 0$ *and* $a_{j_2j_10} < 0$,

$$(a_{j_1j_21}, a_{j_2j_11}) = \underbrace{(3^{\text{rd}}\text{UP}_-, \text{UP}_-)}_{\text{third-order negative saddle}}. \tag{4.54}$$

The equilibrium of $(x^*_{j_2}, x^*_{j_1}) = (a_{j_1j_21}, a_{j_2j_11})$ *is a* $(3^{\text{rd}}\text{UP}_-,\text{UP}_-)$*-third-order negative saddle.*

(vii) If

$$\Delta_{j_1j_2}=0\,\text{and}\,\Delta_{j_2j_1}=0,\,a_{j_2j_1s_1}\neq a_{j_2j_1s_2}, \tag{4.55}$$

then

$$\begin{aligned}
&\dot{x}_{j_1}=a_{j_1j_20}(x_{j_2}-a_{j_1j_2s_1})(x_{j_2}-a_{j_1j_2s_2})^2,\\
&\dot{x}_{j_2}=a_{j_2j_10}(x_{j_1}-a_{j_2j_1l_1})(x_{j_1}-a_{j_2j_1l_2})^2,\\
&j_1,j_2\in\{1,2\};j_1\neq j_2.
\end{aligned} \tag{4.56}$$

The first integral manifold is

$$\begin{aligned}
&a_{j_2j_10}\Big\{\frac{1}{4}\left[(x_{j_1}-a_{j_2j_1l_1})^4-(x_{j_10}-a_{j_2j_1l_1})^4\right]\\
&+\frac{2}{3}(a_{j_2j_1l_1}-a_{j_2j_1l_2})\left[(x_{j_1}-a_{j_2j_1l_1})^3-(x_{j_10}-a_{j_2j_1l_1})^3\right]\\
&+\frac{1}{2}(a_{j_2j_1l_1}-a_{j_2j_1l_2})^2\left[(x_{j_1}-a_{j_2j_1l_1})^2-(x_{j_10}-a_{j_2j_1l_1})^2\right]\Big\}\\
&=a_{j_1j_20}\Big\{\frac{1}{4}\left[(x_{j_1}-a_{j_1j_2s_1})^4-(x_{j_10}-a_{j_1j_2s_1})^4\right]\\
&+\frac{2}{3}(a_{j_1j_2s_1}-a_{j_1j_2s_2})\left[(x_{j_2}-a_{j_1j_2s_1})^3-(x_{j_20}-a_{j_1j_2s_1})^3\right]\\
&+\frac{1}{2}\left[(a_{j_1j_2s_1}-a_{j_1j_2s_2})^2\right]\left[(x_{j_2}-a_{j_1j_2s_1})^2-(x_{j_20}-a_{j_1j_2s_1})^2\right]\Big\}.
\end{aligned} \tag{4.57}$$

(vii_1) The equilibrium of $(x^*_{j_2},x^*_{j_1})=(a_{j_1j_2s_1},a_{j_2j_il_1})(s_1,s_2,l_1,l_2\in\{1,2\},s_1\neq s_2,l_1\neq l_2)$ *has the following properties.*

- *For* $a_{j_1j_20}(a_{j_1j_2s_1}-a_{j_1j_2s_2})^2>0$ *and* $a_{j_2j_10}(a_{j_2j_1l_1}-a_{j_2j_1l_2})^2>0$*,*

$$(a_{j_1j_2s_1},a_{j_2j_1l_1})=\underbrace{(\mathrm{UP}_+,\mathrm{UP}_+)}_{\text{positive saddle}}. \tag{4.58}$$

  The equilibrium of $(x^*_{j_2},x^*_{j_1})=(a_{j_1j_2s_1},a_{j_2j_1l_1})$ *is a* (UP$_+$,UP$_+$)*-positive saddle.*

- *For* $a_{j_1j_20}(a_{j_1j_2s_1}-a_{j_1j_2s_2})^2<0$ *and* $a_{j_2j_10}(a_{j_2j_1l_1}-a_{j_2j_1l_2})^2>0$*,*

$$(a_{j_1j_2s_1},a_{j_2j_1l_1})=\underbrace{(\mathrm{DP}_+,\mathrm{DP}_-)}_{\text{CCW center}}. \tag{4.59}$$

  The equilibrium of $(x^*_{j_2},x^*_{j_1})=(a_{j_1j_2s_1},a_{j_2j_1l_1})$ *is a* (DP$_+$,DP$_-$)*-counter-clockwise center.*

- *For* $a_{j_1j_20}(a_{j_1j_2s_1}-a_{j_1j_2s_2})^2>0$ *and* $a_{j_2j_10}(a_{j_2j_1l_1}-a_{j_2j_1l_2})^2<0$*,*

$$(a_{j_1j_2s_1},a_{j_2j_1l_1})=\underbrace{(\mathrm{DP}_-,\mathrm{DP}_+)}_{\text{CW center}}. \tag{4.60}$$

  The equilibrium of $(x^*_{j_2},x^*_{j_1})=(a_{j_1j_2s_1},a_{j_2j_1l_1})$ *is a* (DP$_-$,DP$_+$)*-clockwise center.*

- *For* $a_{j_1j_20}(a_{j_1j_2s_1} - a_{j_1j_2s_2})^2 < 0$ *and* $a_{j_2j_10}(a_{j_2j_1l_1} - a_{j_2j_1l_2})^2 < 0$,

$$(a_{j_1j_2s_1}, a_{j_2j_1l_1}) = \underbrace{(\mathrm{UP}_-, \mathrm{UP}_-)}_{\text{negative saddle}}. \tag{4.61}$$

The equilibrium of $(x^*_{j_2}, x^*_{j_1}) = (a_{j_1j_2s_1}, a_{j_2j_1l_1})$ *is a* (UP$_-$,UP$_-$)*-negative saddle.*

(vii$_2$) The equilibrium of $(x^*_{j_2}, x^*_{j_1}) = (a_{j_1j_2s_2}, a_{j_2j_1l_1})$ $(s_1, s_2, l_1, l_2 \in \{1,2\}, s_1 \neq s_2, l_1 \neq l_2)$ on the x_{j_2}-direction *has the following properties.*

- *For* $a_{j_1j_20}(a_{j_1j_2s_2} - a_{j_1j_2s_1}) > 0$ *and* $a_{j_2j_10}(a_{j_2j_1l_1} - a_{j_2j_1l_2})^2 > 0$,

$$(a_{j_1j_2s_2}, a_{j_2j_1l_1}) = \underbrace{(\mathrm{US}, \mathrm{UP})}_{\text{up-parabola upper-saddle}}. \tag{4.62}$$

The equilibrium of $(x^*_{j_2}, x^*_{j_1}) = (a_{j_1j_2s_2}, a_{j_2j_1l_1})$ *is a* (US,UP)*-up-parabola upper-saddle.*

- *For* $a_{j_1j_20}(a_{j_1j_2s_2} - a_{j_1j_2s_1}) < 0$ *and* $a_{j_2j_10}(a_{j_2j_1l_1} - a_{j_2j_1l_2})^2 > 0$,

$$(a_{j_1j_2s_2}, a_{j_2j_1l_1}) = \underbrace{(\mathrm{LS}, \mathrm{DP})}_{\text{down-parabola lower-saddle}} \tag{4.63}$$

The equilibrium of $(x^*_{j_2}, x^*_{j_1}) = (a_{j_1j_2s_2}, a_{j_2j_1l_1})$ *is an* (LS,DP)*-down-parabola lower-saddle.*

- *For* $a_{j_1j_20}(a_{j_1j_2s_2} - a_{j_1j_2s_1}) > 0$ *and* $a_{j_2j_10}(a_{j_2j_1l_1} - a_{j_2j_1l_2})^2 < 0$,

$$(a_{j_1j_2s_2}, a_{j_2j_1l_1}) = \underbrace{(\mathrm{US}, \mathrm{DP})}_{\text{down-parabola upper-saddle}}. \tag{4.64}$$

The equilibrium of $(x^*_{j_2}, x^*_{j_1}) = (a_{j_1j_2s_2}, a_{j_2j_1l_1})$ *is a* (US,DP)*-down-parabola upper-saddle.*

- *For* $a_{j_1j_20}(a_{j_1j_2s_2} - a_{j_1j_2s_1}) < 0$ *and* $a_{j_2j_10}(a_{j_2j_1l_1} - a_{j_2j_1l_2})^2 < 0$,

$$(a_{j_1j_2s_2}, a_{j_2j_1l_1}) = \underbrace{(\mathrm{LS}, \mathrm{UP})}_{\text{up-parabola lower-saddle}}. \tag{4.65}$$

The equilibrium of $(x^*_{j_2}, x^*_{j_1}) = (a_{j_1j_2s_2}, a_{j_2j_1l_1})$ *is an* (LS,UP)*-up-parabola lower-saddle.*

(vii$_3$) The equilibrium of $(x^*_{j_2}, x^*_{j_1}) = (a_{j_1j_2s_1}, a_{j_2j_1l_2})$ $(s_1, s_2, l_1, l_2 \in \{1,2\}, s_1 \neq s_2, l_1 \neq l_2)$ on the x_{j_2}-direction *has the following properties.*

- *For* $a_{j_1j_20}(a_{j_1j_2s_1}-a_{j_1j_2s_2})^2>0$ *and* $a_{j_2j_10}(a_{j_2j_1l_2}-a_{j_2j_1l_1})>0$,

$$(a_{j_1j_2s_1},a_{j_2j_1l_2})=\underbrace{(\mathrm{UP},\mathrm{US})}_{\text{up-parabola upper-saddle}}. \tag{4.66}$$

The equilibrium of $(x^*_{j_2},x^*_{j_1})=(a_{j_1j_2s_1},a_{j_2j_1l_2})$ *is a* (UP,US)*-up-parabola upper-saddle.*

- *For* $a_{j_1j_20}(a_{j_1j_2s_1}-a_{j_1j_2s_2})^2<0$ *and* $a_{j_2j_10}(a_{j_2j_1l_2}-a_{j_2j_1l_1})>0$,

$$(a_{j_1j_2s_1},a_{j_2j_1l_2})=\underbrace{(\mathrm{DP},\mathrm{US})}_{\text{down-parabola upper-saddle}}. \tag{4.67}$$

The equilibrium of $(x^*_{j_2},x^*_{j_1})=(a_{j_1j_2s_1},a_{j_2j_1l_2})$ *is a* (DP,US)*-down-parabola upper-saddle.*

- *For* $a_{j_1j_20}(a_{j_1j_2s_1}-a_{j_1j_2s_2})^2>0$ *and* $a_{j_2j_10}(a_{j_2j_1l_2}-a_{j_2j_1l_1})<0$,

$$(a_{j_1j_2s_1},a_{j_2j_1l_2})=\underbrace{(\mathrm{DP},\mathrm{LS})}_{\text{down-parabola lower-saddle}}. \tag{4.68}$$

The equilibrium of $(x^*_{j_2},x^*_{j_1})=(a_{j_1j_2s_1},a_{j_2j_1l_2})$ *is a* (DP,LS)*-down-parabola lower-saddle.*

- *For* $a_{j_1j_20}(a_{j_1j_2s_1}-a_{j_1j_2s_2})^2<0$ *and* $a_{j_2j_10}(a_{j_2j_1l_2}-a_{j_2j_1l_1})<0$,

$$(a_{j_1j_2s_1},a_{j_2j_1l_2})=\underbrace{(\mathrm{UP},\mathrm{LS})}_{\text{up-parabola lower-saddle}}. \tag{4.69}$$

The equilibrium of $(x^*_{j_2},x^*_{j_1})=(a_{j_1j_2s_1},a_{j_2j_1l_2})$ *is a* (UP,LS)*-up-parabola lower-saddle.*

(vii$_4$*) The equilibrium of* $(x^*_{j_2},x^*_{j_1})=(a_{j_1j_2s_2},a_{j_2j_1l_2})(s_1,s_2,l_1,l_2\in\{1,2\},s_1\neq s_2,l_1\neq l_2)$ *has the following properties.*

- *For* $a_{j_1j_20}(a_{j_1j_2s_2}-a_{j_1j_2s_1})>0$ *and* $a_{j_2j_10}(a_{j_2j_1l_2}-a_{j_2j_1l_1})>0$,

$$(a_{j_1j_2s_2},a_{j_2j_1l_2})=\underbrace{(\mathrm{II}_+,\mathrm{II}_+)}_{\text{double-inflection saddle}}. \tag{4.70}$$

The equilibrium of $(x^*_{j_2},x^*_{j_1})=(a_{j_1j_2s_2},a_{j_2j_1l_2})$ *is an* (II$_+$,II$_+$)*-double increasing-inflection saddle.*

- *For* $a_{j_1j_20}(a_{j_1j_2s_2}-a_{j_1j_2s_1})<0$ *and* $a_{j_2j_o0}(a_{j_2j_1l_2}-a_{j_2j_1l_1})>0$,

$$(a_{j_1j_2s_2},a_{j_2j_1l_2})=\underbrace{(\mathrm{DI}_+,\mathrm{DI}_-)}_{\text{double-inflection saddle}}. \tag{4.71}$$

The equilibrium of $(x^*_{j_2},x^*_{j_1})=(a_{j_1j_2s_2},a_{j_2j_1l_2})$ *is a* (DI$_+$,DI$_-$)*-double decreasing-inflection saddle.*

- *For* $a_{j_1j_20}(a_{j_1j_2s_2}-a_{j_1j_2s_1})>0$ *and* $a_{j_2j_10}(a_{j_2j_1l_2}-a_{j_2j_1l_1})<0$,

$$(a_{j_1j_2s_2},a_{j_2j_1l_2})=\underbrace{(\mathrm{DI}_-,\mathrm{DI}_+)}_{\text{double-inflection saddle}}. \tag{4.72}$$

The equilibrium of $(x^*_{j_2},x^*_{j_1})=(a_{j_1j_2s_2},a_{j_2j_1l_2})$ *is a* (DI_-,DI_+)*-double decreasing-inflection saddle.*

- *For* $a_{j_1j_20}(a_{j_1j_2s_2}-a_{j_1j_2s_1})<0$ *and* $a_{j_2j_10}(a_{j_2j_1l_2}-a_{j_2j_1l_1})<0$,

$$(a_{j_1j_2s_2},a_{j_2j_1l_2})=\underbrace{(\mathrm{II}_-,\mathrm{II}_-)}_{\text{double-inflection}}. \tag{4.73}$$

The equilibrium of $(x^*_{j_2},x^*_{j_1})=(a_{j_1j_2s_2},a_{j_2j_1l_2})$ *is an* (II_-,II_-)*-double increasing-inflection saddle.*

(viii) If

$$\begin{aligned}&\Delta_{j_1j_2}=0\text{ and }\Delta_{j_2j_1}=0,\\&a_{j_2j_1s_1}=a_{j_2j_1s_2}=a_{j_2j_11};a_{j_1j_2l_1}\neq a_{j_1j_2l_2};\end{aligned} \tag{4.74}$$

then

$$\begin{aligned}&\dot{x}_{j_1}=a_{j_1j_20}(x_{j_2}-a_{j_1j_21})^3,\\&\dot{x}_{j_2}=a_{j_2j_10}(x_{j_1}-a_{j_2j_1l_1})(x_{j_1}-a_{j_2j_1l_2})^2,\\&j_1,j_2\in\{1,2\};j_1\neq j_2.\end{aligned} \tag{4.75}$$

The first integral manifold is

$$\begin{aligned}&a_{j_2j_10}\{\frac{1}{4}\left[(x_{j_1}-a_{j_2j_1l_1})^4-(x_{j_10}-a_{j_2j_1i_1})^4\right]\\&+\frac{2}{3}(a_{j_2j_1l_1}-a_{j_2j_1l_2})\left[(x_{j_1}-a_{j_2j_1l_1})^3-(x_{j_10}-a_{j_2j_il_1})^3\right]\\&+\frac{1}{2}(a_{j_2j_1l_1}-a_{j_2j_il_2})^2\left[(x_{j_1}-a_{j_2j_il_1})^2-(x_{j_10}-a_{j_2j_1l_1})^2\right]\}\\&=\frac{1}{4}a_{j_1j_20}\left[(x_{j_2}-a_{j_1j_21})^4-(x_{j_20}-a_{j_1j_21})^4\right].\end{aligned} \tag{4.76}$$

(viii$_1$) The equilibrium of $(x^*_{j_2},x^*_{j_1})=(a_{j_1j_21},a_{j_2j_1l_1})$ $(l_1,l_2\in\{1,2\},l_1\neq l_2)$ *has the following properties.*

- *For* $a_{j_1j_20}>0$ *and* $a_{j_2j_10}(a_{j_2j_1l_1}-a_{j_2j_1l_2})^2>0$,

$$(a_{j_1j_21},a_{j_2j_1l_1})=\underbrace{(3^{\mathrm{rd}}\mathrm{UP}_+,\mathrm{UP}_+)}_{\text{third-order positive saddle}}. \tag{4.77}$$

The equilibrium of $(x^*_{j_2},x^*_{j_1})=(a_{j_1j_21},a_{j_2j_1l_1})$ *is a* ($3^{\mathrm{rd}}\mathrm{UP}_+$,$\mathrm{UP}_+$)*-third-order positive saddle.*

- *For* $a_{j_1j_20}<0$ *and* $a_{j_2j_10}(a_{j_2j_1l_1}-a_{j_2j_1l_2})^2>0$,

$$(a_{j_1j_21},a_{j_2j_1l_1})=\underbrace{(3^{\rm rd}{\rm DP}_+,{\rm DP}_-)}_{\text{third-order CCW center}}. \tag{4.78}$$

The equilibrium of $(x^*_{j_2},x^*_{j_1})=(a_{j_1j_21},a_{j_2j_1l_1})$ *is a* ($3^{\rm rd}$DP$_+$,DP$_-$)*-third-order-counter-clockwise center.*

- *For* $a_{j_1j_20}>0$ *and* $a_{j_2j_10}(a_{j_2j_1l_1}-a_{j_2j_1l_2})^2<0$,

$$(a_{j_1j_21},a_{j_2j_1l_1})=\underbrace{(3^{\rm rd}{\rm DP}_-,{\rm DP}_+)}_{\text{third-order CW center}}. \tag{4.79}$$

The equilibrium of $(x^*_{j_2},x^*_{j_1})=(a_{j_1j_21},a_{j_2j_1l_1})$ *is a* ($3^{\rm rd}$DP$_-$,DP$_+$)*-third-order-clockwise center.*

- *For* $a_{j_1j_20}<0$ *and* $a_{j_2j_10}(a_{j_2j_1l_1}-a_{j_2j_1l_2})^2<0$,

$$(a_{j_1j_21},a_{j_2j_1l_1})=\underbrace{(3^{\rm rd}{\rm UP}_-,{\rm UP}_-)}_{\text{third-order negative saddle}}. \tag{4.80}$$

The equilibrium of $(x^*_{j_2},x^*_{j_1})=(a_{j_1j_21},a_{j_2j_1l_1})$ *is a* ($3^{\rm rd}$UP$_-$,UP$_-$)*-third-order negative saddle.*

(viii$_2$) The equilibrium of $(x^*_{j_2},x^*_{j_1})=(a_{j_1j_21},a_{j_2j_1l_2})(l_1,l_2\in\{1,2\},l_1\neq l_2)$ *has the following properties.*

- *For* $a_{j_1j_20}>0$ *and* $a_{j_2j_10}(a_{j_2j_1l_2}-a_{j_2j_1l_1})>0$,

$$(a_{j_1j_21},a_{j_2j_1l_2})=\underbrace{(3^{\rm rd}{\rm UP},{\rm US})}_{(3,2)\text{-up-parabola upper-saddle}}. \tag{4.81}$$

The equilibrium of $(x^*_{j_2},x^*_{j_1})=(a_{j_1j_21},a_{j_2j_1l_2})$ *is a* ($3^{\rm rd}$UP,US)*-third-order up-parabola upper-saddle.*

- *For* $a_{j_1j_20}<0$ *and* $a_{j_2j_10}(a_{j_2j_1l_2}-a_{j_2j_1l_1})>0$,

$$(a_{j_1j_21},a_{j_2j_1l_2})=\underbrace{(3^{\rm rd}{\rm DP},{\rm US})}_{(3,2)\text{-down-parabola upper-saddle}}. \tag{4.82}$$

The equilibrium of $(x^*_{j_2},x^*_{j_1})=(a_{j_1j_21},a_{j_2j_1l_2})$ *is a* ($3^{\rm rd}$DP,US)*-third-order down-parabola upper-saddle.*

- *For* $a_{j_1j_20}>0$ *and* $a_{j_2j_10}(a_{j_2j_1l_2}-a_{j_2j_1l_1})<0$,

$$(a_{j_1j_21},a_{j_2j_1l_2})=\underbrace{(3^{\rm rd}{\rm DP,LS})}_{(3,2)\text{-down-parabola lower-saddle}}. \tag{4.83}$$

The equilibrium of $(x_{j_2}^*,x_{j_1}^*)=(a_{j_1j_21},a_{j_2j_1l_2})$ *is a* ($3^{\rm rd}$DP,LS)*-third-order down-parabola lower-saddle.*

- *For* $a_{j_1j_20}<0$ *and* $a_{j_2j_10}(a_{j_2j_1l_2}-a_{j_2j_1l_1})<0$,

$$(a_{j_1j_21},a_{j_2j_1l_2})=\underbrace{(3^{\rm rd}{\rm UP,LS})}_{(3,2)\text{-up-parabola lower-saddle}}. \tag{4.84}$$

The equilibrium of $(x_{j_2}^*,x_{j_1}^*)=(a_{j_2j_11},a_{j_1j_2l_2})$ *is a* ($3^{\rm rd}$UP,LS)*-third-order up-parabola lower-saddle.*

(ix) If

$$\begin{aligned}&\Delta_{j_1j_2}=0 \text{ and } \Delta_{j_2j_1}=0,\\&a_{j_2j_1s_1}\neq a_{j_2j_1s_2};a_{j_1j_2l_1}=a_{j_1j_2l_2}=a_{j_1j_21};\end{aligned} \tag{4.85}$$

then

$$\begin{aligned}&\dot{x}_{j_1}=a_{j_1j_20}(x_{j_2}-a_{j_1j_2s_1})(x_{j_2}-a_{j_1j_2s_2})^2,\\&\dot{x}_{j_2}=a_{j_2j_10}(x_{j_1}-a_{j_2j_11})^3,\\&j_1,j_2\in\{1,2\};j_1\neq j_2.\end{aligned} \tag{4.86}$$

The first integral manifold is

$$\begin{aligned}&\frac{1}{4}a_{j_2j_10}\left[(x_{j_1}-a_{j_2j_1})^4-(x_{j_10}-a_{j_2j_11})^4\right]\\&=a_{j_1j_20}\Big\{\frac{1}{4}\left[(x_{j_2}-a_{j_1j_2s_1})^4-(x_{j_20}-a_{j_1j_2s_1})^4\right]\\&+\frac{2}{3}(a_{j_1j_2s_1}-a_{j_1j_2s_2})\left[(x_{j_1}-a_{j_1j_2s_1})^3-(x_{j_10}-a_{j_1j_2s_1})^3\right]\\&+\frac{1}{2}(a_{j_1j_2s_1}-a_{j_1j_2s_2})^2\left[(x_{j_1}-a_{j_1j_2s_1})^2-(x_{j_10}-a_{j_1j_2s_1})^2\right]\Big\}.\end{aligned} \tag{4.87}$$

(viii$_I$) The equilibrium of $(x_{j_2}^*,x_{j_1}^*)=(a_{j_1j_2s_1},a_{j_2j_11})(s_1,s_2\in\{1,2\},s_1\neq s_2)$ *has the following properties.*

- *For* $a_{j_1j_20}(a_{j_1j_2s_1}-a_{j_1j_2s_2})^2>0$ *and* $a_{j_2j_10}>0$,

$$(a_{j_1j_2s_1},a_{j_2j_11})=\underbrace{({\rm UP}_+,3^{\rm rd}{\rm UP}_+)}_{\text{third-order positive saddle}}. \tag{4.88}$$

The equilibrium of $(x_{j_2}^*,x_{j_1}^*)=(a_{j_1j_2s_1},a_{j_2j_11})$ *is a* (UP$_+$,$3^{\rm rd}$UP$_+$)*-third-order positive saddle.*

- *For* $a_{j_1j_20}(a_{j_1j_2s_1}-a_{j_1j_2s_2})^2<0$ *and* $a_{j_2j_10}>0$,

$$(a_{j_1j_2s_1},a_{j_2j_11})=\underbrace{(\mathrm{DP}_+,3^{\mathrm{rd}}\mathrm{DP}_-)}_{\text{third-order CCW center}}. \tag{4.89}$$

The equilibrium of $(x^*_{j_2},x^*_{j_1})=(a_{j_1j_2s_1},a_{j_2j_11})$ *is a* $(\mathrm{DP}_+,3^{\mathrm{rd}}\mathrm{DP}_-)$*-third-order-counter-clockwise center.*

- *For* $a_{j_1j_20}(a_{j_1j_2s_1}-a_{j_1j_2s_2})^2>0$ *and* $a_{j_2j_10}<0$,

$$(a_{j_1j_2s_1},a_{j_2j_11})=\underbrace{(\mathrm{DP}_-,3^{\mathrm{rd}}\mathrm{DP}_+)}_{\text{third-order CW center}}. \tag{4.90}$$

The equilibrium of $(x^*_{j_2},x^*_{j_1})=(a_{j_1j_2s_1},a_{j_2j_11})$ *is a* $(\mathrm{DP}_-,3^{\mathrm{rd}}\mathrm{DP}_+)$*-third-order-clockwise center.*

- *For* $a_{j_1j_20}(a_{j_1j_2s_1}-a_{j_1j_2s_2})^2<0$ *and* $a_{j_2j_10}<0$,

$$(a_{j_1j_2s_1},a_{j_2j_11})=\underbrace{(\mathrm{UP}_-,3^{\mathrm{rd}}\mathrm{UP}_-)}_{\text{third-order negative saddle}}. \tag{4.91}$$

The equilibrium of $(x^*_{j_2},x^*_{j_1})=(a_{j_1j_2s_1},a_{j_2j_11})$ *is a* $(\mathrm{UP}_-,3^{\mathrm{rd}}\mathrm{UP}_-)$*-third-order negative saddle.*

(ix$_2$*) The equilibrium of* $(x^*_{j_2},x^*_{j_1})=(a_{j_1j_2s_2},a_{j_2j_11})(s_1,s_2\in\{1,2\},s_1\neq s_2)$ *has the following properties.*

- *For* $a_{j_1j_20}(a_{j_1j_2s_2}-a_{j_1j_2s_1})>0$ *and* $a_{j_2j_10}>0$,

$$(a_{j_1j_2s_2},a_{j_2j_11})=\underbrace{(\mathrm{US},3^{\mathrm{rd}}\mathrm{UP})}_{\text{third-order up-parabola upper-saddle}}. \tag{4.92}$$

The equilibrium of $(x^*_{j_2},x^*_{j_1})=(a_{j_1j_2s_2},a_{j_2j_11})$ *is a* $(\mathrm{US},3^{\mathrm{rd}}\mathrm{UP})$*-third-order up-parabola upper-saddle.*

- *For* $a_{j_1j_20}(a_{j_1j_2s_2}-a_{j_1j_2s_1})<0$ *and* $a_{j_2j_10}>0$,

$$(a_{j_1j_2s_2},a_{j_2j_11})=\underbrace{(\mathrm{LS},3^{\mathrm{rd}}\mathrm{DP})}_{\text{third-order down-parabola lower-saddle}}. \tag{4.93}$$

The equilibrium of $(x^*_{j_2},x^*_{j_1})=(a_{j_1j_2s_2},a_{j_2j_11})$ *is an* $(\mathrm{LS},3^{\mathrm{rd}}\mathrm{DP})$*-third-order down-parabola lower-saddle.*

- *For* $a_{j_1 j_2 0}(a_{j_1 j_2 s_2} - a_{j_1 j_2 s_1}) > 0$ *and* $a_{j_2 j_1 0} < 0$,

$$(a_{j_1 j_2 s_2}, a_{j_2 j_1 1}) = \underbrace{(\text{US}, 3^{\text{rd}}\text{DP})}_{\text{third-order down-parabola upper-saddle}}. \tag{4.94}$$

The equilibrium of $(x^*_{j_2}, x^*_{j_1}) = (a_{j_1 j_2 s_2}, a_{j_2 j_1 1})$ *is a* (US,3rdDP)*-third-order down-parabola upper-saddle.*

- *For* $a_{j_1 j_2 0}(a_{j_1 j_2 s_2} - a_{j_1 j_2 s_1}) < 0$ *and* $a_{j_2 j_1 0} < 0$,

$$(a_{j_1 j_2 s_2}, a_{j_2 j_1 1}) = \underbrace{(\text{LS}, 3^{\text{rd}}\text{UP})}_{\text{third-order up-parabola lower-saddle}}. \tag{4.95}$$

The equilibrium of $(x^*_{j_2}, x^*_{j_1}) = (a_{j_1 j_2 s_2}, a_{j_2 j_1 1})$ *is an* (LS,3rdUP)*-third-order up-parabola lower-saddle.*

(x) If

$$\begin{aligned} &\Delta_{j_1 j_2} = 0 \text{ and } \Delta_{j_2 j_1} = 0, \\ &a_{j_2 j_1 s_1} = a_{j_2 j_1 s_2} = a_{j_2 j_1 1}; a_{j_1 j_2 l_1} = a_{j_1 j_2 l_2} = a_{j_1 j_2 1}; \end{aligned} \tag{4.96}$$

then

$$\begin{aligned} &\dot{x}_{j_1} = a_{j_1 j_2 0}(x_{j_2} - a_{j_1 j_2 1})^3, \\ &\dot{x}_{j_2} = a_{j_2 j_0 0}(x_{j_1} - a_{j_2 j_1 1})^3, \\ &j_1, j_2 \in \{1, 2\}; j_1 \neq j_2. \end{aligned} \tag{4.97}$$

The first integral manifold is

$$\begin{aligned} &\frac{1}{4} a_{j_2 j_1 0}\left[(x_{j_1} - a_{j_2 j_1 1})^4 - (x_{j_1 0} - a_{j_2 j_1 1})^4\right] \\ &= \frac{1}{4} a_{j_1 j_2 0}\left[(x_{j_2} - a_{j_1 j_2 1})^4 - (x_{j_2 0} - a_{j_1 j_2 1})^4\right]. \end{aligned} \tag{4.98}$$

The equilibrium of $(x^*_{j_2}, x^*_{j_1}) = (a_{j_1 j_2 1}, a_{j_2 j_1 1})$ *has the following properties.*

- *For* $a_{j_1 j_2 0} > 0$ *and* $a_{j_2 j_1 0} > 0$,

$$(a_{j_1 j_2 1}, a_{j_2 j_1 1}) = \underbrace{(3^{\text{rd}}\text{UP}_+, 3^{\text{rd}}\text{UP}_+)}_{\text{double third-order positive saddle}}. \tag{4.99}$$

The equilibrium of $(x^*_{j_2}, x^*_{j_1}) = (a_{j_1 j_2 1}, a_{j_2 j_1 1})$ *is a* (3rdUP$_+$,3rdUP$_+$)*-double third-order positive saddle.*

- *For $a_{j_1j_20}<0$ and $a_{j_2j_10}>0$,*

$$(a_{j_1j_21},a_{j_2j_11})=\underbrace{(3^{\rm rd}{\rm DP}_{+},3^{\rm rd}{\rm DP}_{-})}_{\text{double third-order CCW center}}. \tag{4.100}$$

The equilibrium of $(x_{j_2}^*,x_{j_1}^*)=(a_{j_1j_21},a_{j_2j_11})$ is a $(3^{\rm rd}{\rm DP}_{+},3^{\rm rd}{\rm DP}_{-})$-double third-order counter-clockwise center.

- *For $a_{j_1j_20}>0$ and $a_{j_2j_10}<0$,*

$$(a_{j_1j_21},a_{j_2j_11})=\underbrace{(3^{\rm rd}{\rm DP}_{-},3^{\rm rd}{\rm DP}_{+})}_{\text{double third-order CW center}}. \tag{4.101}$$

The equilibrium of $(x_{j_2}^*,x_{j_1}^*)=(a_{j_1j_21},a_{j_2j_11})$ is a $(3^{\rm rd}{\rm DP}_{-},3^{\rm rd}{\rm DP}_{+})$-double third-order clockwise center.

- *For $a_{j_1j_20}<0$ and $a_{j_2j_10}<0$,*

$$(a_{j_1j_21},a_{j_2j_11})=\underbrace{(3^{\rm rd}{\rm UP}_{-},3^{\rm rd}{\rm UP}_{-})}_{\text{double third-order negative saddle}}. \tag{4.102}$$

The equilibrium of $(x_{j_2}^*,x_{j_1}^*)=(a_{j_1j_21},a_{j_2j_11})$ is a $(3^{\rm rd}{\rm UP}_{-},3^{\rm rd}{\rm UP}_{-})$-double third-order negative saddle.

(xi) If

$$\Delta_{j_1j_2}=0\,\text{and}\,\Delta_{j_2j_1}>0, \tag{4.103}$$

then

$$\begin{aligned}&\dot{x}_{j_2}=a_{j_2j_10}(x_{j_1}-a_{j_2j_11})(x_{j_1}-a_{j_2j_12})(x_{j_1}-a_{j_2j_13}),\\&\dot{x}_{j_1}=a_{j_1j_20}(x_{j_2}-a_{j_1j_2s_1})(x_{j_2}-a_{j_1j_2s_2})^2,\\&j_1,j_2\in\{1,2\};j_1\neq j_2.\end{aligned} \tag{4.104}$$

The first integral manifold is

$$\begin{aligned}&a_{j_2j_10}\Big\{\frac{1}{4}\left[(x_{j_1}-a_{j_2j_1l_1})^4-(x_{j_10}-a_{j_2j_1l_1})^4\right]\\&+\frac{1}{3}\sum\nolimits_{l_2-1,l_2\neq l_1}^{3}(a_{j_2j_1l_1}-a_{j_2j_1l_2})\left[(x_{j_1}-a_{j_2j_1l_1})^3-(x_{j_10}-a_{j_2j_1l_1})^3\right]\\&+\frac{1}{2}\prod\nolimits_{l_2=1,l_2\neq l_1}^{3}(a_{j_2j_1l_1}-a_{j_2j_1l_2})\left[(x_{j_1}-a_{j_2j_{l_l}})^2-(x_{j_10}-a_{j_2j_1l_1})^2\right]\Big\}\\&=a_{j_1j_20}\Big\{\frac{1}{4}\left[(x_{j_1}-a_{j_1j_2s_1})^4-(x_{j_10}-a_{j_1j_2s_1})^4\right]\\&+\frac{2}{3}(a_{j_1j_2s_1}-a_{j_1j_2s_2})\left[(x_{j_2}-a_{j_1j_2s_1})^3-(x_{j_20}-a_{j_ij_2s_1})^3\right]\\&+\frac{1}{2}(a_{j_1j_2s_1}-a_{j_1j_2s_2})^2\left[(x_{j_2}-a_{j_1j_2s_1})^2-(x_{j_20}-a_{j_1j_2s_1})^2\right]\Big\}.\end{aligned} \tag{4.105}$$

(xi$_1$) The equilibrium of $(x^*_{j_2}, x^*_{j_1}) = (a_{j_1 j_2 s_1}, a_{j_2 j_1 l_1})(s_1, s_2 \in \{1,2\}, s_1 \neq s_2; l_1, l_2 \in \{1,2,3\}, l_1 \neq l_2)$ *has the following properties.*

- *For* $a_{j_1 j_2 0}(a_{j_1 j_2 s_1} - a_{j_1 j_2 s_2})^2 > 0$ *and* $a_{j_2 j_1 0}\prod_{l_2=1,1,l_2 \neq l_1}^{3}(a_{j_2 j_1 l_1} - a_{j_2 j_1 l_2}) > 0$,

$$(a_{j_1 j_2 s_1}, a_{j_2 j_1 l_1}) = \underbrace{(\mathrm{UP}_+, \mathrm{UP}_+)}_{\text{positive saddle}}. \tag{4.106}$$

The equilibrium of $(x^*_{j_2}, x^*_{j_1}) = (a_{j_1 j_2 s_1}, a_{j_2 j_1 l_1})$ *is a* (UP$_+$,UP$_+$)*-positive saddle.*

- *For* $a_{j_1 j_2 0}(a_{j_1 j_2 s_1} - a_{j_1 j_2 s_2})^2 < 0$ *and* $a_{j_2 j_1 0}\prod_{l_2=1,l_2 \neq l_1}^{3}(a_{j_2 j_1 l_1} - a_{j_2 j_1 l_2}) > 0$,

$$(a_{j_1 j_2 s_1}, a_{j_2 j_1 l_1}) = \underbrace{(\mathrm{DP}_+, \mathrm{DP}_-)}_{\text{CCW center}}. \tag{4.107}$$

The equilibrium of $(x^*_{j_2}, x^*_{j_1}) = (a_{j_i j_2 s_1}, a_{j_2 j_1 l_1})$ *is a* (DP$_+$,DP$_-$)*-counter-clockwise center.*

- *For* $a_{j_1 j_2 0}(a_{j_1 j_2 s_1} - a_{j_1 j_2 s_2})^2 > 0$ *and* $a_{j_2 j_1 0}\prod_{l_2=1,l_2 \neq l_1}^{3}(a_{j_2 j_1 l_1} - a_{j_2 j_1 l_2}) < 0$,

$$(a_{j_1 j_2 s_1}, a_{j_2 j_1 l_1}) = \underbrace{(\mathrm{DP}_-, \mathrm{DP}_+)}_{\text{CW center}}. \tag{4.108}$$

The equilibrium of $(x^*_{j_2}, x^*_{j_1}) = (a_{j_1 j_2 s_1}, a_{j_2 j_1 l_1})$ *is a* (DP$_-$,DP$_+$)*-clockwise center.*

- *For* $a_{j_1 j_2 0}(a_{j_1 j_2 s_1} - a_{j_1 j_2 s_2})^2 < 0$ *and* $a_{j_2 j_1 0}\prod_{l_2=1,l_2 \neq l_1}^{3}(a_{j_2 j_1 l_1} - a_{j_2 j_1 l_2}) < 0$,

$$(a_{j_1 j_2 s_1}, a_{j_2 j_1 l_1}) = \underbrace{(\mathrm{UP}_-, \mathrm{UP}_-)}_{\text{negative saddle}}. \tag{4.109}$$

The equilibrium of $(x^*_{j_2}, x^*_{j_1}) = (a_{j_1 j_2 s_1}, a_{j_2 j_1 l_1})$ *is a* (DP$_-$,UP$_-$)*-negative saddle.*

(xi$_2$) The equilibrium of $(x^*_{j_2}, x^*_{j_1}) = (a_{j_1 j_2 s_2}, a_{j_2 j_1 i_1})(s_1, s_2 \in \{1,2\}, s_1 \neq s_2; l_1, l_2 \in \{1,2,3\}, l_1 \neq l_2)$ *has the following properties.*

- *For* $a_{j_1 j_2 0}(a_{j_1 j_2 s_2} - a_{j_1 j_2 s_1}) > 0$ *and* $a_{j_2 j_1 0}\prod_{l_2=1,l_2 \neq l_1}^{3}(a_{j_2 j_1 l_1} - a_{j_2 j_1 l_2}) > 0$,

$$(a_{j_1 j_2 s_2}, a_{j_2 j_1 l_1}) = \underbrace{(\mathrm{US}, \mathrm{UP})}_{\text{up-pparabola upperer saddle}}. \tag{4.110}$$

The equilibrium of $(x^*_{j_2}, x^*_{j_1}) = (a_{j_1 j_2 s_2}, a_{j_2 j_1 l_1})$ *is a* (US,UP)*-up-parabola upper-saddle.*

- *For* $a_{j_1j_20}(a_{j_1j_2s_2}-a_{j_1j_2s_1})<0$ *and* $a_{j_2j_10}\prod_{l_2=1,l_2\neq l_1}^{3}(a_{j_2j_1l_1}-a_{j_2j_1l_2})>0$,

$$(a_{j_1j_2s_2},a_{j_2j_1l_1})=\underbrace{(\mathrm{LS},\mathrm{DP})}_{\text{down-parabola lower-saddle}}. \tag{4.111}$$

The equilibrium of $(x_{j_2}^*,x_{j_1}^*)=(a_{j_1j_2s_2},a_{j_2j_1l_1})$ *is an* (LS,DP)-*down-parabola lower-saddle.*

- *For* $a_{j_1j_20}(a_{j_1j_2s_2}-a_{j_1j_2s_1})>0$ *and* $a_{j_2j_10}\prod_{l_2=1,l_2\neq l_1}^{3}(a_{j_2j_1l_1}-a_{j_2j_1l_2})<0$,

$$(a_{j_1j_2s_2},a_{j_2j_1l_1})=\underbrace{(\mathrm{US},\mathrm{DP})}_{\text{down-parabola upper-saddle}}. \tag{4.112}$$

The equilibrium of $(x_{j_2}^*,x_{j_1}^*)=(a_{j_1j_2s_2},a_{j_2j_1l_1})$ *is a* (US,DP)-*down-parabola upper-saddle.*

- *For* $a_{j_1j_20}(a_{j_1j_2s_2}-a_{j_1j_2s_1})<0$ *and* $a_{j_2j_10}\prod_{l_2=1,l_2\neq l_1}^{3}(a_{j_2j_1l_1}-a_{j_2j_1l_2})<0$,

$$(a_{j_1j_2s_2},a_{j_2j_1l_1})=\underbrace{(\mathrm{LS},\mathrm{UP})}_{\text{up-parabola lower-saddle}}. \tag{4.113}$$

The equilibrium of $(x_{j_2}^*,x_{j_1}^*)=(a_{j_1j_2s_2},a_{j_2j_1l_1})$ *is an* (LS,UP)-*up-parabola lower-saddle.*

(xii) If

$$\Delta_{j_1j_2}=0\,\text{and}\,\Delta_{j_2j_1}>0,a_{j_1j_2s_1}=a_{j_1j_2s_2}=a_{j_1j_21} \tag{4.114}$$

then

$$\begin{aligned}&\dot{x}_{j_2}=a_{j_2j_10}(x_{j_1}-a_{j_2j_11})(x_{j_1}-a_{j_2j_12})(x_{j_1}-a_{j_2j_13}),\\&\dot{x}_{j_1}=a_{j_1j_20}(x_{j_2}-a_{j_1j_21})^3,\\&j_1,j_2\in\{1,2\};j_1\neq j_2.\end{aligned} \tag{4.115}$$

The first integral manifold is

$$\begin{aligned}&a_{j_2j_10}\{\frac{1}{4}\left[(x_{j_1}-a_{j_2j_1l_1})^4-(x_{j_10}-a_{j_2j_1l_1})^4\right]\\&+\frac{1}{3}\sum_{l_2=1,l_2\neq l_1}^{3}(a_{j_2j_1l_1}-a_{j_2j_1l_2})\left[(x_{j_1}-a_{j_2j_1l_1})^3-(x_{j_10}-a_{j_2j_1l_1})^3\right]\\&+\frac{1}{2}\prod_{l_2=1,l_2\neq l_1}^{3}(a_{j_2j_1l_1}-a_{j_2j_1l_2})\left[(x_{j_1}-a_{j_2j_1l_1})^2-(x_{j_10}-a_{j_2j_1l_1})^2\right]\}\\&=\frac{1}{4}a_{j_1j_20}\left[(x_{j_1}-a_{j_1j_2s_1})^4-(x_{j_10}-a_{j_1j_2s_1})^4\right].\end{aligned} \tag{4.116}$$

The equilibrium of $(x_{j_2}^*,x_{j_1}^*)=(a_{j_1j_21},a_{j_2j_1l_1})$ $(l_1,l_2\in\{1,2,3\},l_1\neq l_2)$ *has the following properties.*

- *For* $a_{j_1j_20}>0$ *and* $a_{j_2j_10}\prod_{l_2=1,l_2\neq l_1}^{3}(a_{j_2j_1l_1}-a_{j_2j_1l_2})>0$,

$$(a_{j_1j_21},a_{j_2j_1l_1})=\underbrace{(3^{\rm rd}{\rm UP}_{+},{\rm UP}_{+})}_{\text{third-order positive saddle}}. \tag{4.117}$$

The equilibrium of $(x_{j_2}^*,x_{j_1}^*)=(a_{j_1j_21},a_{j_2j_1l_1})$ *is a* $(3^{\rm rd}{\rm UP}_+,{\rm UP}_+)$*-third-order positive saddle.*

- *For* $a_{j_1j_20}<0$ *and* $a_{j_2j_10}\prod_{l_2=1,l_2\neq l_1}^{3}(a_{j_2j_1l_1}-a_{j_2j_1l_2})>0$,

$$(a_{j_1j_21},a_{j_2j_1l_1})=\underbrace{(3^{\rm rd}{\rm DP}_{+},{\rm DP}_{-})}_{\text{third-order CCW center}}. \tag{4.118}$$

The equilibrium of $(x_{j_2}^*,x_{j_1}^*)=(a_{j_1j_21},a_{j_2j_1l_1})$ *is a* $(3^{\rm rd}{\rm DP}_+,{\rm DP}_-)$*-third-order-counter-clockwise center.*

- *For* $a_{j_1j_20}>0$ *and* $a_{j_2j_10}\prod_{l_2=1,l_2\neq l_1}^{3}(a_{j_2j_1l_1}-a_{j_2j_1l_2})<0$,

$$(a_{j_1j_21},a_{j_2j_1l_1})=\underbrace{(3^{\rm rd}{\rm DP}_{-},{\rm DP}_{+})}_{\text{third-order CW center}}. \tag{4.119}$$

The equilibrium of $(x_{j_2}^*,x_{j_1}^*)=(a_{j_1j_21},a_{j_2j_1l_1})$ *is a* $(3^{\rm rd}{\rm DP}_-,{\rm DP}_+)$*-third-order-clockwise center.*

- *For* $a_{j_1j_20}<0$ *and* $a_{j_2j_10}\prod_{l_2=1,l_2\neq l_1}^{3}(a_{j_2j_1l_1}-a_{j_2j_1l_2})<0$,

$$(a_{j_1j_21},a_{j_2j_1l_1})=\underbrace{(3^{\rm rd}{\rm UP}_{-},{\rm UP}_{-})}_{\text{third-order negative saddle}}. \tag{4.120}$$

The equilibrium of $(x_{j_2}^*,x_{j_1}^*)=(a_{j_1j_21},a_{j_2j_1l_1})$ *is a* $(3^{\rm rd}{\rm UP}_-,{\rm UP}_-)$*-third-order negative saddle.*

(xiii) If

$$\Delta_{j_1j_2}>0 \text{ and } \Delta_{j_2j_1}<0, \tag{4.121}$$

then

$$\begin{array}{l}\dot{x}_{j_1}=a_{j_1j_20}(x_{j_2}-a_{j_1j_21})(x_{j_2}-a_{j_1j_22})(x_{j_2}-a_{j_1j_23}),\\ \dot{x}_{j_2}=a_{j_2j_10}(x_{j_1}-a_{j_2j_11})\left[(x_{j_1}-a_{j_2j_1})^2+b_{j_2j_1}\right]\\ j_1,j_2\in\{1,2\},j_2\neq j_1\end{array} \tag{4.122}$$

where

$$\begin{aligned}
&\{a_{j_1j_2 1}, a_{j_1j_2 2}, a_{j_1j_2 3}\} = \text{sort}\{b_{j_1j_2 1}, b_{j_1j_2 2}, b_{j_1j_2 3}\}, a_{j_1j_2 1} < a_{j_1j_2 2} < a_{j_1j_2 3};\\
&b_{j_1j_2 2}, b_{j_1j_2 3} = -\frac{1}{2}(B_{j_1j_2} \pm \sqrt{\Delta_{j_1j_2}}), a_{j_1j_2 0} = A_{j_1j_2 0};\\
&a_{j_2j_1 1} = b_{j_2j_1 1}, a_{j_2j_1} = -\frac{1}{2}B_{j_2j_1}, b_{j_2j_1} = \frac{1}{4}(-\Delta_{j_2j_1}).
\end{aligned} \tag{4.123}$$

The first integral manifold is

$$\begin{aligned}
&a_{j_2j_1 0}\Big\{\frac{1}{4}\left[(x_{j_1} - a_{j_2j_1 1})^4 - (x_{j_1 0} - a_{j_2j_1 1})^4\right]\\
&+\frac{2}{3}(a_{j_2j_1 1} - a_{j_2j_1})\left[(x_{j_1} - a_{j_2j_1 1})^3 - (x_{j_1 0} - a_{j_2j_1 1})^3\right]\\
&+\frac{1}{2}\left[(a_{j_2j_1 1} - a_{j_2j_1})^2 + b_{j_2j_1}\right]\left[(x_{j_1} - a_{j_2j_1 1})^2 - (x_{j_1 0} - a_{j_2j_1 1})^2\right]\Big\}\\
&= a_{j_1j_2 0}\Big\{\frac{1}{4}\left[(x_{j_2} - a_{j_1j_2 s_1})^4 - (x_{j_2 0} - a_{j_1j_2 s_1})^4\right]\\
&+\frac{1}{3}\sum\nolimits_{s_2=1, s_2\neq s_1}^{3}(a_{j_1j_2 s_1} - a_{j_1j_2 s_2})\left[(x_{j_2} - a_{j_1j_2 s_1})^3 - (x_{j_2 0} - a_{j_1j_2 s_1})^3\right]\\
&+\frac{1}{2}\prod\nolimits_{s_2=1, s_2\neq s_1}^{3}(a_{j_1j_2 s_1} - a_{j_1j_2 s_2})\left[(x_{j_2} - a_{j_1j_2 s_1})^2 - (x_{j_2 0} - a_{j_1j_2 s_1})^2\right]\Big\}.
\end{aligned} \tag{4.124}$$

The equilibrium of $(x_{j_2}^*, x_{j_1}^*) = (a_{j_1j_2 s_1}, a_{j_2j_1 1})(s_1, s_2 \in \{1,2,3\}, s_1 \neq s_2)$ on the x_{j_2}-direction *has the following properties.*

- *For* $a_{j_1j_2 0}\prod_{s_2=1, s_2\neq s_1}^{3}(a_{j_1j_2 s_1} - a_{j_1j_2 s_2}) > 0$ *and* $a_{j_2j_1 0} > 0$,

$$(a_{j_1j_2 s_1}, a_{j_2j_1 1}) = \underbrace{(\text{UP}_+, \text{UP}_+)}_{\text{positive saddle}}. \tag{4.125}$$

The equilibrium of $(x_{j_2}^*, x_{j_1}^*) = (a_{j_1j_2 s_1}, a_{j_2j_1 1})$ *is a* (UP_+,UP_+)-*positive saddle.*

- *For* $a_{j_1j_2 0}\prod_{s_2=1, s_2\neq s_1}^{3}(a_{j_1j_2 s_1} - a_{j_1j_2 s_2}) < 0$ *and* $a_{j_2j_1 0} > 0$,

$$(a_{j_1j_2 s_1}, a_{j_2j_1 1}) = \underbrace{(\text{DP}_+, \text{DP}_-)}_{\text{CCW center}}. \tag{4.126}$$

The equilibrium of $(x_{j_2}^*, x_{j_1}^*) = (a_{j_1j_2 s_1}, a_{j_2j_1 1})$ *is a* (DP_+,DP_-)-*counter-clockwise center.*

- *For* $a_{j_1j_2 0}\prod_{s_2=1, s_2\neq s_1}^{3}(a_{j_1j_2 s_1} - a_{j_1j_2 s_2}) > 0$ *and* $a_{j_2j_1 0} < 0$,

$$(a_{j_1j_2 s_1}, a_{j_2j_1 1}) = \underbrace{(\text{DP}_-, \text{DP}_+)}_{\text{CW center}}. \tag{4.127}$$

The equilibrium of $(x_{j_2}^*, x_{j_1}^*) = (a_{j_1j_2 s_1}, a_{j_2j_1 1})$ *is a* (DP_-,DP_+)-*clockwise center.*

- *For* $a_{j_1j_20}\prod_{s_2=1,s_2\neq s_1}^{3}(a_{j_1j_2s_1}-a_{j_1j_2s_2})<0$ *and* $a_{j_2j_10}<0$,

$$(a_{j_1j_2s_1},a_{j_2j_11})=\underbrace{(\mathrm{UP}_-,\mathrm{UP}_-)}_{\text{negative saddle}}. \tag{4.128}$$

The equilibrium of $(x_{j_2}^*,x_{j_1}^*)=(a_{j_1j_2s_1},a_{j_2j_11})$ *is a* (UP$_-$,UP$_-$)*-negative saddle.*

(xiv) If

$$\Delta_{j_1j_2}>0\,\text{and}\,\Delta_{j_2j_1}=0,a_{j_2j_1l_1}\neq a_{j_2j_1l_2} \tag{4.129}$$

then

$$\begin{aligned}&\dot{x}_{j_1}=a_{j_1j_20}(x_{j_2}-a_{j_1j_21})(x_{j_2}-a_{j_1j_22})(x_{j_2}-a_{j_1j_23}),\\&\dot{x}_{j_2}=a_{j_2j_10}(x_{j_1}-a_{j_2j_1l_1})(x_{j_1}-a_{j_2j_1l_2})^2\\&j_1,j_2\in\{1,2\},j_2\neq j_1;l_1,l_2\in\{1,2\},l_1\neq l_2.\end{aligned} \tag{4.130}$$

The first integral manifold is

$$\begin{aligned}&a_{j_2j_10}\{\frac{1}{4}\left[(x_{j_1}-a_{j_2j_1l_1})^4-(x_{j_10}-a_{j_2j_1l_1})^4\right]\\&+\frac{2}{3}(a_{j_2j_1l_1}-a_{j_1j_2l_2})\left[(x_{j_1}-a_{j_2j_1l_1})^3-(x_{j_10}-a_{j_2j_1l_1})^3\right]\\&+\frac{1}{2}(a_{j_2j_1l_1}-a_{j_{1j_2l_2}})^2\left[(x_{j_1}-a_{j_2j_1l_1})^2-(x_{j_10}-a_{j_2j_1l_1})^2\right]\}\\&=a_{j_1j_20}\{\frac{1}{4}\left[(x_{j_2}-a_{j_1j_2s_1})^4-(x_{j_20}-a_{j_1j_2s_1})^4\right]\\&+\frac{1}{3}\sum\nolimits_{s_2=1,s_2\neq s_1}^{3}(a_{j_1j_2s_1}-a_{j_1j_2s_2})\left[(x_{j_2}-a_{j_1j_2s_1})^3-(x_{j_20}-a_{j_1j_2s_1})^3\right]\\&+\frac{1}{2}\prod\nolimits_{s_2=1,s_2\neq s_1}^{3}(a_{j_1j_2s_1}-a_{j_1j_2s_2})\left[(x_{j_2}-a_{j_1j_2s_1})^2-(x_{j_20}-a_{j_1j_2s_1})^2\right]\}.\end{aligned} \tag{4.131}$$

(xiv$_1$) The equilibrium of $(x_{j_2}^*,x_{j_1}^*)=(a_{j_1j_2s_1},a_{j_2j_1l_1})(s_1,s_2\in\{1,2,3\},s_1\neq s_2;l_1,l_2\in\{1,2\},l_1\neq l_2)$ *has the following properties.*

- *For* $a_{j_1j_20}\prod_{s_2=1,s_2\neq s_1}^{3}(a_{j_1j_2s_1}-a_{j_1j_2s_2})>0$ *and* $a_{j_2j_10}(a_{j_2j_1l_1}-a_{j_2j_1l_2})^2>0$,

$$(a_{j_1j_2s_1},a_{j_2j_1l_1})=\underbrace{(\mathrm{UP}_+,\mathrm{UP}_+)}_{\text{positive saddle}}. \tag{4.132}$$

  The equilibrium of $(x_{j_2}^*,x_{j_1}^*)=(a_{j_1j_2s_1},a_{j_2j_1l_1})$ *is a* (UP$_+$,UP$_+$)*-positive saddle.*

- *For* $a_{j_1j_20}\prod_{s_2=1,s_2\neq s_1}^{3}(a_{j_1j_2s_1}-a_{j_1j_2s_2})<0$ *and* $a_{j_2j_10}(a_{j_2j_1l_1}-a_{j_2j_1l_2})^2>0$,

$$(a_{j_1j_2s_1},a_{j_2j_1l_1})=\underbrace{(\mathrm{DP}_+,\mathrm{DP}_-)}_{\text{CCW center}}. \tag{4.133}$$

  The equilibrium of $(x_{j_2}^*,x_{j_1}^*)=(a_{j_1j_2s_1},a_{j_2j_1l_1})$ *is a* (DP$_+$,DP$_-$)*-counter-clockwise center.*

- *For* $a_{j_1j_20}\prod_{s_2=1,s_2\neq s_1}^{3}(a_{j_1j_2s_1}-a_{j_1j_2s_2})>0$ *and* $a_{j_2j_10}(a_{j_2j_1l_1}-a_{j_2j_1l_2})^2<0$,

$$(a_{j_1j_2s_1},a_{j_2j_1l_1})=\underbrace{(\mathrm{DP}_-,\mathrm{DP}_+)}_{\text{CW center}}. \tag{4.134}$$

The equilibrium of $(x_{j_2}^*,x_{j_1}^*)=(a_{j_1j_2s_1},a_{j_2j_1l_1})$ *is a* (DP_-,DP_+)*-clockwise center.*

- *For* $a_{j_1j_20}\prod_{s_2=1,s_2\neq s_1}^{3}(a_{j_1j_2s_1}-a_{j_1j_2s_2})<0$ *and* $a_{j_2j_10}(a_{j_2j_1l_1}-a_{j_2j_1l_2})^2<0$,

$$(a_{j_1j_2s_1},a_{j_2j_1l_1})=\underbrace{(\mathrm{UP}_-,\mathrm{UP}_-)}_{\text{negative saddle}}. \tag{4.135}$$

The equilibrium of $(x_{j_2}^*,x_{j_1}^*)=(a_{j_1j_2s_1},a_{j_2j_1l_1})$ *is a* (UP_-,UP_-)*-negative saddle.*

(xiv$_2$*) The equilibrium of* $(x_{j_2}^*,x_{j_1}^*)=(a_{j_1j_2s_1},a_{j_2j_1l_2})(s_1,s_2\in\{1,2,3\},s_1\neq s_2;l_1,l_2\in\{1,2\},l_1\neq l_2)$ *has the following properties.*

- *For* $a_{j_1j_20}\prod_{s_2=1,s_2\neq s_1}^{3}(a_{j_1j_2s_1}-a_{j_1j_2s_2})>0$ *and* $a_{j_2j_10}(a_{j_2j_1l_2}-a_{j_2j_1l_1})>0$,

$$(a_{j_1j_2s_1,},a_{j_2j_1l_2})=\underbrace{(\mathrm{UP},\mathrm{US})}_{\text{up-parabola upper-saddle}}. \tag{4.136}$$

The equilibrium of $(x_{j_2}^*,x_{j_1}^*)=(a_{j_1j_2s_1},a_{j_2j_1l_2})$ *is a* (UP,US)*-up-parabola upper-saddle.*

- *For* $a_{j_1j_20}\prod_{s_2=1,s_2\neq s_1}^{3}(a_{j_1j_2s_1}-a_{j_1j_2s_2})<0$ *and* $a_{j_2j_10}(a_{j_2j_1l_2}-a_{j_2j_1l_1})>0$,

$$(a_{j_1j_2s_1,},a_{j_2j_1l_2})=\underbrace{(\mathrm{DP},\mathrm{US})}_{\text{down-parabola upper-saddle}}. \tag{4.137}$$

The equilibrium of $(x_{j_2}^*,x_{j_1}^*)=(a_{j_1j_2s_1},a_{j_2j_1l_2})$ *is a* (DP,US)*-down-parabola upper-saddle.*

- *For* $a_{j_1j_20}\prod_{s_2=1,s_2\neq s_1}^{3}(a_{j_1j_2s_1}-a_{j_1j_2s_2})>0$ *and* $a_{j_2j_10}(a_{j_2j_1l_2}-a_{j_2j_1l_1})<0$,

$$(a_{j_1j_2s_1},a_{j_2j_1l_2})=\underbrace{(\mathrm{DP},\mathrm{LS})}_{\text{down-parabola lower-saddle}}. \tag{4.138}$$

The equilibrium of $(x_{j_2}^*,x_{j_1}^*)=(a_{j_1j_2s_1},a_{j_2j_1l2})$ *is a* (DP,LS)*-down-parabola lower-saddle.*

- *For* $a_{j_1j_20}\prod_{s_2=1,s_2\neq s_1}^{3}(a_{j_1j_2s_1}-a_{j_1j_2s_2})<0$ *and* $a_{j_2j_10}(a_{j_2j_1l_2}-a_{j_2j_1l_1})<0$,

$$(a_{j_1j_2s_1},a_{j_2j_1l_2})=\underbrace{(\mathrm{UP},\mathrm{LS})}_{\text{up-parabola lower-saddle}}. \tag{4.139}$$

The equilibrium of $(x_{j_2}^*,x_{j_1}^*)=(a_{j_1j_2s_1},a_{j_2j_1l_2})$ *is a* (UP,LS)*-up-parabola lower-saddle.*

(xv) If

$$\Delta_{j_1j_2} > 0 \text{ and } \Delta_{j_2j_1} = 0, a_{j_2j_1l_1} = a_{j_2j_1l_2} = a_{j_2j_1}1, \tag{4.140}$$

then

$$\begin{aligned} &\dot{x}_{j_1} = a_{j_1j_2}0(x_{j_2} - a_{j_1j_2}1)(x_{j_2} - a_{j_1j_2}2)(x_{j_2} - a_{j_1j_2}3),\\ &\dot{x}_{j_2} = a_{j_2j_1}0(x_{j_1} - a_{j_2j_1}1)^3\\ &j_1, j_2 \in \{1,2\}, j_2 \neq j_1. \end{aligned} \tag{4.141}$$

The first integral manifold is

$$\begin{aligned} &\frac{1}{4}a_{j_2j_1}0\left[(x_{j_1} - a_{j_2j_1}1)^4 - (x_{j_1}0 - a_{j_2j_1}1)^4\right]\\ &= a_{j_1j_2}0\Big\{\frac{1}{4}\left[(x_{j_2} - a_{j_1j_2s_1})^4 - (x_{j_2}0 - a_{j_1j_2s_1})^4\right]\\ &+\frac{1}{3}\sum\nolimits_{s_2=1, s_2\neq s_1}^{3}(a_{j_1j_2s_1} - a_{j_1j_2s_2}\left[(x_{j_2} - a_{j_1j_2s_1})^3 - (x_{j_2}0 - a_{j_1j_2s_1})^3\right]\\ &+\frac{1}{2}\prod\nolimits_{s_2=1, s_2\neq s_1}^{3}(a_{j_1j_2s_1} - a_{j_1j_2s_2})\left[(x_{j_2} - a_{j_1j_2s_1})^2 - (x_{j_2}0 - a_{j_1j_2s_1})^2\right]\Big\}. \end{aligned} \tag{4.142}$$

The equilibrium of $(x^*_{j_2}, x^*_{j_1}) = (a_{j_1j_2s_1}, a_{j_2j_1}1)$ $(s_1, s_2 \in \{1,2,3\}, s_1 \neq s_2)$ *has the following properties.*

- *For* $a_{j_1j_2}0\prod\nolimits_{s_2=1, s_2\neq s_1}^{3}(a_{j_1j_2s_1} - a_{j_1j_2s_2}) > 0$ *and* $a_{j_2j_1}0 > 0$,

$$(a_{j_1j_2s_1}, a_{j_2j_1}1) = \underbrace{(\text{UP}_+, 3^{\text{rd}}\text{UP}_+)}_{\text{third-order positive saddle}}. \tag{4.143}$$

  The equilibrium of $(x^*_{j_2}, x^*_{j_1}) = (a_{j_1j_2s_1}, a_{j_2j_1}1)$ *is a* $(\text{UP}_+, 3^{\text{rd}}\text{UP}_+)$*-third-order positive saddle.*
- *For* $a_{j_1j_2}0\prod\nolimits_{s_2=1, s_2\neq s_1}^{3}(a_{j_1j_2s_1} - a_{j_1j_2s_2}) < 0$ *and* $a_{j_2j_1}0 > 0$,

$$(a_{j_1j_2s_1}, a_{j_2j_1}1) = \underbrace{(\text{DP}_+, 3^{\text{rd}}\text{DP}_-)}_{\text{third-order CCW center}}. \tag{4.144}$$

  The equilibrium of $(x^*_{j_2}, x^*_{j_1}) = (a_{j_1j_2s_1}, a_{j_2j_1}1)$ *is a* $(\text{DP}_+, 3^{\text{rd}}\text{DP}_-)$*-third-order counter-clockwise center.*
- *For* $a_{j_1j_2}0\prod\nolimits_{s_2=1, s_2\neq s_1}^{3}(a_{j_1j_2s_1} - a_{j_1j_2s_2}) > 0$ *and* $a_{j_2j_1}0 < 0$,

$$(a_{j_1j_2s_1}, a_{j_2j_1}1) = \underbrace{(\text{DP}_-, 3^{\text{rd}}\text{DP}_+)}_{\text{third-order CCW center}}. \tag{4.145}$$

  The equilibrium of $(x^*_{j_2}, x^*_{j_1}) = (a_{j_1j_2s_1}, a_{j_2j_1}1)$ *is a* $(\text{DP}_-, 3^{\text{rd}}\text{DP}_+)$*-third-order clockwise center.*

- *For* $a_{j_1j_20}\prod_{s_2=1,s_2\neq s_1}^{3}(a_{j_1j_2s_1}-a_{j_1j_2s_2})<0$ *and* $a_{j_2j_10}<0$,

$$(a_{j_1j_2s_1},a_{j_2j_11})=\underbrace{(\mathrm{UP}_-,3^{\mathrm{rd}}\mathrm{UP}_-)}_{\text{third-order negative saddle}}. \tag{4.146}$$

The equilibrium of $(x_{j_2}^*,x_{j_1}^*)=(a_{j_1j_2s_1},a_{j_2j_11})$ *is a* $(\mathrm{UP}_-,3^{\mathrm{rd}}\mathrm{UP}_-)$*-third-order negative saddle.*

(xvi) If

$$\Delta_{j_1j_2}>0\,\text{and}\,\Delta_{j_2j_1}>0, \tag{4.147}$$

then

$$\begin{aligned}&\dot{x}_{j_1}=a_{j_1j_20}(x_{j_2}-a_{j_1j_21})(x_{j_2}-a_{j_1j_22})(x_{j_2}-a_{j_1j_23}),\\&\dot{x}_{j_2}=a_{j_2j_10}(x_{j_1}-a_{j_2j_11})(x_{j_1}-a_{j_2j_12})(x_{j_1}-a_{j_2j_13}),\\&j_1,j_2\in\{1,2\};j_1\neq j_2.\end{aligned} \tag{4.148}$$

The first integral manifold is

$$\begin{aligned}&a_{j_2j_10}\Big\{\frac{1}{4}\left[(x_{j_1}-a_{j_2j_1l_1})^4-(x_{j_10}-a_{j_2j_1l_1})^4\right]\\&+\frac{1}{3}\sum\nolimits_{l_2=1,l_2\neq l_1}^{3}(a_{j_2j_1i_1}-a_{j_2j_1l_2})\left[(x_{j_1}-a_{j_2j_1l_1})^3-(x_{j_10}-a_{j_2j_1l_1})^3\right]\\&+\frac{1}{2}\prod\nolimits_{l_2=1,l_2\neq l_1}^{3}(a_{j_2j_1l_1}-a_{j_2j_1l_2})\left[(x_{j_1}-a_{j_2j_1l_1})^2-(x_{j_10}-a_{j_2j_il_1})\right]\Big\}\\&=a_{j_1j_20}\Big\{\frac{1}{4}\left[(x_{j_2}-a_{j_1j_2s_1})^4-(x_{j_20}-a_{j_1j_2s_1})^4\right]\\&+\frac{1}{3}\sum\nolimits_{s_2=1,s_2\neq s_1}^{3}(a_{j_1j_2s_1}-a_{j_1j_2s_2})\left[(x_{j_2}-a_{j_1j_2s_1})^3-(x_{j_20}-a_{j_1j_2s_1})^3\right]\\&+\frac{1}{2}\prod\nolimits_{s_2=1,s_2\neq s_1}^{3}(a_{j_1j_2s_1}-a_{j_1j_2s_2})\left[(x_{j_2}-a_{j_1j_2s_1})^2-(x_{j_20}-a_{j_1j_2s_1})^2\right]\Big\}.\end{aligned} \tag{4.149}$$

(xvi$_1$) The equilibrium of $(x_{j_2}^*,x_{j_1}^*)=(a_{j_1j_2s_1},a_{j_2j_1l_1})(s_1,s_2,l_1,l_2\in\{1,2,3\},s_1\neq s_2,$ $l_1\neq l_2)$ *has the following properties.*

- *For* $a_{j_1j_20}\prod_{s_2=1,s_2\neq s_1}^{3}(a_{j_1j_2s_1}-a_{j_1j_2s_2})>0$ *and* $a_{j_2j_10}\prod_{l_2=1,l_2\neq l_1}^{3}(a_{j_2j_1l_1}-a_{j_2j_1l_2})>0$,

$$(a_{j_1j_2s_1},a_{j_2j_1l_1})=\underbrace{(\mathrm{UP}_+,\mathrm{UP}_+)}_{\text{positive saddle}}. \tag{4.150}$$

The equilibrium of $(x_{j_2}^*,x_{j_1}^*)=(a_{j_1j_2s_1},a_{j_2j_1l_1})$ *is a* $(\mathrm{UP}_+,\mathrm{UP}_+)$*-positive saddle.*

- *For* $a_{j_1j_20}\prod_{s_2=1,s_2\neq s_1}^{3}(a_{j_1j_2s_1}-a_{j_1j_2s_2})<0$ *and* $a_{j_2j_10}\prod_{l_2-1,l_2\neq l_1}^{3}(a_{j_2j_1l_1}-a_{j_2j_1l_2})>0$,

$$(a_{j_1j_2s_1},a_{j_2j_1l_1})=\underbrace{(\mathrm{DP}_+,\mathrm{DP}_-)}_{\text{CCW center}}. \tag{4.151}$$

The equilibrium of $(x^*_{j_2},x^*_{j_1})=(a_{j_1j_2s_1},a_{j_2j_1l_1})$ *is a* (DP$_+$,DP$_-$)*-counter-clockwise center.*

- *For* $a_{j_1j_20}\prod_{s_2=1,s_2\neq s_1}^{3}(a_{j_1j_2s_1}-a_{j_1j_2s_2})>0$ *and* $a_{j_2j_10}\prod_{l_2=1,l_2\neq l_1}^{3}(a_{j_2j_1l_1}-a_{j_2j_1l_2})<0$,

$$(a_{j_1j_2s_1},a_{j_2j_1l_1})=\underbrace{(\mathrm{DP}_-,\mathrm{DP}_+)}_{\text{CW center}}. \tag{4.152}$$

The equilibrium of $(x^*_{j_2},x^*_{j_1})=(a_{j_1j_2s_1},a_{j_2j_1l_1})$ *is a (DP$_-$,DP$_+$)-clockwise center.*

- *For* $a_{j_1j_20}\prod_{s_2=1,s_2\neq s_1}^{3}(a_{j_1j_2s_1}-a_{j_1j_2s_2})<0$ *and* $a_{j_2j_10}\prod_{l_2=1,l_2\neq l_1}^{3}(a_{j_2j_1l_1}-a_{j_2j_1l_2})<0$,

$$(a_{j_1j_2s_1},a_{j_2j_1l_1})=\underbrace{(\mathrm{UP}_-,\mathrm{UP}_-)}_{\text{negative saddle}}. \tag{4.153}$$

The equilibrium of $(x^*_{j_2},x^*_{j_1})=(a_{j_1j_2s_1},a_{j_2j_1l_1})$ *is a* (UP$_-$,UP$_-$)*-negative saddle.*

(xvi$_2$) For $(x^*_{j_2},x^*_{j_1})=(a_{j_1j_2s_1},a_{j_2j_1l_1})\,(s_1,l_1=1,2,3)$ *the corresponding equilibrium networks have the following properties.*

- *For* $a_{j_1j_20}>0$ *and* $a_{j_2j_10}>0$,

$$\begin{Bmatrix} (a_{j_1j_2}3,a_{j_2j_1}1) & (a_{j_1j_2}3,a_{j_2j_1}2) & (a_{j_1j_2}3,a_{j_2j_1}3) \\ (a_{j_1j_2}2,a_{j_2j_1}1) & (a_{j_1j_2}2,a_{j_2j_1}2) & (a_{j_1j_2}2,a_{j_2j_1}3) \\ (a_{j_1j_2}1,a_{j_2j_1}1) & (a_{j_1j_2}1,a_{j_2j_1}2) & (a_{j_1j_2}1,a_{j_2j_1}3) \end{Bmatrix}$$
$$=\begin{Bmatrix} \underbrace{(\mathrm{UP}_+,\mathrm{UP}_+)}_{\text{positive saddle}} & \underbrace{(\mathrm{DP}_-,\mathrm{DP}_+)}_{\text{CW center}} & \underbrace{(\mathrm{UP}_+,\mathrm{UP}_+)}_{\text{positive saddle}} \\ \underbrace{(\mathrm{DP}_+,\mathrm{DP}_-)}_{\text{CCW center}} & \underbrace{(\mathrm{UP}_-,\mathrm{UP}_-)}_{\text{negative saddle}} & \underbrace{(\mathrm{DP}_+,\mathrm{DP}_-)}_{\text{CCW center}} \\ \underbrace{(\mathrm{UP}_+,\mathrm{UP}_+)}_{\text{positive saddle}} & \underbrace{(\mathrm{DP}_-,\mathrm{DP}_+)}_{\text{CW center}} & \underbrace{(\mathrm{UP}_+,\mathrm{UP}_+)}_{\text{positive saddle}} \end{Bmatrix}. \tag{4.154}$$

- *For* $a_{j_1j_2 0} < 0$ *and* $a_{j_2j_1 0} > 0$,

$$\begin{Bmatrix} (a_{j_1j_2 3}, a_{j_2j_1 1}) & (a_{j_1j_2 3}, a_{j_2j_1 2}) & (a_{j_1j_2 3}, a_{j_2j_1 3}) \\ (a_{j_1j_2 2}, a_{j_2j_1 1}) & (a_{j_1j_2 2}, a_{j_2j_1 2}) & (a_{j_1j_2 2}, a_{j_2j_1 3}) \\ (a_{j_1j_2 1}, a_{j_2j_1 1}) & (a_{j_1j_2 1}, a_{j_2j_1 2}) & (a_{j_1j_2 1}, a_{j_2j_1 3}) \end{Bmatrix}$$
$$= \begin{Bmatrix} \underbrace{(\mathrm{DP}_+, \mathrm{DP}_-)}_{\text{CCW center}} & \underbrace{(\mathrm{UP}_-, \mathrm{UP}_-)}_{\text{negative saddle}} & \underbrace{(\mathrm{DP}_+, \mathrm{DP}_-)}_{\text{CCW center}} \\ \underbrace{(\mathrm{UP}_+, \mathrm{UP}_+)}_{\text{positive saddle}} & \underbrace{(\mathrm{DP}_-, \mathrm{DP}_+)}_{\text{CW center}} & \underbrace{(\mathrm{UP}_+, \mathrm{UP}_+)}_{\text{positive saddle}} \\ \underbrace{(\mathrm{DP}_+, \mathrm{DP}_-)}_{\text{CCW center}} & \underbrace{(\mathrm{UP}_-, \mathrm{UP}_-)}_{\text{negative saddle}} & \underbrace{(\mathrm{DP}_+, \mathrm{DP}_-)}_{\text{CCW center}} \end{Bmatrix}. \tag{4.155}$$

- *For* $a_{j_1j_2 0} > 0$ *and* $a_{j_2j_1 0} < 0$,

$$\begin{Bmatrix} (a_{j_1j_2 3}, a_{j_2j_1 1}) & (a_{j_1j_2 3}, a_{j_2j_1 2}) & (a_{j_1j_2 3}, a_{j_2j_1 3}) \\ (a_{j_1j_2 2}, a_{j_2j_1 1}) & (a_{j_1j_2 2}, a_{j_2j_1 2}) & (a_{j_1j_2 2}, a_{j_2j_1 3}) \\ (a_{j_1j_2 1}, a_{j_2j_1 1}) & (a_{j_1j_2 1}, a_{j_2j_1 2}) & (a_{j_1j_2 1}, a_{j_2j_1 3}) \end{Bmatrix}$$
$$= \begin{Bmatrix} \underbrace{(\mathrm{DP}_-, \mathrm{DP}_+)}_{\text{CW center}} & \underbrace{(\mathrm{UP}_+, \mathrm{UP}_+)}_{\text{positive saddle}} & \underbrace{(\mathrm{DP}_-, \mathrm{DP}_+)}_{\text{CW center}} \\ \underbrace{(\mathrm{UP}_-, \mathrm{UP}_-)}_{\text{negative saddle}} & \underbrace{(\mathrm{DP}_-, \mathrm{DP}_+)}_{\text{CCW center}} & \underbrace{(\mathrm{UP}_-, \mathrm{UP}_-)}_{\text{negative saddle}} \\ \underbrace{(\mathrm{DP}_-, \mathrm{DP}_+)}_{\text{CW center}} & \underbrace{(\mathrm{UP}_+, \mathrm{UP}_+)}_{\text{positive saddle}} & \underbrace{(\mathrm{DP}_-, \mathrm{DP}_+)}_{\text{CW center}} \end{Bmatrix}. \tag{4.156}$$

- *For* $a_{j_1j_2 0} < 0$ *and* $a_{j_2j_1 0} < 0$,

$$\begin{Bmatrix} (a_{j_1j_2 3}, a_{j_2j_1 1}) & (a_{j_1j_2 3}, a_{j_2j_1 2}) & (a_{j_1j_2 3}, a_{j_2j_1 3}) \\ (a_{j_1j_2 2}, a_{j_2j_1 1}) & (a_{j_1j_2 2}, a_{j_2j_1 2}) & (a_{j_1j_2 2}, a_{j_2j_1 3}) \\ (a_{j_1j_2 1}, a_{j_2j_1 1}) & (a_{j_1j_2 1}, a_{j_2j_1 2}) & (a_{j_1j_2}, a_{j_2j_1 3}) \end{Bmatrix}$$
$$= \begin{Bmatrix} \underbrace{(\mathrm{UP}_-, \mathrm{UP}_-)}_{\text{negative saddle}} & \underbrace{(\mathrm{DP}_+, \mathrm{DP}_-)}_{\text{CCW center}} & \underbrace{(\mathrm{UP}_-, \mathrm{UP}_-)}_{\text{negative saddle}} \\ \underbrace{(\mathrm{DP}_-, \mathrm{DP}_+)}_{\text{CW center}} & \underbrace{(\mathrm{UP}_+, \mathrm{UP}_+)}_{\text{positive saddle}} & \underbrace{(\mathrm{DP}_-, \mathrm{DP}_+)}_{\text{CW center}} \\ \underbrace{(\mathrm{UP}_-, \mathrm{UP}_-)}_{\text{negative saddle}} & \underbrace{(\mathrm{DP}_+, \mathrm{DP}_-)}_{\text{CCW center}} & \underbrace{(\mathrm{UP}_-, \mathrm{UP}_-)}_{\text{negative saddle}} \end{Bmatrix}. \tag{4.157}$$

(xvii) Under

$$\Delta_{j_1j_2} < 0 \text{ and } \Delta_{j_2j_1} = 0, \tag{4.158}$$

The standard form is

$$\begin{aligned}
\dot{x}_{j_2} &= a_{j_2j_1 0}(x_{j_1} - a_{j_2j_1l_1})(x_{j_1} - a_{j_2j_1l_2})^2,\\
\dot{x}_{j_1} &= a_{j_1j_2 0}(x_{j_2} - a_{j_1j_2 1})\left[(x_{j_2} - a_{j_1j_2})^2 + b_{j_1j_2}\right],\\
j_1, j_2 &\in \{1,2\}; j_1 \neq j_2.
\end{aligned} \tag{4.159}$$

The equilibrium of $(x^*_{j_2}, x^*_{j_1}) = (a_{j_1j_2 1}, a_{j_2j_1l_2})$ *possesses the following appearing bifurcations.*

- *For* $a_{j_1j_2 0} > 0$ *and* $a_{j_2j_1 0}(a_{j_2j_1l_2} - a_{j_2j_1l_1}) > 0$,

$$\underbrace{(a_{j_1j_2 1}, a_{j_2j_1l_2})}_{(\text{UP},\text{US})\text{-appearing bifurcation}} \rightleftharpoons \left\{\begin{array}{c} \underbrace{(a_{j_1j_2 1}, a_{j_2j_1(l_2)_2})}_{(\text{UP}_+,\text{UP}_+)\text{-positive saddle}} \\ \underbrace{(a_{j_1j_2 1}, a_{j_2j_1(l_2)_1})}_{(\text{DP}_-,\text{DP}_+)\text{-CW center}} \end{array}\right\} \tag{4.160}$$

  The (UP,US)*-up-parabola upper-saddle is a* (UP,US)*-appearing bifurcation for saddle and center.*

- *For* $a_{j_1j_2 0} < 0$ *and* $a_{j_2j_1 0}(a_{j_2j_l l_2} - a_{j_2j_1l_1}) > 0$,

$$\underbrace{(a_{j_1j_2 1}, a_{j_2j_1l_2})}_{(\text{DP},\text{US})\text{-appearing bifurcation}} \rightleftharpoons \left\{\begin{array}{c} \underbrace{(a_{j_1j_2 1}, a_{j_2j_1(l_2)_2})}_{(\text{DP}_+,\text{DP}_-)\text{-CCW center}} \\ \underbrace{(a_{j_1j_2 2}, a_{j_2j_1(l_2)_1})}_{(\text{UP}_-,\text{UP}_-)\text{-negative saddle}} \end{array}\right\}. \tag{4.161}$$

  The (DP,US)*-down-parabola upper-saddle is a* (DP,US)*-appearing bifurcation for center and saddle.*

- *For* $a_{j_1j_2 0} > 0$ *and* $a_{j_2j_1 0}(a_{j_2j_1l_2} - a_{j_2j_1l_1}) < 0$,

$$\underbrace{(a_{j_1j_2 1}, a_{j_2j_1l_2})}_{(\text{DP},\text{LS})\text{-appearing bifurcation}} \rightleftharpoons \left\{\begin{array}{c} \underbrace{(a_{j_1j_2 1}, a_{j_2j_1(l_2)_2})}_{(\text{DP}_-,\text{DP}_+)\text{-CW center}} \\ \underbrace{(a_{j_1j_2 1}, a_{j_2j_1(l_2)_1})}_{(\text{UP}_+,\text{UP}_+)\text{-positive saddle}} \end{array}\right\}. \tag{4.162}$$

  The (DP,LS)*-down-parabola lower-saddle is a* (DP,LS)*-appearing bifurcation for center and* saddle.

- *For* $a_{j_1j_20}<0$ *and* $a_{j_2j_10}(a_{j_2j_2l_2}-a_{j_2j_1l_1})<0$,

$$\underbrace{(a_{j_1j_21},a_{j_2j_1l_2})}_{(\text{UP},\text{LS})\text{-appearing bifurcation}}\rightleftharpoons\left\{\begin{array}{c}\underbrace{(a_{j_1j_21},a_{j_2j_1(l_2)_2})}_{(\text{UP}_-,\text{UP}_+)\text{-negative saddle}}\\ \underbrace{(a_{j_1j_21},a_{j_2j_1(l_2)_1})}_{(\text{DP}_+,\text{DP}_-)\text{-CCW center}}\end{array}\right\}. \tag{4.163}$$

The (UP,LS)*-up-parabola lower-saddle is a* (UP,LS)*-appearing bifurcation for saddle and center.*

– *Such parabola-saddles are also the switching bifurcations for saddle and center.*

(xix) Under

$$\Delta_{j_1j_2}=0 \text{ and } \Delta_{j_2j_1}<0, a_{j_1j_2s_1}\neq a_{j_1j_2s_2}, \tag{4.164}$$

the standard form of dynamical systems is

$$\begin{aligned}&\dot{x}_{j_1}=a_{j_1j_20}(x_{j_2}-a_{j_1j_2s_1})(x_{j_2}-a_{j_1j_2s_2})^2,\\&\dot{x}_{j_2}=a_{j_2j_10}(x_{j_1}-a_{j_2j_11})\left[(x_{j_1}-a_{j_2j_1})^2+b_{j_2j_1}\right]\\&j_1,j_2\in\{1,2\},j_2\neq j_1.\end{aligned} \tag{4.165}$$

The equilibrium state of $x^*_{j_2}=a_{j_1j_2s_2}(s_1,s_2\in\{1,2\},s_1\neq s_2)$ on the x_{j_2}-direction is for *the following appearing bifurcations.*

- *For* $a_{j_1j_20}(a_{j_1j_2s_2}-a_{j_1j_2s_1})>0$ *and* $a_{j_2j_10}>0$,

$$\underbrace{(a_{j_1j_2s_2},a_{j_2j_11})}_{(\text{US},\text{UP})\text{-appearing bifurcation}}\rightleftharpoons\left\{\begin{array}{c}\underbrace{(a_{j_1j_2(s_2)_2},a_{j_2j_11})}_{(\text{UP}_+,\text{UP}_+)\text{-positive saddle}}\\ \underbrace{(a_{j_1j_2(s_2)_1},a_{j_2j_11})}_{(\text{DP}_+,\text{DP}_-)\text{-CCW center}}\end{array}\right\}. \tag{4.166}$$

The (US,UP)*-up-parabola upper-saddle is a* (US,UP)*-appearing bifurcations for saddle and center.*

- *For* $a_{j_1j_20}(a_{j_1j_2s_2}-a_{j_1j_2s_1})<0$ *and* $a_{j_2j_10}>0$,

$$\underbrace{(a_{j_1j_2s_2},a_{j_2j_11})}_{(\text{LS},\text{DP})\text{-appearing bifurcation}}\rightleftharpoons\left\{\begin{array}{c}\underbrace{(a_{j_1j_2(s_2)_2},a_{j_2j_11})}_{(\text{DP}_+,\text{DP}_-)\text{-CCW center}}\\ \underbrace{(a_{j_1j_2(s_2)_1},a_{j_2j_11})}_{(\text{UP}_+,\text{UP}_+)\text{-positive saddle}}\end{array}\right\}. \tag{4.167}$$

The (LS,DP)*-down-parabola lower-saddle is an* (LS,DP)*-appearing bifurcations for saddle and center.*

- *For* $a_{j_1j_20}(a_{j_1j_2s_2}-a_{j_1j_2s_1})>0$ *and* $a_{j_2j_10}<0$,

$$\underbrace{(a_{j_1j_2s_2}, a_{j_2j_1}1)}_{(\text{US},\text{DP})\text{-appearing bifurcation}} \rightleftharpoons \left\{ \begin{array}{c} \underbrace{(a_{j_1j_2(s_2)_2}, a_{j_2j_1}1)}_{(\text{DP}_-,\text{DP}_+)\text{-CW center}} \\ \underbrace{(a_{j_1j_2(s_2)_1}, a_{j_2j_1}1)}_{(\text{UP}_-,\text{UP}_-)\text{-negative saddle}} \end{array} \right\}. \tag{4.168}$$

The (US,DP)*-down-parabola upper-saddle is a* (US,DP)*-appearing bifurcations for saddle and center.*

- *For $a_{j_1j_2 0}(a_{j_1j_2s_2} - a_{j_1j_2s_1}) < 0$ and $a_{j_2j_1 0} < 0$,*

$$\underbrace{(a_{j_1j_2s_2}, a_{j_2j_1}1)}_{(\text{LS},\text{UP})\text{-appearing bifurcation}} \rightleftharpoons \left\{ \begin{array}{c} \underbrace{(a_{j_1j_2(s_2)_2}, a_{j_2j_1}1)}_{(\text{UP}_-,\text{UP}_-)\text{-negative saddle}} \\ \underbrace{(a_{j_1j_2(s_2)_1}, a_{j_2j_1}1)}_{(\text{DP}_-,\text{DP}_+)\text{-CW center}} \end{array} \right\}. \tag{4.169}$$

The (LS,UP)*-up-parabola lower-saddle is an* (LS,UP)*-appearing bifurcations for saddle and center.*

– *Such parabola-saddles are also the switching bifurcations for saddle and center.*

(xx) Under

$$\Delta_{j_1j_2} = 0 \,\text{and}\, \Delta_{j_2j_1} < 0, a_{j_1j_2s_1} = a_{j_1j_2s_2} = a_{j_1j_2}1, \tag{4.170}$$

the standard form is

$$\begin{aligned} \dot{x}_{j_1} &= a_{j_1j_2 0}(x_{j_2} - a_{j_1j_2}1)^3, \\ \dot{x}_{j_2} &= a_{j_2j_1 0}(x_{j_1} - a_{j_2j_1}1)\left[(x_{j_1} - a_{j_2j_1})^2 + b_{j_2j_1}\right] \\ & j_1, j_2 \in \{1,2\}, j_2 \neq j_1. \end{aligned} \tag{4.171}$$

The equilibrium of $(x^*_{j_2}, x^*_{j_1}) = (a_{j_1j_2}1, a_{j_2j_1}1)$ has the following appearing (or switching) bifurcations.

- *For $a_{j_1j_2 0} > 0$ and $a_{j_2j_1 0} > 0$,*

$$\underbrace{(a_{j_1j_2}1, a_{j_2j_1}1)}_{(\text{UP}_+,\text{UP}_+)\text{-positive saddle}} \rightleftharpoons \underbrace{(a_{j_{j_1j_2}}1, a_{j_2j_1}1)}_{(3^{\text{rd}}\text{UP}_+,\text{UP}_+)\text{-appearing bifucation}} \rightleftharpoons \left\{ \begin{array}{c} \underbrace{(a_{j_1j_2}3, a_{j_2j_1}1)}_{(\text{UP}_+,\text{UP}_+)\text{-positive saddle}} \\ \underbrace{(a_{j_1j_2}2, a_{j_2j_1}1)}_{(\text{DP}_+,\text{DP}_-)\text{-CCW center}} \\ \underbrace{(a_{j_1j_2}1, a_{j_2j_1}1)}_{(\text{UP}_+,\text{UP}_+)\text{-positive saddle}} \end{array} \right\}. \tag{4.172}$$

The (3^{rd}UP$_+$,UP$_+$)*-third-order positive saddle is a* (3^{rd}UP$_+$,UP$_+$)*-appearing bifurcation for a saddle to saddle, center, and saddle.*

- *For* $a_{j_1j_2 0} < 0$ *and* $a_{j_2j_1 0} > 0$,

$$\underbrace{(a_{j_1j_1}1, a_{j_2j_1}1)}_{(\mathrm{DP}_+,\mathrm{DP}_-)\text{-CCW center}} \rightleftharpoons \underbrace{(a_{j_1j_2}1, a_{j_2j_1}1)}_{(3^{\mathrm{rd}}\mathrm{DP}_+,\mathrm{DP}_-)\text{-appearing bifucation}} \rightleftharpoons \left\{ \begin{array}{c} \underbrace{(a_{j_1j_2}3, a_{j_2j_1}1)}_{(\mathrm{DP}_+,\mathrm{DP}_-)\text{-CCW center}} \\ \underbrace{(a_{j_1j_2}2, a_{j_2j_1}1)}_{(\mathrm{UP}_+,\mathrm{UP}_+)\text{-positive saddle}} \\ \underbrace{(a_{j_1j_2}1, a_{j_2j_1}1)}_{(\mathrm{DP}_+,\mathrm{DP}_-)\text{-CCW center}} \end{array} \right\}. \tag{4.173}$$

The $(3^{\mathrm{rd}}\mathrm{DP}_+,\mathrm{DP}_-)$*-third-order-counter-clockwise center is a* $(3^{\mathrm{rd}}\mathrm{DP}_+,\mathrm{DP}_-)$*-appearing bifurcation for a center to center, saddle, and center.*

- *For* $a_{j_1j_2 0} > 0$ *and* $a_{j_2j_1 0} < 0$,

$$\underbrace{(a_{j_1j_2}1, a_{j_2j_1}1)}_{(\mathrm{DP}_-,\mathrm{DP}_+)\text{-CW center}} \rightleftharpoons \underbrace{(a_{j_1j_2}1, a_{j_2j_1}1)}_{(3^{\mathrm{rd}}\mathrm{DP}_-,\mathrm{DP}_+)\text{-appearing bifucation}} \rightleftharpoons \left\{ \begin{array}{c} \underbrace{(a_{j_1j_2}3, a_{j_2j_1}1)}_{(\mathrm{DP}_-,\mathrm{DP}_+)\text{-CW center}} \\ \underbrace{(a_{j_1j_2}2, a_{j_2j_1}1)}_{(\mathrm{UP}_-,\mathrm{UP}_-)\text{-negative saddle}} \\ \underbrace{(a_{j_1j_2}1, a_{j_2j_1}1)}_{(\mathrm{DP}_-,\mathrm{DP}_+)\text{-CW center}} \end{array} \right\}. \tag{4.174}$$

The $(3^{\mathrm{rd}}\mathrm{DP}_-,\mathrm{DP}_+)$*-third-order clockwise center is a* $(3^{\mathrm{rd}}\mathrm{DP}_-,\mathrm{DP}_+)$*-appearing bifurcation for a center to center, saddle, and center.*

- *For* $a_{j_1j_2 0} < 0$ *and* $a_{j_2j_1 0} < 0$,

$$\underbrace{(a_{j_1j_2}1, a_{j_2j_1}1)}_{(\mathrm{UP}_-,\mathrm{UP}_-)\text{-negative saddle}} \rightleftharpoons \underbrace{(a_{j_1j_2}1, a_{j_2j_1}1)}_{(3^{\mathrm{rd}}\mathrm{UP}_-,\mathrm{UP}_-)\text{-appearing bifucation}} \rightleftharpoons \left\{ \begin{array}{c} \underbrace{(a_{j_1j_2}3, a_{j_2j_1}1)}_{(\mathrm{UP}_-,\mathrm{UP}_-)\text{-negative saddle}} \\ \underbrace{(a_{j_1j_2}2, a_{j_2j_1}1)}_{(\mathrm{DP}_-,\mathrm{DP}_+)\text{-CW center}} \\ \underbrace{(a_{j_1j_2}1, a_{j_2j_1}1)}_{(\mathrm{UP}_-,\mathrm{UP}_-)\text{-negative saddle}} \end{array} \right\}. \tag{4.175}$$

The (3rdUP$_-$,UP$_-$)*-third-order negative saddle is a* (3rdUP$_-$,UP$_-$)*-appearing bifurcation for a saddle to saddle, center, and saddle.*

- *Such third-order saddle bifurcations are also the switching bifurcations for saddle, center, and saddle.*
- *Such third-order center bifurcations are also the switching bifurcations for center, saddle, and center.*

(xxi) Under

$$\Delta_{j_1j_2} = 0 \text{ and } \Delta_{j_2j_1} = 0, a_{j_2j_1s_1} \neq a_{j_2j_1s_2}, a_{j_1j_2l_1} \neq a_{j_1j_2l_2}, \tag{4.176}$$

the standard form is

$$\begin{aligned} \dot{x}_{j_1} &= a_{j_1j_20}(x_{j_2} - a_{j_1j_2s_1})(x_{j_2} - a_{j_1j_2s_2})^2, \\ \dot{x}_{j_2} &= a_{j_2j_10}(x_{j_1} - a_{j_2j_1l_1})(x_{j_1} - a_{j_2j_2i_2})^2, \\ j_1, j_2 &\in \{1,2\}; j_1 \neq j_2. \end{aligned} \tag{4.177}$$

(xxi$_1$) The equilibrium of $(x^*_{j_2}, x^*_{j_1}) = (a_{j_1j_2s_2}, a_{j_2j_1l_1})$ *has the following appearing bifurcations.*

- *For* $a_{j_1j_20}(a_{j_1j_2s_2} - a_{j_1j_2s_1}) > 0$ *and* $a_{j_2j_10}(a_{j_2j_1l_1} - a_{j_2j_1l_2})^2 > 0$,

$$\underbrace{(a_{j_1j_2s_2}, a_{j_2j_1l_1})}_{(\text{US, UP})\text{-appearing bifurcation}} \rightleftharpoons \left\{ \begin{array}{c} \underbrace{(a_{j_1j_2(s_2)_2}, a_{j_2j_1l_1})}_{(\text{UP}_+, \text{UP}_+)\text{-positive saddle}} \\ \underbrace{(a_{j_1j_2(s_2)_1}, a_{j_2j_1l_1})}_{(\text{DP}_+, \text{DP}_-)\text{-CCW center}} \end{array} \right\}. \tag{4.178}$$

The (US,UP)*-up-parabola upper-saddle is a* (US,UP)*-appearing bifurcation for center and saddle.*

- *For* $a_{j_1j_20}(a_{j_1j_2s_2} - a_{j_1j_2s_1}) < 0$ *and* $a_{j_2j_10}(a_{j_2j_1l_1} - a_{j_2j_1l_2})^2 > 0$,

$$\underbrace{(a_{j_1j_2s_2}, a_{j_2j_1l_1})}_{(\text{LS, DP})\text{-appearing bifurcation}} \rightleftharpoons \left\{ \begin{array}{c} \underbrace{(a_{j_1j_2(s_2)_2}, a_{j_2j_1l_1})}_{(\text{DP}_+, \text{DP}_-)\text{-CCW center}} \\ \underbrace{(a_{j_1j_2(s_2)_2}, a_{j_2j_1l_1})}_{(\text{UP}_+, \text{UP}_+)\text{-positive saddle}} \end{array} \right\} \tag{4.179}$$

The (LS,DP)*-down-parabola lower-saddle is an* (LS,DP)*-appearing bifurcation for center and saddle.*

- *For* $a_{j_1j_20}(a_{j_1j_2s_2} - a_{j_1j_2s_1}) > 0$ *and* $a_{j_2j_10}(a_{j_2j_1l_1} - a_{j_2j_1l_2})^2 < 0$,

$$\underbrace{(a_{j_1j_2s_2}, a_{j_2j_1l_1})}_{(\text{US, DP})\text{-appearing bifurcation}} \rightleftharpoons \left\{ \begin{array}{c} \underbrace{(a_{j_1j_2(s_2)_2}, a_{j_2j_1l_1})}_{(\text{DP}_-, \text{DP}_+)\text{-CW center}} \\ \underbrace{(a_{j_1j_2(s_2)_2}, a_{j_2j_1l_1})}_{(\text{UP}_-, \text{UP}_-)\text{-negative saddle}} \end{array} \right\}. \tag{4.180}$$

The (US,DP)-*down-parabola upper-saddle is a* (US,DP)-*appearing bifurcation for center and saddle.*

- *For* $a_{j_1j_20}(a_{j_1j_2s_2} - a_{j_1j_2s_1}) < 0$ *and* $a_{j_2j_10}(a_{j_2j_1l_1} - a_{j_2j_1l_2})^2 < 0$,

$$\underbrace{(a_{j_1j_2s_2}, a_{j_2j_1l_1})}_{(\text{LS, UP})\text{-appearing bifurcation}} \rightleftharpoons \left\{ \begin{array}{c} \underbrace{(a_{j_1j_2(s_2)_2}, a_{j_2j_1l_1})}_{(\text{UP}_-, \text{UP}_-)\text{-negative saddle}} \\ \underbrace{(a_{j_1j_2(s_2)_1}, a_{j_2j_1l_1})}_{(\text{DP}_-, \text{DP}_+)\text{-CW center}} \end{array} \right\}. \tag{4.181}$$

The (LS,UP)-*up-parabola lower-saddle is an* (LS,UP)-*appearing bifurcation for center and saddle.*

– *Such parabola-saddle bifurcations are also the appearing bifurcations for saddle and center.*

(xxi$_2$) The equilibrium of $(x^*_{j_2}, x^*_{j_1}) = (a_{j_1j_2s_1}, a_{j_2j_1}1)$ *has the following appearing bifurcations.*

- *For* $a_{j_1j_20}(a_{j_1j_2s_1} - a_{j_1j_2s_2})^2 > 0$ *and* $a_{j_2j_10}(a_{j_2j_1l_2} - a_{j_2j_1l_1}) > 0$,

$$\underbrace{(a_{j_1j_2s_1}, a_{j_2j_1l_2})}_{(\text{UP, US})\text{-appearing bifurcation}} \rightleftharpoons \left\{ \begin{array}{c} \underbrace{(a_{j_1j_2s_1}, a_{j_2j_1(l_2)_2})}_{(\text{UP}_+, \text{UP}_+)\text{-positive saddle}} \\ \underbrace{(a_{j_1j_2s_1}, a_{j_2j_1(l_2)_1})}_{(\text{DP}_-, \text{DP}_+)\text{-CW center}} \end{array} \right\}. \tag{4.182}$$

The (UP,US)-*up-parabola upper-saddle is a* (UP,US)-*appearing bifurcation for center and saddle.*

- *For* $a_{j_1j_20}(a_{j_1j_2s_1} - a_{j_1j_2s_2})^2 < 0$ *and* $a_{j_2j_10}(a_{j_2j_1l_2} - a_{j_2j_1l_1}) > 0$,

$$\underbrace{(a_{j_1j_2s_1}, a_{j_2j_1l_2})}_{(\text{DP, US})\text{-appearing bifurcation}} \rightleftharpoons \left\{ \begin{array}{c} \underbrace{(a_{j_1j_2s_1}, a_{j_2j_1(l_2)_2})}_{(\text{DP}_+, \text{DP}_-)\text{-CCW center}} \\ \underbrace{(a_{j_1j_2s_1}, a_{j_2j_1(l_2)_1})}_{(\text{UP}_-, \text{UP}_-)\text{-negative saddle}} \end{array} \right\}. \tag{4.183}$$

The (DP,US)-*down-parabola upper-saddle is a* (DP,US)-*appearing bifurcation for center and saddle.*

- *For* $a_{j_1j_2 0}(a_{j_1j_2s_1}-a_{j_1j_2s_2})^2>0$ *and* $a_{j_2j_1 0}(a_{j_2j_1l_2}-a_{j_2j_1l_1})^2<0$,

$$\underbrace{(a_{j_1j_2s_1},a_{j_2j_1l_2})}_{(\mathrm{DP},\mathrm{LS})\text{-appearing bifurcation}} \rightleftharpoons \left\{\begin{array}{c}\underbrace{(a_{j_1j_2s_1},a_{j_2j_1(l_2)_2})}_{(\mathrm{DP}_-,\mathrm{DP}_+)\text{-CW center}}\\ \underbrace{(a_{j_1j_2s_1},a_{j_2j_1(l_2)_1})}_{(\mathrm{UP}_+,\mathrm{UP}_+)\text{-positive saddle}}\end{array}\right\}. \tag{4.184}$$

The (DP,LS)*-down-parabola lower-saddle is a* (DP,LS)*-appearing bifurcation for center and saddle.*

- *For* $a_{j_1j_2 0}(a_{j_1j_2s_1}-a_{j_1j_2s_2})^2<0$ *and* $a_{j_2j_1 0}(a_{j_2j_1l_2}-a_{j_2j_1l_1})<0$,

$$\underbrace{(a_{j_1j_2s_1},a_{j_2j_1l_2})}_{(\mathrm{UP},\mathrm{LS})\text{-appearing bifurcation}} \rightleftharpoons \left\{\begin{array}{c}\underbrace{(a_{j_1j_2s_1},a_{j_2j_1(l_2)_2})}_{(\mathrm{UP}_-,\mathrm{UP}_-)\text{-negative saddle}}\\ \underbrace{(a_{j_1j_2s_1},a_{j_2j_1(l_2)_1})}_{(\mathrm{DP}_+,\mathrm{DP}_-)\text{-CCW center}}\end{array}\right\}. \tag{4.185}$$

The (UP,LS)*-up-parabola lower-saddle is a* (UP,LS)*-appearing bifurcation for center and saddle.*

– *Such parabola-saddle bifurcations are also the switching bifurcations for saddle and center.*

(xxi$_3$*) The equilibrium of* $(x^*_{j_2},x^*_{j_1})=(a_{j_1j_2s_2},a_{j_2j_1l_2})$ *has the following bifurcations.*

- *For* $a_{j_1j_2 0}(a_{j_1j_2s_2}-a_{j_1j_2s_1})>0$ *and* $a_{j_2j_1 0}(a_{j_2j_1l_2}-a_{j_2j_1l_1})>0$,

$$\underbrace{(a_{j_1j_2s_2},a_{j_2j_2l_2})}_{(\mathrm{II}_+,\mathrm{II}_+)\text{-appearing bifurcation}} \rightleftharpoons \left\{\begin{array}{cc}\underbrace{(a_{j_1j_2(s_2)_2},a_{j_2j_1(l_2)_1})}_{(\mathrm{DP}_-,\mathrm{DP}_+)\text{-CW center}} & \underbrace{(a_{j_1j_2(s_2)_2},a_{j_2j_1(l_2)_2})}_{(\mathrm{UP}_+,\mathrm{UP}_+)\text{-positive saddle}}\\ \underbrace{(a_{j_1j_2(s_2)_1},a_{j_2j_1(l_2)_1})}_{(\mathrm{UP}_-,\mathrm{UP}_-)\text{-negative saddle}} & \underbrace{(a_{j_1j_2(s_2)_1},a_{j_2j_1(l_2)_2})}_{(\mathrm{DP}_+,\mathrm{DP}_-)\text{-CCW center}}\end{array}\right\} \tag{4.186}$$

The (II$_+$,II$_+$)*-double increasing-inflection saddle is an* (II$_+$,II$_+$)*-appearing bifurcation for a network of saddles and centers.*

- *For* $a_{j_1j_2 0}(a_{j_1j_2s_2}-a_{j_1j_2s_1})<0$ *and* $a_{j_2j_1 0}(a_{j_2j_1l_2}-a_{j_2j_1l_1})>0$,

$$\underbrace{(a_{j_1j_2s_2},a_{j_2j_1l_2})}_{(\mathrm{DI}_+,\mathrm{DI}_-)\text{-appearing bifurcation}} \rightleftharpoons \left\{\begin{array}{cc}\underbrace{(a_{j_1j_2(s_2)_2},a_{j_2j_1(l_2)_1})}_{(\mathrm{UP}_-,\mathrm{UP}_-)\text{-negative saddle}} & \underbrace{(a_{j_1j_2(s_2)_2},a_{j_2j_1(l_2)_2})}_{(\mathrm{DP}_+,\mathrm{DP}_-)\text{-CCW center}}\\ \underbrace{(a_{j_1j_2(s_2)_1},a_{j_2j_1(l_2)_1})}_{(\mathrm{DP}_-,\mathrm{DP}_+)\text{-CW center}} & \underbrace{(a_{j_1j_2(s_2)_1},a_{j_2j_1(l_2)_2})}_{(\mathrm{UP}_+,\mathrm{UP}_+)\text{-positive saddle}}\end{array}\right\}. \tag{4.187}$$

The (DI$_+$,DI$_-$)*-double decreasing-inflection saddle is a* (DI$_+$,DI$_-$)*-appearing bifurcation for a network of saddles and centers.*

- *For* $a_{j_1j_20}(a_{j_1j_2s_2}-a_{j_1j_2s_1})>0$ *and* $a_{j_2j_10}(a_{j_2j_1l_2}-a_{j_2j_1l_1})<0$,

$$\underbrace{(a_{j_1j_2s_2},a_{j_2j_2l_2})}_{(\mathrm{DI}_-,\mathrm{DI}_+)\text{-appearing bifurcation}} \rightleftharpoons \left\{\begin{array}{cc} \underbrace{(a_{j_1j_2(s_2)_2},a_{j_2j_1(l_2)_1})}_{(\mathrm{UP}_+,\mathrm{UP}_+)\text{-positive saddle}} & \underbrace{(a_{j_1j_2(s_2)_2},a_{j_2j_1(l_2)_2})}_{(\mathrm{DP}_-,\mathrm{DP}_+)\text{-CW center}} \\ \underbrace{(a_{j_1j_2(s_2)_1},a_{j_2j_1(l_2)_1})}_{(\mathrm{DP}_+,\mathrm{DP}_-)\text{-CCW center}} & \underbrace{(a_{j_1j_2(s_2)_1},a_{j_2j_1(l_2)_2})}_{(\mathrm{UP}_-,\mathrm{UP}_-)\text{-negative saddle}} \end{array}\right\}. \tag{4.188}$$

The (DI$_-$,DI$_+$)*-double decreasing-inflection saddle is a* (DI$_-$,DI$_+$)*-appearing bifurcation for a network of saddles and centers.*

- *For* $a_{j_1j_20}(a_{j_1j_2s_2}-a_{j_1j_2s_1})<0$ *and* $a_{j_2j_10}(a_{j_2j_1l_2}-a_{j_2j_1l_1})<0$,

$$\underbrace{(a_{j_1j_2s_2},a_{j_2j_1l_2})}_{(\mathrm{II}_-,\mathrm{II}_-)\text{-appearing bifurcation}} \rightleftharpoons \left\{\begin{array}{cc} \underbrace{(a_{j_1j_2(s_2)_2},a_{j_2j_1(l_2)_1})}_{(\mathrm{DP}_+,\mathrm{DP}_-)\text{-CCW center}} & \underbrace{(a_{j_1j_2(s_2)_2},a_{j_2j_1(l_2)_2})}_{(\mathrm{UP}_-,\mathrm{UP}_-)\text{-negative saddle}} \\ \underbrace{(a_{j_1j_2(s_2)_1},a_{j_2j_1(l_2)_1})}_{(\mathrm{UP}_+,\mathrm{UP}_+)\text{-positive saddle}} & \underbrace{(a_{j_1j_2(s_2)_1},a_{j_2j_1(l_2)_2})}_{(\mathrm{DP}_-,\mathrm{DP}_+)\text{-CW center}} \end{array}\right\}. \tag{4.189}$$

The (II$_-$,II$_-$)*-double increasing-inflection saddle is an* (II$_-$,II$_-$)*-appearing bifurcation for a network of saddles and centers.*

– *Such double inflection-saddles are also the switching bifurcations for networks of saddles and centers.*

(xxii) Under

$$\Delta_{j_1j_2}=0 \text{ and } \Delta_{j_2j_1}=0; a_{j_2j_1s_1}=a_{j_2j_1s_2}=a_{j_2j_11}, a_{j_1j_2l_1}\neq a_{j_1j_2l_2}, \tag{4.190}$$

the standard form is

$$\begin{aligned} \dot{x}_{j_1} &= a_{j_1j_20}(x_{j_2}-a_{j_1j_21})^3, \\ \dot{x}_{j_2} &= a_{j_2j_10}(x_{j_1}-a_{j_2j_1l_1})(x_{j_1}-a_{j_2j_1l_2})^2, \\ & j_1,j_2 \in \{1,2\}; j_1 \neq j_2. \end{aligned} \tag{4.191}$$

(xxii$_1$) The equilibrium of $(x_{j_2}^*,x_{j_1}^*)=(a_{j_1j_21},a_{j_2j_1l_1})$ *has the following bifurcations.*

- *For* $a_{j_1j_20}>0$ *and* $a_{j_2j_10}(a_{j_2j_1l_1}-a_{j_2j_1l_2})^2>0$,

$$\underbrace{(a_{j_1j_2}1, a_{j_2j_1l_1})}_{(\mathrm{UP}_+,\mathrm{UP}_+)\text{-positive saddle}} \rightleftharpoons \underbrace{(a_{j_1j_2}1, a_{j_2j_1l_1})}_{(3^{\mathrm{rd}}\mathrm{UP}_+,\mathrm{UP}_+)\text{-appearing bifucation}} \rightleftharpoons \left\{ \begin{array}{c} \underbrace{(a_{j_1j_2}3, a_{j_2j_ll_1})}_{(\mathrm{UP}_+,\mathrm{UP}_+)\text{-positive saddle}} \\ \underbrace{(a_{j_1j_2}2, a_{j_2j_1l_1})}_{(\mathrm{DP}_+,\mathrm{DP}_-)\text{-CCW center}} \\ \underbrace{(a_{j_1j_2}1, a_{j_2j_ll_1})}_{(\mathrm{UP}_+,\mathrm{UP}_+)\text{-positive saddle}} \end{array} \right\}. \tag{4.192}$$

The $(3^{\mathrm{rd}}\mathrm{UP}_+,\mathrm{UP}_+)$*-third-order positive saddle is a* $(3^{\mathrm{rd}}\mathrm{UP}_+,\mathrm{UP}_+)$*-appearing bifurcation from a positive saddle to the positive saddle, clockwise center, and positive saddle.*

- *For* $a_{j_1j_2}0 < 0$ *and* $a_{j_2j_0}0(a_{j_2j_1l_1} - a_{j_2j_1l_2})^2 > 0$,

$$\underbrace{(a_{j_1j_2}1, a_{j_2j_1l_1})}_{(\mathrm{DP}_+,\mathrm{DP}_-)\text{-CCW center}} \rightleftharpoons \underbrace{(a_{j_1j_2}1, a_{j_2j_1l_1})}_{(3^{\mathrm{rd}}\mathrm{DP}_+,\mathrm{DP}_-)\text{-appearing bifucation}} \rightleftharpoons \left\{ \begin{array}{c} \underbrace{(a_{j_1j_2}3, a_{j_2j_ll_1})}_{(\mathrm{DP}_+,\mathrm{DP}_-)\text{-CCW center}} \\ \underbrace{(a_{j_1j_2}2, a_{j_2j_1l_1})}_{(\mathrm{UP}_+,\mathrm{UP}_+)\text{-positive saddle}} \\ \underbrace{(a_{j_1j_2}1, a_{j_2j_ll_1})}_{(\mathrm{DP}_+,\mathrm{DP}_-)\text{-CCW center}} \end{array} \right\}. \tag{4.193}$$

The $(3^{\mathrm{rd}}\mathrm{DP}_+,\mathrm{DP}_-)$*-third-order clockwise center is a* $(3^{\mathrm{rd}}\mathrm{DP}_+,\mathrm{DP}_-)$*-appearing bifurcation from a center to the center, saddle, clockwise center.*

- *For* $a_{j_1j_2}0 > 0$ *and* $a_{j_2j_1}0(a_{j_2j_1l_1} - a_{j_2j_1l_2})^2 < 0$,

$$\underbrace{(a_{j_1j_2}1, a_{j_2j_1l_1})}_{(\mathrm{DP}_-,\mathrm{DP}_+)\text{-CW center}} \rightleftharpoons \underbrace{(a_{j_1j_2}1, a_{j_2j_1l_1})}_{(3^{\mathrm{rd}}\mathrm{DP}_-,\mathrm{DP}_+)\text{-appearing bifucation}} \rightleftharpoons \left\{ \begin{array}{c} \underbrace{(a_{j_1j_2}3, a_{j_2j_ll_1})}_{(\mathrm{DP}_-,\mathrm{DP}_+)\text{-CW center}} \\ \underbrace{(a_{j_1j_2}2, a_{j_2j_1l_1})}_{(\mathrm{UP}_-,\mathrm{UP}_-)\text{-negative saddle}} \\ \underbrace{(a_{j_1j_2}1, a_{j_2j_ll_1})}_{(\mathrm{DP}_-,\mathrm{DP}_+)\text{-CW center}} \end{array} \right\}. \tag{4.194}$$

The $(3^{\mathrm{rd}}\mathrm{DP}_-,\mathrm{DP}_+)$*-third-order clockwise center is a* $(3^{\mathrm{rd}}\mathrm{DP}_-,\mathrm{DP}_+)$*-appearing bifurcation from a clockwise center to center, saddle, and center.*

- *For* $a_{j_1j_20}<0$ *and* $a_{j_2j_10}(a_{j_2j_1l_1}-a_{j_2j_1l_2})<0$,

$$\underbrace{(a_{j_1j_21},a_{j_2j_1l_1})}_{(\mathrm{UP}_-,\mathrm{UP}_-)\text{-negative saddle}} \rightleftharpoons \underbrace{(a_{j_1j_21},a_{j_2j_1l_1})}_{(3^{\mathrm{rd}}\mathrm{UP}_-,\mathrm{UP}_-)\text{-appearing bifucation}} \rightleftharpoons \left\{\begin{array}{c}\underbrace{(a_{j_1j_23},a_{j_2j_1l_1})}_{(\mathrm{UP}_-,\mathrm{UP}_-)\text{-negative saddle}}\\ \underbrace{(a_{j_1j_22},a_{j_2j_1l_1})}_{(\mathrm{DP}_+,\mathrm{DP}_-)\text{-CCW center}}\\ \underbrace{(a_{j_1j_21},a_{j_2j_1l_1})}_{(\mathrm{UP}_-,\mathrm{UP}_-)\text{-negative saddle}}\end{array}\right\}. \tag{4.195}$$

The (3^rd^UP_,UP_)*-third-order negative saddle is a* (3^rd^UP_,UP_)*-appearing bifurcation from a saddle to saddle, center, and saddle.*

– *The third-order saddles are also the switching bifurcations of saddle, center, and saddle.*
– *The third-order centers are also the switching bifurcations of center, saddle, and center.*

(xxii$_2$*) The equilibrium of* $(x^*_{j_2},x^*_{j_1})=(a_{j_1j_21},a_{j_2j_1l_2})$ *has the following appearing bifurcations.*

- *For* $a_{j_1j_20}>0$ *and* $a_{j_2j_10}(a_{j_2j_1l_2}-a_{j_2j_1l_1})>0$,

$$\underbrace{(a_{j_1j_21},a_{j_2j_1l_2})}_{(3^{\mathrm{rd}}\mathrm{UP},\mathrm{US})\text{-appearing bifurcation}} \rightleftharpoons \left\{\begin{array}{cc}\underbrace{(a_{j_1j_23},a_{j_2j_1(l_2)_1})}_{(\mathrm{DP}_-,\mathrm{DP}_+)\text{-CW center}} & \underbrace{(a_{j_1j_23},a_{j_2j_1(l_2)_2})}_{(\mathrm{UP}_+,\mathrm{UP}_+)\text{-positive saddle}}\\ \underbrace{(a_{j_1j_22},a_{j_2j_1(l_2)_1})}_{(\mathrm{UP}_-,\mathrm{UP}_-)\text{-negative saddle}} & \underbrace{(a_{j_1j_22},a_{j_2j_1(l_2)_2})}_{(\mathrm{DP}_+,\mathrm{DP}_-)\text{-CCW center}}\\ \underbrace{(a_{j_1j_21},a_{j_2j_1(l_2)_1})}_{(\mathrm{DP}_-,\mathrm{DP}_+)\text{-CW center}} & \underbrace{(a_{j_1j_21},a_{j_2j_1(l_2)_2})}_{(\mathrm{UP}_+,\mathrm{UP}_+)\text{-positive saddle}}\end{array}\right\} \tag{4.196}$$

The (3^rd^UP,US)-third-order-*up-parabola upper-saddle is a* (3^rd^UP,US)-*appearing bifurcation for a network of saddles and centers.*

- *For* $a_{j_1j_20}<0$ *and* $a_{j_2j_10}(a_{j_2j_1l_2}-a_{j_2j_1l_1})>0$,

$$\underbrace{(a_{j_1j_2}1, a_{j_2j_1l_2})}_{(3^{\rm rd}\mathrm{DP},\,\mathrm{US})\text{-appearing bifurcation}} \rightleftharpoons \left\{ \begin{array}{cc} \underbrace{(a_{j_1j_2}3, a_{j_2j_1(l_2)_1})}_{(\mathrm{UP}_-,\,\mathrm{UP}_-)\text{-negative saddle}} & \underbrace{(a_{j_1j_2}3, a_{j_2j_1(l_2)_2})}_{(\mathrm{DP}_+,\,\mathrm{DP}_-)\text{-CCW center}} \\ \underbrace{(a_{j_1j_2}2, a_{j_2j_1(l_2)_1})}_{(\mathrm{DP}_-,\,\mathrm{DP}_+)\text{-CW center}} & \underbrace{(a_{j_1j_2}2, a_{j_2j_1(l_2)_2})}_{(\mathrm{UP}_+,\,\mathrm{UP}_+)\text{-positive saddle}} \\ \underbrace{(a_{j_1j_2}1, a_{j_2j_1(l_2)_1})}_{(\mathrm{UP}_-,\,\mathrm{UP}_-)\text{-negative saddle}} & \underbrace{(a_{j_1j_2}1, a_{j_2j_1(l_2)_2})}_{(\mathrm{DP}_+,\,\mathrm{DP}_-)\text{-CCW center}} \end{array} \right\}. \tag{4.197}$$

The ($3^{\rm rd}$DP,US)-*third-order-down-parabola upper-saddle is a* ($3^{\rm rd}$DP, US)-*appearing bifurcation for a network of saddles and centers.*

- *For* $a_{j_1j_2}0 > 0$ *and* $a_{j_2j_1}0(a_{j_2j_1l_2} - a_{j_2j_1l_1}) < 0$,

$$\underbrace{(a_{j_1j_2}1, a_{j_2j_1l_2})}_{(3^{\rm rd}\mathrm{DP},\,\mathrm{LS})\text{-appearing bifurcation}} \rightleftharpoons \left\{ \begin{array}{cc} \underbrace{(a_{j_1j_2}3, a_{j_2j_1(l_2)_1})}_{(\mathrm{UP}_+,\,\mathrm{UP}_+)\text{-positive saddle}} & \underbrace{(a_{j_1j_3}3, a_{j_2j_1(l_2)_2})}_{(\mathrm{DP}_-,\,\mathrm{DP}_+)\text{-CW center}} \\ \underbrace{(a_{j_1j_2}2, a_{j_2j_1(l_2)_1})}_{(\mathrm{DP}_+,\,\mathrm{DP}_-)\text{-CCW center}} & \underbrace{(a_{j_1j_2}2, a_{j_2j_1(l_2)_2})}_{(\mathrm{UP}_-,\,\mathrm{UP}_-)\text{-negative saddle}} \\ \underbrace{(a_{j_1j_2}1, a_{j_2j_1(l_2)_1})}_{(\mathrm{UP}_+,\,\mathrm{UP}_+)\text{-positive saddle}} & \underbrace{(a_{j_1j_2}1, a_{j_2j_1(l_2)_2})}_{(\mathrm{DP}_-,\,\mathrm{DP}_+)\text{-CW center}} \end{array} \right\}. \tag{4.198}$$

The ($3^{\rm rd}$DP,LS)-*third-order down-parabola lower-saddle is a* ($3^{\rm rd}$DP,LS)-*appearing bifurcation for a network of saddles and centers.*

- *For* $a_{j_1j_2}0 < 0$ *and* $a_{j_2j_1}0(a_{j_2j_1l_2} - a_{j_2j_1l_1}) < 0$,

$$\underbrace{(a_{j_1j_2}1, a_{j_2j_1l_2})}_{(3^{\rm rd}\mathrm{UP},\,\mathrm{LS})\text{-appearing bifurcation}} \rightleftharpoons \left\{ \begin{array}{cc} \underbrace{(a_{j_1j_2}3, a_{j_2j_1(l_2)_1})}_{(\mathrm{DP}_+,\,\mathrm{DP}_-)\text{-CCW center}} & \underbrace{(a_{j_1j_3}3, a_{j_2j_1(l_2)_2})}_{(\mathrm{UP}_-,\,\mathrm{UP}_-)\text{-negative saddle}} \\ \underbrace{(a_{j_1j_2}2, a_{j_2j_1(l_2)_1})}_{(\mathrm{UP}_+,\,\mathrm{UP}_+)\text{-positive saddle}} & \underbrace{(a_{j_1j_2}2, a_{j_2j_1(l_2)_2})}_{(\mathrm{DP}_-,\,\mathrm{DP}_+)\text{-CW center}} \\ \underbrace{(a_{j_1j_2}1, a_{j_2j_1(l_2)_1})}_{(\mathrm{DP}_+,\,\mathrm{DP}_-)\text{-CCW center}} & \underbrace{(a_{j_1j_2}1, a_{j_2j_1(l_2)_2})}_{(\mathrm{UP}_-,\,\mathrm{UP}_-)\text{-negative saddle}} \end{array} \right\}. \tag{4.199}$$

The ($3^{\rm rd}$UP,LS)-*third-order-up-parabola lower-saddle is a* ($3^{\rm rd}$UP,LS)-*appearing bifurcation for a network of saddles and centers.*

The (3,2)-parabola-saddles are also the switching bifurcations for networks of saddles and centers.
(xxiii) Under

$$\Delta_{j_1j_2}=0\,\text{and}\,\Delta_{j_2j_1}=0;\, a_{j_2j_1s_1}=a_{j_2j_1s_2}=a_{j_2j_1}1,\, a_{j_1j_2l_1}=a_{j_1j_2l_2}=a_{j_1j_2}1, \tag{4.200}$$

the standard form is

$$\begin{aligned}
&\dot{x}_{j_1}=a_{j_1j_2}0(x_{j_2}-a_{j_1j_2}1)^3,\\
&\dot{x}_{j_2}=a_{j_2j_1}0(x_{j_1}-a_{j_2j_1})^3,\\
&j_1,j_2\in\{1,2\};j_1\neq j_2.
\end{aligned} \tag{4.201}$$

The equilibrium of $(x_{j_2}^*,x_{j_1}^*)=(a_{j_1j_2}1,a_{j_2j_1}1)$ *has the following appearing bifurcations.*

- *For* $a_{j_1j_2}0>0$ *and* $a_{j_2j_1}0>0$,

$$\underbrace{(a_{j_1j_2}1,a_{j_2j_1}1)}_{(\text{UP}_+,\text{UP}_+)\text{-positive saddle}} \rightleftharpoons \underbrace{(a_{j_1j_2}1,a_{j_2j_1}1)}_{(3^{\text{rd}}\text{UP}_+,3^{\text{rd}}\text{UP}_+)\text{-appearing bifurcation}}$$
$$\rightleftharpoons \left\{\begin{array}{ccc}
\underbrace{(a_{j_1j_2}3,a_{j_2j_1}1)}_{(\text{UP}_+,\text{UP}_+)\text{-positive saddle}} & \underbrace{(a_{j_1j_2}3,a_{j_2j_1}2)}_{(\text{DP}_-,\text{DP}_+)\text{-CW center}} & \underbrace{(a_{j_1j_2}3,a_{j_2j_1}3)}_{(\text{UP}_+,\text{UP}_+)\text{-positive saddle}}\\
\underbrace{(a_{j_1j_2}2,a_{j_2j_1}1)}_{(\text{DP}_+,\text{DP}_-)\text{-CCW center}} & \underbrace{(a_{j_1j_2}2,a_{j_2j_1}2)}_{(\text{UP}_-,\text{UP}_-)\text{-negative saddle}} & \underbrace{(a_{j_1j_2}2,a_{j_2j_1}3)}_{(\text{DP}_+,\text{DP}_-)\text{-CCW center}}\\
\underbrace{(a_{j_1j_2}1,a_{j_2j_1}1)}_{(\text{UP}_+,\text{UP}_+)\text{-positive saddle}} & \underbrace{(a_{j_1j_2}1,a_{j_2j_1}2)}_{(\text{DP}_-,\text{DP}_+)\text{-CW center}} & \underbrace{(a_{j_1j_2}1,a_{j_2j_1}3)}_{(\text{UP}_+,\text{UP}_+)\text{-positive saddle}}
\end{array}\right\}. \tag{4.202}$$

The (3^{rd}UP$_+$,3rd UP$_+$)*-double third-order positive saddle is a* (3^{rd}UP$_+$,3rd UP$_+$)*-appearing bifurcation for a saddle to a network of saddles and centers.*

- *For* $a_{j_1j_2}0<0$ *and* $a_{j_2j_1}0>0$,

$$\underbrace{(a_{j_1j_2}1,a_{j_2j_1}1)}_{(\text{DP}_+,\text{DP}_-)\text{-CCW center}} \rightleftharpoons \underbrace{(a_{j_1j_2}1,a_{j_2j_1}1)}_{(3^{\text{rd}}\text{DP}_+,3^{\text{rd}}\text{DP}_-)\text{-appearing bifurcation}}$$
$$\rightleftharpoons \left\{\begin{array}{ccc}
\underbrace{(a_{j_1j_2}3,a_{j_2j_1}1)}_{(\text{DP}_+,\text{DP}_-)\text{-CCW center}} & \underbrace{(a_{j_1j_2}3,a_{j_2j_1}2)}_{(\text{UP}_-,\text{UP}_-)\text{-negative saddle}} & \underbrace{(a_{j_1j_2}3,a_{j_2j_1}3)}_{(\text{DP}_+,\text{DP}_-)\text{-CCW center}}\\
\underbrace{(a_{j_1j_2}2,a_{j_2j_1}1)}_{(\text{UP}_+,\text{UP}_+)\text{-positive saddle}} & \underbrace{(a_{j_1j_2}2,a_{j_2j_1}2)}_{(\text{DP}_-,\text{DP}_+)\text{-CW center}} & \underbrace{(a_{j_1j_2}2,a_{j_2j_1}3)}_{(\text{UP}_+,\text{UP}_+)\text{-positive saddle}}\\
\underbrace{(a_{j_1j_2}1,a_{j_2j_1}1)}_{(\text{DP}_+,\text{DP}_-)\text{-CCW center}} & \underbrace{(a_{j_1j_2}1,a_{j_2j_1}2)}_{(\text{UP}_-,\text{UP}_-)\text{-negative saddle}} & \underbrace{(a_{j_1j_2}1,a_{j_2j_1}3)}_{(\text{DP}_+,\text{UP}_-)\text{-CCW center}}
\end{array}\right\}. \tag{4.203}$$

The (3^{rd}DP$_+$,3^{rd}DP$_-$)*-double third-order counter clockwise center is a* (3^{rd}DP$_+$,3^{rd}DP$_-$)*-appearing bifurcation for a center to a network of saddles and centers.*

- *For* $a_{j_1j_2}0>0$ *and* $a_{j_2j_1}0<0$,

$$\underbrace{(a_{j_1j_2}1, a_{j_2j_1}1)}_{(\text{DP}_-, \text{DP}_+)\text{-CW center}} \rightleftharpoons \underbrace{(a_{j_1j_2}1, a_{j_2j_1}1)}_{(3^{\text{rd}}\text{DP}_+, 3^{\text{rd}}\text{DP}_-)\text{-appearing bifurcation}}$$
$$\rightleftharpoons \left\{ \begin{array}{ccc} \underbrace{(a_{j_1j_2}3, a_{j_2j_1}1)}_{(\text{DP}_-, \text{DP}_+)\text{-CW center}} & \underbrace{(a_{j_1j_2}3, a_{j_2j_1}2)}_{(\text{UP}_+, \text{UP}_+)\text{-positive saddle}} & \underbrace{(a_{j_1j_2}3, a_{j_2j_1}3)}_{(\text{DP}_-, \text{DP}_+)\text{-CW center}} \\ \underbrace{(a_{j_1j_2}2, a_{j_2j_1}1)}_{(\text{UP}_-, \text{UP}_-)\text{-negative saddle}} & \underbrace{(a_{j_1j_2}2, a_{j_2j_1}2)}_{(\text{DP}_+, \text{DP}_-)\text{-CCW center}} & \underbrace{(a_{j_1j_2}2, a_{j_2j_1}3)}_{(\text{UP}_-, \text{UP}_-)\text{-negative saddle}} \\ \underbrace{(a_{j_1j_2}1, a_{j_2j_1}1)}_{(\text{DP}_-, \text{DP}_+)\text{-CW center}} & \underbrace{(a_{j_1j_2}1, a_{j_2j_1}2)}_{(\text{UP}_+, \text{UP}_+)\text{-positive saddle}} & \underbrace{(a_{j_1j_2}1, a_{j_2j_1}3)}_{(\text{DP}_-, \text{DP}_+)\text{-CW center}} \end{array} \right\} \tag{4.204}$$

The (3^{rd}DP$_+$, 3^{rd}DP$_-$)*-double third-order clockwise center is a* (3^{rd}DP$_+$, 3^{rd}DP$_-$)*-switching bifurcation for a center to a network of saddles and centers.*

- *For* $a_{j_1j_2}0 < 0$ *and* $a_{j_2j_1}0 < 0$,

$$\underbrace{(a_{j_1j_2}1, a_{j_2j_1}1)}_{(\text{UP}_-, \text{UP}_-)\text{-negative saddle}} \rightleftharpoons \underbrace{(a_{j_1j_2}1, a_{j_2j_1}1)}_{(3^{\text{rd}}\text{UP}_-, 3^{\text{rd}}\text{UP}_-)\text{-appearing bifurcation}}$$
$$\rightleftharpoons \left\{ \begin{array}{ccc} \underbrace{(a_{j_1j_2}3, a_{j_2j_1}1)}_{(\text{UP}_-, \text{UP}_-)\text{-negative saddle}} & \underbrace{(a_{j_1j_2}3, a_{j_2j_1}2)}_{(\text{DP}_+, \text{DP}_-)\text{-CCW center}} & \underbrace{(a_{j_1j_2}3, a_{j_2j_1}3)}_{(\text{UP}_-, \text{UP}_-)\text{-negative saddle}} \\ \underbrace{(a_{j_1j_2}2, a_{j_2j_1}1)}_{(\text{DP}_-, \text{DP}_+)\text{-CW center}} & \underbrace{(a_{j_1j_2}2, a_{j_2j_1}2)}_{(\text{UP}_+, \text{UP}_+)\text{-positive saddle}} & \underbrace{(a_{j_1j_2}2, a_{j_2j_1}3)}_{(\text{DP}_-, \text{DP}_+)\text{-CW center}} \\ \underbrace{(a_{j_1j_2}1, a_{j_2j_1}1)}_{(\text{UP}_-, \text{UP}_-)\text{-negative saddle}} & \underbrace{(a_{j_1j_2}1, a_{j_2j_1}2)}_{(\text{DP}_+, \text{DP}_-)\text{-CCW center}} & \underbrace{(a_{j_1j_2}1, a_{j_2j_1}3)}_{(\text{UP}_-, \text{UP}_-)\text{-negative saddle}} \end{array} \right\}. \tag{4.205}$$

The (3^{rd}UP$_-$, 3^{rd}UP$_-$)*-double third-order negative saddle is a* (3^{rd}UP$_-$, 3^{rd}UP)*-appearing bifurcation for a saddle to a network of saddles and centers.*

– *The double third-order saddles and centers are also the switching bifurcations of two networks of saddles and centers.*

(xxiii) Under

$$\Delta_{j_1j_2} = 0 \,\text{and}\, \Delta_{j_2j_1} > 0, a_{j_1j_2s_1} \neq a_{j_1j_2s_2}, \tag{4.206}$$

the standard form is

$$\begin{aligned}
&\dot{x}_{j_2} = a_{j_2 j_1 0}(x_{j_1} - a_{j_2 j_1 1})(x_{j_1} - a_{j_2 j_1 2})(x_{j_1} - a_{j_2 j_1 3}),\\
&\dot{x}_{j_1} = a_{j_1 j_2 0}(x_{j_2} - a_{j_1 j_2 s_1})(x_{j_2} - a_{j_1 j_2 s_2})^2,\\
&j_1, j_2 \in \{1,2\}; j_1 \neq j_2.
\end{aligned} \tag{4.207}$$

The equilibrium of $(x^*_{j_2}, x^*_{j_1}) = (a_{j_1 j_2 s_2}, a_{j_2 j_1 l_1})(s_1, s_2 \in \{1,2\}, s_1 \neq s_2; l_1, l_2 \in \{1,2,3\}$ $l_1 \neq l_2)$ *has the following appearing bifurcation properties.*

- *For* $a_{j_1 j_2 0}(a_{j_1 j_2 s_2} - a_{j_1 j_2 s_1}) > 0$ *and* $a_{j_2 j_1 0}\prod_{l_2=1, l_2 \neq l_1}^{3}(a_{j_2 j_1 l_1} - a_{j_2 j_1 l_2}) > 0$,

$$\underbrace{(a_{j_1 j_2 s_2}, a_{j_2 j_1 l_1})}_{(\text{US},\,\text{UP})\text{-appearing bifurcation}} \rightleftharpoons \left\{ \begin{array}{c} \underbrace{(a_{j_1 j_2 (s_2)_2}, a_{j_2 j_1 l_1})}_{(\text{UP}_+,\,\text{UP}_+)\text{-positive saddle}} \\ \underbrace{(a_{j_1 j_2 (s_2)_1}, a_{j_2 j_1 l_1})}_{(\text{DP}_+,\,\text{DP}_-)\text{-CCW center}} \end{array} \right\}. \tag{4.208}$$

  The (US,UP)*-up-parabola upper-saddle is a* (US,UP)*-appearing bifurcation for saddle and center.*

- *For* $a_{j_1 j_2 0}(a_{j_1 j_2 s_2} - a_{j_1 j_2 s_1}) < 0$ *and* $a_{j_2 j_1 0}\prod_{l_2=1, l_2 \neq l_1}^{3}(a_{j_2 j_1 l_1} - a_{j_2 j_1 l_2}) > 0$,

$$\underbrace{(a_{j_1 j_2 s_2}, a_{j_2 j_1 l_1})}_{(\text{LS},\,\text{DP})\text{-appearing bifurcation}} \left\{ \begin{array}{c} \underbrace{(a_{j_1 j_2 (s_2)_2}, a_{j_2 j_1 l_1})}_{(\text{DP}_+,\,\text{DP}_-)\text{-CCW center}} \\ \underbrace{(a_{j_1 j_2 (s_2)_1}, a_{j_2 j_1 l_1})}_{(\text{UP}_+,\,\text{UP}_+)\text{-positive saddle}} \end{array} \right\}. \tag{4.209}$$

  The (LS,DP)*-down-parabola lower-saddle is an* (LS,DP)*-appearing bifurcation for saddle and center.*

- *For* $a_{j_1 j_2 0}(a_{j_1 j_2 s_2} - a_{j_1 j_2 s_1}) > 0$ *and* $a_{j_2 j_1 0}\prod_{l_2=1, l_2 \neq l_1}^{3}(a_{j_2 j_1 l_1} - a_{j_2 j_1 l_2}) < 0$,

$$\underbrace{(a_{j_1 j_2 s_2}, a_{j_2 j_1 l_1})}_{(\text{US},\,\text{DP})\text{-appearing bifurcation}} \rightleftharpoons \left\{ \begin{array}{c} \underbrace{(a_{j_1 j_2 (s_2)_2}, a_{j_2 j_1 l_1})}_{(\text{DP}_-,\,\text{DP}_+)\text{-CW center}} \\ \underbrace{(a_{j_1 j_2 (s_2)_1}, a_{j_2 j_1 l_1})}_{(\text{UP}_-,\,\text{UP}_-)\text{-negative saddle}} \end{array} \right\}. \tag{4.210}$$

  The (US,DP)*-down-parabola upper-saddle is a* (US,DP)*-appearing bifurcation for saddle and center.*

- *For* $a_{j_1j_20}(a_{j_1j_2s_2} - a_{j_1j_2s_1}) < 0$ *and* $a_{j_2j_10}\prod_{l_2=1,l_2\neq l_1}^{3}(a_{j_2j_1l_1} - a_{j_2j_1l_2}) < 0$,

$$\underbrace{(a_{j_1j_2s_2}, a_{j_2j_1l_1})}_{\text{(LS, UP)-appearing bifurcation}} \left\{ \begin{array}{c} \underbrace{(a_{j_1j_2(s_2)_2}, a_{j_2j_1l_1})}_{\text{(UP}_-\text{, UP}_-\text{)-negative saddle}} \\ \underbrace{(a_{j_1j_2(s_2)_1}, a_{j_2j_1l_1})}_{\text{(DP}_-\text{, DP}_+\text{)-CW center}} \end{array} \right\}. \tag{4.211}$$

The (LS,UP)*-up-parabola lower-saddle is an* (LS,UP)*-appearing bifurcation for saddle and center.*

The parabola-saddles are also the switching bifurcations of saddle and center.

(xxiv) Under

$$\Delta_{j_1j_2} = 0 \text{ and } \Delta_{j_2j_1} > 0, a_{j_1j_2s_1} = a_{j_1j_2s_2} = a_{j_1j_21}, \tag{4.212}$$

the standard form is

$$\begin{array}{l} \dot{x}_{j_2} = a_{j_2j_10}(x_{j_1} - a_{j_2j_11})(x_{j_1} - a_{j_2j_12})(x_{j_1} - a_{j_2j_13}), \\ \dot{x}_{j_1} = a_{j_1j_20}(x_{j_2} - a_{j_1j_21})^3, \\ j_1, j_2 \in \{1,2\}; j_1 \neq j_2. \end{array} \tag{4.213}$$

The equilibrium of $(x_{j_2}^*, x_{j_1}^*) = (a_{j_1j_21}, a_{j_2j_1l_1})(l_1, l_2 \in \{1,2,3\}, l_1 \neq l_2)$ *has the following appearing bifurcations.*

- *For* $a_{j_1j_20} > 0$ *and* $a_{j_2j_10}\prod_{l_2=1,l_2\neq l_1}^{3}(a_{j_2j_1i_1} - a_{j_2j_1l_2}) > 0$,

$$\underbrace{(a_{j_1j_21}, a_{j_2j_1l_1})}_{\text{(UP}_+\text{, UP}_+\text{)-positive saddle}} \rightleftharpoons \underbrace{(a_{j_1j_21}, a_{j_2j_1l_1})}_{\text{(3}^{\text{rd}}\text{UP}_+\text{, UP}_+\text{)-appearing bifucation}} \rightleftharpoons \left\{ \begin{array}{c} \underbrace{(a_{j_1j_23}, a_{j_2j_1l_1})}_{\text{(UP}_+\text{, UP}_+\text{)-positive saddle}} \\ \underbrace{(a_{j_1j_22}, a_{j_2j_1l_1})}_{\text{(DP}_+\text{, DP}_-\text{)-CCW center}} \\ \underbrace{(a_{j_1j_21}, a_{j_2j_1l_1})}_{\text{(UP}_+\text{, UP}_+\text{)-positive saddle}} \end{array} \right\}. \tag{4.214}$$

The (3[rd]UP₊,UP₊)*-third-order positive saddle is a* (3[rd]UP₊,UP₊)*-appearing bifurcation from a saddle to saddle, center, and saddle.*

- *For* $a_{j_1j_20}<0$ *and* $a_{j_2j_10}\prod_{l_2=1,l_2\neq l_1}^{3}(a_{j_2j_1l_1}-a_{j_2j_1l_2})>0$,

$$
\underbrace{(a_{j_1j_2}1,a_{j_2j_1l_1})}_{(\mathrm{DP}_+,\mathrm{DP}_-)\text{-CCW center}} \rightleftharpoons \underbrace{(a_{j_1j_2}1,a_{j_2j_1l_1})}_{(3^{\mathrm{rd}}\mathrm{DP}_+,\mathrm{DP}_-)\text{-appearing bifucation}} \rightleftharpoons \left\{\begin{array}{c}\underbrace{(a_{j_1j_2}3,a_{j_2j_1l_1})}_{(\mathrm{DP}_+,\mathrm{DP}_-)\text{-CCW center}}\\ \underbrace{(a_{j_1j_2}2,a_{j_2j_1l_1})}_{(\mathrm{UP}_+,\mathrm{UP}_+)\text{-positive saddle}}\\ \underbrace{(a_{j_1j_2}1,a_{j_2j_1l_1})}_{(\mathrm{DP}_+,\mathrm{DP}_-)\text{-CCW center}}\end{array}\right\}. \tag{4.215}
$$

The $(3^{\mathrm{rd}}\mathrm{DP}_+,\mathrm{DP}_-)$*-third-order-counter-clockwise center is a* $(3^{\mathrm{rd}}\mathrm{DP}_+,\mathrm{DP}_-)$*-appearing bifurcation from a center to center, saddle, and center.*

- *For* $a_{j_1j_20}>0$ *and* $a_{j_2j_10}\prod_{l_2=1,l_2\neq l_1}^{3}(a_{j_2j_1l_1}-a_{j_2j_1l_2})<0$,

$$
\underbrace{(a_{j_1j_2}1,a_{j_2j_1l_1})}_{(\mathrm{DP}_-,\mathrm{DP}_+)\text{-CW center}} \rightleftharpoons \underbrace{(a_{j_1j_1}1,a_{j_2j_1l_1})}_{(3^{\mathrm{rd}}\mathrm{DP}_-,\mathrm{DP}_+)\text{-appearing bifucation}} \rightleftharpoons \left\{\begin{array}{c}\underbrace{(a_{j_1j_2}3,a_{j_2j_1l_1})}_{(\mathrm{DP}_-,\mathrm{DP}_+)\text{-CW center}}\\ \underbrace{(a_{j_1j_2}2,a_{j_2j_1l_1})}_{(\mathrm{UP}_-,\mathrm{UP}_-)\text{-negative saddle}}\\ \underbrace{(a_{j_1j_2}1,a_{j_2j_1l_1})}_{(\mathrm{DP}_-,\mathrm{DP}_+)\text{-CW center}}\end{array}\right\}. \tag{4.216}
$$

The $(3^{\mathrm{rd}}\mathrm{DP}_-,\mathrm{DP}_+)$*-third-order clockwise center is a* $(3^{\mathrm{rd}}\mathrm{DP}_-,\mathrm{DP}_+)$*-appearing bifurcation from a center to center, saddle, and center.*

- *For* $a_{j_1j_20}<0$ *and* $a_{j_2j_10}\prod_{l_2=1,l_2\neq l_1}^{3}(a_{j_2j_1l_1}-a_{j_2j_1l_2})<0$,

$$
\underbrace{(a_{j_1j_2}1,a_{j_2j_1l_1})}_{(\mathrm{UP}_-,\mathrm{UP}_-)\text{-negative saddle}} \rightleftharpoons \underbrace{(a_{j_1j_1}1,a_{j_2j_1l_1})}_{(3^{\mathrm{rd}}\mathrm{UP}_-,\mathrm{UP}_-)\text{-appearing bifucation}} \rightleftharpoons \left\{\begin{array}{c}\underbrace{(a_{j_1j_2}3,a_{j_2j_1l_1})}_{(\mathrm{UP}_-,\mathrm{UP}_-)\text{-negative saddle}}\\ \underbrace{(a_{j_1j_2}2,a_{j_2j_1l_1})}_{(\mathrm{DP}_-,\mathrm{DP}_+)\text{-CW center}}\\ \underbrace{(a_{j_1j_2}1,a_{j_2j_1l_1})}_{(\mathrm{UP}_-,\mathrm{UP}_-)\text{-negative saddle}}\end{array}\right\}. \tag{4.217}
$$

The $(3^{\text{rd}}\text{UP}_{-},\text{UP}_{-})$*-third-order negative saddle is a* $(3^{\text{rd}}\text{UP}_{-},\text{UP}_{-})$*-appearing bifurcation from a saddle to saddle, center, and saddle.*

- *The third-order saddles are also the switching bifurcations of saddle, center, and saddle.*
- *The third-order centers are also the switching bifurcations of center, saddle, and center.*

4.2 Proof of Theorem 4.1

Consider a two-dimensional, dynamical system as

$$\begin{aligned}
\dot{x}_{j_1} &= f_{j_1}(\mathbf{x},\mathbf{p}) = A_{j_1j_2 3} + A_{j_1j_2 2}x_{j_2} + A_{j_1j_2 1}x_{j_2}^2 + A_{j_1j_2 0}x_{j_2}^3,\\
\dot{x}_{j_2} &= f_{j_2}(\mathbf{x},\mathbf{p}) = A_{j_2j_1 3} + A_{j_2j_1 2}x_{j_1} + A_{j_2j_1 1}x_{j_1}^2 + A_{j_2j_1 0}x_{j_1}^3.
\end{aligned}$$

The 2-diemnsional dynamical system becomes

$$\begin{aligned}
\dot{x}_{j_2} &= a_{j_2j_1 0}(x_{j_1} - b_{j_2j_1 1})(x_{j_1}^2 + B_{j_2j_1}x_{j_1} + C_{j_2j_1}),\\
\dot{x}_{j_1} &= a_{j_1j_2 0}(x_{j_2} - b_{j_1j_2 1})(x_{j_2}^2 + B_{j_1j_2}x_{j_2} + C_{j_1j_2})
\end{aligned}$$

with

$$\begin{aligned}
&\Delta_{j_2j_1} = (B_{j_2j_1})^2 - 4C_{j_2j_1},\ \Delta_{j_1j_2} = (B_{j_1j_2})^2 - 4C_{j_1j_2}\\
&\text{for } j_1, j_2 \in \{1,2\}, j_1 \neq j_2.
\end{aligned}$$

(i) If

$$\Delta_{j_1j_2} < 0 \text{ and } \Delta_{j_2j_1} < 0,$$

then

$$\begin{aligned}
&\dot{x}_{j_1} = a_{j_1j_2 0}(x_{j_2} - a_{j_1j_2 1})\left[(x_{j_2} - a_{j_1j_2})^2 + b_{j_1j_2}\right],\\
&\dot{x}_{j_2} = a_{j_2j_1 0}(x_{j_1} - a_{j_2j_1 1})\left[(x_{j_1} - a_{j_2j_1})^2 + b_{j_2j_1}\right],\\
&j_1, j_2 \in \{1,2\}, j_2 \neq j_1.
\end{aligned}$$

In phase space,

$$\frac{dx_{j_1}}{dx_{j_2}} = \frac{a_{j_1j_2 0}}{a_{j_2j_1 0}} \frac{(x_{j_2} - a_{j_1j_2 1})\left[(x_{j_2} - a_{j_1j_2})^2 + b_{j_1j_2}\right]}{(x_{j_1} - a_{j_2j_1 1})\left[(x_{j_1} - a_{j_2j_1})^2 + b_{j_2j_1}\right]}$$

and

$$
\begin{aligned}
&a_{j_2j_10}\big\{(x_{j_1}-a_{j_2j_11})^3+2(a_{j_2j_11}-a_{j_2j_1})(x_{j_1}-a_{j_2j_11})^3\\
&+\big[(a_{j_2j_11}-a_{j_2j_1})^2+b_{j_2j_1}\big](x_{j_1}-a_{j_2j_11})\big\}dx_{j_1}\\
&=a_{j_1j_20}\big\{(x_{j_2}-a_{j_1j_21})^3+2(a_{j_1j_21}-a_{j_1j_2})(x_{j_2}-a_{j_1j_21})^2\\
&+\big[(a_{j_1j_21}-a_{j_1j_2})^2+b_{j_1j_2}\big](x_{j_2}-a_{j_1j_21})\big\}dx_{j_2}.
\end{aligned}
$$

With an initial condition (x_{j_10}, x_{j_20}), the first integral manifold is generated, i.e.,

$$
\begin{aligned}
&a_{j_2j_10}\Big\{\frac{1}{4}\big[(x_{j_1}-a_{j_2j_11})^4-(x_{j_10}-a_{j_2j_1})^4\big]\\
&+\frac{2}{3}(a_{j_2j_11}-a_{j_2j_1})\big[(x_{j_1}-a_{j_2j_11})^3-(x_{j_10}-a_{j_2j_11})^3\big]\\
&+\frac{1}{2}\big[(a_{j_2j_11}-a_{j_2j_1})^2+b_{j_2j_1}\big]\big[(x_{j_1}-a_{j_2j_1})^2-(x_{j_10}-a_{j_2j_1})^2\big]\Big\}\\
&=a_{j_1j_20}\Big\{\frac{1}{4}\big[(x_{j_2}-a_{j_1j_21})^4-(x_{j_20}-a_{j_1j_21})^4\big]\\
&+\frac{2}{3}(a_{j_1j_21}-a_{j_1j_2})\big[(x_{j_2}-a_{j_1j_21})^3-(x_{j_20}-a_{j_1j_21})^3\big]\\
&+\frac{1}{2}\big[(a_{j_1j_21}-a_{j_1j_2})^2+b_{j_1j_2}\big]\big[(x_{j_2}-a_{j_1j_21})^2-(x_{j_20}-a_{j_1j_21})^2\big]\Big\}.
\end{aligned}
$$

Consider two cases (I) $x^*_{j_2}=a_{j_1j_21}$ and (II) $x^*_{j_1}=a_{j_2j_11}$.

(I) At $x^*_{j_2}=a_{j_1j_21}$ and $\overline{x}_{j_1}\neq a_{j_2j_11}$, in phase space,

$$
\left.\frac{dx_{j_1}}{dx_{j_2}}\right|_{x^*_{j_2}=a_{j_1j_21}}=\left.\frac{a_{j_1j_20}}{a_{j_2j_10}}\frac{(x_{j_2}-a_{j_1j_21})\big[(x_{j_2}-a_{j_1j_2})^2+b_{j_1j_2}\big]}{(\overline{x}_{j_1}-a_{j_2j_11})\big[(\overline{x}_{j_1}-a_{j_2j_1})^2+b_{j_2j_1}\big]}\right|_{x^*_{j_2}=a_{j_1j_21}}=0.
$$

If

$$
\left.\frac{d^2x_{j_1}}{dx^2_{j_2}}\right|_{x^*_{j_2}=a_{j_1j_21}}=\frac{a_{j_1j_20}}{a_{j_2j_10}}\frac{\big[(a_{j_1j_21}-a_{j_1j_2})^2+b_{j_1j_2}\big]}{(\overline{x}_{j_1}-a_{j_2j_11})\big[(\overline{x}_{j_1}-a_{j_2j_1})^2+b_{j_2j_1}\big]}>0,
$$

there is an up-parabola flow at $x^*_{j_2}=a_{j_1j_21}$ on the x_{j_2}-direction. If

$$
\left.\frac{d^2x_{j_1}}{dx^2_{j_2}}\right|_{x^*_{j_2}=a_{j_1j_21}}=\frac{a_{j_1j_20}}{a_{j_2j_10}}\frac{\big[(a_{j_1j_21}-a_{j_1j_2})^2+b_{j_1j_2}\big]}{(\overline{x}_{j_1}-a_{j_2j_11})\big[(\overline{x}_{j_1}-a_{j_2j_1})^2+b_{j_2j_1}\big]}<0,
$$

there is a down-parabola flow at $x_{j_2}^* = a_{j_1j_2 1}$ on the x_{j_2}-direction.

Because

$$\dot{x}_{j_2} = a_{j_2j_1 0}(\overline{x}_{j_1} - a_{j_2j_1 1})\left[(\overline{x}_{j_1} - a_{j_2j_1})^2 + b_{j_2j_1}\right]$$

and

$$(\overline{x}_{j_1} - a_{j_2j_1})^2 + b_{j_2j_1} > 0 \text{ and } (a_{j_1j_2 1} - a_{j_1j_2})^2 + b_{j_1j_2} > 0,$$

the parabola flows at $x_{j_2}^* = a_{j_1j_2 1}$ are positive and negative in the x_{j_2}-direction for $a_{j_2j_1 0}(\overline{x}_{j_1} - a_{j_2j_1 1}) > 0$ and $a_{j_2j_1 0}(\overline{x}_{j_1} - a_{j_2j_1 1}) < 0$, respectively.

Therefore, the equilibrium of $x_{j_2}^* = a_{j_1j_2 1}$ on the x_{j_2}-direction has the following properties.

- For $a_{j_1j_2 0} > 0$ and $a_{j_2j_1 0}(\overline{x}_{j_1} - a_{j_2j_1 1}) > 0$,

$$(a_{j_1j_2 1}, \dot{x}_{j_2}) = \underbrace{(\text{UP}, \text{pF})}_{\text{up-parabola flow } (+)}.$$

- For $a_{j_1j_2 0} < 0$ and $a_{j_2j_1 0}(\overline{x}_{j_1} - a_{j_2j_1 1}) > 0$,

$$(a_{j_1j_2 1}, \dot{x}_{j_2}) = \underbrace{(\text{DP}, \text{pF})}_{\text{down-parabola flow } (+)}.$$

- For $a_{j_1j_2 0} > 0$ and $a_{j_2j_1 0}(\overline{x}_{j_1} - a_{j_2j_1 1}) < 0$,

$$(a_{j_1j_2 1}, \dot{x}_{j_2}) = \underbrace{(\text{DP}, \text{nF})}_{\text{down-parabola flow } (-)}.$$

- For $a_{j_1j_2 0} < 0$ and $a_{j_2j_1 0}(\overline{x}_{j_1} - a_{j_2j_1 1}) < 0$,

$$(a_{j_1j_2 1}, \dot{x}_{j_2}) = \underbrace{(\text{UP}, \text{nF})}_{\text{up-parabola flow } (-)}.$$

(II) At $x_{j_1}^* = a_{j_2j_1 1}$ and $\overline{x}_{j_2} \neq a_{j_1j_2 1}$, in phase space,

$$\left.\frac{dx_{j_2}}{dx_{j_1}}\right|_{x_{j_1}^* = a_{j_2j_1 1}} = \frac{a_{j_2j_1 0}}{a_{j_1j_2 0}} \left.\frac{(x_{j_1} - a_{j_2j_1 1})\left[(x_{j_1} - a_{j_2j_1})^2 + b_{j_2j_1}\right]}{(\overline{x}_{j_2} - a_{j_1j_2 1})\left[(\overline{x}_{j_2} - a_{j_1j_2})^2 + b_{j_1j_2}\right]}\right|_{x_{j_1}^* = a_{j_2j_1 1}} = 0.$$

If

$$\left.\frac{d^2x_{j_2}}{dx_{j_1}^2}\right|_{x_{j_2}^* = a_{j_1j_2 1}} = \frac{a_{j_2j_1 0}}{a_{j_1j_2 0}} \frac{(a_{j_2j_1 1} - a_{j_2j_1})^2 + b_{j_2j_1}}{(\overline{x}_{j_2} - a_{j_1j_2 1})\left[(\overline{x}_{j_2} - a_{j_1j_2})^2 + b_{j_1j_2}\right]} > 0,$$

there is an up-parabola flow at $x_{j_1}^* = a_{j_2 j_1 1}$ on the x_{j_1}-direction. If

$$\left.\frac{d^2 x_{j_2}}{dx_{j_1}^2}\right|_{x_{j_2}^* = a_{j_1 j_2 1}} = \frac{a_{j_2 j_1 0}}{a_{j_1 j_2 0}} \frac{(a_{j_2 j_1 1} - a_{j_2 j_1})^2 + b_{j_2 j_1}}{(\overline{x}_{j_2} - a_{j_1 j_2 1})\left[(\overline{x}_{j_2} - a_{j_1 j_2})^2 + b_{j_1 j_2}\right]} < 0,$$

there is a down-parabola flow at $x_{j_1}^* = a_{j_2 j_1 1}$ on the x_{j_1}-direction.

Because,

$$\dot{x}_{j_1} = a_{j_1 j_2 0}(\overline{x}_{j_2} - a_{j_1 j_2 1})\left[(\overline{x}_{j_2} - a_{j_1 j_2})^2 + b_{j_1 j_2}\right]$$

and

$$(\overline{x}_{j_1} - a_{j_2 j_1})^2 + b_{j_2 j_1} > 0 \text{ and } (a_{j_1 j_2 1} - a_{j_1 j_2})^2 + b_{j_1 j_2} > 0,$$

the parabola flows at $x_{j_1}^* = a_{j_2 j_1 1}$ are positive and negative in the x_{j_1}-direction for $a_{j_1 j_2 0}(\overline{x}_{j_2} - a_{j_1 j_2 1}) > 0$ and $a_{j_1 j_2 0}(\overline{x}_{j_2} - a_{j_1 j_2 1}) < 0$, respectively.

Therefore, the equilibrium of $x_{j_1}^* = a_{j_2 j_1 1}$ on the x_{j_2}-direction has the following properties.

- For $a_{j_1 j_2 0}(\overline{x}_{j_2} - a_{j_1 j_2 1}) > 0$ and $a_{j_2 j_1 0} > 0$,

$$(\dot{x}_{j_1}, a_{j_2 j_1 1}) = \underbrace{(\text{pF}, \text{UP})}_{\text{up-parabola flow } (+)}.$$

- For $a_{j_1 j_2 0}(\overline{x}_{j_2} - a_{j_1 j_2 1}) < 0$ and $a_{j_2 j_1 0} > 0$,

$$(\dot{x}_{j_1}, a_{j_2 j_1 1}) = \underbrace{(\text{nF}, \text{DP})}_{\text{down-parabola flow } (-)}.$$

- For $a_{j_1 j_2 0}(\overline{x}_{j_2} - a_{j_1 j_2 1}) > 0$ and $a_{j_2 j_1 0} < 0$,

$$(\dot{x}_{j_1}, a_{j_2 j_1 1}) = \underbrace{(\text{pF}, \text{DP})}_{\text{down-parabola flow } (+)}.$$

- For $a_{j_1 j_2 0}(\overline{x}_{j_2} - a_{j_1 j_2 1}) < 0$ and $a_{j_2 j_1 0} < 0$,

$$(\dot{x}_{j_1}, a_{j_2 j_1 1}) = \underbrace{(\text{nF}, \text{UP})}_{\text{up-parabola flow } (-)}.$$

From cases (I) and (II), thus, the equilibrium of $(x_{j_2}^*, x_{j_1}^*) = (a_{j_1 j_2 1}, a_{j_2 j_1 1})$ has the following properties as in Eqs. (4.7)–(4.10).

- For $a_{j_1j_20} > 0$ and $a_{j_2j_10} > 0$,

$$(a_{j_1j_21}, a_{j_2j_11}) = \underbrace{(\mathrm{UP}_+, \mathrm{UP}_+)}_{\text{positive saddle}}.$$

  The equilibrium of $(x_{j_2}^*, x_{j_1}^*) = (a_{j_1j_21}, a_{j_2j_11})$ is a $(\mathrm{UP}_+,\mathrm{UP}_+)$-positive saddle.
- For $a_{j_1j_20} < 0$ and $a_{j_2j_10} > 0$,

$$(a_{j_1j_21}, a_{j_2j_11}) = \underbrace{(\mathrm{DP}_+, \mathrm{DP}_-)}_{\text{CCW center}}.$$

  The equilibrium of $(x_{j_2}^*, x_{j_1}^*) = (a_{j_1j_21}, a_{j_2j_11})$ is a $(\mathrm{DP}_+,\mathrm{DP}_-)$-counter-clockwise center.
- For $a_{j_1j_20} > 0$ and $a_{j_2j_10} < 0$,

$$(a_{j_1j_21}, a_{j_2j_11}) = \underbrace{(\mathrm{DP}_-, \mathrm{DP}_+)}_{\text{CW center}}.$$

  The equilibrium of $(x_{j_2}^*, x_{j_1}^*) = (a_{j_1j_21}, a_{j_2j_11})$ is a $(\mathrm{DP}_-,\mathrm{DP}_+)$-clockwise center.
- For $a_{j_1j_20} < 0$ and $a_{j_2j_10} < 0$,

$$(a_{j_1j_21}, a_{j_2j_11}) = \underbrace{(\mathrm{UP}_-, \mathrm{UP}_-)}_{\text{negative saddle}}.$$

  The equilibrium of $(x_{j_2}^*, x_{j_1}^*) = (a_{j_1j_21}, a_{j_2j_11})$ is a $(\mathrm{UP}_-,\mathrm{UP}_-)$-negative saddle.

(ii) If

$$\Delta_{j_1j_2} < 0 \text{ and } \Delta_{j_2j_1} = 0, a_{j_2j_1l_1} \neq a_{j_2j_1l_2}$$

then

$$\begin{aligned}
&\dot{x}_{j_2} = a_{j_2j_10}(x_{j_1} - a_{j_2j_1l_1})(x_{j_1} - a_{j_2j_1l_2})^2,\\
&\dot{x}_{j_1} = a_{j_1j_20}(x_{j_2} - a_{j_1j_21})\left[(x_{j_2} - a_{j_1j_2})^2 + b_{j_1j_2}\right],\\
&j_1, j_2 \in \{1,2\}; j_1 \neq j_2.
\end{aligned}$$

In phase space,

$$\frac{dx_{j_2}}{dx_{j_1}} = \frac{a_{j_1j_20}(x_{j_1} - a_{j_2j_1l_1})(x_{j_1} - a_{j_2j_1l_2})^2}{a_{j_2j_10}(x_{j_2} - a_{j_1j_21})\left[(x_{j_1} - a_{j_2j_11})^2 + b_{j_2j_1}\right]},$$

and

$$\begin{aligned}
&a_{j_2j_10}\left[(x_{j_1}-a_{j_2j_1l_1})^3+2(a_{j_2j_1l_1}-a_{j_2j_1l_2})(x_{j_1}-a_{j_2j_1l_1})^2\right.\\
&\left.+(a_{j_2j_1l_1}-a_{j_2j_1l_2})^2(x_{j_1}-a_{j_2j_1l_1})\right]dx_{j_1}\\
&=a_{j_1j_20}\{(x_{j_2}-a_{j_1j_21})^3+2(a_{j_1j_21}-a_{j_1j_2})(x_{j_2}-a_{j_1j_21})^2\\
&+\left[(a_{j_1j_21}-a_{j_1j_2})^2+b_{j_1j_2}\right](x_{j_2}-a_{j_1j_21})\}dx_{j_2}.
\end{aligned}$$

With an initial condition (x_{j_10},x_{j_20}), the first integral manifold is generated, i.e.,

$$\begin{aligned}
&a_{j_2j_10}\{\frac{1}{4}\left[(x_{j_1}-a_{j_2j_1l_1})^4-(x_{j_10}-a_{j_2j_1l_1})^4\right]\\
&+\frac{2}{3}(a_{j_2j_1l_1}-a_{j_2j_1l_2})\left[(x_{j_1}-a_{j_2j_1l_1})^3-(x_{j_10}-a_{j_2j_1l_1})^3\right]\\
&+\frac{1}{2}(a_{j_2j_1l_1}-a_{j_2j_1l_2})^2\left[(x_{j_1}-a_{j_2j_1l_1})^2-(x_{j_10}-a_{j_2j_1l_1})^2\right]\}\\
&=a_{j_1j_20}\{\frac{1}{4}\left[(x_{j_1}-a_{j_1j_21})^4-(x_{j_10}-a_{j_1j_21})^4\right]\\
&+\frac{2}{3}(a_{j_1j_21}-a_{j_1j_2})\left[(x_{j_2}-a_{j_1j_21})^3-(x_{j_20}-a_{j_1j_21})^3\right]\\
&+\frac{1}{2}\left[(a_{j_1j_21}-a_{j_1j_2})^2+b_{j_1j_2}\right]\left[(x_{j_2}-a_{j_1j_21})^2-(x_{j_20}-a_{j_1j_21})^2\right]\}.
\end{aligned}$$

(ii_1) Consider two cases (I) $x^*_{j_2}=a_{j_1j_21}$ and (II) $x^*_{j_1}=a_{j_2j_1l_1}$.

(I) For $x^*_{j_2}=a_{j_1j_21}$ and $\bar{x}_{j_1}\neq a_{j_2j_1l_1},a_{j_2j_1l_2}(l_1,l_2\in\{1,2\},l_1\neq l_2)$, we have

$$\left.\frac{dx_{j_1}}{dx_{j_2}}\right|_{x^*_{j_2}=a_{j_1j_21}}=\left.\frac{a_{j_1j_20}(x_{j_2}-a_{j_1j_21})\left[(x_{j_2}-a_{j_1j_2})^2+b_{j_1j_2}\right]}{a_{j_1j_20}(\bar{x}_{j_1}-a_{j_2j_1l_1})(\bar{x}_{j_1}-a_{j_2j_2l_2})^2}\right|_{x^*_{j_2}=a_{j_1j_21}}=0.$$

If

$$\left.\frac{d^2x_{j_1}}{dx^2_{j_2}}\right|_{x^*_{j_2}=a_{j_1j_21}}=\frac{a_{j_1j_20}}{a_{j_2j_10}}\frac{(a_{j_1j_21}-a_{j_1j_2})^2+b_{j_1j_2}}{(\bar{x}_{j_1}-a_{j_2j_1l_1})(\bar{x}_{j_1}-a_{j_2j_1l_2})^2}>0,$$

there is an up-parabola flow at $x^*_{j_2}=a_{j_1j_21}$ in on the x_{j_2}-direction. If

$$\left.\frac{d^2x_{j_1}}{dx^2_{j_2}}\right|_{x^*_{j_2}=a_{j_1j_21}}=\frac{a_{j_1j_20}}{a_{j_2j_10}}\frac{(a_{j_1j_21}-a_{j_1j_2})^2+b_{j_1j_2}}{(\bar{x}_{j_1}-a_{j_2j_1l_1})(\bar{x}_{j_1}-a_{j_2j_1l_2})^2}<0,$$

there is a down-parabola flow at $x_{j_2}^* = a_{j_1 j_2 1}$ in on the x_{j_2}-direction. Because

$$\dot{x}_{j_2} = a_{j_2 j_1 0}(\overline{x}_{j_1} - a_{j_2 j_1 l_1})(\overline{x}_{j_1} - a_{j_2 j_1 l_2})^2$$

and

$$(\overline{x}_{j_1} - a_{j_2 j_1 l_2})^2 > 0 \text{ and} (a_{j_1 j_2 1} - a_{j_1 j_2})^2 + b_{j_1 j_2} > 0,$$

the flows at $x_{j_2}^* = a_{j_1 j_2 1}$ are positive and negative in the x_{j_2}-direction for $\dot{x}_{j_2} > 0$ and $\dot{x}_{j_2} < 0$, respectively.

Thus, the equilibrium of $x_{j_2}^* = a_{j_1 j_2 1}$ on the x_{j_2}-direction has the following properties.

- For $a_{j_1 j_2 0} > 0$ and $a_{j_2 j_1 0}(\overline{x}_{j_1} - a_{j_2 j_1 l_1})(\overline{x}_{j_1} - a_{j_2 j_1 l_2})^2 > 0$,

$$(a_{j_1 j_2 1}, \dot{x}_{j_2}) = \underbrace{(\text{UP}, \text{pF})}_{\text{up-parabola flow } (+)}.$$

- For $a_{j_1 j_2 0} < 0$ and $a_{j_2 j_1 0}(\overline{x}_{j_1} - a_{j_2 j_1 l_1})(\overline{x}_{j_1} - a_{j_2 j_1 l_2})^2 > 0$,

$$(a_{j_1 j_2 1}, \dot{x}_{j_2}) = \underbrace{(\text{DP}, \text{pF})}_{\text{down-parabola flow } (+)}.$$

- For $a_{j_1 j_2 0} > 0$ and $a_{j_2 j_1 0}(\overline{x}_{j_1} - a_{j_2 j_1 1})(\overline{x}_{j_1} - a_{j_2 j_1 l_2})^2 < 0$,

$$(a_{j_1 j_2 1}, \dot{x}_{j_2}) = \underbrace{(\text{DP}, \text{nF})}_{\text{up-parabola flow } (-)}.$$

- For $a_{j_1 j_2 0} < 0$ and $a_{j_2 j_1 0}(\overline{x}_{j_1} - a_{j_2 j_1 l_1})(\overline{x}_{j_1} - a_{j_2 j_1 l_2})^2 < 0$,

$$(a_{j_1 j_2 1}, \dot{x}_{j_2}) = \underbrace{(\text{UP}, \text{nF})}_{\text{up-parabola flow } (-)}.$$

(II) At $x_{j_1}^* = a_{j_2 j_1 l_1}$ and $\overline{x}_{j_2} \neq a_{j_1 j_2 1}$, in phase space,

$$\left.\frac{dx_{j_2}}{dx_{j_1}}\right|_{x_{j_1}^* = a_{j_2 j_1 l_1}} = \frac{a_{j_2 j_1 0}}{a_{j_1 j_2 0}} \left.\frac{(x_{j_1} - a_{j_2 j_1 l_1})(x_{j_1} - a_{j_2 j_1 l_2})^2}{(\overline{x}_{j_2} - a_{j_1 j_2 1})\left[(\overline{x}_{j_2} - a_{j_1 j_2})^2 + b_{j_1 j_2}\right]}\right|_{x_{j_1}^* = a_{j_2 j_1 l_1}} = 0.$$

If

$$\left.\frac{d^2 x_{j_2}}{dx_{j_1}^2}\right|_{x_{j_1}^* = a_{j_2 j_1 l_1}} = \frac{a_{j_2 j_1 0}}{a_{j_1 j_2 0}} \frac{(a_{j_2 j_1 l_1} - a_{j_2 j_1 l_2})^2}{(\overline{x}_{j_2} - a_{j_1 j_2 1})\left[(\overline{x}_{j_2} - a_{j_1 j_2})^2 + b_{j_1 j_2}\right]} > 0,$$

there is an up-parabola flow at $x_{j_1}^* = a_{j_2 j_1 l_1}$ on the x_{j_1}-direction. If

$$\left.\frac{d^2 x_{j_2}}{dx_{j_1}^2}\right|_{x_{j_1}^* = a_{j_2 j_1 l_1}} = \frac{a_{j_2 j_1 0}}{a_{j_1 j_2 0}} \frac{(a_{j_2 j_1 l_1} - a_{j_2 j_1 l_2})^2}{(\overline{x}_{j_2} - a_{j_1 j_2 1})\left[(\overline{x}_{j_2} - a_{j_1 j_2})^2 + b_{j_1 j_2}\right]} < 0,$$

there is a down-parabola flow at $x_{j_1}^* = a_{j_2 j_1 l_1}$ on the x_{j_1}-direction.

Because

$$\dot{x}_{j_1} = a_{j_1 j_2 0}(\overline{x}_{j_2} - a_{j_1 j_2 1})\left[(\overline{x}_{j_2} - a_{j_1 j_2})^2 + b_{j_1 j_2}\right]$$

and

$$(\overline{x}_{j_1} - a_{j_2 j_1})^2 + b_{j_2 j_1} > 0 \,\text{and}\, (a_{j_2 j_1 l_1} - a_{j_2 j_1 l_2})^2 > 0,$$

the parabola flows at $x_{j_1}^* = a_{j_2 j_1 l_1}$ are positive and negative in the x_{j_1}-direction for $a_{j_1 j_2 0}(\overline{x}_{j_2} - a_{j_1 j_2 1}) > 0$ and $a_{j_1 j_2 0}(\overline{x}_{j_2} - a_{j_1 j_2 1}) < 0$, respectively.

Thus, the equilibrium of $x_{j_1}^* = a_{j_2 j_1 l_1}$ on the x_{j_2}-direction has the following properties.

- For $a_{j_1 j_2 0}(\overline{x}_{j_2} - a_{j_1 j_2 1}) > 0$ and $a_{j_2 j_1 0}(a_{j_2 j_1 l_1} - a_{j_2 j_1 l_2})^2 > 0$,

$$(\dot{x}_{j_1}, a_{j_2 j_1 l_1}) = \underbrace{(\text{pF}, \text{UP})}_{\text{up-parabola flow } (+)}.$$

- For $a_{j_1 j_2 0}(\overline{x}_{j_2} - a_{j_1 j_2 1}) < 0$ and $a_{j_2 j_1 0}(a_{j_2 j_1 l_1} - a_{j_2 j_1 l_2})^2 > 0$,

$$(\dot{x}_{j_1}, a_{j_2 j_1 l_1}) = \underbrace{(\text{nF}, \text{DP})}_{\text{down-parabola flow } (-)}.$$

- For $a_{j_1 j_2 0}(\overline{x}_{j_2} - a_{j_1 j_2 1}) > 0$ and $a_{j_2 j_1 0}(a_{j_2 j_1 l_1} - a_{j_2 j_1 l_2})^2 < 0$,

$$(\dot{x}_{j_1}, a_{j_2 j_1 l_1}) = \underbrace{(\text{pF}, \text{DP})}_{\text{down-parabola flow } (+)}.$$

- For $a_{j_1j_20}(\overline{x}_{j_2} - a_{j_1j_21}) < 0$ and $a_{j_2j_10}(a_{j_2j_1l_1} - a_{j_2j_1l_2})^2 < 0$,

$$(\dot{x}_{j_1}, a_{j_2j_1l_1}) = \underbrace{(\text{nF}, \text{UP})}_{\text{up-parabola flow }(-)}.$$

Therefore, from cases (I) and (II), the equilibrium of $(x_{j_2}^*, x_{j_1}^*) = (a_{j_1j_21}, a_{j_2j_1l_1})$ $(l_1, l_2 \in \{1, 2\}, l_1 \neq l_2)$ has the following properties as in Eqs. (4.14)–(4.17).

- For $a_{j_1j_20} > 0$ and $a_{j_2j_10}(a_{j_2j_1l_1} - a_{j_2j_1l_2})^2 > 0$,

$$(a_{j_1j_21}, a_{j_2j_1l_1}) = \underbrace{(\text{UP}_+, \text{UP}_+)}_{\text{positive saddle}}.$$

  The equilibrium of $(x_{j_2}^*, x_{j_1}^*) = (a_{j_1j_21}, a_{j_2j_1l_1})$ is a $(\text{UP}_+,\text{UP}_+)$-positive saddle.
- For $a_{j_1j_20} < 0$ and $a_{j_2j_10}(a_{j_2j_1l_1} - a_{j_2j_1l_2})^2 > 0$,

$$(a_{j_1j_21}, a_{j_2j_1l_1}) = \underbrace{(\text{DP}_+, \text{DP}_-)}_{\text{CCW center}}.$$

  The equilibrium of $(x_{j_2}^*, x_{j_1}^*) = (a_{j_1j_21}, a_{j_2j_1l_1})$ is a $(\text{DP}_+,\text{DP}_-)$-counter-clockwise center.
- For $a_{j_1j_20} > 0$ and $a_{j_2j_10}(a_{j_2j_1l_1} - a_{j_2j_1l_2})^2 < 0$,

$$(a_{j_1j_21}, a_{j_2j_1l_1}) = \underbrace{(\text{DP}_-, \text{DP}_+)}_{\text{CW center}}.$$

  The equilibrium of $(x_{j_2}^*, x_{j_1}^*) = (a_{j_1j_21}, a_{j_2j_1l_1})$ is a $(\text{DP}_-,\text{DP}_+)$-clockwise center.
- For $a_{j_1j_20} < 0$ and $a_{j_2j_10}(a_{j_2j_1l_1} - a_{j_2j_1l_2})^2 < 0$,

$$(a_{j_1j_21}, a_{j_2j_1l_1}) = \underbrace{(\text{UP}_-, \text{UP}_-)}_{\text{negative saddle}}.$$

  The equilibrium of $(x_{j_2}^*, x_{j_1}^*) = (a_{j_1j_21}, a_{j_2j_1l_1})$ is a $(\text{UP}_-,\text{UP}_-)$-negative saddle.

(ii$_2$) For $x_{j_2}^* = a_{j_2j_1l_2}$ and $\overline{x}_{j_2} \neq a_{j_1j_21}$, we have

$$\left.\frac{dx_{j_2}}{dx_{j_1}}\right|_{x_{j_1}^* = a_{j_2j_1l_2}} = \left.\frac{a_{j_1j_20}(x_{j_1} - a_{j_2j_1l_1})(x_{j_1} - a_{j_2j_1l_2})^2}{a_{j_2j_10}(\overline{x}_{j_2} - a_{j_1j_21})\left[(\overline{x}_{j_2} - a_{j_1j_2})^2 + b_{j_1j_2}\right]}\right|_{x_{j_1}^* = a_{j_2j_1l_2}} = 0,$$

$$\left.\frac{d^3x_{j_2}}{dx_{j_1}^3}\right|_{x_{j_1}^* = a_{j_2j_1l_2}} = 0.$$

If

$$\left.\frac{d^3x_{j_2}}{dx_{j_1}^3}\right|_{x_{j_1}^*=a_{j_2j_1l_2}} = \frac{2a_{j_1j_20}(a_{j_2j_1l_2}-a_{j_2j_1l_1})}{a_{j_1j_20}(\bar{x}_{j_2}-a_{j_1j_21})\left[(\bar{x}_{j_2}-a_{j_1j_2})^2+b_{j_1j_2}\right]} > 0,$$

the flow at $x_{j_2}^* = a_{j_2j_1l_2}$ is an increasing-inflection flow. If

$$\left.\frac{d^3x_{j_2}}{dx_{j_1}^3}\right|_{x_{j_1}^*=a_{j_2j_1l_2}} = \frac{2a_{j_1j_20}(a_{j_2j_1l_2}-a_{j_2j_1l_1})}{a_{j_1j_20}(\bar{x}_{j_2}-a_{j_1j_21})\left[(\bar{x}_{j_2}-a_{j_1j_2})^2+b_{j_1j_2}\right]} < 0,$$

the flow at $x_{j_1}^* = a_{j_2j_1l_2}$ is a decreasing-inflection flow. Because of

$$\dot{x}_{j_1} = a_{j_1j_20}(\bar{x}_{j_2}-a_{j_1j_21})\left[(\bar{x}_{j_2}-a_{j_1j_2})^2+b_{j_1j_2}\right],$$

the inflection flows at $x_{j_1}^* = a_{j_2j_2l_2}$ are positive and negative in the x_{j_1}-direction for $a_{j_1j_20}(\bar{x}_{j_2}-a_{j_1j_21}) > 0$ and $a_{j_1j_20}(\bar{x}_{j_2}-a_{j_1j_21}) < 0$.

Thus, the equilibrium of $x_{j_1}^* = a_{j_2j_1l_2}$ on the x_{j_2}-direction for the x_{j_1}-direction has the following properties.

- For $a_{j_1j_20}(\bar{x}_{j_2}-a_{j_1j_21}) > 0$ and $a_{j_2j_10} > 0$,

$$(\dot{x}_{j_2}, a_{j_1j_2l_2}) = \underbrace{(\text{pF}, \text{II})}_{\text{increasing-infection flow } (+)}.$$

- For $a_{j_1j_20}(\bar{x}_{j_2}-a_{j_1j_21}) < 0$ and $a_{j_2j_10} > 0$,

$$(\dot{x}_{j_2}, a_{j_2j_1l_2}) = \underbrace{(\text{nF}, \text{DI})}_{\text{decreasing-inflection flow } (-)}$$

- For $a_{j_1j_20}(\bar{x}_{j_2}-a_{j_1j_21}) > 0$ and $a_{j_2j_10} < 0$,

$$(\dot{x}_{j_2}, a_{j_2j_1l_2}) = \underbrace{(\text{pF}, \text{DI})}_{\text{decreasing-inflection flow } (+)}.$$

- For $a_{j_1j_20}(\bar{x}_{j_2}-a_{j_1j_21}) < 0$ and $a_{j_2j_10} < 0$,

$$(\dot{x}_{j_2}, a_{j_2j_1l_2}) = \underbrace{(\text{nF}, \text{II})}_{\text{increasing-inflection flow } (-)}.$$

Therefore, from case (I) for case (ii$_1$) and the above analysis, the equilibrium of $(x_{j_2}^*, x_{j_1}^*) = (a_{j_1j_21}, a_{j_2j_1l_2})$ $(l_1, l_2 \in \{1,2\}, l_1 \neq l_2)$ has the following properties as in Eqs. (4.18)–(4.21).

- For $a_{j_1j_20} > 0$ and $a_{j_2j_10}(a_{j_2j_1l_2} - a_{j_2j_1l_1}) > 0$,

$$(a_{j_1j_21}, a_{j_2j_1l_2}) = \underbrace{(\mathrm{UP}, \mathrm{US})}_{\text{up-parabola upper-saddle}}.$$

The equilibrium of $(x_{j_2}^*, x_{j_1}^*) = (a_{j_1j_21}, a_{j_2j_1l_2})$ is a (UP,US)-up-parabola upper-saddle.

- For $a_{j_1j_20} < 0$ and $a_{j_2j_10}(a_{j_2j_1l_2} - a_{j_2j_1l_1}) > 0$,

$$(a_{j_1j_21}, a_{j_2j_1l_2}) = \underbrace{(\mathrm{DP}, \mathrm{US})}_{\text{down-parabola upper-saddle}}.$$

The equilibrium of $(x_{j_2}^*, x_{j_1}^*) = (a_{j_1j_21}, a_{j_2j_1l_2})$ is a (DP,US)-down-parabola upper-saddle.

- For $a_{j_1j_20} > 0$ and $a_{j_2j_10}(a_{j_2j_1l_2} - a_{j_2j_1l_1}) < 0$,

$$(a_{j_1j_21}, a_{j_2j_1l_2}) = \underbrace{(\mathrm{DP}, \mathrm{LS})}_{\text{down-parabola lower-saddle}}.$$

The equilibrium of $(x_{j_2}^*, x_{j_1}^*) = (a_{j_1j_21}, a_{j_2j_1l_2})$ is a (DP,LS)-down-parabola lower-saddle.

- For $a_{j_1j_20} < 0$ and $a_{j_2j_10}(a_{j_2j_1l_2} - a_{j_2j_1l_1}) < 0$,

$$(a_{j_1j_21}, a_{j_2j_1l_2}) = \underbrace{(\mathrm{UP}, \mathrm{LS})}_{\text{up-parabola lower-saddle}}.$$

The equilibrium of $(x_{j_2}^*, x_{j_1}^*) = (a_{j_1j_21}, a_{j_2j_1l_2})$ is a (UP,LS)-up-parabola lower-saddle.

(iii) If

$$\Delta_{j_1j_2} < 0 \text{ and } \Delta_{j_2j_1} = 0, a_{j_2j_1l_1} = a_{j_2j_1l_2} = a_{j_2j_11},$$

then

$$\begin{aligned}
&\dot{x}_{j_2} = a_{j_2j_10}(x_{j_1} - a_{j_2j_11})^3,\\
&\dot{x}_{j_1} = a_{j_1j_20}(x_{j_2} - a_{j_1j_21})\left[(x_{j_2} - a_{j_1j_2})^2 + b_{j_1j_2}\right],\\
&j_1, j_2, l_1, l_2 \in \{1,2\}; j_1 \neq j_2; l_1 \neq l_2.
\end{aligned}$$

In phase space,

$$\frac{dx_{j_2}}{dx_{j_1}} = \frac{a_{j_1j_20}(x_{j_1} - a_{j_2j_11})^3}{a_{j_2j_10}(x_{j_2} - a_{j_1j_21})\left[(x_{j_1} - a_{j_2j_11})^2 + b_{j_2j_1}\right]},$$

and

$$
\begin{aligned}
& a_{j_2 j_1 0}(x_{j_1} - a_{j_2 j_1 1})^3 dx_{j_1} \\
& = a_{j_1 j_2 0}\{(x_{j_2} - a_{j_1 j_2 1})^3 + 2(a_{j_1 j_2 1} - a_{j_1 j_2})(x_{j_2} - a_{j_1 j_2 1})^2 \\
& \quad + [(a_{j_1 j_2 1} - a_{j_1 j_2})^2 + b_{j_1 j_2}](x_{j_2} - a_{j_1 j_2 1})\} dx_{j_2}.
\end{aligned}
$$

With an initial condition $(x_{j_1 0}, x_{j_2 0})$, the first integral manifold is

$$
\begin{aligned}
& \frac{1}{4} a_{j_2 j_1 0}\left[(x_{j_1} - a_{j_2 j_1 1})^4 - (x_{j_1 0} - a_{j_2 j_1 1})^4\right] \\
& = a_{j_1 j_2 0}\{\frac{1}{4}\left[(x_{j_1} - a_{j_1 j_2 1})^4 - (x_{j_1 0} - a_{j_1 j_2 1})^4\right] \\
& + \frac{2}{3}(a_{j_1 j_2 1} - a_{j_1 j_2})\left[(x_{j_2} - a_{j_1 j_2 1})^3 - (x_{j_2 0} - a_{j_1 j_2 1})^3\right] \\
& + \frac{1}{2}\left[(a_{j_1 j_2 1} - a_{j_1 j_2})^2 + b_{j_1 j_2}\right]\left[(x_{j_2} - a_{j_1 j_2 1})^2 - (x_{j_2 0} - a_{j_1 j_2 1})^2\right]\}.
\end{aligned}
$$

Consider two cases (I) $x_{j_2}^* = a_{j_1 j_2 1}$ and (II) $x_{j_1}^* = a_{j_2 j_1 1}$.

(I) For $x_{j_2}^* = a_{j_1 j_2 1}$ and $x_{j_1}^* \neq a_{j_2 j_1 1}$, similar to the proof of case (I) for (ii$_1$),

$$
\left.\frac{dx_{j_1}}{dx_{j_2}}\right|_{x_{j_2}^* = a_{j_1 j_2 1}} = \frac{a_{j_1 j_2 0}}{a_{j_1 j_2 0}} \left.\frac{(x_{j_2} - a_{j_1 j_2 1})\left[(x_{j_2} - a_{j_1 j_2})^2 + b_{j_1 j_2}\right]}{(\bar{x}_{j_1} - a_{j_2 j_1 1})^3}\right|_{x_{j_2}^* = a_{j_1 j_2 1}} = 0.
$$

If

$$
\left.\frac{d^2 x_{j_1}}{dx_{j_2}^2}\right|_{x_{j_2}^* = a_{j_1 j_2 1}} = \frac{a_{j_1 j_2 0}}{a_{j_2 j_1 0}} \frac{(a_{j_1 j_2 1} - a_{j_1 j_2})^2 + b_{j_1 j_2}}{(\bar{x}_{j_1} - a_{j_2 j_1 1})^3} > 0,
$$

there is an up-parabola flow at $x_{j_2}^* = a_{j_1 j_2 1}$ in on the x_{j_2}-direction. If

$$
\left.\frac{d^2 x_{j_1}}{dx_{j_2}^2}\right|_{x_{j_2}^* = a_{j_1 j_2 1}} = \frac{a_{j_1 j_2 0}}{a_{j_2 j_1 0}} \frac{(a_{j_1 j_2 1} - a_{j_1 j_2})^2 + b_{j_1 j_2}}{(\bar{x}_{j_1} - a_{j_2 j_1 1})^3} < 0,
$$

there is a down-parabola flow at $x_{j_2}^* = a_{j_1 j_2 1}$ in on the x_{j_2}-direction. Because of

$$
\dot{x}_{j_2} = a_{j_2 j_1 0}(\bar{x}_{j_1} - a_{j_2 j_1 1})^3
$$

and

$$(a_{j_1j_2 1}-a_{j_1j_2})^2+b_{j_1j_2}>0,$$

the flows at $x_{j_2}^*=a_{j_1j_2 1}$ are positive and negative in the x_{j_2}-direction for $\dot{x}_{j_2}>0$ and $\dot{x}_{j_2}<0$, respectively.

Thus, the equilibrium of $x_{j_2}^*=a_{j_1j_2 1}$ on the x_{j_2}-direction has the following properties.

- For $a_{j_1j_2 0}>0$ and $a_{j_2i_1 0}(\overline{x}_{j_1}-a_{j_2j_1 1})^3>0$,

$$(a_{j_1j_2 1},,\dot{x}_{j_2})=\underbrace{(\text{UP},\text{pF})}_{\text{up-parabola flow }(+)}.$$

- For $a_{j_1j_2 0}<0$ and $a_{j_2j_1 0}(\overline{x}_{j_1}-a_{j_2j_1 1})^3>0$,

$$(a_{j_1j_2 1},\dot{x}_{j_2})=\underbrace{(\text{DP},\text{pF})}_{\text{down-parabola flow }(+)}.$$

- For $a_{j_1j_2 0}>0$ and $a_{j_2j_1,0}(\overline{x}_{j_1}-a_{j_2j_1 1})^3<0$,

$$(a_{j_1j_2 1},\dot{x}_{j_2})=\underbrace{(\text{DP},\text{nF})}_{\text{down-parabola flow }(-)}.$$

- For $a_{j_1j_2 0}<0$ and $a_{j_2j_1 0}(\overline{x}_{j_1}-a_{j_2j_1 1})^3<0$,

$$(a_{j_1j_2 1},\dot{x}_{j_2})=\underbrace{(\text{UP},\text{nF})}_{\text{up-parabola flow }(-)}.$$

(II) For $x_{j_2}^*=a_{j_2j_1 1}\left.\frac{dx_{j_2}}{dx_{j_1}}\right|_{x_{j_1}^*=a_{j_2j_1 1}}$ and $\overline{x}_{j_2}\neq a_{j_1j_2 1}$, we have

$$\left.\frac{dx_{j_2}}{dx_{j_1}}\right|_{x_{j_1}^*=a_{j_2j_1 1}}=\frac{a_{j_2j_1 0}}{a_{j_1j_2 0}}\left.\frac{(x_{j_1}-a_{j_2j_1 1})^3}{(\overline{x}_{j_2}-a_{j_1j_2 1})\left[(\overline{x}_{j_2}-a_{j_1j_2})^2+b_{j_1j_2}\right]}\right|_{x_{j_1}^*=a_{j_2j_1 1}}=0,$$

$$\left.\frac{d^2x_{j_2}}{dx_{j_1}^2}\right|_{x_{j_1}^*=a_{j_2j_1 1}}=\left.\frac{d^3x_{j_2}}{dx_{j_1}^3}\right|_{x_{j_1}^*=a_{j_2j_1 1}}=0.$$

If

$$\left.\frac{d^4x_{j_2}}{dx_{j_1}^4}\right|_{x_{j_1}^*=a_{j_2j_1 1}}=\frac{6a_{j_2j_1 0}}{a_{j_1j_2 0}(\overline{x}_{j_2}-a_{j_1j_2 1})\left[(\overline{x}_{j_2}-a_{j_1j_2})^2+b_{j_1j_2}\right]}>0,$$

the flow at $x_{j_2}^* = a_{j_2 j_1 1}$ is a third-order up-parabola flow. If

$$\left.\frac{d^4 x_{j_2}}{dx_{j_1}^4}\right|_{x_{j_1}^* = a_{j_2 j_1 1}} = \frac{6a_{j_2 j_1 0}}{a_{j_1 j_2 0}(\overline{x}_{j_2} - a_{j_1 j_2 1})\left[(\overline{x}_{j_2} - a_{j_1 j_2})^2 + b_{j_1 j_2}\right]} < 0,$$

the flow at $x_{j_1}^* = a_{j_2 j_1 1}$ is a third-order down-parabola flow. Because of

$$\dot{x}_{j_1} = a_{j_1 j_2 0}(\overline{x}_{j_2} - a_{j_1 j_2 1})\left[(\overline{x}_{j_2} - a_{j_1 j_2})^2 + b_{j_1 j_2}\right],$$

the third-order parabola flows at $x_{j_1}^* = a_{j_2 j_1 1}$ are positive and negative in the x_{j_1}-direction for $a_{j_1 j_2 0}$ $(\overline{x}_{j_2} - a_{j_1 j_2 1}) > 0$ and $a_{j_1 j_2 0}(\overline{x}_{j_2} - a_{j_1 j_2 1}) < 0$.

Thus, the equilibrium of $x_{j_1}^* = a_{j_2 j_1 l_2}$ on the x_{j_2}-direction for the x_{j_1}-direction has the following properties.

- For $a_{j_1 j_2 0}(\overline{x}_{j_2} - a_{j_1 j_2 1}) > 0$ and $a_{j_2 j_1 0} > 0$,

$$(\dot{x}_{j_2}, a_{j_2 j_1 1}) = \underbrace{(\text{pF}, 3^{\text{rd}}\text{UP})}_{\text{third-order up-parabola flow } (+)}.$$

- For $a_{j_1 j_2 0}(\overline{x}_{j_2} - a_{j_1 j_2 1}) < 0$ and $a_{j_2 j_1 0} > 0$,

$$(\dot{x}_{j_2}, a_{j_2 j_1 1}) = \underbrace{(\text{nF}, 3^{\text{rd}}\text{DP})}_{\text{third-order down-parabola flow } (-)}.$$

- For $a_{j_1 j_2 0}(\overline{x}_{j_2} - a_{j_1 j_2 1}) > 0$ and $a_{j_2 j_1 0} < 0$,

$$(\dot{x}_{j_2}, a_{j_2 j_1 1}) = \underbrace{(\text{pF}, 3^{\text{rd}}\text{DP})}_{\text{third-order down-parabola flow } (+)}.$$

- For $a_{j_1 j_2 0}(\overline{x}_{j_2} - a_{j_1 j_2 1}) < 0$ and $a_{j_2 j_1 0} < 0$,

$$(\dot{x}_{j_2}, a_{j_2 j_1 1}) = \underbrace{(\text{nF}, 3^{\text{rd}}\text{UP})}_{\text{third-order up-parabola flow } (-)}.$$

Therefore, from cases (I) and (II), the equilibrium of $(x_{j_2}^*, x_{j_1}^*) = (a_{j_1 j_2 1}, a_{j_2 j_1 1})$ has the flowing properties as in Eqs. (4.25)–(4.28):

- For $a_{j_1 j_2 0} > 0$ and $a_{j_2 j_1 0} > 0$,

$$(a_{j_1 j_2 1}, a_{j_2 j_1 1}) = \underbrace{(\text{UP}_+, 3^{\text{rd}}\text{UP}_+)}_{\text{third-order positive saddle}}.$$

The equilibrium of $(x_{j_2}^*, x_{j_1}^*) = (a_{j_1 j_2 1}, a_{j_2 j_1 1})$ is a $(\text{UP}_+, 3^{\text{rd}}\text{UP}_+)$-third-order positive saddle.

- For $a_{j_1 j_2 0} < 0$ and $a_{j_2 j_1 0} > 0$,

$$(a_{j_1 j_2 1}, a_{j_2 j_1 1}) = \underbrace{(\text{DP}_+, 3^{\text{rd}}\text{DP}_-)}_{\text{third-order CCW center}}.$$

The equilibrium of $(x_{j_2}^*, x_{j_1}^*) = (a_{j_1 j_2 1}, a_{j_2 j_1 1})$ is a $(\text{DP}_+, 3^{\text{rd}}\text{DP}_-)$-third-order counter-clockwise center.

- For $a_{j_1 j_2 0} > 0$ and $a_{j_2 j_1 0} < 0$,

$$(a_{j_1 j_2 1}, a_{j_2 j_1 1}) = \underbrace{(\text{DP}_-, 3^{\text{rd}}\text{DP}_+)}_{\text{third-order CW center}}.$$

The equilibrium of $(x_{j_2}^*, x_{j_1}^*) = (a_{j_1 j_2 1}, a_{j_2 j_1 1})$ is a $(\text{DP}_-, \text{DP}_+)$-third-order clockwise center.

- For $a_{j_1 j_2 0} < 0$ and $a_{j_2 j_1 0} < 0$,

$$(a_{j_1 j_2 1}, a_{j_2 j_1 1}) = \underbrace{(\text{UP}_-, 3^{\text{rd}}\text{UP}_-)}_{\text{third-order positive saddle}}.$$

The equilibrium of $(x_{j_2}^*, x_{j_1}^*) = (a_{j_1 j_2 1}, a_{j_2 j_1 1})$ is a $(\text{UP}_-, 3^{\text{rd}}\text{UP}_-)$-third-order negative saddle.

(iv) If

$$\Delta_{j_1 j_2} < 0 \text{ and } \Delta_{j_2 j_1} > 0,$$

then

$$\begin{aligned}
\dot{x}_{j_2} &= a_{j_1 j_2 0}(x_{j_1} - a_{j_2 j_1 1})(x_{j_1} - a_{j_2 j_1 2})(x_{j_1} - a_{j_2 j_1 3}),\\
\dot{x}_{j_1} &= a_{j_2 j_1 0}(x_{j_2} - a_{j_1 j_2 1})\left[(x_{j_2} - a_{j_1 j_2})^2 + b_{j_1 j_2}\right],\\
& j_1, j_2 \in \{1, 2\}; j_1 \neq j_2
\end{aligned}$$

where

$$\begin{aligned}
&\{a_{j_2 j_1 1}, a_{j_2 j_1 2}, a_{j_2 j_1 3}\} = \text{sort}\{b_{j_2 j_1 1}, b_{j_2 j_1 2}, b_{j_2 j_1 3}\}, a_{j_2 j_1 1} < a_{j_2 j_1 2} < a_{j_2 j_1 3};\\
&b_{j_2 j_1 2}, b_{j_2 j_1 3} = -\frac{1}{2}(B_{j_2 j_1} \pm \sqrt{\Delta_{j_2 j_1}}), a_{j_2 j_1 0} = A_{j_2 j_1 0};\\
&a_{j_1 j_2} = -\frac{1}{2} B_{j_1 j_2}, b_{j_1 j_2} = \frac{1}{4}(-\Delta_{j_1 j_2}).
\end{aligned}$$

In phase space,

$$\frac{dx_{j_2}}{dx_{j_1}} = \frac{a_{j_2j_10}(x_{j_1} - a_{j_2j_11})(x_{j_1} - a_{j_2j_12})(x_{j_1} - a_{j_2j_13})}{a_{j_1j_20}(x_{j_2} - a_{j_1j_21})\left[(x_{j_2} - a_{j_1j_2})^2 + b_{j_1j_2}\right]},$$

and

$$\begin{aligned}
&a_{j_2j_10}\left[(x_{j_1} - a_{j_2j_1l_1})^3 + \sum\nolimits_{l_2=1,l_2\neq l_1}^{3}(a_{j_2j_1l_1} - a_{j_2j_1l_2})(x_{j_1} - a_{j_2j_11})^2\right.\\
&\left.+\prod\nolimits_{l_2=1,l_2\neq l_1}^{3}(a_{j_2j_1l_1} - a_{j_2j_1l_2})(x_{j_1} - a_{j_2j_11})\right]dx_{j_1}\\
&= a_{j_1j_20}\left\{(x_{j_2} - a_{j_1j_21})^3 + 2(a_{j_1j_21} - a_{j_1j_2})(x_{j_2} - a_{j_1j_21})^2\right.\\
&\left.+\left[(a_{j_1j_21} - a_{j_1j_2})^2 + b_{j_1j_2}\right](x_{j_2} - a_{j_1j_2})\right\}dx_{j_2}.
\end{aligned}$$

With an initial condition (x_{j_10}, x_{j_20}), the first integral manifold is generated, i.e.,

$$\begin{aligned}
&a_{j_2j_10}\left\{\frac{1}{4}\left[(x_{j_1} - a_{j_2j_1l_1})^4 - (x_{j_10} - a_{j_2j_1l_1})^4\right]\right.\\
&+\frac{1}{3}\sum\nolimits_{l_2=1,l_2\neq l_1}^{3}(a_{j_2j_1l_1} - a_{j_2j_1l_2})\left[(x_{j_1} - a_{j_2j_1l_1})^3 - (x_{j_10} - a_{j_2j_1l_1})^3\right]\\
&\left.+\frac{1}{2}\prod\nolimits_{l_2=1,l_2\neq l_1}^{3}(a_{j_2j_1l_1} - a_{j_2j_1l_2})(x_{j_1} - a_{j_2j_1l_1})^2 - (x_{j_10} - a_{j_2j_1l_1})^2]\right\}\\
&= a_{j_1j_20}\left\{\frac{1}{4}\left[(x_{j_2} - a_{j_1j_21})^4 - (x_{j_20} - a_{j_1j_21})^4\right]\right.\\
&+\frac{2}{3}(a_{j_1j_21} - a_{j_1j_2})\left[(x_{j_2} - a_{j_1j_21})^3 - (x_{j_20} - a_{j_1j_21})^3\right]\\
&\left.+\frac{1}{2}\left[(a_{j_1j_21} - a_{j_1j_2})^2 + b_{j_1j_2}\right]\left[(x_{j_2} - a_{j_1j_21})^2 - (x_{j_20} - a_{j_1j_21})^2\right]\right\}.
\end{aligned}$$

Consider two cases (I) $x^*_{j_2} = a_{j_1j_21}$ and (II) $x^*_{j_1} = a_{j_2j_1l_1}\,(l_1 = 1,2,3)$.

(I) For $x^*_{j_2} = a_{j_1j_21}$ and $\bar{x}_{j_1} \neq a_{j_2j_11}, a_{j_2j_12}, a_{j_2j_13}$, we have

$$\left.\frac{dx_{j_1}}{dx_{j_2}}\right|_{x^*_{j_2} = a_{j_1j_21}} = \left.\frac{a_{j_1j_20}(x_{j_2} - a_{j_1j_21})\left[(x_{j_2} - a_{j_1j_2})^2 + b_{j_1j_2}\right]}{a_{j_1j_20}\prod_{l_1=1}^{3}(\bar{x}_{j_1} - a_{j_2j_1l_1})}\right|_{x^*_{j_2} = a_{j_1j_21}} = 0.$$

If

$$\left.\frac{d^2x_{j_1}}{dx^2_{j_2}}\right|_{x^*_{j_2} = a_{j_1j_21}} = \frac{a_{j_1j_20}}{a_{j_2j_10}}\,\frac{(a_{j_1j_21} - a_{j_1j_2})^2 + b_{j_1j_2}}{\prod_{l_1=1}^{3}(\bar{x}_{j_1} - a_{j_2j_1l_1})} > 0,$$

there is an up-parabola flow at $x^*_{j_2} = a_{j_1 j_2 1}$ in on the x_{j_2}-direction. If

$$\left.\frac{d^2 x_{j_1}}{dx^2_{j_2}}\right|_{x^*_{j_2} = a_{j_1 j_2 1}} = \frac{a_{j_1 j_2 0}}{a_{j_2 j_1 0}} \frac{(a_{j_1 j_2 1} - a_{j_1 j_2})^2 + b_{j_1 j_2}}{\prod_{l_1=1}^{3} (\overline{x}_{j_1} - a_{j_2 j_1 l_1})} < 0,$$

there is a down-parabola flow at $x^*_{j_2} = a_{j_1 j_2 1}$ in on the x_{j_2}-direction. Because of

$$\dot{x}_{j_2} = a_{j_2 j_1 0} \prod_{l_1=1}^{3} (\overline{x}_{j_1} - a_{j_2 j_1 l_1})$$

and

$$(a_{j_1 j_2 1} - a_{j_1 j_2})^2 + b_{j_1 j_2} > 0,$$

the flows at $x^*_{j_2} = a_{j_1 j_2 1}$ are positive and negative in the x_{j_2}-direction for $\dot{x}_{j_2} > 0$ and $\dot{x}_{j_2} < 0$, respectively.

Thus, the equilibrium of $x^*_{j_2} = a_{j_1 j_2 1}$ with on $\overline{x}_{j_1} \neq a_{j_2 j_1 l_1} (l_1 = 1,2,3)$ the x_{j_2}-direction has the following properties.

- For $a_{j_1 j_2 0} > 0$ and $a_{j_2 j_1 0} \prod_{l_1=1}^{3} (\overline{x}_{j_1} - a_{j_2 j_1 l_1}) > 0$,

$$(a_{j_1 j_2 1}, \dot{x}_{j_2}) = \underbrace{(\mathrm{UP}, \mathrm{pF})}_{\text{up-parabola flow }(+)}.$$

- For $a_{j_1 j_2 0} < 0$ and $a_{j_2 j_1 0} \prod_{l_1=1}^{3} (\overline{x}_{j_1} - a_{j_2 j_1 1}) > 0$,

$$(a_{j_1 j_2 1}, \dot{x}_{j_2}) = \underbrace{(\mathrm{DP}, \mathrm{pF})}_{\text{down-parabola flow }(+)}.$$

- For $a_{j_1 j_2 0} > 0$ and $a_{j_2 j_1 0} \prod_{l_1=1}^{3} (\overline{x}_{j_1} - a_{j_2 j_1 l_1}) < 0$,

$$(a_{j_1 j_2 1}, \dot{x}_{j_2}) = \underbrace{(\mathrm{DP}, \mathrm{nF})}_{\text{down-parabola flow }(-)}.$$

- For $a_{j_1 j_2 0} < 0$ and $a_{j_2 j_1 0} \prod_{l_1=1}^{3} (\overline{x}_{j_1} - a_{j_2 j_1 l_1}) < 0$,

$$(a_{j_1 j_2 1}, \dot{x}_{j_2}) = \underbrace{(\mathrm{UP}, \mathrm{nF})}_{\text{up-parabola flow }(-)}.$$

(II) For $x_{j_2}^* = a_{j_2 j_1 l_1} (l_1, l_2 \in \{1,2,3\}, l_1 \neq l_2)$ and $\overline{x}_{j_2} \neq a_{j_1 j_2 1}$, we have

$$\left.\frac{dx_{j_2}}{dx_{j_1}}\right|_{x_{j_1}^* = a_{j_2 j_1 l_1}} = \left.\frac{a_{j_1 j_2 0}(x_{j_1} - a_{j_2 j_1 l_1})\prod_{l_1=1, l_2 \neq l_1}^{3}(x_{j_1} - a_{j_2 j_1 l_2})}{a_{j_2 j_1 0}(\overline{x}_{j_2} - a_{j_1 j_2 1})\left[(\overline{x}_{j_2} - a_{j_1 j_2})^2 + b_{j_1 j_2}\right]}\right|_{x_{j_1}^* = a_{j_2 j_1 l_1}} = 0.$$

If

$$\left.\frac{d^2 x_{j_2}}{dx_{j_1}^2}\right|_{x_{j_1}^* = a_{j_2 j_1 l_1}} = \frac{a_{j_1 j_2 0}\prod_{l_2=1, l_2 \neq l_1}^{3}(a_{j_2 j_1 l_1} - a_{j_2 j_1 l_2})}{a_{j_2 j_1 0}(\overline{x}_{j_2} - a_{j_1 j_2 1})\left[(\overline{x}_{j_2} - a_{j_1 j_2})^2 + b_{j_1 j_2}\right]} > 0,$$

the flow at $x_{j_2}^* = a_{j_2 j_1 l_1}$ is an up-parabola flow. If

$$\left.\frac{d^2 x_{j_2}}{dx_{j_1}^2}\right|_{x_{j_1}^* = a_{j_2 j_1 l_1}} = \frac{a_{j_1 j_2 0}\prod_{l_2=1, l_2 \neq l_1}^{3}(a_{j_2 j_1 l_1} - a_{j_2 j_1 l_2})}{a_{j_2 j_1 0}(\overline{x}_{j_2} - a_{j_1 j_2 1})\left[(\overline{x}_{j_2} - a_{j_1 j_2})^2 + b_{j_1 j_2}\right]} < 0,$$

the flow at $x_{j_2}^* = a_{j_2 j_1 l_1}$ is a down-parabola flow. Because of

$$\dot{x}_{j_1} = a_{j_1 j_2 0}(\overline{x}_{j_2} - a_{j_1 j_2 1})\left[(\overline{x}_{j_2} - a_{j_1 j_2})^2 + b_{j_1 j_2}\right],$$

the parabola flows at $x_{j_2}^* = a_{j_2 j_1 l_1}$ are positive and negative in the x_{j_1}-direction for $a_{j_1 j_2 0}(\overline{x}_{j_2} - a_{j_1 j_2 1}) > 0$ and $a_{j_1 j_2 0}(\overline{x}_{j_2} - a_{j_1 j_2 1}) < 0$.

Therefore, the equilibrium of $x_{j_2}^* = a_{j_2 j_1 l_1}$ with $\overline{x}_{j_2} \neq a_{j_1 j_2 1}$ on the x_{j_2}-direction for the x_{j_1}-direction has the following properties.

- For $a_{j_1 j_2 0}(\overline{x}_{j_2} - a_{j_1 j_2 1}) > 0$ and $a_{j_2 j_1 0}\prod_{l_2=1, l_2 \neq l_1}^{3}(a_{j_2 j_1 l_1} - a_{j_2 j_1 l_2}) > 0$,

$$(\dot{x}_{j_2}, a_{j_2 j_1 l_1}) = \underbrace{(\text{pF}, \text{UP})}_{\text{up-parabola flow } (+)}.$$

- For $a_{j_1 j_2 0}(\overline{x}_{j_2} - a_{j_1 j_2 1}) < 0$ and $a_{j_2 j_1 0}\prod_{l_2=1, l_2 \neq l_1}^{3}(a_{j_2 j_1 l_1} - a_{j_2 j_1 l_2}) > 0$,

$$(\dot{x}_{j_2}, a_{j_2 j_1 l_1}) = \underbrace{(\text{nF}, \text{DP})}_{\text{down-parabola flow } (-)}.$$

- For $a_{j_1 j_2 0}(\overline{x}_{j_2} - a_{j_1 j_2 1}) > 0$ and $a_{j_2 j_1 0}\prod_{l_2=1, l_2 \neq l_1}^{3}(a_{j_2 j_1 l_1} - a_{j_2 j_1 l_2}) < 0$,

$$(\dot{x}_{j_2}, a_{j_2 j_1 l_1}) = \underbrace{(\text{pF}, \text{DP})}_{\text{down-parabola flow } (+)}.$$

- For $a_{j_1j_20}(\bar{x}_{j_2}-a_{j_1j_21})<0$ and $a_{j_2j_10}\prod_{l_2=1,l_2\neq l_1}^{3}(a_{j_2j_1l_1}-a_{j_2j_1l_2})<0$,

$$(\dot{x}_{j_2},a_{j_2j_1l_1})=\underbrace{(\text{nF},\text{UP})}_{\text{up-parabola flow }(-)}.$$

Therefore, from cases (I) and (II) the equilibrium of $(x_{j_2}^*,x_{j_1}^*)=(a_{j_1j_21},a_{j_2j_1l_1})$ (l_1, $l_2\in\{1,2,3\},l_1\neq l_2$) has the following properties as in Eqs. (4.33)–(4.36).

- For $a_{j_1j_20}>0$ and $a_{j_2j_10}\prod_{l_2=1,l_2\neq l_1}^{3}(a_{j_2j_1l_1}-a_{j_2j_1l_2})>0$,

$$(a_{j_1j_21},a_{j_2j_1l_1})=\underbrace{(\text{UP}_+,\text{UP}_+)}_{\text{positive saddle}}.$$

  The equilibrium of $(x_{j_2}^*,x_{j_1}^*)=(a_{j_1j_21},a_{j_2j_1l_1})$ is a $(\text{UP}_+,\text{UP}_+)$-positive saddle.
- For $a_{j_1j_20}<0$ and $a_{j_2j_10}\prod_{l_2=1,l_2\neq l_1}^{3}(a_{j_2j_1l_1}-a_{j_2j_il_2})>0$,

$$(a_{j_1j_21},a_{j_2j_1l_1})=\underbrace{(\text{DP}_+,\text{DP}_-)}_{\text{CCW center}}.$$

  The equilibrium of $(x_{j_2}^*,x_{j_1}^*)=(a_{j_1j_21},a_{j_2j_1l_1})$ is a $(\text{DP}_+,\text{DP}_-)$-counter-clockwise center.
- For $a_{j_1j_20}>0$ and $a_{j_2j_10}\prod_{l_2=1,l_2\neq l_1}^{3}(a_{j_2j_1l_1}-a_{j_2j_1l_2})<0$,

$$(a_{j_1j_21},a_{j_2j_1l_1})=\underbrace{(\text{DP}_-,\text{DP}_+)}_{\text{CW center}}$$

  The equilibrium of $(x_{j_2}^*,x_{j_1}^*)=(a_{j_1j_21},a_{j_2j_1l_1})$ is a $(\text{DP}_-,\text{DP}_+)$-clockwise center.
- For $a_{j_1j_20}<0$ and $a_{j_2j_10}\prod_{l_2=1,l_2\neq l_1}^{3}(a_{j_2j_1l_1}-a_{j_2j_1l_2})<0$,

$$(a_{j_1j_21},a_{j_2j_1l_1})=\underbrace{(\text{UP}_-,\text{UP}_-)}_{\text{negative saddle}}.$$

  The equilibrium of $(x_{j_2}^*,x_{j_1}^*)=(a_{j_1j_21},a_{j_2j_1l_1})$ is a $(\text{UP}_-,\text{UP}_+)$-negative saddle.

(v) If

$$\Delta_{j_1j_2}=0 \text{ and } \Delta_{j_2j_1}<0, a_{j_1j_2s_1}\neq a_{j_1j_2s_2},$$

then

$$\begin{aligned}
&\dot{x}_{j_1} = a_{j_1j_20}(x_{j_2} - a_{j_1j_2s_1})(x_{j_2} - a_{j_1j_2s_2})^2,\\
&\dot{x}_{j_2} = a_{j_2j_10}(x_{j_1} - a_{j_2j_11})\left[(x_{j_1} - a_{j_2j_1})^2 + b_{j_2j_1}\right]\\
&j_1, j_2 \in \{1,2\}, j_2 \neq j_1.
\end{aligned}$$

In phase space,

$$\frac{dx_{j_1}}{dx_{j_2}} = \frac{a_{j_1j_20}}{a_{j_2j_10}} \frac{(x_{j_2} - a_{j_1j_2s_1})(x_{j_2} - a_{j_1j_2s_2})^2}{(x_{j_1} - a_{j_2j_11})\left[(x_{j_1} - a_{j_2j_1})^2 + b_{j_2j_1}\right]},$$

and

$$\begin{aligned}
&a_{j_2j_10}(x_{j_1} - a_{j_2j_11})\left[(x_{j_1} - a_{j_2j_1})^2 + b_{j_2j_1}\right]dx_{j_1}\\
&= a_{j_1j_20}\left[(x_{j_2} - a_{j_1j_2s_1})^3 + 2(a_{j_1j_2s_1} - a_{j_1j_2s_2})(x_{j_2} - a_{j_1j_2s_1})^2\right.\\
&\left.+(a_{j_1j_2s_1} - a_{j_1j_2s_2})^2(x_{j_2} - a_{j_1j_2s_1})\right]dx_{j_2}.
\end{aligned}$$

With an initial condition (x_{j_10}, x_{j_20}), the first integral manifold is generated, i.e.,

$$\begin{aligned}
&a_{j_2j_10}\Big\{\frac{1}{4}\left[(x_{j_1} - a_{j_2j_11})^4 - (x_{j_10} - a_{j_2j_11})^4\right]\\
&+\frac{2}{3}(a_{j_2j_11} - a_{j_2j_1})\left[(x_{j_1} - a_{j_2j_11})^3 - (x_{j_10} - a_{j_2j_11})^3\right]\\
&+\frac{1}{2}\left[(a_{j_2j_11} - a_{j_2j_1})^2 + b_{j_2j_1}\right]\left[(x_{j_1} - a_{j_2j_11})^2 - (x_{j_10} - a_{j_2j_11})^2\right]\Big\}\\
&= a_{j_1j_20}\Big\{\frac{1}{4}\left[(x_{j_2} - a_{j_1j_2s_1})^4 - (x_{j_20} - a_{j_1j_2s_1})^4\right]\\
&+\frac{2}{3}(a_{j_1j_2s_1} - a_{j_1j_2s_2})\left[(x_{j_2} - a_{j_1j_2s_1})^3 - (x_{j_20} - a_{j_1j_2s_1})^3\right]\\
&+\frac{1}{2}(a_{j_1j_2s_1} - a_{j_1j_2s_2})^2\left[(x_{j_2} - a_{j_1j_2s_1})^2 - (x_{j_20} - a_{j_1j_2s_1})^2\right]\Big\}.
\end{aligned}$$

(v_1) Consider two cases (I) $x^*_{j_2} = a_{j_1j_2s_1}$ and (II) $x^*_{j_1} = a_{j_2j_11}$.

(I) In phase space, at $x^*_{j_2} = a_{j_1j_2s_1}(s_1, s_2 \in \{1,2\}, s_1 \neq s_2)$ with $\overline{x}_{j_1} \neq a_{j_2j_11}$, we have

$$\left.\frac{dx_{j_1}}{dx_{j_2}}\right|_{x^*_{j_2} = a_{j_1j_2s_1}} = \frac{a_{j_1j_20}}{a_{j_2j_10}} \frac{(x_{j_2} - a_{j_1j_2s_1})(x_{j_2} - a_{j_1j_2s_2})^2}{(\overline{x}_{j_1} - a_{j_2j_11})\left[(\overline{x}_{j_1} - a_{j_2j_1})^2 + b_{j_2j_1}\right]} = 0.$$

If

$$\left.\frac{d^2x_{j_1}}{dx^2_{j_2}}\right|_{x^*_{j_2} = a_{j_1j_2s_1}} = \frac{a_{j_1j_20}}{a_{j_2j_10}} \frac{(a_{j_1j_2s_1} - a_{j_1j_2s_2})^2}{(\overline{x}_{j_1} - a_{j_2j_11})\left[(\overline{x}_{j_1} - a_{j_2j_1})^2 + b_{j_2j_1}\right]} > 0,$$

there is an up-parabola flow at $x_{j_2}^* = a_{j_1 j_2 s_1}$ in the x_{j_1}-direction. If

$$\left.\frac{d^2 x_{j_1}}{dx_{j_2}^2}\right|_{x_{j_2}^* = a_{j_1 j_2 s_1}} = \frac{a_{j_1 j_2 0}}{a_{j_2 j_1 0}} \frac{(a_{j_1 j_2 s_1} - a_{j_1 j_2 s_2})^2}{(\overline{x}_{j_1} - a_{j_2 j_1 1})\left[(\overline{x}_{j_1} - a_{j_2 j_1})^2 + b_{j_2 j_1}\right]} < 0,$$

there is a down-parabola flow at $x_{j_2}^* = a_{j_1 j_2 s_1}$ in the x_{j_1}-direction. Because of

$$\dot{x}_{j_2} = a_{j_2 j_1 0}(\overline{x}_{j_1} - a_{j_2 j_1 1})\left[(\overline{x}_{j_1} - a_{j_2 j_1})^2 + b_{j_2 j_1}\right],$$

the parabola flows at $x_{j_2}^* = a_{j_1 j_2 s_1}$ in the x_{j_2}-direction are positive and negative for $a_{j_2 j_1 0}(\overline{x}_{j_1} - a_{j_2 j_1 1}) > 0$ and $a_{j_2 j_1 0}(\overline{x}_{j_1} - a_{j_2 j_1 1}) < 0$, respectively.

Thus, the equilibrium of $x_{j_2}^* = a_{j_1 j_2 s_1} (s_1, s_2 \in \{1, 2\}, s_1 \neq s_2)$ on the x_{j_1}-direction has the following properties.

- For $a_{j_1 j_2 0}(a_{j_1 j_2 s_1} - a_{j_1 j_2 s_2})^2 > 0$ and $a_{j_2 j_1 0}(\overline{x}_{j_1} - a_{j_2 j_1 1}) > 0$,

$$(a_{j_1 j_2 s_1}, \dot{x}_{j_2}) = \underbrace{(\text{UP}, \text{pF})}_{\text{up-parabola flow }(+)}.$$

  The flow at $x_{j_2}^* = a_{j_1 j_2 s_1}$ is a (UP,pF)-positive up-parabola flow.

- For $a_{j_1 j_2 0}(a_{j_1 j_2 s_1} - a_{j_1 j_2 s_2})^2 < 0$ and $a_{j_2 j_1 0}(\overline{x}_{j_1} - a_{j_2 j_1 1}) > 0$,

$$(a_{j_1 j_2 s_1}, \dot{x}_{j_2}) = \underbrace{(\text{DP}, \text{pF})}_{\text{down-parabola flow }(+)}.$$

  The flow at $x_{j_2}^* = a_{j_1 j_2 s_1}$ is a (DP,pF)-positive down-parabola flow.

- For $a_{j_1 j_2 0}(a_{j_1 j_2 s_1} - a_{j_1 j_2 s_2})^2 > 0$ and $a_{j_2 j_1 0}(\overline{x}_{j_1} - a_{j_2 j_1 1}) < 0$,

$$(a_{j_1 j_2 s_1}, \dot{x}_{j_2}) = \underbrace{(\text{DP}, \text{nF})}_{\text{down-parabola flow }(-)}.$$

  The flow at $x_{j_2}^* = a_{j_1 j_2 s_1}$ is a (DP,nF)-negative down-parabola flow.

- For $a_{j_1 j_2 0}(a_{j_1 j_2 s_1} - a_{j_1 j_2 s_2})^2 < 0$ and $a_{j_2 j_1 0}(\overline{x}_{j_1} - a_{j_2 j_1 1}) < 0$,

$$(a_{j_1 j_2 s_1}, \dot{x}_{j_2}) = \underbrace{(\text{UP}, \text{nF})}_{\text{up-parabola flow }(-)}.$$

  The flow at $x_{j_2}^* = a_{j_1 j_2 s_1}$ is a (UP,nF)-negative up-parabola flow.

(II) At $x_{j_1}^* = a_{j_2 j_1 1}$ and $\overline{x}_{j_2} \neq a_{j_1 j_2 s_1}, a_{j_1 j_2 s_2}$, in phase space,

$$\left.\frac{dx_{j_2}}{dx_{j_1}}\right|_{x_{j_1}^* = a_{j_2 j_1 1}} = \frac{a_{j_2 j_1 0}}{a_{j_1 j_2 0}} \left.\frac{(x_{j_1} - a_{j_2 j_1 1})\left[(x_{j_1} - a_{j_2 j_1})^2 + b_{j_2 j_1}\right]}{(\overline{x}_{j_2} - a_{j_1 j_2 s_1})(\overline{x}_{j_2} - a_{j_1 j_2 s_2})^2}\right|_{x_{j_1}^* = a_{j_2 j_1 1}} = 0.$$

If

$$\left.\frac{d^2 x_{j_2}}{dx_{j_1}^2}\right|_{x_{j_2}^* = a_{j_1 j_2 1}} = \frac{a_{j_2 j_1 0}}{a_{j_1 j_2 0}} \frac{(a_{j_2 j_1 1} - a_{j_2 j_1})^2 + b_{j_2 j_1}}{(\overline{x}_{j_2} - a_{j_1 j_2 s_1})(\overline{x}_{j_2} - a_{j_1 j_2 s_2})^2} > 0,$$

there is an up-parabola flow at $x_{j_1}^* = a_{j_2 j_1 1}$ on the x_{j_1}-direction. If

$$\left.\frac{d^2 x_{j_2}}{dx_{j_1}^2}\right|_{x_{j_2}^* = a_{j_1 j_2 1}} = \frac{a_{j_2 j_1 0}}{a_{j_1 j_2 0}} \frac{(a_{j_2 j_1 1} - a_{j_2 j_1})^2 + b_{j_2 j_1}}{(\overline{x}_{j_2} - a_{j_1 j_2 s_1})(\overline{x}_{j_2} - a_{j_1 j_2 s_2})^2} < 0,$$

there is a down-parabola flow at $x_{j_1}^* = a_{j_2 j_1 1}$ on the x_{j_1}-direction.

Because of

$$\dot{x}_{j_1} = a_{j_1 j_2 0}(\overline{x}_{j_2} - a_{j_1 j_2 s_1})(\overline{x}_{j_2} - a_{j_1 j_2 s_2})^2$$

with

$$(\overline{x}_{j_1} - a_{j_2 j_1})^2 + b_{j_2 j_1} > 0,$$

the parabola flows at $x_{j_1}^* = a_{j_2 j_1 1}$ are positive and negative in the x_{j_1}-direction for $a_{j_1 j_2 0}(\overline{x}_{j_2} - a_{j_1 j_2 1}) > 0$ and $a_{j_1 j_2 0}(\overline{x}_{j_2} - a_{j_1 j_2 1}) < 0$, respectively.

Thus, the equilibrium of $x_{j_1}^* = a_{j_2 j_1 1}$ on the x_{j_2}-direction has the following properties.

- For $a_{j_1 j_2 0}(\overline{x}_{j_2} - a_{j_1 j_2 s_1})(\overline{x}_{j_2} - a_{j_1 j_2 s_2})^2 > 0$ and $a_{j_2 j_1 0} > 0$,

$$(\dot{x}_{j_1}, a_{j_2 j_1 1}) = \underbrace{(\text{pF}, \text{UP})}_{\text{up-parabola flow } (+)}.$$

- For $a_{j_1 j_2 0}(\overline{x}_{j_2} - a_{j_1 j_2 s_1})(\overline{x}_{j_2} - a_{j_1 j_2 s_2})^2 < 0$ and $a_{j_2 j_1 0} > 0$,

$$(\dot{x}_{j_1}, a_{j_2 j_1 1}) = \underbrace{(\text{nF}, \text{DP})}_{\text{down-parabola flow } (-)}.$$

- For $a_{j_1j_20}(\overline{x}_{j_2} - a_{j_1j_2s_1})(\overline{x}_{j_2} - a_{j_1j_2s_2})^2 > 0$ and $a_{j_2j_10} < 0$,

$$(\dot{x}_{j_1}, a_{j_2j_11}) = \underbrace{(\text{pF}, \text{DP})}_{\text{down-parabola flow } (+)}.$$

- For $a_{j_1j_20}(\overline{x}_{j_2} - a_{j_1j_2s_1})(\overline{x}_{j_2} - a_{j_1j_2s_2})^2 < 0$ and $a_{j_2j_10} < 0$,

$$(\dot{x}_{j_1}, a_{j_2j_11}) = \underbrace{(\text{nF}, \text{UP})}_{\text{up-parabola flow } (-)}.$$

Therefore, from case (I) and (II), the equilibrium of $(x^*_{j_2}, x^*_{j_1}) = (a_{j_1j_2s_1}, a_{j_1j_21})$ $(s_1, s_2 \in \{1,2\}, s_1 \neq s_2)$ has the following properties as in Eqs. (4.40)–(4.43).

- For $a_{j_1j_20}(a_{j_1j_2s_1} - a_{j_1j_2s_2})^2 > 0$ and $a_{j_2j_10} > 0$,

$$(a_{j_1j_2s_1}, a_{j_1j_21}) = \underbrace{(\text{UP}_+, \text{UP}_+)}_{\text{positive saddle}}.$$

  The equilibrium of $(x^*_{j_2}, x^*_{j_1}) = (a_{j_1j_2s_1}, a_{j_1j_21})$ is a $(\text{UP}_+,\text{UP}_+)$-positive saddle.
- For $a_{j_1j_20}(a_{j_1j_2s_1} - a_{j_1j_2s_2})^2 < 0$ and $a_{j_2j_10} > 0$,

$$(a_{j_1j_2s_1}, a_{j_1j_21}) = \underbrace{(\text{DP}_+, \text{DP}_-)}_{\text{CCW center}}.$$

  The equilibrium of $(x^*_{j_2}, x^*_{j_1}) = (a_{j_1j_2s_1}, a_{j_1j_21})$ is a $(\text{DP}_+,\text{DP}_-)$-counter-clockwise center.
- For $a_{j_1j_20}(a_{j_1j_2s_1} - a_{j_1j_2s_2})^2 > 0$ and $a_{j_2j_10} < 0$,

$$(a_{j_1j_2s_1}, a_{j_1j_21}) = \underbrace{(\text{DP}_-, \text{DP}_+)}_{\text{CW center}}.$$

  The equilibrium of $(x^*_{j_2}, x^*_{j_1}) = (a_{j_1j_2s_1}, a_{j_1j_21})$ is a $(\text{DP}_-,\text{DP}_+)$-clockwise center.
- For $a_{j_1j_20}(a_{j_1j_2s_1} - a_{j_1j_2s_2})^2 < 0$ and $a_{j_2j_10} < 0$,

$$(a_{j_1j_2s_1}, a_{j_1j_21}) = \underbrace{(\text{UP}_-, \text{UP}_-)}_{\text{negative saddle}}.$$

  The equilibrium of $(x^*_{j_2}, x^*_{j_1}) = (a_{j_1j_2s_1}, a_{j_1j_21})$ is a $(\text{UP}_-,\text{UP}_-)$-negative saddle.

(v_2) In phase space, at $x_{j_2}^* = a_{j_1 j_2 s_2} (s_1, s_2 \in \{1,2\}, s_1 \neq s_2)$, we have

$$\left.\frac{dx_{j_1}}{dx_{j_2}}\right|_{x_{j_2}^* = a_{j_1 j_2 s_2}} = \frac{a_{j_1 j_2 0}}{a_{j_2 j_1 0}} \left.\frac{(x_{j_2} - a_{j_1 j_2 s_1})(x_{j_2} - a_{j_1 j_2 s_2})^2}{(\overline{x}_{j_1} - a_{j_2 j_1 1})\left[(\overline{x}_{j_1} - a_{j_2 j_1})^2 + b_{j_2 j_1}\right]}\right|_{x_{j_2}^* = a_{j_1 j_2 s_2}} = 0,$$

$$\left.\frac{d^2 x_{j_1}}{dx_{j_2}^2}\right|_{x_{j_2}^* = a_{j_1 j_2 s_2}} = 0.$$

If

$$\left.\frac{d^3 x_{j_1}}{dx_{j_2}^3}\right|_{x_{j_2}^* = a_{j_1 j_2 s_2}} = \frac{a_{j_1 j_2 0}}{a_{j_2 j_1 0}} \frac{2(a_{j_1 j_2 s_2} - a_{j_1 j_2 s_1})}{(\overline{x}_{j_1} - a_{j_2 j_1 1})\left[(\overline{x}_{j_1} - a_{j_2 j_1})^2 + b_{j_2 j_1}\right]} > 0,$$

there is an increasing-inflection flow at $x_{j_2}^* = a_{j_1 j_2 s_2}$ in the x_{j_1}-direction. If

$$\left.\frac{d^3 x_{j_1}}{dx_{j_2}^3}\right|_{x_{j_2}^* = a_{j_1 j_2 s_2}} = \frac{a_{j_1 j_2 0}}{a_{j_2 j_1 0}} \frac{2(a_{j_1 j_2 s_2} - a_{j_1 j_2 s_1})}{(\overline{x}_{j_1} - a_{j_2 j_1 1})\left[(\overline{x}_{j_1} - a_{j_2 j_1})^2 + b_{j_2 j_1}\right]} < 0,$$

there is a decreasing-inflection flow at $x_{j_2}^* = a_{j_1 j_2 s_2}$ in the x_{j_1}-direction. Because of

$$\dot{x}_{j_2} = a_{j_2 j_1 0}(\overline{x}_{j_1} - a_{j_2 j_1 1})\left[(\overline{x}_{j_1} - a_{j_2 j_1})^2 + b_{j_2 j_1}\right],$$

the inflection flows at $x_{j_2}^* = a_{j_1 j_2 s_2}$ in the x_{j_2}-direction are positive and negative for $a_{j_2 j_1 0}(\overline{x}_{j_1} - a_{j_2 j_1 1}) > 0$ and $a_{j_2 j_1 0}(\overline{x}_{j_1} - a_{j_2 j_1 1}) < 0$, respectively.

Thus, the equilibrium of $x_{j_2}^* = a_{j_1 j_2 s_2} (s_1, s_2 \in \{1,2\}, s_1 \neq s_2)$ on the x_{j_2}-direction has the following properties:

- For $a_{j_1 j_2 0}(a_{j_1 j_2 s_2} - a_{j_1 j_2 s_1}) > 0$ and $a_{j_2 j_1 0}(\overline{x}_{j_1} - a_{j_2 j_1 1}) > 0$,

$$(a_{j_1 j_2 s_2}, \dot{x}_{j_2}) = \underbrace{(\mathrm{II}, \mathrm{pF})}_{\text{increasing-inflection flow } (+)}.$$

- For $a_{j_1 j_2 0}(a_{j_1 j_2 s_2} - a_{j_1 j_2 s_1}) < 0$ and $a_{j_2 j_1 0}(\overline{x}_{j_1} - a_{j_2 j_1 1}) > 0$,

$$(a_{j_1 j_2 s_2}, \dot{x}_{j_2}) = \underbrace{(\mathrm{DI}, \mathrm{pF})}_{\text{decreasing-inflection flow } (+)}.$$

- For $a_{j_1j_20}(a_{j_1j_2s_2} - a_{j_1j_2s_1}) > 0$ and $a_{j_2j_10}(\overline{x}_{j_1} - a_{j_2j_1}) < 0$,

$$(a_{j_1j_2s_2}, \dot{x}_{j_2}) = \underbrace{(\text{DI}, \text{nF})}_{\text{decreasing-inflection flow }(-)}.$$

- For $a_{j_1j_20}(a_{j_1j_2s_2} - a_{j_1j_2s_1}) < 0$ and $a_{j_2j_10}(\overline{x}_{j_1} - a_{j_2j_11}) < 0$,

$$(a_{j_1j_2s_2}, \dot{x}_{j_2}) = \underbrace{(\text{II}, \text{nF})}_{\text{increasing-inflection flow }(-)}.$$

Therefore, from the proof (II) of case (v_1) and the above analysis, the equilibrium of $(x_{j_2}^*, x_{j_1}^*) = (a_{j_1j_2s_2}, a_{j_2j_11})\,(s_1, s_2 \in \{1,2\}, s_1 \neq s_2)$ on the x_{j_2}-direction has the following properties as in Eqs. (4.44)–(4.47).

- For $a_{j_1j_20}(a_{j_1j_2s_2} - a_{j_1j_2s_1}) > 0$ and $a_{j_2j_10} > 0$,

$$(a_{j_1j_2s_2}, a_{j_2j_11}) = \underbrace{(\text{US}, \text{UP})}_{\text{up-parabola upper-saddle}}.$$

  The equilibrium $(x_{j_2}^*, x_{j_1}^*) = (a_{j_1j_2s_2}, a_{j_2j_11})$ is a (US,UP)- up-parabola upper-saddle.
- For $a_{j_1j_20}(a_{j_1j_2s_2} - a_{j_1j_2s_1}) < 0$ and $a_{j_2j_10} > 0$,

$$(a_{j_1j_2s_2}, a_{j_2j_11}) = \underbrace{(\text{LS}, \text{DP})}_{\text{down-parabola lower-saddle}}.$$

  The equilibrium $(x_{j_2}^*, x_{j_1}^*) = (a_{j_1j_2s_2}, a_{j_2j_11})$ is an (LS,DP)-down-parabola lower-saddle.
- For $a_{j_1j_20}(a_{j_1j_2s_2} - a_{j_1j_2s_1}) > 0$ and $a_{j_2j_10} < 0$,

$$(a_{j_1j_2s_2}, a_{j_2j_11}) = \underbrace{(\text{US}, \text{DP})}_{\text{down-parabola upper-saddle}}.$$

  The equilibrium $(x_{j_2}^*, x_{j_1}^*) = (a_{j_1j_2s_2}, a_{j_2j_11})$ is an (LS,DP)-down-parabola upper-saddle.
- For $a_{j_1j_20}(a_{j_1j_2s_2} - a_{j_1j_2s_1}) < 0$ and $a_{j_2j_10} < 0$,

$$(a_{j_1j_2s_2}, a_{j_2j_11}) = \underbrace{(\text{LS}, \text{UP})}_{\text{up-parabola lower-saddle}}.$$

  The equilibrium $(x_{j_2}^*, x_{j_1}^*) = (a_{j_1j_2s_2}, a_{j_2j_11})$ is an (LS,UP)-up-parabola lower-saddle.

(vi) If

$$\Delta_{j_1 j_2} = 0 \text{ and } \Delta_{j_2 j_1} < 0, a_{j_1 j_2 s_1} = a_{j_1 j_2 s_2} = a_{j_1 j_2 1},$$

then

$$\begin{aligned}
&\dot{x}_{j_1} = a_{j_1 j_2 0}(x_{j_2} - a_{j_1 j_2 1})^3,\\
&\dot{x}_{j_2} = a_{j_2 j_1 0}(x_{j_1} - a_{j_2 j_1 1})\left[(x_{j_1} - a_{j_2 j_1})^2 + b_{j_2 j_1}\right]\\
&j_1, j_2 \in \{1,2\}, j_2 \neq j_1.
\end{aligned}$$

In phase space,

$$\frac{dx_{j_1}}{dx_{j_2}} = \frac{a_{j_1 j_2 0}}{a_{j_2 j_1 0}} \frac{(x_{j_2} - a_{j_1 j_2 1})^3}{(x_{j_1} - a_{j_2 j_1 1})\left[(x_{j_1} - a_{j_2 j_1})^2 + b_{j_2 j_1}\right]}$$

and

$$\begin{aligned}
&a_{j_2 j_1 0}\{(x_{j_1} - a_{j_2 j_1 1})^3 + 2(a_{j_2 j_1 1} - a_{j_2 j_1})(x_{j_1} - a_{j_2 j_1 1})^2\\
&+\left[(a_{j_2 j_1 1} - a_{j_2 j_1})^2 + b_{j_2 j_1}\right](x_{j_1} - a_{j_2 j_1 1})\}dx_{j_1}\\
&= a_{j_1 j_2 0}(x_{j_2} - a_{j_1 j_2 1})^3 dx_{j_2}.
\end{aligned}$$

With an initial condition $(x_{j_1 0}, x_{j_2 0})$, the first integral manifold is obtained, i.e.,

$$\begin{aligned}
&a_{j_2 j_1 0}\{\frac{1}{4}\left[(x_{j_1} - a_{j_2 j_1 1})^4 - (x_{j_1 0} - a_{j_2 j_1 1})^4\right]\\
&+\frac{2}{3}(a_{j_2 j_1 1} - a_{j_2 j_1})\left[(x_{j_1} - a_{j_2 j_1 1})^3 - (x_{j_1 0} - a_{j_2 j_1 1})^3\right]\\
&+\frac{1}{2}\left[(a_{j_2 j_1 1} - a_{j_2 j_1})^2 + b_{j_2 j_1}\right]\left[(x_{j_1} - a_{j_2 j_1 1})^2 - (x_{j_1 0} - a_{j_2 j_1 1})^2\right]\}\\
&= \frac{1}{4} a_{j_1 j_2 0}\left[(x_{j_2} - a_{j_1 j_2 1})^4 - (x_{j_2 0} - a_{j_1 j_2 1})^4\right].
\end{aligned}$$

In phase space, at $x_{j_2}^* = a_{j_1 j_2 1}$ with $\overline{x}_{j_1} \neq a_{j_2 j_1 1}$,

$$\begin{aligned}
&\left.\frac{dx_{j_1}}{dx_{j_2}}\right|_{x_{j_2}^* = a_{j_1 j_2 1}} = \left.\frac{a_{j_1 j_2 0}}{a_{j_2 j_1 0}} \frac{(x_{j_2} - a_{j_1 j_2 1})^3}{(\overline{x}_{j_1} - a_{j_2 j_1 1})\left[(\overline{x}_{j_1} - a_{j_2 j_1})^2 + b_{j_2 j_1}\right]}\right|_{x_{j_2}^* = a_{j_1 j_2 1}} = 0,\\
&\left.\frac{d^2 x_{j_1}}{dx_{j_2}^2}\right|_{x_{j_2}^* = a_{j_1 j_2 1}} = \left.\frac{d^3 x_{j_1}}{dx_{j_2}^3}\right|_{x_{j_2}^* = a_{j_1 j_2 1}} = 0.
\end{aligned}$$

If

$$\left.\frac{d^4 x_{j_1}}{dx_{j_2}^4}\right|_{x_{j_2}^* = a_{j_1 j_2 1}} = \frac{a_{j_1 j_2 0}}{a_{j_2 j_1 0}} \frac{6}{(\bar{x}_{j_1} - a_{j_2 j_1 1})\left[(\bar{x}_{j_1} - a_{j_2 j_1})^2 + b_{j_2 j_1}\right]} > 0,$$

there is a third-order up-parabola flow at $x_{j_2}^* = a_{j_1 j_2 1}$ in the x_{j_1}-direction. If

$$\left.\frac{d^4 x_{j_1}}{dx_{j_2}^4}\right|_{x_{j_2}^* = a_{j_1 j_2 1}} = \frac{a_{j_1 j_2 0}}{a_{j_2 j_1 0}} \frac{6}{(\bar{x}_{j_1} - a_{j_2 j_1 1})\left[(\bar{x}_{j_1} - a_{j_2 j_1})^2 + b_{j_2 j_1}\right]} < 0,$$

there is a third-order down-parabola flow at $x_{j_2}^* = a_{j_1 j_2 1}$ in the x_{j_1}-direction. Because of

$$\dot{x}_{j_2} = a_{j_2 j_1 0}(\bar{x}_{j_1} - a_{j_2 j_1 1})\left[(\bar{x}_{j_1} - a_{j_2 j_1})^2 + b_{j_2 j_1}\right],$$

the third-order parabola flows at $x_{j_2}^* = a_{j_1 j_2 1}$ in the x_{j_2}-direction are positive and negative for $a_{j_2 j_1 0}(\bar{x}_{j_1} - a_{j_2 j_1 1}) > 0$ and $a_{j_2 j_1 0}(\bar{x}_{j_1} - a_{j_2 j_1 1}) < 0$, respectively.

Thus, the equilibrium of $x_{j_2}^* = a_{j_1 j_2 1}$ on the x_{j_2}-direction has the following properties.

- For $a_{j_1 j_2 0} > 0$ and $a_{j_2 j_1 0}(\bar{x}_{j_1} - a_{j_2 j_1 1}) > 0$,

$$(a_{j_1 j_2 1}, \dot{x}_{j_2}) = \underbrace{(3^{\text{rd}}\text{UP, pF})}_{\text{third-order up-parabola flow } (+)}.$$

- For $a_{j_1 j_2 0} < 0$ and $a_{j_2 j_1 0}(\bar{x}_{j_1} - a_{j_2 j_1 1}) > 0$,

$$(a_{j_1 j_2 1}, \dot{x}_{j_2}) = \underbrace{(3^{\text{rd}}\text{DP, pF})}_{\text{third-order down-parabola flow } (+)}.$$

- For $a_{j_1 j_2 0} > 0$ and $a_{j_2 j_1 0}(\bar{x}_{j_1} - a_{j_2 j_1 1}) < 0$,

$$(a_{j_1 j_2 1}, \dot{x}_{j_2}) = \underbrace{(3^{\text{rd}}\text{DP, nF})}_{\text{third-order down-parabola flow } (-)}.$$

- For $a_{j_1 j_2 0} < 0$ and $a_{j_2 j_1 0}(\bar{x}_{j_1} - a_{j_2 j_1 1}) < 0$,

$$(a_{j_1 j_2 1}, \dot{x}_{j_2}) = \underbrace{(3^{\text{rd}}\text{UP, nF})}_{\text{third-order up-parabola flow } (-)}.$$

(II) Similarly, at $x_{j_2}^* = a_{j_1j_2 1}$ and $x_{j_2}^* \neq a_{j_1j_2 1}$, in phase space,

$$\left.\frac{dx_{j_2}}{dx_{j_1}}\right|_{x_{j_1}^* = a_{j_2j_1 1}} = \frac{a_{j_2j_1 0}}{a_{j_1j_2 0}} \left.\frac{(x_{j_1} - a_{j_2j_1 1})\left[(x_{j_1} - a_{j_2j_1})^2 + b_{j_2j_1}\right]}{(\overline{x}_{j_2} - a_{j_1j_2 1})^3}\right|_{x_{j_1}^* = a_{j_2j_1 1}} = 0.$$

If

$$\left.\frac{d^2x_{j_2}}{dx_{j_1}^2}\right|_{x_{j_2}^* = a_{j_1j_2 1}} = \frac{a_{j_2j_1 0}}{a_{j_1j_2 0}} \frac{(a_{j_2j_1 1} - a_{j_2j_1})^2 + b_{j_2j_1}}{(\overline{x}_{j_2} - a_{j_1j_2 1})^3} > 0,$$

there is an up-parabola flow at $x_{j_1}^* = a_{j_2j_1 1}$ on the x_{j_1}-direction. If

$$\left.\frac{d^2x_{j_2}}{dx_{j_1}^2}\right|_{x_{j_2}^* = a_{j_1j_2 1}} = \frac{a_{j_2j_1 0}}{a_{j_1j_2 0}} \frac{(a_{j_2j_1 1} - a_{j_2j_1})^2 + b_{j_2j_1}}{(\overline{x}_{j_2} - a_{j_1j_2 1})^3} < 0,$$

there is a down-parabola flow at $x_{j_1}^* = a_{j_2j_1 1}$ on the x_{j_1}-direction.

Because of

$$\dot{x}_{j_1} = a_{j_1j_2 0}(\overline{x}_{j_2} - a_{j_1j_2 1})^3$$

and

$$(\overline{x}_{j_1} - a_{j_2j_1})^2 + b_{j_2j_1} > 0,$$

the parabola flows at $x_{j_1}^* = a_{j_2j_1 1}$ are positive and negative in the x_{j_1}-direction for $a_{j_1j_2 0}(\overline{x}_{j_2} - a_{j_1j_2 1})^3 > 0$ and $a_{j_1j_2 0}(\overline{x}_{j_2} - a_{j_1j_2 1})^3 < 0$, respectively.

Thus, the equilibrium of $x_{j_1}^* = a_{j_2j_1 1}$ on the x_{j_2}-direction has the following properties.

- For $a_{j_1j_2 0}(\overline{x}_{j_2} - a_{j_1j_2 1})^3 > 0$ and $a_{j_2j_1 0} > 0$,

$$(\dot{x}_{j_1}, a_{j_2j_1 1}) = \underbrace{(\text{pF}, \text{UP})}_{\text{up-parabola flow } (+)}.$$

- For $a_{j_1j_2 0}(\overline{x}_{j_2} - a_{j_1j_2 1})^3 < 0$ and $a_{j_2j_1 0} > 0$,

$$(\dot{x}_{j_1}, a_{j_2j_1 1}) = \underbrace{(\text{nF}, \text{DP})}_{\text{down-parabola flow } (-)}.$$

- For $a_{j_1 j_2 0}(\overline{x}_{j_2} - a_{j_1 j_2 1})^3 > 0$ and $a_{j_2 j_1 0} < 0$,

$$(\dot{x}_{j_1}, a_{j_2 j_1 1}) = \underbrace{(\text{pF}, \text{DP})}_{\text{down-parabola flow } (+)}.$$

- For $a_{j_1 j_2 0}(\overline{x}_{j_2} - a_{j_1 j_2 1})^3 < 0$ and $a_{j_2 j_1 0} < 0$,

$$(\dot{x}_{j_1}, a_{j_2 j_1 1}) = \underbrace{(\text{nF}, \text{UP})}_{\text{up-parabola flow } (-)}.$$

Therefore, from the proofs of cases (I) and (II), the equilibrium of $(x_{j_2}^*, x_{j_1}^*) = (a_{j_1 j_2 1}, a_{j_2 j_1 1})$ has the following properties as in Eqs. (4.51)–(4.54).

- For $a_{j_1 j_2 0} > 0$ and $a_{j_2 j_1 0} > 0$,

$$(a_{j_1 j_2 1}, a_{j_2 j_1 1}) = \underbrace{(3^{\text{rd}}\text{UP}_+, \text{UP}_+)}_{\text{third-order positive saddle}}.$$

The equilibrium of $(x_{j_2}^*, x_{j_1}^*) = (a_{j_1 j_2 1}, a_{j_2 j_1 1})$ is a $(3^{\text{rd}}\text{UP}_+,\text{UP}_+)$-third-order positive saddle.

- For $a_{j_1 j_2 0} < 0$ and $a_{j_2 j_1 0} > 0$,

$$(a_{j_1 j_2 1}, a_{j_2 j_1 1}) = \underbrace{(3^{\text{rd}}\text{DP}_+, \text{DP}_-)}_{\text{third-order CCW center}}.$$

The equilibrium of $(x_{j_2}^*, x_{j_1}^*) = (a_{j_1 j_2 1}, a_{j_2 j_1 1})$ is a $(3^{\text{rd}}\text{DP}_+,\text{DP}_-)$-third-order counter-clockwise center.

- For $a_{j_1 j_2 0} > 0$ and $a_{j_2 j_1 0} < 0$,

$$(a_{j_1 j_2 1}, a_{j_2 j_1 1}) = \underbrace{(3^{\text{rd}}\text{DP}_-, \text{DP}_+)}_{\text{third-order CW center}}.$$

The equilibrium of $(x_{j_2}^*, x_{j_1}^*) = (a_{j_1 j_2 1}, a_{j_2 j_1 1})$ is a $(3^{\text{rd}}\text{DP}_-,\text{DP}_+)$-third-order clockwise center.

- For $a_{j_1 j_2 0} < 0$ and $a_{j_2 j_1 0} < 0$,

$$(a_{j_1 j_2 1}, a_{j_2 j_1 1}) = \underbrace{(3^{\text{rd}}\text{UP}_-, \text{UP}_-)}_{\text{third-order negative saddle}}.$$

The equilibrium of $(x_{j_2}^*, x_{j_1}^*) = (a_{j_1 j_2 1}, a_{j_2 j_1 1})$ is a $(3^{\text{rd}}\text{UP}_-,\text{UP}_-)$-third-order negative saddle.

(vii) If

$$\Delta_{j_1 j_2} = 0 \text{ and } \Delta_{j_2 j_1} = 0, a_{j_2 j_1 s_1} \neq a_{j_2 j_1 s_2},$$

then

$$\begin{aligned}
&\dot{x}_{j_1} = a_{j_1 j_2 0}(x_{j_2} - a_{j_1 j_2 s_1})(x_{j_2} - a_{j_1 j_2 s_2})^2,\\
&\dot{x}_{j_2} = a_{j_2 j_1 j_0}(x_{j_1} - a_{j_2 j_1 l_1})(x_{j_1} - a_{j_2 j_1 l_2})^2,\\
&j_1, j_2 \in \{1,2\}; j_1 \neq j_2.
\end{aligned}$$

In phase space,

$$\frac{dx_{j_1}}{dx_{j_2}} = \frac{a_{j_1 j_2 0}}{a_{j_2 j_1 0}} \frac{(x_{j_2} - a_{j_1 j_2 s_1})(x_{j_2} - a_{j_1 j_2 s_2})^2}{(x_{j_1} - a_{j_2 j_1 l_1})(x_{j_1} - a_{j_2 j_1 l_2})^2},$$

and

$$\begin{aligned}
&a_{j_2 j_1 0}\left[(x_{j_1} - a_{j_2 j_1 l_1})^3 + 2(a_{j_2 j_1 l_1} - a_{j_2 j_1 l_2})(x_{j_1} - a_{j_2 j_1 l_1})^2\right.\\
&\left.+(a_{j_2 j_1 l_1} - a_{j_2 j_1 l_2})^2(x_{j_1} - a_{j_2 j_1 l_1})\right] dx_{j_1}\\
&= a_{j_1 j_2 0}\left[(x_{j_2} - a_{j_1 j_2 s_1})^3 + 2(a_{j_1 j_2 s_1} - a_{j_1 j_2 s_2})(x_{j_2} - a_{j_1 j_2 s_1})^2\right.\\
&\left.+(a_{j_1 j_2 s_1} - a_{j_1 j_2 s_2})^2(x_{j_2} - a_{j_1 j_2 s_1})\right] dx_{j_2}.
\end{aligned}$$

With an initial condition $(x_{j_1 0}, x_{j_2 0})$, the first integral manifold is obtained, i.e.,

$$\begin{aligned}
&a_{j_2 j_1 0}\Big\{\frac{1}{4}\left[(x_{j_1} - a_{j_2 j_1 l_1})^4 - (x_{j_1 0} - a_{j_2 j_1 l_1})^4\right]\\
&+\frac{2}{3}(a_{j_2 j_1 l_1} - a_{j_2 j_1 l_2})\left[(x_{j_1} - a_{j_2 j_1 l_1})^3 - (x_{j_1 0} - a_{j_2 j_1 l_1})^3\right]\\
&+\frac{1}{2}(a_{j_2 j_1 l_1} - a_{j_2 j_1 l_2})^2\left[(x_{j_1} - a_{j_2 j_1 l_1})^2 - (x_{j_1 0} - a_{j_2 j_1 l_1})^2\right]\Big\}\\
&= a_{j_1 j_2 0}\Big\{\frac{1}{4}\left[(x_{j_2} - a_{j_1 j_2 s_1})^4 - (x_{j_2 0} - a_{j_1 j_2 s_1})^4\right]\\
&+\frac{2}{3}(a_{j_1 j_2 s_1} - a_{j_1 j_2 s_2})\left[(x_{j_2} - a_{j_1 j_2 s_1})^3 - (x_{j_2 0} - a_{j_1 j_2 s_1})^3\right]\\
&+\frac{1}{2}(a_{j_1 j_2 s_1} - a_{j_1 j_2 s_2})^2\left[(x_{j_2} - a_{j_1 j_2 s_1})^2 - (x_{j_2 0} - a_{j_1 j_2 s_1})^2\right]\Big\}.
\end{aligned}$$

(vii$_1$) Consider two cases (I) $x_{j_2}^* = a_{j_1 j_2 s_1}$ and (II) $x_{j_1}^* = a_{j_2 j_1 l_1}$

(I) In phase space, at $x_{j_2}^* = a_{j_1 j_2 s_1} (s_1, s_2 \in \{1,2\}, s_1 \neq s_2)$ and $\bar{x}_{j_1} \neq a_{j_2 j_1 l_1}, a_{j_2 j_1 l_2}$, we have

$$\left.\frac{dx_{j_1}}{dx_{j_2}}\right|_{x_{j_2}^* = a_{j_1 j_2 s_1}} = \left.\frac{a_{j_1 j_2 0}}{a_{j_2 j_1 0}} \frac{(x_{j_2} - a_{j_1 j_2 s_1})(x_{j_2} - a_{j_1 j_2 s_2})^2}{(\bar{x}_{j_1} - a_{j_2 j_1 l_1})(\bar{x}_{j_1} - a_{j_2 j_1 l_2})^2}\right|_{x_{j_2}^* = a_{j_1 j_2 s_1}} = 0.$$

If

$$\left.\frac{d^2 x_{j_1}}{dx_{j_2}^2}\right|_{x_{j_2}^* = a_{j_1 j_2 s_1}} = \frac{a_{j_1 j_2 0}}{a_{j_2 j_1 0}} \frac{(a_{j_1 j_2 s_1} - a_{j_1 j_2 s_2})^2}{(\bar{x}_{j_1} - a_{j_2 j_1 l_1})(\bar{x}_{j_1} - a_{j_2 j_1 l_2})^2} > 0,$$

there is an up-parabola flow at $x_{j_2}^* = a_{j_1 j_2 s_1}$ in the x_{j_1}-direction. If

$$\left.\frac{d^2 x_{j_1}}{dx_{j_2}^2}\right|_{x_{j_2}^* = a_{j_1 j_2 s_1}} = \frac{a_{j_1 j_2 0}}{a_{j_2 j_1 0}} \frac{(a_{j_1 j_2 s_1} - a_{j_1 j_2 s_2})^2}{(\bar{x}_{j_1} - a_{j_2 j_1 l_1})(\bar{x}_{j_1} - a_{j_2 j_1 l_2})^2} < 0,$$

there is a down-parabola flow at $x_{j_2}^* = a_{j_1 j_2 s_1}$ in the x_{j_1}-direction. Because of

$$\dot{x}_{j_2} = a_{j_2 j_1 0} (\bar{x}_{j_1} - a_{j_2 j_1 l_1})(\bar{x}_{j_1} - a_{j_2 j_1 l_2})^2,$$

the flows at $x_{j_2}^* = a_{j_1 j_2 s_1}$ in the x_{j_2}-direction are positive and negative for $\dot{x}_{j_2} > 0$ and $\dot{x}_{j_2} < 0$, respectively.

Thus, the equilibrium of $x_{j_2}^* = a_{j_1 j_2 s_1} (s_1, s_2 \in \{1,2\}, s_1 \neq s_2)$ on the x_{j_2}-direction has the following properties.

- For $a_{j_1 j_2 0}(a_{j_1 j_2 s_1} - a_{j_1 j_2 s_2})^2 > 0$ and $a_{j_2 j_1 0}(\bar{x}_{j_1} - a_{j_2 j_1 l_1})(\bar{x}_{j_1} - a_{j_2 j_1 l_2})^2 > 0$,

$$(a_{j_1 j_2 s_1}, \dot{x}_{j_2}) = \underbrace{(\text{UP}, \text{pF})}_{\text{up-parabola flow } (+)}.$$

- For $a_{j_1 j_2 0}(a_{j_1 j_2 s_1} - a_{j_1 j_2 s_2})^2 < 0$ and $a_{j_2 j_1 0}(\bar{x}_{j_1} - a_{j_2 j_1 l_1})(\bar{x}_{j_1} - a_{j_2 j_1 l_2})^2 > 0$,

$$(a_{j_1 j_2 s_1}, \dot{x}_{j_2}) = \underbrace{(\text{DP}, \text{pF})}_{\text{down-parabola flow } (+)}.$$

- For $a_{j_1 j_2 0}(a_{j_1 j_2 s_1} - a_{j_1 j_2 s_2})^2 > 0$ and $a_{j_2 j_1 0}(\bar{x}_{j_1} - a_{j_2 j_1 l_1})(\bar{x}_{j_1} - a_{j_2 j_1 l_2})^2 < 0$,

$$(a_{j_1 j_2 s_1}, \dot{x}_{j_2}) = \underbrace{(\text{DP}, \text{nF})}_{\text{down-parabola flow } (-)}.$$

- For $a_{j_1j_20}(a_{j_1j_2s_1}-a_{j_1j_2s_2})^2<0$ and $a_{j_2j_10}(\bar{x}_{j_1}-a_{j_2j_1l_1})(\bar{x}_{j_1}-a_{j_2j_1l_2})^2<0$,

$$(a_{j_1j_2s_1},\dot{x}_{j_2})=\underbrace{(\mathrm{UP},\mathrm{nF})}_{\text{up-parabola flow }(-)}.$$

(II) In phase space, at $x^*_{j_1}=a_{j_2j_1l_1}\neq a_{j_2j_1l_2}$ and $\bar{x}_{j_2}\neq a_{j_1j_2s_1},a_{j_1j_2s_2}(s_1,s_2\in\{1,2\},s_1\neq s_2)$, we have

$$\left.\frac{dx_{j_2}}{dx_{j_1}}\right|_{x^*_{j_1}=a_{j_2j_1l_1}}=\left.\frac{a_{j_2j_10}}{a_{j_1j_20}}\frac{(x_{j_1}-a_{j_2j_1l_1})(x_{j_1}-a_{j_2j_1l_2})^2}{(\bar{x}_{j_2}-a_{j_1j_2s_1})(\bar{x}_{j_2}-a_{j_1j_2s_2})^2}\right|_{x^*_{j_1}=a_{j_2j_1l_1}}=0.$$

If

$$\left.\frac{d^2x_{j_2}}{dx^2_{j_1}}\right|_{x^*_{j_1}=a_{j_2j_1l_1}}=\frac{a_{j_1j_20}}{a_{j_2j_10}}\frac{(a_{j_2j_1l_1}-a_{j_2j_1l_2})^2}{(\bar{x}_{j_2}-a_{j_1j_2s_1})(\bar{x}_{j_2}-a_{j_1j_2s_2})^2}>0,$$

there is an up-parabola flow at $x^*_{j_1}=a_{j_2j_1l_1}$ in the x_{j_2}-direction. If

$$\left.\frac{d^2x_{j_2}}{dx^2_{j_1}}\right|_{x^*_{j1}=a_{j_2j_1l_1}}=\frac{a_{j_1j_20}}{a_{j_2j_10}}\frac{(a_{j_2j_1l_1}-a_{j_2j_1l_2})^2}{(\bar{x}_{j_2}-a_{j_1j_2s_1})(\bar{x}_{j_2}-a_{j_1j_2s_2})^2}<0,$$

there is a down-parabola flow at $x^*_{j_1}=a_{j_2j_1l_1}$ in the x_{j_1}-direction. Because of

$$\dot{x}_{j_1}=a_{j_1j_20}(\bar{x}_{j_2}-a_{j_1j_2s_1})(\bar{x}_{j_2}-a_{j_1j_2s_2})^2,$$

the flows at $x^*_{j_1}=a_{j_2j_1l_1}$ in the x_{j_1}-direction are positive and negative for $\dot{x}_{j_1}>0$ and $\dot{x}_{j_1}<0$, respectively.

Thus, the equilibrium of $x^*_{j_1}=a_{j_2j_1l_1}$ and $\bar{x}_{j_2}\neq a_{j_1j_2s_1},a_{j_1j_2s_2}(s_1,s_2,l_1,l_2\in\{1,2\}s_1\neq s_2,l_1\neq l_2)$ on the x_{j_2}-direction has the following properties.

- For $a_{j_1j_20}(\bar{x}_{j_2}-a_{j_1j_2s_1})(\bar{x}_{j_2}-a_{j_1j_2s_2})^2>0$ and $a_{j_2j_10}(a_{j_2j_1l_1}-a_{j_2j_1j_2})^2>0$,

$$(\dot{x}_{j_1},a_{j_2j_1l_1})=\underbrace{(\mathrm{pF},\mathrm{UP})}_{\text{up-parabola flow }(+)}.$$

- For $a_{j_1j_20}(\bar{x}_{j_2}-a_{j_1j_2s_1})(\bar{x}_{j_2}-a_{j_1j_2s_2})^2<0$ and $a_{j_2j_10}(a_{j_2j_1l_1}-a_{j_2j_1l_2})^2>0$,

$$(\dot{x}_{j_1},a_{j_2j_1l_1})=\underbrace{(\mathrm{nF},\mathrm{DP})}_{\text{down-parabola flow }(-)}.$$

- For $a_{j_1j_20}(\bar{x}_{j_2} - a_{j_1j_2s_1})(\bar{x}_{j_2} - a_{j_1j_2s_2})^2 > 0$ and $a_{j_2j_10}(a_{j_2j_1l_1} - a_{j_2j_1l_2})^2 < 0$,

$$(\dot{x}_{j_1}, a_{j_2j_1l_1}) = \underbrace{(\text{pF}, \text{DP})}_{\text{down-parabola flow } (+)}.$$

- For $a_{j_1j_20}(\bar{x}_{j_2} - a_{j_1j_2s_1})(\bar{x}_{j_2} - a_{j_1j_2s_2})^2 < 0$ and $a_{j_2j_10}(a_{j_2j_1l_1} - a_{j_2j_1l_2})^2 < 0$,

$$(\dot{x}_{j_1}, a_{j_2j_1l_1}) = \underbrace{(\text{nF}, \text{UP})}_{\text{up-parabola flow } (-)}.$$

Therefore, from the proofs of cases (I) and (II), the equilibrium of $(x_{j_2}^*, x_{j_1}^*) = (a_{j_1j_2s_1}, a_{j_2j_1l_1})(s_1, s_2, l_1, l_2 \in \{1,2\}, s_1 \neq s_2, l_1 \neq l_2)$ has the following properties as in Eqs. (4.58)–(4.61).

- For $a_{j_1j_20}(a_{j_1j_2s_1} - a_{j_1j_2s_2})^2 > 0$ and $a_{j_2j_10}(a_{j_2j_1l_1} - a_{j_2j_1l_2})^2 > 0$,

$$(a_{j_1j_2s_1}, a_{j_2j_1l_1}) = \underbrace{(\text{UP}_+, \text{UP}_+)}_{\text{positive saddle}}.$$

  The equilibrium of $(x_{j_2}^*, x_{j_1}^*) = (a_{j_1j_2s_1}, a_{j_2j_1l_1})$ is a $(\text{UP}_+,\text{UP}_+)$-positive saddle.
- For $a_{j_1j_20}(a_{j_1j_2s_1} - a_{j_1j_2s_2})^2 < 0$ and $a_{j_2j_10}(a_{j_2j_1l_1} - a_{j_2j_1l_2})^2 > 0$,

$$(a_{j_1j_2s_1}, a_{j_2j_1l_1}) = \underbrace{(\text{DP}_+, \text{DP}_-)}_{\text{CCW center}}.$$

  The equilibrium of $(x_{j_2}^*, x_{j_1}^*) = (a_{j_1j_2s_1}, a_{j_2j_1l_1})$ is a $(\text{DP}_+,\text{DP}_-)$-counter-clockwise center.
- For $a_{j_1j_20}(a_{j_1j_2s_1} - a_{j_1j_2s_2})^2 > 0$ and $a_{j_2j_10}(a_{j_2j_1l_1} - a_{j_2j_1l_2})^2 < 0$,

$$(a_{j_1j_2s_1}, a_{j_2j_1l_1}) = \underbrace{(\text{DP}_-, \text{DP}_+)}_{\text{CW center}}.$$

  The equilibrium of $(x_{j_2}^*, x_{j_1}^*) = (a_{j_1j_2s_1}, a_{j_2j_1l_1})$ is a $(\text{DP}_-,\text{DP}_+)$-clockwise center.
- For $a_{j_1j_20}(a_{j_1j_2s_1} - a_{j_1j_2s_2})^2 < 0$ and $a_{j_2j_10}(a_{j_2j_1l_1} - a_{j_2j_1l_2})^2 < 0$,

$$(a_{j_1j_2s_1}, a_{j_2j_1l_1}) = \underbrace{(\text{UP}_-, \text{UP}_-)}_{\text{negative saddle}}.$$

  The equilibrium of $(x_{j_2}^*, x_{j_1}^*) = (a_{j_1j_2s_1}, a_{j_2j_1l_1})$ is a $(\text{UP}_-,\text{UP}_-)$-negative saddle.

(vii$_2$) In phase space, at $x_{j_2}^* = a_{j_1 j_2 s_2} (s_1, s_2 \in \{1,2\}, s_1 \neq s_2)$ and $\bar{x}_{j_1} \neq a_{j_2 j_1 l_1}, a_{j_2 j_1 l_2}$, we have

$$\left.\frac{dx_{j_1}}{dx_{j_2}}\right|_{x_{j_2}^* = a_{j_1 j_2 s_2}} = \left.\frac{a_{j_1 j_2 0}}{a_{j_2 j_1 0}} \frac{(x_{j_2} - a_{j_1 j_2 s_1})(x_{j_2} - a_{j_1 j_2 s_2})^2}{(\bar{x}_{j_1} - a_{j_2 j_1 l_1})(\bar{x}_{j_1} - a_{j_2 j_1 l_2})^2}\right|_{x_{j_2}^* = a_{j_1 j_2 s_2}} = 0,$$

$$\left.\frac{d^2 x_{j_1}}{dx_{j_2}^2}\right|_{x_{j_2}^* = a_{j_1 j_2 s_2}} = 0.$$

If

$$\left.\frac{d^3 x_{j_1}}{dx_{j_2}^3}\right|_{x_{j_2}^* = a_{j_1 j_2 s_2}} = \frac{a_{j_1 j_2 0}}{a_{j_2 j_1 0}} \frac{2(a_{j_1 j_2 s_2} - a_{j_1 j_2 s_1})}{(\bar{x}_{j_1} - a_{j_2 j_1 l_1})(\bar{x}_{j_1} - a_{j_2 j_1 l_2})^2} > 0,$$

there is an increasing-inflection flow at $x_{j_2}^* = a_{j_1 j_2 s_2}$ in the x_{j_1}-direction. If

$$\left.\frac{d^3 x_{j_1}}{dx_{j_2}^3}\right|_{x_{j_2}^* = a_{j_1 j_2 s_2}} = \frac{a_{j_1 j_2 0}}{a_{j_2 j_1 0}} \frac{2(a_{j_1 j_2 s_2} - a_{j_1 j_2 s_1})}{(\bar{x}_{j_1} - a_{j_2 j_1 l_1})(\bar{x}_{j_1} - a_{j_2 j_1 l_2})^2} < 0,$$

there is a decreasing-inflection flow at $x_{j_2}^* = a_{j_1 j_2 s_2}$ in the x_{j_1}-direction. Because of

$$\dot{x}_{j_2} = a_{j_2 j_1 0}(\bar{x}_{j_1} - a_{j_2 j_1 l_1})(\bar{x}_{j_1} - a_{j_2 j_1 j_2})^2,$$

the flows at $x_{j_2}^* = a_{j_1 j_2 s_2}$ in the x_{j_2}-direction are positive and negative for $\dot{x}_{j_2} > 0$ and $\dot{x}_{j_2} < 0$, respectively.

Thus, the equilibrium of $x_{j_2}^* = a_{j_1 j_2 s_2} (s_1, s_2 \in \{1,2\}, s_1 \neq s_2)$ and $x_{j_1} \neq a_{j_2 j_1 1}$ on the x_{j_2}-direction has the following properties.

- For $a_{j_1 j_2 0}(a_{j_1 j_2 s_2} - a_{j_1 j_2 s_1}) > 0$ and $a_{j_2 j_1 0}(\bar{x}_{j_1} - a_{j_2 j_1 l_1})(\bar{x}_{j_1} - a_{j_2 j_1 l_2})^2 > 0$,

$$(a_{j_1 j_2 s_2}, \dot{x}_{j_2}) = \underbrace{(\text{II}, \text{pF})}_{\text{increasing-inflection flow } (+)}$$

- For $a_{j_1 j_2 0}(a_{j_1 j_2 s_2} - a_{j_1 j_2 s_1}) < 0$ and $a_{j_2 j_1 0}(\bar{x}_{j_1} - a_{j_2 j_1 l_1})(\bar{x}_{j_1} - a_{j_2 j_1 l_2})^2 > 0$,

$$(a_{j_1 j_2 s_2}, \dot{x}_{j_2}) = \underbrace{(\text{DI}, \text{pF})}_{\text{decreasing-inflection flow } (+)}.$$

- For $a_{j_1j_20}(a_{j_1j_2s_2}-a_{j_1j_2s_1})>0$ and $a_{j_2j_10}(\overline{x}_{j_1}-a_{j_2j_1l_1})(\overline{x}_{j_1}-a_{j_2j_1l_2})^2<0$,

$$(a_{j_1j_2s_2},\dot{x}_{j_2})=\underbrace{(\text{DI, nF})}_{\text{decreasing-inflection flow }(-)}.$$

- For $a_{j_1j_20}(a_{j_1j_2s_2}-a_{j_1j_2s_1})<0$ and $a_{j_2j_10}(\overline{x}_{j_1}-a_{j_2j_1l_1})(\overline{x}_{j_1}-a_{j_2j_1l_2})^2<0$,

$$(a_{j_1j_2s_2},\dot{x}_{j_2})=\underbrace{(\text{II, nF})}_{\text{increasing-inflection flow }(-)}$$

Therefore, from the proof (II) of case (vii$_1$) and the foregoing analysis, the equilibrium of $(x_{j_2}^*,x_{j_1}^*)=(a_{j_1j_2s_2},a_{j_2j_1l_1})(s_1,s_2,l_1,l_2\in\{1,2\},s_1\neq s_2,l_1\neq l_2)$ on the x_{j_2}-direction has the following properties as in Eqs. (4.62)–(4.65).

- For $a_{j_1j_20}(a_{j_1j_2s_2}-a_{j_1j_2s_1})>0$ and $a_{j_2j_10}(a_{j_2j_1l_1}-a_{j_2j_1l_2})^2>0$,

$$(a_{j_1j_2s_2},a_{j_2j_1l_1})=\underbrace{(\text{US, UP})}_{\text{up-parabola upper-saddle}}$$

  The equilibrium of $(x_{j_2}^*,x_{j_1}^*)=(a_{j_1j_2s_2},a_{j_2j_1l_1})$ is a (US,UP)-up-parabola upper-saddle.
- For $a_{j_1j_20}(a_{j_1j_2s_2}-a_{j_1j_2s_1})<0$ and $a_{j_2j_10}(a_{j_2j_1l_1}-a_{j_2j_1l_2})^2>0$,

$$(a_{j_1j_2s_2},a_{j_2j_1l_1})=\underbrace{(\text{LS, DP})}_{\text{down-parabola lower-saddle}}$$

  The equilibrium of $(x_{j_2}^*,x_{j_1}^*)=(a_{j_1j_2s_2},a_{j_2j_1l_1})$ is an (LS,DP)-down-parabola lower-saddle.
- For $a_{j_1j_20}(a_{j_1j_2s_2}-a_{j_1j_2s_1})>0$ and $a_{j_2j_10}(a_{j_2j_1l_1}-a_{j_2j_1l_2})^2<0$,

$$(a_{j_1j_2s_2},a_{j_2j_1l_1})=\underbrace{(\text{US, DP})}_{\text{down-parabola upper-saddle}}$$

  The equilibrium of $(x_{j_2}^*,x_{j_1}^*)=(a_{j_1j_2s_2},a_{j_2j_1l_1})$ is a (US,DP)-down-parabola upper-saddle.
- For $a_{j_1j_20}(a_{j_1j_2s_2}-a_{j_1j_2s_1})<0$ and $a_{j_2j_10}(a_{j_2j_1l_1}-a_{j_2j_1l_2})^2<0$,

$$(a_{j_1j_2s_2},a_{j_2j_1l_1})=\underbrace{(\text{LS, UP})}_{\text{up-parabola-lower-saddle}}.$$

  The equilibrium of $(x_{j_2}^*,x_{j_1}^*)=(a_{j_1j_2s_2},a_{j_2j_1l_1})$ is an (LS,UP)-up-parabola lower-saddle.

(vii$_3$) In phase space, at $x_{j_1}^* = a_{j_2j_1l_2}$ and $\overline{x}_{j_2} \neq a_{j_1j_2s_1}, a_{j_1j_2s_2} (s_1, s_2 \in \{1,2\}, s_1 \neq s_2)$, we have

$$\left.\frac{dx_{j_2}}{dx_{j_1}}\right|_{x_{j_1}^* = a_{j_2j_1l_2}} = \left.\frac{a_{j_2j_10}}{a_{j_1j_20}} \frac{(x_{j_1} - a_{j_2j_1l_1})(x_{j_1} - a_{j_2j_1l_2})^2}{(\overline{x}_{j_2} - a_{j_1j_2s_1})(\overline{x}_{j_2} - a_{j_1j_2s_2})^2}\right|_{x_{j_1}^* = a_{j_2j_1l_2}} = 0,$$

$$\left.\frac{d^2x_{j_2}}{dx_{j_1}^2}\right|_{x_{j_1}^* = a_{j_2j_1l_2}} = 0.$$

If

$$\left.\frac{d^3x_{j_2}}{dx_{j_1}^3}\right|_{x_{j_1}^* = a_{j_2j_1l_2}} = \frac{a_{j_2j_10}}{a_{j_1j_20}} \frac{(a_{j_2j_1l_2} - a_{j_2j_1l_1})}{(\overline{x}_{j_2} - a_{j_1j_2s_1})(\overline{x}_{j_2} - a_{j_1j_2s_2})^2} > 0,$$

there is an increasing-inflection flow at $x_{j_1}^* = a_{j_2j_1l_2}$ in the x_{j_2}-direction. If

$$\left.\frac{d^3x_{j_2}}{dx_{j_1}^3}\right|_{x_{j_1}^* = a_{j_2j_1l_2}} = \frac{a_{j_2j_10}}{a_{j_1j_20}} \frac{(a_{j_2j_1l_2} - a_{j_2j_1l_1})}{(\overline{x}_{j_2} - a_{j_1j_2s_1})(\overline{x}_{j_2} - a_{j_1j_2s_2})^2} < 0,$$

there is a decreasing-inflection flow at $x_{j_1}^* = a_{j_2j_1l_2}$ in the x_{j_2}-direction. Because of

$$\dot{x}_{j_1} = a_{j_1j_20}(\overline{x}_{j_2} - a_{j_1j_2s_1})(\overline{x}_{j_2} - a_{j_1j_2s_2})^2,$$

the flows at $x_{j_1}^* = a_{j_2j_1l_2}$ in the x_{j_1}-direction are positive and negative for $\dot{x}_{j_1} > 0$ and $\dot{x}_{j_1} < 0$, respectively.

Thus, the equilibrium of $x_{j_1}^* = a_{j_2j_1l_2}$ and $\overline{x}_{j_2} \neq a_{j_1j_2s_1}, a_{j_1j_2s_2} (s_1, s_2 \in \{1,2\} s_1 \neq s_2)$ on the x_{j_2}-direction has the following properties.

- For $a_{j_1j_20}(\overline{x}_{j_2} - a_{j_1j_2s_1})(\overline{x}_{j_2} - a_{j_1j_2s_2})^2 > 0$ and $a_{j_2j_10}(a_{j_2j_1l_2} - a_{j_2j_1i_1}) > 0$,

$$(\dot{x}_{j_1}, a_{j_2j_1l_2}) = \underbrace{(\text{pF}, \text{II})}_{\text{increasing-infection flow } (+)}.$$

- For $a_{j_1j_20}(\overline{x}_{j_2} - a_{j_1j_2s_1})(\overline{x}_{j_2} - a_{j_1j_2s_2})^2 < 0$ and $a_{j_2j_10}(a_{j_2j_1l_2} - a_{j_2j_1l_1}) > 0$,

$$(\dot{x}_{j_1}, a_{j_2j_1l_2}) = \underbrace{(\text{nF}, \text{DI})}_{\text{decreasing-inflection flow } (-)}.$$

- For $a_{j_1j_20}(\overline{x}_{j_2}-a_{j_1j_2s_1})(\overline{x}_{j_2}-a_{j_1j_2s_2})^2>0$ and $a_{j_2j_10}(a_{j_2j_1l_2}-a_{j_2j_1l_1})<0$,

$$(\dot{x}_{j_1},a_{j_2j_1l_2})=\underbrace{(\text{pF},\text{DI})}_{\text{decreasing-inflection flow }(+)}.$$

- For $a_{j_1j_20}(\overline{x}_{j_2}-a_{j_1j_2s_1})(\overline{x}_{j_2}-a_{j_1j_2s_2})^2<0$ and $a_{j_2j_10}(a_{j_2j_1l_2}-a_{j_2j_1l_1})<0$,

$$(\dot{x}_{j_1},a_{j_2j_1l_2})=\underbrace{(\text{nF},\text{II})}_{\text{increasing-inflection flow }(-)}.$$

Therefore, from the proof (I) of case (vii$_1$) and the foregoing analysis, the equilibrium of $(x_{j_2}^*,x_{j_1}^*)=(a_{j_1j_2s_1},a_{j_2j_1l_2})$ has the following properties as in Eqs. (4.66)–(4.69).

- For $a_{j_1j_20}(a_{j_1j_2s_1}-a_{j_1j_2s_2})^2>0$ and $a_{j_2j_10}(a_{j_2j_1l_2}-a_{j_2j_1l_1})>0$,

$$(a_{j_1j_2s_1},a_{j_2j_1l_2})=\underbrace{(\text{UP},\text{US})}_{\text{up-parabola upper-saddle}}$$

  The equilibrium of $(x_{j_2}^*,x_{j_1}^*)=(a_{j_1j_2s_1},a_{j_2j_1l_2})$ is a (UP,US)-up-parabola upper-saddle.
- For $a_{j_1j_20}(a_{j_1j_2s_1}-a_{j_1j_2s_2})^2<0$ and $a_{j_2j_10}(a_{j_2j_1l_2}-a_{j_2j_1l_1})>0$,

$$(a_{j_1j_2s_1},a_{j_2j_1l_2})=\underbrace{(\text{DP},\text{US})}_{\text{down-parabola upper-saddle}}$$

  The equilibrium of $(x_{j_2}^*,x_{j_1}^*)=(a_{j_1j_2s_1},a_{j_2j_1l_2})$ is a (DP,US)-down-parabola upper-saddle.
- For $a_{j_1j_20}(a_{j_1j_2s_1}-a_{j_1j_2s_2})^2>0$ and $a_{j_2j_10}(a_{j_2j_1l_2}-a_{j_2j_1l_1})<0$,

$$(a_{j_1j_2s_1},a_{j_2j_1l_2})=\underbrace{(\text{DP},\text{LS})}_{\text{down-parabola lower-saddle}}.$$

  The equilibrium of $(x_{j_2}^*,x_{j_1}^*)=(a_{j_1j_2s_1},a_{j_2j_1l_2})$ is a (DP,LS)-down-parabola lower-saddle.
- For $a_{j_1j_20}(a_{j_1j_2s_1}-a_{j_1j_2s_2})^2<0$ and $a_{j_2j_10}(a_{j_2j_1l_2}-a_{j_2j_1l_1})<0$,

$$(a_{j_1j_2s_1},a_{j_2j_1l_2})=\underbrace{(\text{UP},\text{LS})}_{\text{up-parabola lower-saddle}}.$$

  The equilibrium of $(x_{j_2}^*,x_{j_1}^*)=(a_{j_1j_2s_1},a_{j_2j_1l_2})$ is a (UP,LS)-up-parabola lower-saddle.

(vii$_4$) From the proofs of case (vii$_2$) and (vii$_3$), the equilibrium of $(x^*_{j_2}, x^*_{j_1}) = (a_{j_1 j_2 s_2}, a_{j_2 j_1 l_2})$ has the following properties as in Eqs. (4.70)–(4.73).

- For $a_{j_1 j_2 0}(a_{j_1 j_2 s_2} - a_{j_1 j_2 s_1}) > 0$ and $a_{j_2 j_1 0}(a_{j_2 j_1 l_2} - a_{j_2 j_1 l_1}) > 0$,

$$(a_{j_1 j_2 s_2}, a_{j_2 j_1 l_2}) = \underbrace{(\mathrm{II}_+, \mathrm{II}_+)}_{\text{double inflection saddle}}.$$

The equilibrium of $(x^*_{j_2}, x^*_{j_1}) = (a_{j_1 j_2 s_2}, a_{j_2 j_1 l_2})$ is an (II$_+$,II$_+$)-double increasing-inflection saddle.

- For $a_{j_1 j_2 0}(a_{j_1 j_2 s_2} - a_{j_1 j_2 s_1}) < 0$ and $a_{j_2 j_1 0}(a_{j_2 j_1 l_2} - a_{j_2 j_1 l_1}) > 0$,

$$(a_{j_1 j_2 s_2}, a_{j_2 j_1 l_2}) = \underbrace{(\mathrm{DI}_+, \mathrm{DI}_-)}_{\text{double inflection saddle}}.$$

The equilibrium of $(x^*_{j_2}, x^*_{j_1}) = (a_{j_1 j_2 s_2}, a_{j_2 j_1 l_2})$ is a (DI$_+$,DI$_-$)-double decreasing-inflection saddle.

- For $a_{j_1 j_2 0}(a_{j_1 j_2 s_2} - a_{j_1 j_2 s_1}) > 0$ and $a_{j_2 j_1 0}(a_{j_2 j_1 l_2} - a_{j_2 j_1 l_1}) < 0$,

$$(a_{j_1 j_2 s_2}, a_{j_2 j_1 l_2}) = \underbrace{(\mathrm{DI}_-, \mathrm{DI}_+)}_{\text{double inflection saddle}}.$$

The equilibrium of $(x^*_{j_2}, x^*_{j_1}) = (a_{j_1 j_2 s_2}, a_{j_2 j_1 l_2})$ is a (DI$_-$,DI$_+$)-double decreasing-inflection saddle.

- For $a_{j_1 j_2 0}(a_{j_1 j_2 s_2} - a_{j_1 j_2 s_1}) < 0$ and $a_{j_2 j_1 0}(a_{j_2 j_1 l_2} - a_{j_2 j_1 l_1}) < 0$,

$$(a_{j_1 j_2 s_2}, a_{j_2 j_1 l_2}) = \underbrace{(\mathrm{II}_-, \mathrm{II}_-)}_{\text{double inflection saddle}}.$$

The equilibrium of $(x^*_{j_2}, x^*_{j_1}) = (a_{j_1 j_2 s_2}, a_{j_2 j_1 l_2})$ is an (II$_-$,II$_-$)-double increasing -inflection saddle.

(viii) If

$$\begin{aligned} &\Delta_{j_1 j_2} = 0 \text{ and } \Delta_{j_2 j_1} = 0, \\ &a_{j_2 j_1 s_1} = a_{j_2 j_1 s_2} = a_{j_2 j_1 1}, a_{j_1 j_2 l_1} \neq a_{j_1 j_2 l_2}; \end{aligned}$$

then

$$\begin{aligned} &\dot{x}_{j_1} = a_{j_1 j_2 0}(x_{j_2} - a_{j_1 j_2 1})^3, \\ &\dot{x}_{j_2} = a_{j_2 j_1 0}(x_{j_1} - a_{j_2 j_1 l_1})(x_{j_1} - a_{j_2 j_1 l_2})^2, \\ &j_1, j_2 \in \{1, 2\}; j_1 \neq j_2. \end{aligned}$$

In phase space,

$$\frac{dx_{j_1}}{dx_{j_2}} = \frac{a_{j_1j_20}}{a_{j_2j_10}} \frac{(x_{j_2} - a_{j_1j_21})^3}{(x_{j_1} - a_{j_2j_1l_1})(x_{j_1} - a_{j_2j_1l_2})^2},$$

and

$$\begin{aligned}
&a_{j_2j_10}\left[(x_{j_1} - a_{j_2j_1l_1})^3 + 2(a_{j_2j_1l_1} - a_{j_2j_1l_2})(x_{j_1} - a_{j_2j_1l_1})^2\right.\\
&\left.+(a_{j_2j_1l_1} - a_{j_2j_1l_2})^2(x_{j_1} - a_{j_2j_1l_1})\right]dx_{j_1}\\
&= a_{j_1j_20}(x_{j_2} - a_{j_1j_21})^3 dx_{j_2}.
\end{aligned}$$

With an initial condition (x_{j_10}, x_{j_20}), the first integral manifold is obtained, i.e.,

$$\begin{aligned}
&a_{j_2j_10}\Big\{\frac{1}{4}\left[(x_{j_1} - a_{j_2j_1l_1})^4 - (x_{j_10} - a_{j_2j_1l_1})^4\right]\\
&+\frac{2}{3}(a_{j_2j_1l_1} - a_{j_2j_1l_2})\left[(x_{j_1} - a_{j_2j_1l_1})^3 - (x_{j_10} - a_{j_2j_1l_1})^3\right]\\
&+\frac{1}{2}(a_{j_2j_1l_1} - a_{j_2j_1l_2})^2\left[(x_{j_1} - a_{j_2j_1l_1})^2 - (x_{j_10} - a_{j_2j_1l_1})^2\right]\Big\}\\
&= \frac{1}{4}a_{j_1j_20}\left[(x_{j_2} - a_{j_1j_21})^4 - (x_{j_20} - a_{j_1j_21})^4\right].
\end{aligned}$$

(viii_1) Consider two cases (I) $x_{j_2}^* = a_{j_1j_21}$ and (II) $x_{j_1}^* = a_{j_2j_1l_1}$.

(I) In phase space, at $x_{j_2}^* = a_{j_1j_21}$ and $\overline{x}_{j_1} \neq a_{j_2j_1l_1}, a_{j_2j_1l_2}$, we have

$$\begin{aligned}
&\left.\frac{dx_{j_1}}{dx_{j_2}}\right|_{x_{j_2}^* = a_{j_1j_21}} = \frac{a_{j_1j_20}}{a_{j_2j_10}} \left.\frac{(x_{j_2} - a_{j_1j_21})^3}{(\overline{x}_{j_1} - a_{j_2j_1l_1})(\overline{x}_{j_1} - a_{j_2j_1l_2})^2}\right|_{x_{j_2}^* = a_{j_1j_21}} = 0,\\
&\left.\frac{d^2x_{j_1}}{dx_{j_2}^2}\right|_{x_{j_2}^* = a_{j_1j_21}} = \left.\frac{d^3x_{j_1}}{dx_{j_2}^3}\right|_{x_{j_2}^* = a_{j_1j_21}} = 0.
\end{aligned}$$

If

$$\left.\frac{d^4x_{j_1}}{dx_{j_2}^4}\right|_{x_{j_2}^* = a_{j_1j_21}} = \frac{a_{j_1j_20}}{a_{j_2j_10}} \frac{6}{(\overline{x}_{j_1} - a_{j_2j_1l_1})(\overline{x}_{j_1} - a_{j_2j_1l_2})^2} > 0,$$

there is a third-order up-parabola flow at $x_{j_2}^* = a_{j_1j_21}$ in the x_{j_1}-direction. If

$$\left.\frac{d^4x_{j_1}}{dx_{j_2}^4}\right|_{x_{j_2}^* = a_{j_1j_21}} = \frac{a_{j_1j_20}}{a_{j_2j_10}} \frac{6}{(\overline{x}_{j_1} - a_{j_2j_1l_1})(\overline{x}_{j_1} - a_{j_2j_1l_2})^2} < 0,$$

there is a third-order down-parabola flow at $x_{j_2}^* = a_{j_1j_2 1}$ in the x_{j_1}-direction. Because of

$$\dot{x}_{j_2} = a_{j_2j_1 0}(\bar{x}_{j_1} - a_{j_2j_1 l_1})(\bar{x}_{j_1} - a_{j_2j_1 l_2})^2,$$

the flows at $x_{j_2}^* = a_{j_1j_2 1}$ in the x_{j_2}-direction are positive and negative for $\dot{x}_{j_2} > 0$ and $\dot{x}_{j_2} < 0$, respectively.

Thus, the equilibrium of $x_{j_2}^* = a_{j_1j_2 1}$ and $\bar{x}_{j_1} \neq a_{j_2j_1 l_1}, a_{j_2j_1 l_2}$ on the x_{j_1}-direction has the following properties.

- For $a_{j_1j_2 0}(a_{j_1j_2 s_2} - a_{j_1j_2 s_1}) > 0$ and $a_{j_2j_1 0}(\bar{x}_{j_1} - a_{j_2j_1 l_1})(\bar{x}_{j_1} - a_{j_2j_1 l_2})^2 > 0$,

$$(a_{j_1j_2 1}, \dot{x}_{j_2}) = \underbrace{(3^{\text{rd}}\text{UP}, \text{pF})}_{\text{third-order up-parabola flow } (+)}.$$

- For $a_{j_1j_2 0}(a_{j_1j_2 s_2} - a_{j_1j_2 s_1}) < 0$ and $a_{j_2j_1 0}(\bar{x}_{j_1} - a_{j_2j_1 l_1})(\bar{x}_{j_1} - a_{j_2j_1 l_2})^2 > 0$,

$$(a_{j_1j_2 1}, \dot{x}_{j_2}) = \underbrace{(3^{\text{rd}}\text{DP}, \text{pF})}_{\text{third-order down-parabola flow } (+)}.$$

- For $a_{j_1j_2 0}(a_{j_1j_2 s_2} - a_{j_1j_2 s_1}) > 0$ and $a_{j_2j_1 0}(\bar{x}_{j_1} - a_{j_2j_1 l_1})(\bar{x}_{j_1} - a_{j_2j_1 l_2})^2 < 0$,

$$(a_{j_1j_2 1}, \dot{x}_{j_2}) = \underbrace{(3^{\text{rd}}\text{DP}, \text{nF})}_{\text{third-order down-parabola flow } (-)}.$$

- For $a_{j_1j_2 0}(a_{j_1j_2 s_2} - a_{j_1j_2 s_1}) < 0$ and $a_{j_2j_1 0}(\bar{x}_{j_1} - a_{j_2j_1 l_1})(\bar{x}_{j_1} - a_{j_2j_1 l_2})^2 < 0$,

$$(a_{j_1j_2 1}, \dot{x}_{j_2}) = \underbrace{(3^{\text{rd}}\text{UP}, \text{nF})}_{\text{third-order up-parabola flow } (-)}.$$

(II) In phase space, at $x_{j_1}^* = a_{j_2j_1 l_1}$ and $\bar{x}_{j_2} \neq a_{j_1j_2 s_1}, a_{j_1j_2 s_2} (s_1, s_2 \in \{1, 2\}, s_1 \neq s_2)$, we have

$$\frac{dx_{j_2}}{dx_{j_1}}\bigg|_{x_{j_1}^* = a_{j_2j_1 l_1}} = \frac{a_{j_2j_1 0}}{a_{j_1j_2 0}} \frac{(x_{j_1} - a_{j_2j_1 l_1})(x_{j_1} - a_{j_2j_1 l_2})^2}{(\bar{x}_{j_2} - a_{j_1j_2 1})^3}\bigg|_{x_{j_1}^* = a_{j_2j_1 l_1}} = 0.$$

If

$$\frac{d^2x_{j_2}}{dx_{j_1}^2}\bigg|_{x_{j_1}^* = a_{j_2j_1 l_1}} = \frac{a_{j_1j_2 0}}{a_{j_2j_1 0}} \frac{(a_{j_2j_1 l_1} - a_{j_2j_1 l_2})^2}{(\bar{x}_{j_2} - a_{j_1j_2 1})^3} > 0,$$

there is an up-parabola flow at $x_{j_1}^* = a_{j_2 j_1 l_1}$ in the x_{j_2}-direction. If

$$\left.\frac{d^2 x_{j_2}}{dx_{j_1}^2}\right|_{x_{j_1}^* = a_{j_2 j_1 l_1}} = \frac{a_{j_1 j_2 0}}{a_{j_2 j_1 0}} \frac{(a_{j_2 j_1 l_1} - a_{j_2 j_1 l_2})^2}{(\overline{x}_{j_2} - a_{j_1 j_2 1})^3} < 0,$$

there is a down-parabola flow at $x_{j_1}^* = a_{j_2 j_1 l_1}$ in the x_{j_1}-direction. Because of

$$\dot{x}_{j_1} = a_{j_1 j_2 0}(\overline{x}_{j_2} - a_{j_1 j_2 1})^3,$$

the flows at $x_{j_1}^* = a_{j_2 j_1 l_1}$ in the x_{j_1}-direction are positive and negative for $\dot{x}_{j_1} > 0$ and $\dot{x}_{j_1} < 0$, respectively.

Thus, the equilibrium of $x_{j_1}^* = a_{j_2 j_1 l_1}$ and $\overline{x}_{j_2} \neq a_{j_1 j_2 1} (l_1, l_2 \in \{1, 2\}, l_1 \neq l_2)$ on the x_{j_2}-direction has the following properties.

- For $a_{j_1 j_2 0}(\overline{x}_{j_2} - a_{j_1 j_2 1})^3 > 0$ and $a_{j_2 j_1 0}(a_{j_2 j_1 l_1} - a_{j_2 j_1 l_2})^2 > 0$,

$$(\dot{x}_{j_1}, a_{j_2 j_1 l_1}) = \underbrace{(\text{pF}, \text{UP})}_{\text{up-parabola flow } (+)}.$$

- For $a_{j_1 j_2 0}(\overline{x}_{j_2} - a_{j_1 j_2 1})^3 < 0$ and $a_{j_2 j_1 0}(a_{j_2 j_1 l_1} - a_{j_2 j_1 l_2})^2 > 0$,

$$(\dot{x}_{j_1}, a_{j_2 j_1 l_1}) = \underbrace{(\text{nF}, \text{DP})}_{\text{down-parabola flow } (-)}.$$

- For $a_{j_1 j_2 0}(\overline{x}_{j_2} - a_{j_1 j_2 1})^3 > 0$ and $a_{j_2 j_1 0}(a_{j_2 j_1 l_1} - a_{j_2 j_1 l_2})^2 < 0$,

$$(\dot{x}_{j_1}, a_{j_2 j_1 l_1}) = \underbrace{(\text{pF}, \text{DP})}_{\text{down-parabola flow } (+)}.$$

- For $a_{j_1 j_2 0}(\overline{x}_{j_2} - a_{j_1 j_2 1})^3 < 0$ and $a_{j_2 j_1 0}(a_{j_2 j_1 l_1} - a_{j_2 j_1 l_2})^2 < 0$,

$$(\dot{x}_{j_1}, a_{j_2 j_1 l_1}) = \underbrace{(\text{nF}, \text{UP})}_{\text{up-parabola flow } (-)}.$$

Therefore, from cases (I) and (II), the equilibrium of $(x_{j_2}^*, x_{j_1}^*) = (a_{j_1 j_2 1}, a_{j_2 j_1 l_1})$ $(l_1, l_2 \in \{1, 2\}, l_1 \neq l_2)$ has the following properties.

- For $a_{j_1 j_2 0} > 0$ and $a_{j_2 j_1 0}(a_{j_2 j_1 l_1} - a_{j_2 j_1 l_2})^2 > 0$,

$$(a_{j_1 j_2 1}, a_{j_2 j_1 l_1}) = \underbrace{(3^{\text{rd}}\text{UP}_+, \text{UP}_+)}_{\text{third-order positive saddle}}.$$

The equilibrium of $(x_{j_2}^*, x_{j_1}^*) = (a_{j_1 j_2 1}, a_{j_2 j_1 l_1})$ is a $(3^{\text{rd}}\text{UP}_+,\text{UP}_+)$-third-order positive saddle.

- For $a_{j_1j_20}<0$ and $a_{j_2j_10}(a_{j_2j_1l_1}-a_{j_2j_1l_2})^2>0$,

$$(a_{j_1j_21},a_{j_2j_1l_1})=\underbrace{(3^{\rm rd}\mathrm{DP}_+,\mathrm{DP}_-)}_{\text{third-order CCW center}}.$$

The equilibrium of $(x^*_{j_2},x^*_{j_1})=(a_{j_1j_21},a_{j_2j_1l_1})$ is a $(3^{\rm rd}\mathrm{DP}_+,\mathrm{DP}_-)$-third-order-counter-clockwise center.

- For $a_{j_1j_20}>0$ and $a_{j_2j_10}(a_{j_2j_1l_1}-a_{j_2j_1l_2})^2<0$,

$$(a_{j_1j_21},a_{j_2j_1l_1})=\underbrace{(3^{\rm rd}\mathrm{DP}_-,\mathrm{DP}_+)}_{\text{third-order CW center}}.$$

The equilibrium of $(x^*_{j_2},x^*_{j_1})=(a_{j_1j_21},a_{j_2j_1l_1})$ is a $(3^{\rm rd}\mathrm{DP}_-,\mathrm{DP}_+)$-third-order-clockwise center.

- For $a_{j_1j_20}<0$ and $a_{j_2j_10}(a_{j_2j_1l_1}-a_{j_2j_1l_2})^2<0$,

$$(a_{j_1j_21},a_{j_2j_1l_1})=\underbrace{(3^{\rm rd}\mathrm{UP}_-,\mathrm{UP}_-)}_{\text{third-order negative saddle}}.$$

The equilibrium of $(x^*_{j_2},x^*_{j_1})=(a_{j_1j_21},a_{j_2j_1l_1})$ is a $(3^{\rm rd}\mathrm{UP}_-,\mathrm{UP}_-)$-third-order negative saddle.

(viii$_2$) In phase space, at $x^*_{j_1}=a_{j_2j_1l_2}$ and $\overline{x}_{j_2}\neq a_{j_1j_2s_1},a_{j_1j_2s_2}(s_1,s_2\in\{1,2\},s_1\neq s_2)$, we have

$$\frac{dx_{j_2}}{dx_{j_1}}\Big|_{x^*_{j_1}=a_{j_2j_1l_2}}=\frac{a_{j_2j_10}}{a_{j_1j_20}}\frac{(x_{j_1}-a_{j_2j_1l_1})(x_{j_1}-a_{j_2j_1l_2})^2}{(\overline{x}_{j_2}-a_{j_1j_21})^3}\Big|_{x^*_{j_1}=a_{j_2j_1l_2}}=0,$$
$$\frac{d^2x_{j_2}}{dx^2_{j_1}}\Big|_{x^*_{j_1}=a_{j_2j_1l_2}}=0.$$

If

$$\frac{d^3x_{j_2}}{dx^3_{j_1}}\Big|_{x^*_{j_1}=a_{j_2j_1l_2}}=\frac{a_{j_2j_10}}{a_{j_1j_20}}\frac{(a_{j_2j_1l_2}-a_{j_2j_1l_1})}{(\overline{x}_{j_2}-a_{j_1j_21})^3}>0,$$

there is an increasing-inflection flow at $x^*_{j_1}=a_{j_2j_1l_2}$ in the x_{j_2}-direction. If

$$\frac{d^3x_{j_2}}{dx^3_{j_1}}\Big|_{x^*_{j_1}=a_{j_2j_1l_2}}=\frac{a_{j_2j_10}}{a_{j_1j_20}}\frac{(a_{j_2j_1l_2}-a_{j_2j_1l_1})}{(\overline{x}_{j_2}-a_{j_1j_21})^3}<0,$$

there is a decreasing-inflection flow at $x_{j_1}^* = a_{j_2 j_1 l_2}$ in the x_{j_2}-direction. Because of

$$\dot{x}_{j_1} = a_{j_1 j_2 0}(\overline{x}_{j_2} - a_{j_1 j_2 1})^3,$$

the flows at $x_{j_1}^* = a_{j_2 j_1 l_2}$ in the x_{j_1}-direction are positive and negative for $\dot{x}_{j_1} > 0$ and $\dot{x}_{j_1} < 0$, respectively.

Thus, the equilibrium of $x_{j_1}^* = a_{j_2 j_1 l_2}$ and $\overline{x}_{j_2} \neq a_{j_1 j_2 s_1}, a_{j_1 j_2 s_2} (s_1, s_2 \in \{1,2\}, s_1 \neq s_2)$ on the x_{j_2}-direction has the following properties.

- For $a_{j_1 j_2 0}(\overline{x}_{j_2} - a_{j_1 j_2 1})^3 > 0$ and $a_{j_2 j_1 0}(a_{j_2 j_1 l_2} - a_{j_2 j_1 l_1}) > 0$,

$$(\dot{x}_{j_1}, a_{j_2 j_1 l_2}) = \underbrace{(\text{pF}, \text{II})}_{\text{increasing-inflection flow } (+)}.$$

- For $a_{j_1 j_2 0}(\overline{x}_{j_2} - a_{j_1 j_2 1})^3 < 0$ and $a_{j_2 j_1 0}(a_{j_2 j_1 l_2} - a_{j_2 j_1 l_1}) > 0$,

$$(\dot{x}_{j_1}, a_{j_2 j_1 l_2}) = \underbrace{(\text{nF}, \text{DI})}_{\text{decreasing-inflection flow } (-)}.$$

- For $a_{j_1 j_2 0}(\overline{x}_{j_2} - a_{j_1 j_2 1})^3 > 0$ and $a_{j_2 j_1 0}(a_{j_2 j_1 l_2} - a_{j_2 j_1 l_1}) < 0$,

$$(\dot{x}_{j_1}, a_{j_2 j_1 l_2}) = \underbrace{(\text{pF}, \text{DI})}_{\text{decreasing-inflection flow } (+)}.$$

- For $a_{j_1 j_2 0}(\overline{x}_{j_2} - a_{j_1 j_2 1})^3 < 0$ and $a_{j_2 j_1 0}(a_{j_2 j_1 l_2} - a_{j_2 j_1 l_1}) < 0$,

$$(\dot{x}_{j_1}, a_{j_2 j_1 l_2}) = \underbrace{(\text{nF}, \text{II})}_{\text{increasing-inflection flow } (-)}.$$

Therefore, from the proof (II) of case (viii$_{1a}$) and the foregoing analysis, the equilibrium of $(x_{j_2}^*, x_{j_1}^*) = (a_{j_1 j_2 1}, a_{j_2 j_1 l_2})$ has the following properties as in Eqs. (4.81)–(4.84).

- For $a_{j_1 j_2 0} > 0$ and $a_{j_2 j_1 0}(a_{j_2 j_1 l_2} - a_{j_2 j_1 l_1}) > 0$,

$$(a_{j_1 j_2 1}, a_{j_2 j_1 l_2}) = \underbrace{(3^{\text{rd}}\text{UP}, \text{US})}_{\text{third-order up-parabola upper-saddle}}.$$

  The equilibrium of $(x_{j_2}^*, x_{j_1}^*) = (a_{j_1 j_2 1}, a_{j_2 j_1 l_2})$ is a (3^{rd}UP,US)-third-order-up-parabola upper-saddle.

- For $a_{j_1j_20} < 0$ and $a_{j_2j_10}(a_{j_2j_1l_2} - a_{j_2j_1l_1}) > 0$,

$$(a_{j_1j_21}, a_{j_2j_1l_2}) = \underbrace{(3^{\rm rd}{\rm DP, US})}_{\text{third-order down-parabola upper-saddle}}.$$

The equilibrium of $(x^*_{j_2}, x^*_{j_1}) = (a_{j_1j_21}, a_{j_2j_1l_2})$ is a ($3^{\rm rd}$DP,US)-third-order down-parabola upper-saddle.

- For $a_{j_1j_20} > 0$ and $a_{j_2j_10}(a_{j_2j_1l_2} - a_{j_2j_1l_1}) < 0$,

$$(a_{j_1j_21}, a_{j_2j_1l_2}) = \underbrace{(3^{\rm rd}{\rm DP, LS})}_{\text{third-order down-parabola lower-saddle}}.$$

The equilibrium of $(x^*_{j_2}, x^*_{j_1}) = (a_{j_1j_21}, a_{j_2j_1l_2})$ is a ($3^{\rm rd}$DP,LS)-third-order down-parabola lower-saddle.

- For $a_{j_1j_20} < 0$ and $a_{j_2j_10}(a_{j_2j_1l_2} - a_{j_2j_1l_1}) < 0$,

$$(a_{j_1j_21}, a_{j_2j_1l_2}) = \underbrace{(3^{\rm rd}{\rm UP, LS})}_{\text{third-order up-parabola lower-saddle}}.$$

The equilibrium of $(x^*_{j_2}, x^*_{j_1}) = (a_{j_1j_21}, a_{j_2j_1l_2})$ is a ($3^{\rm rd}$UP,LS)-third-order-up-parabola lower-saddle.

(ix) If

$$\begin{aligned}&\Delta_{j_1j_2} = 0 \text{ and } \Delta_{j_2j_1} = 0,\\ &a_{j_2j_1s_1} \neq a_{j_2j_1s_2}; a_{j_1j_2l_1} = a_{j_1j_2l_2} = a_{j_1j_21};\end{aligned}$$

then

$$\begin{aligned}&\dot{x}_{j_1} = a_{j_1j_20}(x_{j_2} - a_{j_1j_2s_1})(x_{j_2} - a_{j_1j_2s_2})^2,\\ &\dot{x}_{j_2} = a_{j_2j_10}(x_{j_1} - a_{j_2j_11})^3,\\ &j_1, j_2 \in \{1, 2\}; j_1 \neq j_2.\end{aligned}$$

In phase space,

$$\frac{dx_{j_1}}{dx_{j_2}} = \frac{a_{j_1j_20}}{a_{j_2j_10}} \frac{(x_{j_2} - a_{j_1j_2s_1})(x_{j_2} - a_{j_1j_2s_2})^2}{(x_{j_1} - a_{j_2j_11})^3}$$

and

$$\begin{aligned}&a_{j_2j_10}(x_{j_1} - a_{j_2j_11})^3 dx_{j_1}\\ &= a_{j_1j_20}\big[(x_{j_2} - a_{j_1j_2s_1})^3 + 2(a_{j_1j_2s_1} - a_{j_1j_2s_2})(x_{j_2} - a_{j_1j_2s_1})^2\\ &+(a_{j_1j_2s_1} - a_{j_1j_2s_2})^2(x_{j_2} - a_{j_1j_2s_1})\big] dx_{j_2}.\end{aligned}$$

With an initial condition (x_{j_10}, x_{j_20}), the first integral manifold is obtained, i.e.,

$$\begin{aligned}
&\frac{1}{4}a_{j_2j_10}\left[(x_{j_1}-a_{j_2j_11})^4-(x_{j_10}-a_{j_2j_11})^4\right]\\
&=a_{j_1j_20}\Big\{\frac{1}{4}\left[(x_{j_2}-a_{j_1j_2s_1})^4-(x_{j_20}-a_{j_1j_2s_1})^4\right]\\
&+\frac{2}{3}(a_{j_1j_2s_1}-a_{j_1j_2s_2})\left[(x_{j_1}-a_{j_1j_2s_1})^3-(x_{j_10}-a_{j_1j_2s_1})^3\right]\\
&+\frac{1}{2}(a_{j_1j_2s_1}-a_{j_1j_2s_2})^2\left[(x_{j_1}-a_{j_1j_2s_1})^2-(x_{j_10}-a_{j_1j_2s_1})^2\right]\Big\}.
\end{aligned}$$

(ix$_1$) Consider two cases (I) $x^*_{j_2}=a_{j_1j_2s_1}$ and (II) $x^*_{j_1}=a_{j_2j_11}$.

(I) At $x^*_{j_2}=a_{j_1j_2s_1}(s_1,s_2\in\{1,2\},s_1\neq s_2)$ and $x_{j_1}\neq a_{j_2j_11}$, in phase space, we have

$$\frac{dx_{j_1}}{dx_{j_2}}\bigg|_{x^*_{j_2}=a_{j_1j_2s_1}}=\frac{a_{j_1j_20}}{a_{j_2j_10}}\frac{(x_{j_2}-a_{j_1j_2s_1})(x_{j_2}-a_{j_1j_2s_2})^2}{(\overline{x}_{j_1}-a_{j_2j_11})^3}\bigg|_{x^*_{j_2}=a_{j_1j_2s_1}}=0.$$

If

$$\frac{d^2x_{j_1}}{dx^2_{j_2}}\bigg|_{x^*_{j_2}=a_{j_1j_2s_1}}=\frac{a_{j_1j_20}}{a_{j_2j_10}}\frac{(a_{j_1j_2s_1}-a_{j_1j_2s_2})^2}{(\overline{x}_{j_1}-a_{j_2j_11})^3}>0,$$

there is an up-parabola flow at $x^*_{j_2}=a_{j_1j_2s_1}$ in the x_{j_1}-direction. If

$$\frac{d^2x_{j_1}}{dx^2_{j_2}}\bigg|_{x^*_{j_2}=a_{j_1j_2s_1}}=\frac{a_{j_1j_20}}{a_{j_2j_10}}\frac{(a_{j_1j_2s_1}-a_{j_1j_2s_2})^2}{(\overline{x}_{j_1}-a_{j_2j_11})^3}<0,$$

there is a down-parabola flow at $x^*_{j_2}=a_{j_1j_2s_1}$ in the x_{j_1}-direction. Because of

$$\dot{x}_{j_2}=a_{j_2j_10}(\overline{x}_{j_1}-a_{j_2j_11})^3,$$

the flows at $x^*_{j_2}=a_{j_1j_2s_1}$ in the x_{j_2}-direction are positive and negative for $\dot{x}_{j_2}>0$ and $\dot{x}_{j_2}<0$, respectively.

Thus, the equilibrium of $x^*_{j_2}=a_{j_1j_2s_1}(s_1,s_2\in\{1,2\},s_1\neq s_2)$ on the x_{j_1}-direction has the following properties.

- For $a_{j_1j_20}(a_{j_1j_2s_1}-a_{j_1j_2s_2})^2>0$ and $a_{j_2j_10}(\overline{x}_{j_1}-a_{j_2j_11})^3>0$,

$$(a_{j_1j_2s_1},\dot{x}_{j_2})=\underbrace{(\text{UP},\text{pF})}_{\text{up-parabola flow }(+)}.$$

- For $a_{j_1j_20}(a_{j_1j_2s_1}-a_{j_1j_2s_2})^2<0$ and $a_{j_2j_10}(\overline{x}_{j_1}-a_{j_2j_11})^3>0$,

$$(a_{j_1j_2s_1},\dot{x}_{j_2}) = \underbrace{(\text{DP},\text{pF})}_{\text{down-parabola flow }(+)}.$$

- For $a_{j_1j_20}(a_{j_1j_2s_1}-a_{j_1j_2s_2})^2>0$ and $a_{j_2j_10}(\overline{x}_{j_1}-a_{j_2j_11})^3<0$,

$$(a_{j_1j_2s_1},\dot{x}_{j_2}) = \underbrace{(\text{DP},\text{nF})}_{\text{down-parabola flow }(-)}.$$

- For $a_{j_1j_20}(a_{j_1j_2s_1}-a_{j_1j_2s_2})^2<0$ and $a_{j_2j_10}(\overline{x}_{j_1}-a_{j_2j_11})^3<0$,

$$(a_{j_1j_2s_1},\dot{x}_{j_2}) = \underbrace{(\text{UP},\text{nF})}_{\text{up-parabola flow }(-)}.$$

(II) In phase space, at $x_{j_1}^*=a_{j_2j_1l_2}$ and $\overline{x}_{j_2}\neq a_{j_1j_2s_1},a_{j_1j_2s_2}(s_1,s_2\in\{1,2\},s_1\neq s_2)$, we have

$$\begin{aligned}
&\frac{dx_{j_2}}{dx_{j_1}}\Big|_{x_{j_1}^*=a_{j_2j_11}} = \frac{a_{j_2j_10}}{a_{j_1j_20}}\frac{(x_{j_1}-a_{j_2j_11})^3}{(\overline{x}_{j_2}-a_{j_1j_2s_1})(\overline{x}_{j_2}-a_{j_1j_2s_2})^2}\Big|_{x_{j_1}^*=a_{j_2j_11}}=0,\\
&\frac{d^2x_{j_2}}{dx_{j_1}^2}\Big|_{x_{j_1}^*=a_{j_2j_11}} = \frac{d^3x_{j_2}}{dx_{j_1}^3}\Big|_{x_{j_1}^*=a_{j_2j_11}}=0.
\end{aligned}$$

If

$$\frac{d^4x_{j_2}}{dx_{j_1}^4}\Big|_{x_{j_1}^*=a_{j_2j_11}} = \frac{a_{j_2j_10}}{a_{j_1j_20}}\frac{6}{(\overline{x}_{j_2}-a_{j_1j_2s_1})(\overline{x}_{j_2}-a_{j_1j_2s_2})^2}>0,$$

there is a third-order up-parabola flow at $x_{j_1}^*=a_{j_2j_1l_2}$ in the x_{j_2}-direction. If

$$\frac{d^4x_{j_2}}{dx_{j_1}^4}\Big|_{x_{j_1}^*=a_{j_2j_11}} = \frac{a_{j_2j_10}}{a_{j_1j_20}}\frac{6}{(\overline{x}_{j_2}-a_{j_1j_2s_1})(\overline{x}_{j_2}-a_{j_1j_2s_2})^2}<0,$$

there is a third-order down-parabola flow at $x_{j_1}^*=a_{j_2j_1l_2}$ in the x_{j_2}-direction. Because of

$$\dot{x}_{j_1}=a_{j_1j_20}(\overline{x}_{j_2}-a_{j_1j_2s_1})(\overline{x}_{j_2}-a_{j_1j_2s_2})^2,$$

the flows at $x_{j_1}^*=a_{j_2j_1l_2}$ in the x_{j_1}-direction are positive and negative for $\dot{x}_{j_1}>0$ and $\dot{x}_{j_1}<0$, respectively.

Thus, the equilibrium of $x_{j_1}^*=a_{j_2j_1l_2}$ and $\overline{x}_{j_2}\neq a_{j_1j_2s_1},a_{j_1j_2s_2}(s_1,s_2\in\{1,2\},s_1\neq s_2)$ on the x_{j_2}-direction has the following properties.

- For $a_{j_1j_20}(\overline{x}_{j_2} - a_{j_1j_2s_1})(\overline{x}_{j_2} - a_{j_1j_2s_2})^2 > 0$ and $a_{j_2j_10} > 0$,

$$(\dot{x}_{j_1}, a_{j_2j_11}) = \underbrace{(\text{pF}, 3^{\text{rd}}\text{UP})}_{\text{third-order up-parabola flow }(+)}.$$

- For $a_{j_1j_20}(\overline{x}_{j_2} - a_{j_1j_2s_1})(\overline{x}_{j_2} - a_{j_1j_2s_2})^2 < 0$ and $a_{j_2j_10} > 0$,

$$(\dot{x}_{j_1}, a_{j_2j_11}) = \underbrace{(\text{nF}, 3^{\text{rd}}\text{DP})}_{\text{third-order down-parabola flow }(-)}.$$

- For $a_{j_1j_20}(\overline{x}_{j_2} - a_{j_1j_2s_1})(\overline{x}_{j_2} - a_{j_1j_2s_2})^2 > 0$ and $a_{j_2j_10} < 0$,

$$(\dot{x}_{j_1}, a_{j_2j_11}) = \underbrace{(\text{pF}, 3^{\text{rd}}\text{DP})}_{\text{third-order down-parabola flow }(+)}.$$

- For $a_{j_1j_20}(\overline{x}_{j_2} - a_{j_1j_2s_1})(\overline{x}_{j_2} - a_{j_1j_2s_2})^2 < 0$ and $a_{j_2j_10} < 0$,

$$(\dot{x}_{j_1}, a_{j_2j_11}) = \underbrace{(\text{nF}, 3^{\text{rd}}\text{UP})}_{\text{third-order up-parabola flow }(-)}.$$

Therefore, from cases (I) and (II), the equilibrium of $(x_{j_2}^*, x_{j_1}^*) = (a_{j_1j_2s_1}, a_{j_2j_11})$ $(s_1, s_2 \in \{1,2\}, s_1 \neq s_2)$ has the following properties as in Eqs. (4.88)–(4.91).

- For $a_{j_1j_20}(a_{j_1j_2s_1} - a_{j_1j_2s_2})^2 > 0$ and $a_{j_2j_10} > 0$,

$$(a_{j_1j_2s_1}, a_{j_2j_11}) = \underbrace{(\text{UP}_+, 3^{\text{rd}}\text{UP}_+)}_{\text{third-order positive saddle}}.$$

  The equilibrium of $(x_{j_2}^*, x_{j_1}^*) = (a_{j_1j_2s_1}, a_{j_2j_11})$ is a $(\text{UP}_+, 3^{\text{rd}}\text{UP}_+)$-third-order positive saddle.
- For $a_{j_1j_20}(a_{j_1j_2s_1} - a_{j_1j_2s_2})^2 < 0$ and $a_{j_2j_10} > 0$,

$$(a_{j_1j_2s_1}, a_{j_2j_11}) = \underbrace{(\text{DP}_+, 3^{\text{rd}}\text{DP}_-)}_{\text{third-order CCW center}}.$$

  The equilibrium of $(x_{j_2}^*, x_{j_1}^*) = (a_{j_1j_2s_1}, a_{j_2j_11})$ is a $(\text{DP}_+, 3^{\text{rd}}\text{DP}_-)$-third-order-counter-clockwise center.

- For $a_{j_1j_20}(a_{j_1j_2s_1} - a_{j_1j_2s_2})^2 > 0$ and $a_{j_2j_10} < 0$,

$$(a_{j_1j_2s_1}, a_{j_2j_11}) = \underbrace{(\mathrm{DP}_-, 3^{\mathrm{rd}}\mathrm{DP}_+)}_{\text{third-order CW center}}.$$

The equilibrium of $(x_{j_2}^*, x_{j_1}^*) = (a_{j_1j_2s_1}, a_{j_2j_11})$ is a $(\mathrm{DP}_-, 3^{\mathrm{rd}}\mathrm{DP}_+)$-third-order-clockwise center.

- For $a_{j_1j_20}(a_{j_1j_2s_1} - a_{j_1j_2s_2})^2 < 0$ and $a_{j_2j_10} < 0$,

$$(a_{j_1j_2s_1}, a_{j_2j_11}) = \underbrace{(\mathrm{UP}_-, 3^{\mathrm{rd}}\mathrm{UP}_-)}_{\text{third-order negative saddle}}.$$

The equilibrium of $(x_{j_2}^*, x_{j_1}^*) = (a_{j_1j_2s_1}, a_{j_2j_11})$ is a $(\mathrm{UP}_-, 3^{\mathrm{rd}}\mathrm{UP}_-)$-third-order negative saddle.

(ix$_2$) In phase space, at $x_{j_2}^* = a_{j_1j_2s_2} (s_1, s_2 \in \{1,2\}, s_1 \neq s_2)$ and $x_{j_1} \neq a_{j_2j_11}$, we have

$$\frac{dx_{j_1}}{dx_{j_2}}\bigg|_{x_{j_2}^* = a_{j_1j_2s_2}} = \frac{a_{j_1j_20}}{a_{j_2j_10}} \frac{(x_{j_2} - a_{j_1j_2s_1})(x_{j_2} - a_{j_1j_2s_2})^2}{(\overline{x}_{j_1} - a_{j_2j_11})^3}\bigg|_{x_{j_2}^* = a_{j_1j_2s_2}} = 0,$$

$$\frac{d^2x_{j_1}}{dx_{j_2}^2}\bigg|_{x_{j_2}^* = a_{j_1j_2s_2}} = 0.$$

If

$$\frac{d^3x_{j_1}}{dx_{j_2}^3}\bigg|_{x_{j_2}^* = a_{j_1j_2s_2}} = \frac{a_{j_1j_20}}{a_{j_2j_10}} \frac{2(a_{j_1j_2s_2} - a_{j_1j_2s_1})}{(\overline{x}_{j_1} - a_{j_2j_11})^3} > 0,$$

there is an increasing-inflection flow at $x_{j_2}^* = a_{j_1j_2s_2}$ in the x_{j_1}-direction. If

$$\frac{d^3x_{j_1}}{dx_{j_2}^3}\bigg|_{x_{j_2}^* = a_{j_1j_2s_2}} = \frac{a_{j_1j_20}}{a_{j_2j_10}} \frac{2(a_{j_1j_2s_2} - a_{j_1j_2s_1})}{(\overline{x}_{j_1} - a_{j_2j_11})^3} < 0,$$

there is a decreasing-inflection flow at $x_{j_2}^* = a_{j_1j_2s_2}$ in the x_{j_1}-direction. Because of

$$\dot{x}_{j_2} = a_{j_2j_10}(\overline{x}_{j_1} - a_{j_2j_11})^3,$$

the flows at $x_{j_2}^* = a_{j_1j_2s_2}$ in the x_{j_2}-direction are positive and negative for $\dot{x}_{j_2} > 0$ and $\dot{x}_{j_2} < 0$, respectively.

Thus, the equilibrium of $x_{j_2}^* = a_{j_1j_2s_2} (s_1, s_2 \in \{1,2\}, s_1 \neq s_2)$ and $x_{j_1} \neq a_{j_2j_11}$ on the x_{j_2}-direction has the following properties.

- For $a_{j_1j_20}(a_{j_1j_2s_2} - a_{j_1j_2s_1}) > 0$ and $a_{j_2j_10}(\overline{x}_{j_1} - a_{j_2j_11})^3 > 0$,

$$(a_{j_1j_2s_2}, \dot{x}_{j_2}) = \underbrace{(\text{II}, \text{pF})}_{\text{increasing-inflection flow } (+)}.$$

- For $a_{j_1j_20}(a_{j_1j_2s_2} - a_{j_1j_2s_1}) < 0$ and $a_{j_2j_10}(\overline{x}_{j_1} - a_{j_2j_11})^3 > 0$,

$$(a_{j_1j_2s_2}, \dot{x}_{j_2}) = \underbrace{(\text{DI}, \text{pF})}_{\text{decreasing-inflection flow } (+)}.$$

- For $a_{j_1j_20}(a_{j_1j_2s_2} - a_{j_1j_2s_1}) > 0$ and $a_{j_2j_10}(\overline{x}_{j_1} - a_{j_2j_11})^3 < 0$,

$$(a_{j_1j_2s_2}, \dot{x}_{j_2}) = \underbrace{(\text{DI}, \text{nF})}_{\text{decreasing-inflection flow } (-)}.$$

- For $a_{j_1j_20}(a_{j_1j_2s_2} - a_{j_1j_2s_1}) < 0$ and $a_{j_2j_10}(\overline{x}_{j_1} - a_{j_2j_11})^3 < 0$,

$$(a_{j_1j_2s_2}, \dot{x}_{j_2}) = \underbrace{(\text{II}, \text{nF})}_{\text{increasing-inflection flow } (-)}.$$

Therefore, from the proof (II) of case (ix$_1$) and the foregoing analysis, the equilibrium of $(x^*_{j_2}, x^*_{j_1}) = (a_{j_1j_2s_2}, a_{j_2j_11})$ has the following properties as in (4.92)–(4.95).

- For $a_{j_1j_20}(a_{j_1j_2s_2} - a_{j_1j_2s_1}) > 0$ and $a_{j_2j_10} > 0$,

$$(a_{j_1j_2s_2}, a_{j_2j_11}) = \underbrace{(\text{US}, 3^{\text{rd}}\text{UP})}_{\text{third-order up-parabola upper-saddle}}.$$

  The equilibrium of $(x^*_{j_2}, x^*_{j_1}) = (a_{j_1j_2s_2}, a_{j_2j_11})$ is a (US,3rdUP)-third-order-up-parabola upper-saddle.
- For $a_{j_1j_20}(a_{j_1j_2s_2} - a_{j_1j_2s_1}) < 0$ and $a_{j_2j_10} > 0$,

$$(a_{j_1j_2s_2}, a_{j_2j_11}) = \underbrace{(\text{LS}, 3^{\text{rd}}\text{DP})}_{\text{third-order down-parabola lower-saddle}}.$$

  The equilibrium of $(x^*_{j_2}, x^*_{j_1}) = (a_{j_1j_2s_2}, a_{j_2j_11})$ is an (LS,3rdDP)-third-order down-parabola lower-saddle.

- For $a_{j_1j_20}(a_{j_1j_2s_2} - a_{j_1j_2s_1}) > 0$ and $a_{j_2j_10} < 0$,

$$(a_{j_1j_2s_2}, a_{j_2j_11}) = \underbrace{(\text{US}, 3^{\text{rd}}\text{DP})}_{\text{third-order down-parabola upper-saddle}}.$$

The equilibrium of $(x^*_{j_2}, x^*_{j_1}) = (a_{j_1j_2s_2}, a_{j_2j_11})$ is a (US,3rdDP)-third-order down-parabola upper-saddle.

- For $a_{j_1j_20}(a_{j_1j_2s_2} - a_{j_1j_2s_1}) < 0$ and $a_{j_2j_10} < 0$,

$$(a_{j_1j_2s_2}, a_{j_2j_11}) = \underbrace{(\text{LS}, 3^{\text{rd}}\text{UP})}_{\text{third-order up-parabola lower-saddle}}.$$

The equilibrium of $(x^*_{j_2}, x^*_{j_1}) = (a_{j_1j_2s_2}, a_{j_2j_11})$ is an (LS,3rdUP)-third-order-up-parabola lower-saddle.

(x) If

$$\begin{aligned}
&\Delta_{j_1j_2} = 0 \text{ and } \Delta_{j_2j_1} = 0,\\
&a_{j_2j_1s_1} = a_{j_2j_1s_2} = a_{j_2j_11}; a_{j_1j_2l_1} = a_{j_1j_2l_2} = a_{j_1j_21};
\end{aligned}$$

then

$$\begin{aligned}
&\dot{x}_{j_1} = a_{j_1j_20}(x_{j_2} - a_{j_1j_21})^3,\\
&\dot{x}_{j_2} = a_{j_2j_10}(x_{j_1} - a_{j_2j_11})^3,\\
&j_1, j_2 \in \{1,2\}; j_1 \neq j_2.
\end{aligned}$$

In phase space,

$$\frac{dx_{j_1}}{dx_{j_2}} = \frac{a_{j_1j_20}}{a_{j_2j_10}} \frac{(x_{j_2} - a_{j_1j_21})^3}{(x_{j_1} - a_{j_2j_11})^3},$$

and

$$a_{j_2j_10}(x_{j_1} - a_{j_2j_11})^3 dx_{j_1} = a_{j_1j_20}(x_{j_2} - a_{j_1j_21})^3 dx_{j_2}.$$

With an initial condition (x_{j_10}, x_{j_20}), the first integral manifold is obtained, i.e.,

$$\begin{aligned}
&\frac{1}{4} a_{j_2j_10}\left[(x_{j_1} - a_{j_2j_11})^4 - (x_{j_10} - a_{j_2j_11})^4\right]\\
&= \frac{1}{4} a_{j_1j_20}\left[(x_{j_2} - a_{j_1j_21})^4 - (x_{j_20} - a_{j_1j_21})^4\right].
\end{aligned}$$

Consider two cases (I) $x^*_{j_2} = a_{j_1j_21}$ and (II) $x^*_{j_1} = a_{j_2j_11}$.

(I) In phase space, at $x_{j_2}^* = a_{j_1 j_2 1}$ and $x_{j_1} \neq a_{j_2 j_1 1}$, we have

$$\frac{dx_{j_1}}{dx_{j_2}}\Big|_{x_{j_2}^* = a_{j_1 j_2 1}} = \frac{a_{j_1 j_2 0}\,(x_{j_2} - a_{j_1 j_2 1})^3}{a_{j_2 j_1 0}\,(x_{j_1} - a_{j_2 j_1 1})^3}\Big|_{x_{j_2}^* = a_{j_1 j_2 1}} = 0,$$
$$\frac{d^2 x_{j_1}}{dx_{j_2}^2}\Big|_{x_{j_2}^* = a_{j_1 j_2 1}} = \frac{d^3 x_{j_1}}{dx_{j_2}^3}\Big|_{x_{j_2}^* = a_{j_1 j_2 1}} = 0.$$

If

$$\frac{d^4 x_{j_1}}{dx_{j_2}^4}\Big|_{x_{j_2}^* = a_{j_1 j_2 1}} = \frac{a_{j_1 j_2 0}}{a_{j_2 j_1 0}} \frac{6}{(\bar{x}_{j_1} - a_{j_2 j_1 1})^3} > 0,$$

there is a third-order up-parabola flow at $x_{j_2}^* = a_{j_1 j_2 1}$ in the x_{j_1}-direction. If

$$\frac{d^4 x_{j_1}}{dx_{j_2}^4}\Big|_{x_{j_2}^* = a_{j_1 j_2 1}} = \frac{a_{j_1 j_2 0}}{a_{j_2 j_1 0}} \frac{6}{(\bar{x}_{j_1} - a_{j_2 j_1 1})^3} < 0,$$

there is a third-order down-parabola flow at $x_{j_2}^* = a_{j_1 j_2 1}$ in the x_{j_1}-direction. Because of

$$\dot{x}_{j_2} = a_{j_2 j_1 0}(\bar{x}_{j_1} - a_{j_2 j_1 1})^3,$$

the flows at $x_{j_2}^* = a_{j_1 j_2 1}$ in the x_{j_2}-direction are positive and negative for $\dot{x}_{j_2} > 0$ and $\dot{x}_{j_2} < 0$, respectively.

Thus, the equilibrium of $x_{j_2}^* = a_{j_1 j_2 1}$ on the x_{j_1}-direction has the following properties.

- For $a_{j_1 j_2 0} > 0$ and $a_{j_2 j_1 0}(\bar{x}_{j_1} - a_{j_2 j_1 1})^3 > 0$,

$$(a_{j_1 j_2 s_1}, \dot{x}_{j_2}) = \underbrace{(3^{\text{rd}}\text{UP}, \text{pF})}_{\text{third-order up-parabola flow }(+)}.$$

- For $a_{j_1 j_2 0} < 0$ and $a_{j_2 j_1 0}(\bar{x}_{j_1} - a_{j_2 j_1 1})^3 > 0$,

$$(a_{j_1 j_2 s_1}, \dot{x}_{j_2}) = \underbrace{(3^{\text{rd}}\text{DP}, \text{pF})}_{\text{third-order down-parabola flow }(+)}.$$

- For $a_{j_1j_20}>0$ and $a_{j_2j_10}(\overline{x}_{j_1}-a_{j_2j_11})^3<0$,

$$(a_{j_1j_2s_1},\dot{x}_{j_2})=\underbrace{(3^{\text{rd}}\text{DP},\text{nF})}_{\text{third-order down-parabola flow }(-)}.$$

- For $a_{j_1j_20}<0$ and $a_{j_2j_10}(\overline{x}_{j_1}-a_{j_2j_11})^3<0$,

$$(a_{j_1j_2s_1},\dot{x}_{j_2})=\underbrace{(3^{\text{rd}}\text{UP},\text{nF})}_{\text{third-order up-parabola flow }(-)}.$$

(II) In phase space, at $x^*_{j_1}=a_{j_2j_11}$ and $\overline{x}_{j_2}\neq a_{j_1j_21}$, we have

$$\begin{aligned}
&\frac{dx_{j_2}}{dx_{j_1}}\bigg|_{x^*_{j_1}=a_{j_2j_11}}=\frac{a_{j_2j_10}\,(x_{j_1}-a_{j_2j_11})^3}{a_{j_1j_20}\,(x_{j_2}-a_{j_1j_21})^3}\bigg|_{x^*_{j_1}=a_{j_2j_11}}=0,\\
&\frac{d^2x_{j_2}}{dx^2_{j_1}}\bigg|_{x^*_{j_1}=a_{j_2j_11}}=\frac{d^3x_{j_2}}{dx^3_{j_1}}\bigg|_{x^*_{j_1}=a_{j_2j_11}}=0.
\end{aligned}$$

If

$$\frac{d^4x_{j_2}}{dx^4_{j_1}}\bigg|_{x^*_{j_1}=a_{j_2j_11}}=\frac{a_{j_1j_20}}{a_{j_2j_10}}\frac{6}{(\overline{x}_{j_2}-a_{j_1j_2s_2})^3}>0,$$

there is a third-order up-parabola flow at $x^*_{j_1}=a_{j_2j_11}$ in the x_{j_2}-direction. If

$$\frac{d^4x_{j_2}}{dx^4_{j_1}}\bigg|_{x^*_{j_1}=a_{j_2j_11}}=\frac{a_{j_1j_20}}{a_{j_2j_10}}\frac{6}{(\overline{x}_{j_2}-a_{j_1j_2s_2})^3}<0,$$

there is a third-order down-parabola flow at $x^*_{j_1}=a_{j_2j_11}$ in the x_{j_2}-direction. Because of

$$\dot{x}_{j_1}=a_{j_1j_20}(\overline{x}_{j_2}-a_{j_1j_21})^3,$$

the flows at $x^*_{j_1}=a_{j_2j_11}$ in the x_{j_1}-direction are positive and negative for $\dot{x}_{j_1}>0$ and $\dot{x}_{j_1}<0$, respectively.

Thus, the equilibrium of $x^*_{j_1}=a_{j_2j_11}$ and $\overline{x}_{j_2}\neq a_{j_1j_21}$ on the x_{j_2}-direction has the following properties.

- For $a_{j_1j_2 0}(\overline{x}_{j_2} - a_{j_1j_2 1})^3 > 0$ and $a_{j_2j_1 0} > 0$,

$$(\dot{x}_{j_1}, a_{j_2j_1 1}) = \underbrace{(\text{pF}, 3^{\text{rd}}\text{UP})}_{\text{third-order up-parabola flow } (+)}.$$

- For $a_{j_1j_2 0}(\overline{x}_{j_2} - a_{j_1j_2 1})^3 < 0$ and $a_{j_2j_1 0} > 0$,

$$(\dot{x}_{j_1}, a_{j_2j_1 1}) = \underbrace{(\text{nF}, 3^{\text{rd}}\text{DP})}_{\text{third-order down-parabola flow } (-)}.$$

- For $a_{j_1j_2 0}(\overline{x}_{j_2} - a_{j_1j_2 1})^3 > 0$ and $a_{j_2j_1 0} < 0$,

$$(\dot{x}_{j_1}, a_{j_2j_1 1}) = \underbrace{(\text{pF}, 3^{\text{rd}}\text{DP})}_{\text{third-order down-parabola flow } (+)}.$$

- For $a_{j_1j_2 0}(\overline{x}_{j_2} - a_{j_1j_2 1})^3 < 0$ and $a_{j_2j_1 0} < 0$,

$$(\dot{x}_{j_1}, a_{j_2j_1 1}) = \underbrace{(\text{nF}, 3^{\text{rd}}\text{UP})}_{\text{third-order up-parabola flow } (-)}.$$

Therefore, from cases (I) and (II), the equilibrium of $(x^*_{j_2}, x^*_{j_1}) = (a_{j_1j_2 1}, a_{j_2j_1 1})$ has the following properties as in Eqs. (4.99)–(4.102).

- For $a_{j_1j_2 0} > 0$ and $a_{j_2j_1 0} > 0$,

$$(a_{j_1j_2 1}, a_{j_2j_1 1}) = \underbrace{(3^{\text{rd}}\text{UP}_+, 3^{\text{rd}}\text{UP}_+)}_{\text{double third-order positive saddle}}.$$

  The equilibrium of $(x^*_{j_2}, x^*_{j_1}) = (a_{j_1j_2 1}, a_{j_2j_1 1})$ is a $(3^{\text{rd}}\text{UP}_+, 3^{\text{rd}}\text{UP}_+)$-double third-order positive saddle.
- For $a_{j_1j_2 0} < 0$ and $a_{j_2j_1 0} > 0$,

$$(a_{j_1j_2 1}, a_{j_2j_1 1}) = \underbrace{(3^{\text{rd}}\text{DP}_+, 3^{\text{rd}}\text{DP}_-)}_{\text{doubel third-order CCW center}}.$$

  The equilibrium of $(x^*_{j_2}, x^*_{j_1}) = (a_{j_1j_2 1}, a_{j_2j_1 1})$ is a $(3^{\text{rd}}\text{DP}_+, 3^{\text{rd}}\text{DP}_-)$-double third-order-counter-clockwise center.

- For $a_{j_1j_20} > 0$ and $a_{j_2j_10} < 0$,

$$(a_{j_1j_21}, a_{j_2j_11}) = \underbrace{(3^{\text{rd}}\text{DP}_-, 3^{\text{rd}}\text{DP}_+)}_{\text{double third-order CW center}}.$$

The equilibrium of $(x^*_{j_2}, x^*_{j_1}) = (a_{j_1j_21}, a_{j_2j_11})$ is a $(3^{\text{rd}}\text{DP}_-, 3^{\text{rd}}\text{DP}_+)$-double third-order-clockwise center.

- For $a_{j_1j_20} < 0$ and $a_{j_2j_10} < 0$,

$$(a_{j_1j_21}, a_{j_2j_11}) = \underbrace{(3^{\text{rd}}\text{UP}_-, 3^{\text{rd}}\text{UP}_-)}_{\text{double third-order negative saddle}}.$$

The equilibrium of $(x^*_{j_2}, x^*_{j_1}) = (a_{j_1j_21}, a_{j_2j_11})$ is a $(3^{\text{rd}}\text{UP}_-, 3^{\text{rd}}\text{UP}_-)$-double third-order negative saddle.

(xi) If

$$\Delta_{j_1j_2} = 0 \text{ and } \Delta_{j_2j_1} > 0, a_{j_1j_2s_1} \neq a_{j_1j_2s_2},$$

then

$$\begin{aligned}
&\dot{x}_{j_2} = a_{j_2j_10}(x_{j_1} - a_{j_2j_11})(x_{j_1} - a_{j_2j_12})(x_{j_1} - a_{j_2j_13}),\\
&\dot{x}_{j_1} = a_{j_1j_20}(x_{j_2} - a_{j_1j_2s_1})(x_{j_2} - a_{j_1j_2s_2})^2,\\
&j_1, j_2 \in \{1,2\}; j_1 \neq j_2.
\end{aligned}$$

In phase space,

$$\frac{dx_{j_1}}{dx_{j_2}} = \frac{a_{j_1j_20}}{a_{j_2j_10}} \frac{(x_{j_2} - a_{j_1j_2s_1})(x_{j_2} - a_{j_1j_2s_2})^2}{(x_{j_1} - a_{j_2j_1l_1})(x_{j_1} - a_{j_2j_1l_2})(x_{j_1} - a_{j_2j_1l_3})},$$

and

$$\begin{aligned}
&a_{j_2j_10}\left[(x_{j_1} - a_{j_2j_1l_1})^3 + \sum\nolimits_{l_2=1, l_2\neq l_1}^{3} (a_{j_2j_1l_1} - a_{j_2j_1l_2})(x_{j_1} - a_{j_2j_11})^2\right.\\
&\left.+\prod\nolimits_{l_2=1, l_2\neq l_1}^{3} (a_{j_2j_1l_1} - a_{j_2j_1l_2})(x_{j_1} - a_{j_2j_11})\right] dx_{j_1}\\
&= a_{j_1j_20}\left[(x_{j_2} - a_{j_1j_2s_1})^3 + 2(a_{j_1j_2s_1} - a_{j_1j_2s_2})(x_{j_2} - a_{j_1j_2s_1})^2\right.\\
&\left.+(a_{j_1j_2s_1} - a_{j_1j_2s_2})^2(x_{j_2} - a_{j_1j_2s_1})\right] dx_{j_2}.
\end{aligned}$$

With an initial condition (x_{j_10}, x_{j_20}), the first integral manifold is generated, i.e.,

$$
\begin{aligned}
&a_{j_2j_10}\Big\{\frac{1}{4}\left[(x_{j_1}-a_{j_2j_1l_1})^4-(x_{j_10}-a_{j_2j_1l_1})^4\right]\\
&+\frac{1}{3}\sum\nolimits_{l_2=1,l_2\neq l_1}^{3}(a_{j_2j_1l_1}-a_{j_2j_1l_2})\left[(x_{j_1}-a_{j_2j_1l_1})^3-(x_{j_10}-a_{j_2j_1l_1})^3\right]\\
&+\frac{1}{2}\prod\nolimits_{l_2=1,l_2\neq l_1}^{3}(a_{j_2j_1l_1}-a_{j_2j_1l_2})\left[(x_{j_1}-a_{j_2j_1l_1})^2-(x_{j_10}-a_{j_2j_1l_1})^2\right]\Big\}\\
&=a_{j_1j_20}\Big\{\frac{1}{4}\left[(x_{j_1}-a_{j_1j_2s_1})^4-(x_{j_10}-a_{j_1j_2s_1})^4\right]\\
&+\frac{2}{3}(a_{j_1j_2s_1}-a_{j_1j_2s_2})\left[(x_{j_2}-a_{j_1j_2s_1})^3-(x_{j_20}-a_{j_1j_2s_1})^3\right]\\
&+\frac{1}{2}\left[(a_{j_1j_2s_1}-a_{j_1j_2s_2})^2\right]\left[(x_{j_2}-a_{j_1j_2s_1})^2-(x_{j_20}-a_{j_1j_2s_1})^2\right]\Big\}.
\end{aligned}
$$

(xi_1) Consider two cases (I) $x^*_{j_2}=a_{j_1j_2s_1}$ and (II)$x^*_{j_1}=a_{j_2j_1l_1}$.

(I) In phase space, at $x^*_{j_2}=a_{j_1j_2s_1}(s_1,s_2\in\{1,2\},s_1\neq s_2)$ and $\bar{x}_{j_1}\neq a_{j_2j_1l_1}(l_1=1,2,3)$, we have

$$
\frac{dx_{j_1}}{dx_{j_2}}\Big|_{x^*_{j_2}=a_{j_1j_2s_1}}=\frac{a_{j_1j_20}}{a_{j_2j_10}}\frac{(x_{j_2}-a_{j_1j_2s_1})(x_{j_2}-a_{j_1j_2s_2})^2}{\prod_{l_1=1}^{3}(\bar{x}_{j_1}-a_{j_2j_1l_1})}\Big|_{x^*_{j_2}=a_{j_1j_2s_1}}=0.
$$

If

$$
\frac{d^2x_{j_1}}{dx^2_{j_2}}\Big|_{x^*_{j_2}=a_{j_1j_2s_1}}=\frac{a_{j_1j_20}}{a_{j_2j_10}}\frac{(a_{j_1j_2s_1}-a_{j_1j_2s_2})^2}{\prod_{l_1=1}^{3}(\bar{x}_{j_1}-a_{j_2j_1l_1})}>0,
$$

there is an up-parabola flow at $x^*_{j_2}=a_{j_1j_2s_1}$ in the x_{j_1}-direction. If

$$
\frac{d^2x_{j_1}}{dx^2_{j_2}}\Big|_{x^*_{j_2}=a_{j_1j_2s_1}}=\frac{a_{j_1j_20}}{a_{j_2j_10}}\frac{(a_{j_1j_2s_1}-a_{j_1j_2s_2})^2}{\prod_{l_1=1}^{3}(\bar{x}_{j_1}-a_{j_2j_1l_1})}<0,
$$

there is a down-parabola flow at $x^*_{j_2}=a_{j_1j_2s_1}$ in the x_{j_1}-direction. Because of

$$
\dot{x}_{j_2}=a_{j_2j_10}\prod\nolimits_{l_1=1}^{3}(\bar{x}_{j_1}-a_{j_2j_1l_1}),
$$

the flows at $x^*_{j_2}=a_{j_1j_2s_1}$ in the x_{j_2}-direction are positive and negative for $\dot{x}_{j_2}>0$ and $\dot{x}_{j_2}<0$, respectively.

Thus, the equilibrium of $x^*_{j_2}=a_{j_1j_2s_1}(s_1,s_2\in\{1,2\},s_1\neq s_2)$ on the x_{j_2}-direction has the following properties.

- For $a_{j_1j_20}(a_{j_1j_2s_1}-a_{j_1j_2s_2})^2>0$ and $a_{j_2j_10}\prod_{l_1=1}^{3}(\bar{x}_{j_1}-a_{j_2j_1l_1})>0$,

$$
(a_{j_1j_2s_1},\dot{x}_{j_2})=\underbrace{(\mathrm{UP},\mathrm{pF})}_{\text{up-parabola flow }(+)}.
$$

- For $a_{j_1j_20}(a_{j_1j_2s_1}-a_{j_1j_2s_2})^2<0$ and $a_{j_2j_10}\prod_{l_1=1}^{3}(\overline{x}_{j_1}-a_{j_2j_1l_1})>0$,

$$(a_{j_1j_2s_1},\dot{x}_{j_2})=\underbrace{(\text{DP},\text{pF})}_{\text{down-parabola flow }(+)}.$$

- For $a_{j_1j_20}(a_{j_1j_2s_1}-a_{j_1j_2s_2})^2>0$ and $a_{j_2j_10}\prod_{l_1=1}^{3}(\overline{x}_{j_1}-a_{j_2j_1l_1})<0$,

$$(a_{j_1j_2s_1},\dot{x}_{j_2})=\underbrace{(\text{DP},\text{nF})}_{\text{down-parabola flow }(-)}.$$

- For $a_{j_1j_20}(a_{j_1j_2s_1}-a_{j_1j_2s_2})^2<0$ and $a_{j_2j_10}\prod_{l_1=1}^{3}(\overline{x}_{j_1}-a_{j_2j_1l_1})<0$,

$$(a_{j_1j_2s_1},\dot{x}_{j_2})=\underbrace{(\text{UP},\text{nF})}_{\text{up-parabola flow }(-)}.$$

(II) In phase space, at $x_{j_1}^*=a_{j_2j_1l_1}(l_1,l_2\in\{1,2,3\},l_1\neq l_2)$ and $\overline{x}_{j_2}\neq a_{j_1j_2s_1},a_{j_1j_2s_2}(s_1,s_2\in\{1,2\},s_1\neq s_2)$, we have

$$\frac{dx_{j_2}}{dx_{j_1}}\bigg|_{x_{j_1}^*=a_{j_2j_1l_1}}=\frac{a_{j_2j_10}}{a_{j_1j_20}}\frac{(x_{j_1}-a_{j_2j_1l_1})\prod_{l_2=1,l_2\neq l_1}^{3}(x_{j_1}-a_{j_2j_1l_2})}{(\overline{x}_{j_2}-a_{j_1j_2s_1})(\overline{x}_{j_2}-a_{j_1j_2s_2})^2}\bigg|_{x_{j_1}^*=a_{j_2j_1l_1}}=0.$$

If

$$\frac{d^2x_{j_2}}{dx_{j_1}^2}\bigg|_{x_{j_1}^*=a_{j_2j_1l_1}}=\frac{a_{j_2j_10}}{a_{j_1j_20}}\frac{\prod_{l_2=1,l_2\neq l_1}^{3}(a_{j_2j_1l_1}-a_{j_2j_1l_2})}{(\overline{x}_{j_2}-a_{j_1j_2s_1})(\overline{x}_{j_2}-a_{j_1j_2s_2})^2}>0,$$

there is an up-parabola flow at $x_{j_1}^*=a_{j_2j_1l_1}$ in the x_{j_2}-direction. If

$$\frac{d^2x_{j_2}}{dx_{j_1}^2}\bigg|_{x_{j_1}^*=a_{j_2j_1l_1}}=\frac{a_{j_2j_10}}{a_{j_1j_20}}\frac{\prod_{l_2=1,l_2\neq l_1}^{3}(a_{j_2j_1l_1}-a_{j_2j_1l_2})}{(\overline{x}_{j_2}-a_{j_1j_2s_1})(\overline{x}_{j_2}-a_{j_1j_2s_2})^2}<0,$$

there is a down-parabola flow at $x_{j_1}^*=a_{j_2j_1l_1}$ in the x_{j_1}-direction. Because of

$$\dot{x}_{j_1}=a_{j_1j_20}(\overline{x}_{j_2}-a_{j_1j_2s_1})(\overline{x}_{j_2}-a_{j_1j_2s_2})^2,$$

the flows at $x_{j_1}^*=a_{j_2j_1l_1}$ in the x_{j_1}-direction are positive and negative for $\dot{x}_{j_1}>0$ and $\dot{x}_{j_1}<0$, respectively.

Thus, the equilibrium of $x_{j_1}^*=a_{j_2j_1l_1}(l_1,l_2\in\{1,2,3\},l_1\neq l_2)$ and $\overline{x}_{j_2}\neq a_{j_1j_2s_1}$, $a_{j_1j_2s_2}(s_1,s_2\in\{1,2\},s_1\neq s_2)$ on the x_{j_2}-direction has the following properties.

- For $a_{j_1j_20}(\overline{x}_{j_2}-a_{j_1j_2s_1})(\overline{x}_{j_2}-a_{j_1j_2s_2})^2>0$ and $a_{j_2j_10}\prod_{l_2=1,l_2\neq l_1}^{3}(a_{j_2j_1l_1}-a_{j_2j_1l_2})>0$,

$$(\dot{x}_{j_1},a_{j_2j_1l_1})=\underbrace{(\text{pF},\text{UP})}_{\text{up-parabola flow }(+)}.$$

- For $a_{j_1j_20}(\overline{x}_{j_2}-a_{j_1j_2s_1})(\overline{x}_{j_2}-a_{j_1j_2s_2})^2<0$ and $a_{j_2j_10}\prod_{l_2=1,l_2\neq l_1}^{3}(a_{j_2j_1l_1}-a_{j_2j_1l_2})>0$,

$$(\dot{x}_{j_1},a_{j_2j_1l_1})=\underbrace{(\text{nF},\text{DP})}_{\text{down-parabola flow }(-)}.$$

- For $a_{j_1j_20}(\overline{x}_{j_2}-a_{j_1j_2s_1})(\overline{x}_{j_2}-a_{j_1j_2s_2})^2>0$ and $a_{j_2j_10}\prod_{l_2=1,l_2\neq l_1}^{3}(a_{j_2j_1l_1}-a_{j_2j_1l_2})<0$,

$$(\dot{x}_{j_1},a_{j_2j_1l_1})=\underbrace{(\text{pF},\text{DP})}_{\text{down-parabola flow }(+)}.$$

- For $a_{j_1j_20}(\overline{x}_{j_2}-a_{j_1j_2s_1})(\overline{x}_{j_2}-a_{j_1j_2s_2})^2<0$ and $a_{j_2j_10}\prod_{l_2=1,l_2\neq l_1}^{3}(a_{j_2j_1l_1}-a_{j_2j_1l_2})<0$,

$$(\dot{x}_{j_1},a_{j_2j_1l_1})=\underbrace{(\text{nF},\text{UP})}_{\text{up-parabola flow }(-)}.$$

Therefore, from the two cases (I) and (II), the equilibrium of $(x_{j_2}^*,x_{j_1}^*)=(a_{j_1j_2s_1},a_{j_2j_1l_1})$ $(s_1,s_2\in\{1,2\},s_1\neq s_2;l_1,l_2\in\{1,2,3\};l_1\neq l_2)$ has the following properties as in Eqs. (4.106)–(4.109).

- For $a_{j_1j_20}(a_{j_1j_2s_1}-a_{j_1j_2s_2})^2>0$ and $a_{j_2j_10}\prod_{l_2=1,l_2\neq l_1}^{3}(a_{j_2j_1l_1}-a_{j_2j_1l_2})>0$,

$$(a_{j_1j_2s_1},a_{j_2j_1l_1})=\underbrace{(\text{UP}_+,\text{UP}_+)}_{\text{positive saddle}}.$$

The equilibrium of $(x_{j_2}^*,x_{j_1}^*)=(a_{j_1j_2s_1},a_{j_2j_1l_1})$ is a $(\text{UP}_+,\text{UP}_+)$-positive saddle.

- For $a_{j_1j_20}(a_{j_1j_2s_1}-a_{j_1j_2s_2})^2<0$ and $a_{j_2j_10}\prod_{l_2=1,l_2\neq l_1}^{3}(a_{j_2j_1l_1}-a_{j_2j_1l_2})>0$,

$$(a_{j_1j_2s_1},a_{j_2j_1l_1})=\underbrace{(\text{DP}_+,\text{DP}_-)}_{\text{CCW center}}.$$

The equilibrium of $(x_{j_2}^*,x_{j_1}^*)=(a_{j_1j_2s_1},a_{j_2j_1l_1})$ is a $(\text{DP}_+,\text{DP}_-)$-counter-clockwise center.

- For $a_{j_1j_20}(a_{j_1j_2s_1}-a_{j_1j_2s_2})^2>0$ and $a_{j_2j_10}\prod_{l_2=1,l_2\neq l_1}^{3}(a_{j_2j_1l_1}-a_{j_2j_1l_2})<0$,

$$(a_{j_1j_2s_1},a_{j_2j_1l_1})=\underbrace{(\mathrm{DP}_-,\mathrm{DP}_+)}_{\text{CW center}}.$$

The equilibrium of $(x_{j_2}^*,x_{j_1}^*)=(a_{j_1j_2s_1},a_{j_2j_1l_1})$ is a (DP$_-$,DP$_+$)-clockwise center.

- For $a_{j_1j_20}(a_{j_1j_2s_1}-a_{j_1j_2s_2})^2<0$ and $a_{j_2j_10}\prod_{l_2=1,l_2\neq l_1}^{3}(a_{j_2j_1l_1}-a_{j_2j_1l_2})<0$,

$$(a_{j_1j_2s_1},a_{j_2j_1l_1})=\underbrace{(\mathrm{UP}_-,\mathrm{UP}_-)}_{\text{negative saddle}}.$$

The equilibrium of $(x_{j_2}^*,x_{j_1}^*)=(a_{j_1j_2s_1},a_{j_2j_1l_1})$ is a (DP$_-$,UP$_-$)-negative saddle.

(viii$_2$) In phase space, at $x_{j_2}^*=a_{j_1j_2s_2}$ $(s_1,\ s_2\in\{1,2\},\ s_1\neq s_2)$ and $\bar{x}_{j_1}\neq a_{j_2j_1l_1}$ $(l_1\in\{1,2,3\})$, we have

$$\frac{dx_{j_1}}{dx_{j_2}}\Big|_{x_{j_2}^*=a_{j_1j_2s_2}}=\frac{a_{j_1j_20}}{a_{j_2j_10}}\frac{(x_{j_2}-a_{j_1j_2s_1})(x_{j_2}-a_{j_1j_2s_2})^2}{\prod_{l_1=1}^{3}(\bar{x}_{j_1}-a_{j_2j_1l_1})}\Big|_{x_{j_2}^*=a_{j_1j_2s_2}}=0,$$

$$\frac{d^2x_{j_1}}{dx_{j_2}^2}\Big|_{x_{j_2}^*=a_{j_1j_2s_2}}=0.$$

If

$$\frac{d^3x_{j_1}}{dx_{j_2}^3}\Big|_{x_{j_2}^*=a_{j_1j_2s_2}}=\frac{a_{j_1j_20}}{a_{j_2j_10}}\frac{2(a_{j_1j_2s_2}-a_{j_1j_2s_1})}{\prod_{l_1=1}^{3}(\bar{x}_{j_1}-a_{j_2j_1l_1})}>0,$$

there is an increasing-inflection flow at $x_{j_2}^*=a_{j_1j_2s_2}$ in the x_{j_1}-direction. If

$$\frac{d^3x_{j_1}}{dx_{j_2}^3}\Big|_{x_{j_2}^*=a_{j_1j_2s_2}}=\frac{a_{j_1j_20}}{a_{j_2j_10}}\frac{2(a_{j_1j_2s_2}-a_{j_1j_2s_1})}{\prod_{l_1=1}^{3}(\bar{x}_{j_1}-a_{j_2j_1l_1})}<0,$$

there is a decreasing-inflection flow at $x_{j_2}^*=a_{j_1j_2s_2}$ in the x_{j_1}-direction. Because of

$$\dot{x}_{j_2}=a_{j_2j_10}\prod_{l_1=1}^{3}(\bar{x}_{j_1}-a_{j_2j_1l_1}),$$

the flows at $x_{j_2}^*=a_{j_1j_2s_2}$ in the x_{j_2}-direction are positive and negative for $a_{j_2j_10}>0$ and $a_{j_2j_10}<0$, respectively.

Thus, the equilibrium of $x_{j_2}^*=a_{j_1j_2s_2}(s_1,\ s_2\in\{1,2\},\ s_1\neq s_2\ )$ and $\bar{x}_{j_1}\neq a_{j_2j_11}$ on the x_{j_2}-direction has the following properties.

- For $a_{j_1j_20}(a_{j_1j_2s_2} - a_{j_1j_2s_1}) > 0$ and $a_{j_2j_10}\prod_{l_1=1}^{3}(\overline{x}_{j_1} - a_{j_2j_1l_1}) > 0$,

$$(a_{j_1j_2s_2}, \dot{x}_{j_2}) = \underbrace{(\text{II}, \text{pF})}_{\text{increasing-inflection flow }(+)}.$$

- For $a_{j_1j_20}(a_{j_1j_2s_2} - a_{j_1j_2s_1}) < 0$ and $a_{j_2j_10}\prod_{l_1=1}^{3}(\overline{x}_{j_1} - a_{j_2j_1l_1}) > 0$,

$$(a_{j_1j_2s_2}, \dot{x}_{j_2}) = \underbrace{(\text{DI}, \text{pF})}_{\text{decreasing-inflection flow }(+)}.$$

- For $a_{j_1j_20}(a_{j_1j_2s_2} - a_{j_1j_2s_1}) > 0$ and $a_{j_2j_10}\prod_{l_1=1}^{3}(\overline{x}_{j_1} - a_{j_2j_1l_1}) < 0$,

$$(a_{j_1j_2s_2}, \dot{x}_{j_2}) = \underbrace{(\text{DI}, \text{nF})}_{\text{decreasing-inflection flow }(-)}.$$

- For $a_{j_1j_20}(a_{j_1j_2s_2} - a_{j_1j_2s_1}) < 0$ and $a_{j_2j_10}\prod_{l_1=1}^{3}(\overline{x}_{j_1} - a_{j_2j_1l_1}) < 0$,

$$(a_{j_1j_2s_2}, \dot{x}_{j_2}) = \underbrace{(\text{II}, \text{nF})}_{\text{increasing-inflection flow }(-)}.$$

Therefore, from the proof (II) of case (xi$_1$) and the above analysis, the equilibrium of $(x_{j_2}^*, x_{j_1}^*) = (a_{j_1j_2s_2}, a_{j_2j_1l_1})$ ($s_1, s_2 \in \{1, 2\}, s_1 \neq s_2; l_1, l_2 \in \{1, 2, 3\}, l_1 \neq l_2$) has the following properties as in Eqs. (4.110)–(4.113).

- For $a_{j_1j_20}(a_{j_1j_2s_2} - a_{j_1j_2s_1}) > 0$ and $a_{j_2j_10}\prod_{l_2=1,l_2\neq l_1}^{3}(a_{j_2j_1l_1} - a_{j_2j_1l_2}) > 0$,

$$(a_{j_1j_2s_2}, a_{j_2j_1l_1}) = \underbrace{(\text{US}, \text{UP})}_{\text{up-parabola upper-saddle}}.$$

  The equilibrium of $(x_{j_2}^*, x_{j_1}^*) = (a_{j_1j_2s_2}, a_{j_2j_1l_1})$ is a (US,UP)-up-parabola upper-saddle.

- For $a_{j_1j_20}(a_{j_1j_2s_2} - a_{j_1j_2s_1}) < 0$ and $a_{j_2j_10}\prod_{l_2=1,l_2\neq l_1}^{3}(a_{j_2j_1l_1} - a_{j_2j_1l_2}) > 0$,

$$(a_{j_1j_2s_2}, a_{j_2j_1l_1}) = \underbrace{(\text{LS}, \text{DP})}_{\text{down-parabola lower-saddle}}.$$

  The equilibrium of $(x_{j_2}^*, x_{j_1}^*) = (a_{j_1j_2s_2}, a_{j_2j_1l_1})$ is an (LS,DP)-down-parabola lower-saddle.

- For $a_{j_1j_20}(a_{j_1j_2s_2}-a_{j_1j_2s_1})>0$ and $a_{j_2j_10}\prod_{l_2=1,l_2\neq l_1}^{3}(a_{j_2j_1l_1}-a_{j_2j_1l_2})<0$,

$$(a_{j_1j_2s_2},a_{j_2j_1l_1})=\underbrace{(\mathrm{US},\mathrm{DP})}_{\text{down-parabola upper-saddle}}.$$

The equilibrium of $(x_{j_2}^*,x_{j_1}^*)=(a_{j_1j_2s_2},a_{j_2j_1l_1})$ is a (US,DP)-down-parabola upper-saddle.

- For $a_{j_1j_20}(a_{j_1j_2s_2}-a_{j_1j_2s_1})<0$ and $a_{j_2j_10}\prod_{l_2=1,l_2\neq l_1}^{3}(a_{j_2j_1l_1}-a_{j_2j_1l_2})<0$,

$$(a_{j_1j_2s_2},a_{j_2j_1l_1})=\underbrace{(\mathrm{LS},\mathrm{UP})}_{\text{up-parabola lower-saddle}}.$$

The equilibrium of $(x_{j_2}^*,x_{j_1}^*)=(a_{j_1j_2s_2},a_{j_2j_1l_1})$ is an (LS,UP)-up-parabola lower-saddle.

(xii) If

$$\Delta_{j_1j_2}=0 \text{ and } \Delta_{j_2j_1}>0, a_{j_1j_2s_1}=a_{j_1j_2s_2}=a_{j_1j_21}$$

then

$$\begin{aligned}
&\dot{x}_{j_2}=a_{j_2j_10}(x_{j_1}-a_{j_2j_11})(x_{j_1}-a_{j_2j_12})(x_{j_1}-a_{j_2j_13}),\\
&\dot{x}_{j_1}=a_{j_1j_20}(x_{j_2}-a_{j_1j_21})^3,\\
&j_1,j_2\in\{1,2\};j_1\neq j_2.
\end{aligned}$$

In phase space, for $a_{j_2j_11}\neq a_{j_2j_12}$ (l_1, $l_2\in\{1,2,3\}$, $l_1\neq l_2$),

$$\frac{dx_{j_1}}{dx_{j_2}}=\frac{a_{j_1j_20}}{a_{j_2j_10}}\frac{(x_{j_2}-a_{j_1j_21})^3}{(x_{j_1}-a_{j_2j_11})(x_{j_1}-a_{j_2j_12})(x_{j_1}-a_{j_2j_13})},$$

and

$$\begin{aligned}
&a_{j_2j_10}\big[(x_{j_1}-a_{j_2j_1l_1})^3+\textstyle\sum_{l_2=1,l_2\neq l_1}^{3}(a_{j_2j_1l_1}-a_{j_2j_1l_2})(x_{j_1}-a_{j_2j_11})^2\\
&+\textstyle\prod_{l_2=1,l_2\neq l_1}^{3}(a_{j_2j_1l_1}-a_{j_2j_1l_2})(x_{j_1}-a_{j_2j_11})\big]dx_{j_1}\\
&=a_{j_1j_20}(x_{j_2}-a_{j_1j_21})^3dx_{j_2}.
\end{aligned}$$

With an initial condition (x_{j_10},x_{j_20}), the first integral manifold is generated, i.e.,

$$\begin{aligned}
&a_{j_2j_10}\Big\{\frac{1}{4}\big[(x_{j_1}-a_{j_2j_1l_1})^4-(x_{j_10}-a_{j_2j_1l_1})^4\big]\\
&+\frac{1}{3}\textstyle\sum_{l_2=1,l_2\neq l_1}^{3}(a_{j_2j_1l_1}-a_{j_2j_1l_2})\big[(x_{j_1}-a_{j_2j_1l_1})^3-(x_{j_10}-a_{j_2j_1l_1})^3\big]\\
&+\frac{1}{2}\textstyle\prod_{l_2=1,l_2\neq l_1}^{3}(a_{j_2j_1l_1}-a_{j_2j_1l_2})\big[(x_{j_1}-a_{j_2j_1l_1})^2-(x_{j_10}-a_{j_2j_1l_1})^2\big]\Big\}\\
&=\frac{1}{4}a_{j_1j_20}\big[(x_{j_1}-a_{j_1j_2s_1})^4-(x_{j_10}-a_{j_1j_2s_1})^4\big].
\end{aligned}$$

Consider two cases (I) $x_{j_2}^* = a_{j_1 j_2 1}$ and (II) $x_{j_1}^* = a_{j_2 j_1 l_1}$

(I) In phase space, at $x_{j_2}^* = a_{j_1 j_2 1}$ and $\bar{x}_{j_1} \neq a_{j_2 j_1 l_1}$ ($l_1 = 1, 2, 3$), we have

$$\frac{dx_{j_1}}{dx_{j_2}}\Big|_{x_{j_2}^* = a_{j_1 j_2 1}} = \frac{a_{j_1 j_2 0}}{a_{j_2 j_1 0}} \frac{(x_{j_2} - a_{j_1 j_2 1})^3}{\prod_{l_1=1}^{3}(\bar{x}_{j_1} - a_{j_2 j_1 l_1})}\Big|_{x_{j_2}^* = a_{j_1 j_2 1}} = 0,$$

$$\frac{d^2 x_{j_1}}{dx_{j_2}^2}\Big|_{x_{j_2}^* = a_{j_1 j_2 1}} = \frac{d^3 x_{j_1}}{dx_{j_2}^3}\Big|_{x_{j_2}^* = a_{j_1 j_2 1}} = 0.$$

If

$$\frac{d^4 x_{j_1}}{dx_{j_2}^4}\Big|_{x_{j_2}^* = a_{j_1 j_2 1}} = \frac{a_{j_1 j_2 0}}{a_{j_2 j_1 0}} \frac{6}{\prod_{l_1=1}^{3}(\bar{x}_{j_1} - a_{j_2 j_1 l_1})} > 0,$$

there is a third-order up-parabola flow at $x_{j_2}^* = a_{j_1 j_2 1}$ in the x_{j_1}-direction. If

$$\frac{d^4 x_{j_1}}{dx_{j_2}^4}\Big|_{x_{j_2}^* = a_{j_1 j_2 1}} = \frac{a_{j_1 j_2 0}}{a_{j_2 j_1 0}} \frac{6}{\prod_{l_1=1}^{3}(\bar{x}_{j_1} - a_{j_2 j_1 l_1})} < 0,$$

there is a third-order down-parabola flow at $x_{j_2}^* = a_{j_1 j_2 1}$ in the x_{j_1}-direction. Because of

$$\dot{x}_{j_2} = a_{j_2 j_1 0} \prod_{l_1=1}^{3}(\bar{x}_{j_1} - a_{j_2 j_1 l_1}),$$

the flows at $x_{j_2}^* = a_{j_1 j_2 1}$ in the x_{j_2}-direction are positive and negative for $\dot{x}_{j_2} > 0$ and $\dot{x}_{j_2} < 0$, respectively.

Thus, the equilibrium of $x_{j_2}^* = a_{j_1 j_2 1}$ on the x_{j_2}-direction has the following properties.

- For $a_{j_1 j_2 0} > 0$ and $a_{j_2 j_1 0} \prod_{l_1=1}^{3}(\bar{x}_{j_1} - a_{j_2 j_1 l_1}) > 0$,

$$(a_{j_1 j_2 1}, \dot{x}_{j_2}) = \underbrace{(3^{\text{rd}}\text{UP}, \text{pF})}_{\text{third-order up-parabola flow } (+)}.$$

- For $a_{j_1 j_2 0} < 0$ and $a_{j_2 j_1 0} \prod_{l_1=1}^{3}(\bar{x}_{j_1} - a_{j_2 j_1 l_1}) > 0$,

$$(a_{j_1 j_2 1}, \dot{x}_{j_2}) = \underbrace{(3^{\text{rd}}\text{DP}, \text{pF})}_{\text{third-order down-parabola flow } (+)}.$$

- For $a_{j_1j_20}>0$ and $a_{j_2j_10}\prod_{l_1=1}^{3}(\overline{x}_{j_1}-a_{j_2j_1l_1})<0$,

$$(a_{j_1j_21},\dot{x}_{j_2})=\underbrace{(3^{\text{rd}}\text{DP},\text{nF})}_{\text{third-order down-parabola flow }(-)}.$$

- For $a_{j_1j_20}<0$ and $a_{j_2j_10}\prod_{l_1=1}^{3}(\overline{x}_{j_1}-a_{j_2j_1l_1})<0$,

$$(a_{j_1j_21},\dot{x}_{j_2})=\underbrace{(3^{\text{rd}}\text{UP},\text{nF})}_{\text{third-order up-parabola flow }(-)}.$$

(II) In phase space, at $x^*_{j_1}=a_{j_2j_1l_1}(l_1,\,l_2\in\{1,2,3\},\,l_1\neq l_2)$ and $\overline{x}_{j_2}\neq a_{j_1j_21}$, we have

$$\frac{dx_{j_2}}{dx_{j_1}}\bigg|_{x^*_{j_1}=a_{j_2j_1l_1}}=\frac{a_{j_2j_10}}{a_{j_1j_20}}\frac{(x_{j_1}-a_{j_2j_1l_1})\prod_{l_2=1,l_2\neq l_1}^{3}(x_{j_1}-a_{j_2j_1l_2})}{(\overline{x}_{j_2}-a_{j_1j_21})^3}\bigg|_{x^*_{j_1}=a_{j_2j_1l_1}}=0.$$

If

$$\frac{d^2x_{j_2}}{dx^2_{j_1}}\bigg|_{x^*_{j_1}=a_{j_2j_11}}=\frac{a_{j_2j_10}}{a_{j_1j_20}}\frac{\prod_{l_2=1,l_2\neq l_1}^{3}(a_{j_2j_1l_1}-a_{j_2j_1l_2})}{(\overline{x}_{j_2}-a_{j_1j_21})^3}>0,$$

there is an up-parabola flow at $x^*_{j_1}=a_{j_2j_1l_1}$ in the x_{j_2}-direction. If

$$\frac{d^2x_{j_2}}{dx^2_{j_1}}\bigg|_{x^*_{j_1}=a_{j_2j_11}}=\frac{a_{j_2j_10}}{a_{j_1j_20}}\frac{\prod_{l_2=1,l_2\neq l_1}^{3}(a_{j_2j_1l_1}-a_{j_2j_1l_2})}{(\overline{x}_{j_2}-a_{j_1j_21})^3}<0,$$

there is a down-parabola flow at $x^*_{j_1}=a_{j_2j_1l_1}$ in the x_{j_1}-direction. Because of

$$\dot{x}_{j_1}=a_{j_1j_20}(\overline{x}_{j_2}-a_{j_1j_21})^3,$$

the flows at $x^*_{j_1}=a_{j_2j_1l_1}$ in the x_{j_1}-direction are positive and negative for $\dot{x}_{j_1}>0$ and $\dot{x}_{j_1}<0$, respectively.

Thus, the equilibrium of $x^*_{j_1}=a_{j_2j_1l_1}(l_1,\,l_2\in\{1,2,3\},\,l_1\neq l_2)$ and $\overline{x}_{j_2}\neq a_{j_1j_21}$ on the x_{j_2}-direction has the following properties.

- For $a_{j_1j_20}(\overline{x}_{j_2}-a_{j_1j_21})^3>0$ and $a_{j_2j_10}\prod_{l_2=1,l_2\neq l_1}^{3}(a_{j_2j_1l_1}-a_{j_2j_1l_2})>0$,

$$(\dot{x}_{j_1},a_{j_2j_1l_1})=\underbrace{(\text{pF},\text{UP})}_{\text{up-parabola flow }(+)}.$$

- For $a_{j_1j_20}(\bar{x}_{j_2}-a_{j_1j_21})^3<0$ and $a_{j_2j_10}\prod_{l_2=1,l_2\neq l_1}^{3}(a_{j_2j_1l_1}-a_{j_2j_1l_2})>0$,

$$(\dot{x}_{j_1},a_{j_2j_1l_1})=\underbrace{(\text{nF},\text{DP})}_{\text{down-parabola flow }(-)}.$$

- For $a_{j_1j_20}(\bar{x}_{j_2}-a_{j_1j_21})^3>0$ and $a_{j_2j_10}\prod_{l_2=1,l_2\neq l_1}^{3}(a_{j_2j_1l_1}-a_{j_2j_1l_2})<0$,

$$(\dot{x}_{j_1},a_{j_2j_1l_1})=\underbrace{(\text{pF},\text{DP})}_{\text{down-parabola flow }(+)}.$$

- For $a_{j_1j_20}(\bar{x}_{j_2}-a_{j_1j_21})^3<0$ and $a_{j_2j_10}\prod_{l_2=1,l_2\neq l_1}^{3}(a_{j_2j_1l_1}-a_{j_2j_1l_2})<0$,

$$(\dot{x}_{j_1},a_{j_2j_1l_1})=\underbrace{(\text{nF},\text{UP})}_{\text{up-parabola flow }(-)}.$$

Therefore, from case (I) and (II), the equilibrium of $(x_{j_2}^*,x_{j_1}^*)=(a_{j_1j_21},a_{j_2j_1l_1})$ ($l_1,l_2\in\{1,2,3\}$, $l_1\neq l_2$) has the following properties as in Eqs. (4.117)–(4.120)

- For $a_{j_1j_20}>0$ and $a_{j_2j_10}\prod_{l_2=1,l_2\neq l_1}^{3}(a_{j_2j_1l_1}-a_{j_2j_1l_2})>0$,

$$(a_{j_1j_21},a_{j_2j_1l_1})=\underbrace{(3^{\text{rd}}\text{UP}_+,\text{UP}_+)}_{\text{third-order positive saddle}}.$$

  The equilibrium of $(x_{j_2}^*,x_{j_1}^*)=(a_{j_1j_21},a_{j_2j_1l_1})$ is a (3^{rd}UP_+,UP_+)-third-order positive saddle.
- For $a_{j_1j_20}<0$ and $a_{j_2j_10}\prod_{l_2=1,l_2\neq l_1}^{3}(a_{j_2j_1l_1}-a_{j_2j_1l_2})>0$,

$$(a_{j_1j_21},a_{j_2j_1l_1})=\underbrace{(3^{\text{rd}}\text{DP}_+,\text{DP}_-)}_{\text{third-order CCW center}}.$$

  The equilibrium of $(x_{j_2}^*,x_{j_1}^*)=(a_{j_1j_21},a_{j_2j_1l_1})$ is a (3^{rd}DP_+,DP_-)-third-order-counter-clockwise center.
- For $a_{j_1j_20}>0$ and $a_{j_2j_10}\prod_{l_2=1,l_2\neq l_1}^{3}(a_{j_2j_1l_1}-a_{j_2j_1l_2})<0$,

$$(a_{j_1j_21},a_{j_2j_1l_1})=\underbrace{(3^{\text{rd}}\text{DP}_-,\text{DP}_+)}_{\text{third-order CW center}}.$$

  The equilibrium of $(x_{j_2}^*,x_{j_1}^*)=(a_{j_1j_21},a_{j_2j_1l_1})$ is a (3^{rd}DP_-,DP_+)-third-order-clockwise center.

- For $a_{j_1 j_2 0} < 0$ and $a_{j_2 j_1 0}\prod_{l_2=1,l_2\neq l_1}^{3}(a_{j_2 j_1 l_1} - a_{j_2 j_1 l_2}) < 0$,

$$(a_{j_1 j_2 1}, a_{j_2 j_1 l_1}) = \underbrace{(3^{\text{rd}}\text{UP}_-, \text{UP}_-)}_{\text{third-order negative saddle}}.$$

The equilibrium of $(x_{j_2}^*, x_{j_1}^*) = (a_{j_1 j_2 1}, a_{j_2 j_1 l_1})$ is a $(3^{\text{rd}}\text{UP}_-,\text{UP}_-)$-third-order negative saddle.

(xiii) If

$$\Delta_{j_1 j_2} > 0 \text{ and } \Delta_{j_2 j_1} < 0,$$

then

$$\begin{aligned}
&\dot{x}_{j_1} = a_{j_1 j_2 0}(x_{j_2} - a_{j_1 j_2 1})(x_{j_2} - a_{j_1 j_2 2})(x_{j_2} - a_{j_1 j_2 3}),\\
&\dot{x}_{j_2} = a_{j_2 j_1 0}(x_{j_1} - a_{j_2 j_1 1})\left[(x_{j_1} - a_{j_2 j_1})^2 + b_{j_2 j_1}\right]\\
&j_1, j_2 \in \{1, 2\}, j_2 \neq j_1
\end{aligned}$$

where

$$\begin{aligned}
&\{a_{j_1 j_2 1}, a_{j_1 j_2 2}, a_{j_1 j_2 3}\} = \text{sort}\{b_{j_1 j_2 1}, b_{j_1 j_2 2}, b_{j_1 j_2 3}\}, a_{j_1 j_2 1} < a_{j_1 j_2 2} < a_{j_1 j_2 3};\\
&b_{j_1 j_2 2}, b_{j_1 j_2 3} = -\frac{1}{2}(B_{j_1 j_2} \pm \sqrt{\Delta_{j_1 j_2}}), a_{j_1 j_2 0} = A_{j_1 j_2 0};\\
&a_{j_2 j_1 1} = b_{j_2 j_1 1}, a_{j_2 j_1} = -\frac{1}{2}B_{j_2 j_1}, b_{j_2 j_1} = \frac{1}{4}(-\Delta_{j_2 j_1}).
\end{aligned}$$

In phase space, for $a_{j_1 j_2 s_1} \neq a_{j_1 j_2 s_2}$ (s_1, $s_2 \in \{1, 2, 3\}$, $s_1 \neq s_2$),

$$\frac{dx_{j_1}}{dx_{j_2}} = \frac{a_{j_1 j_2 0}}{a_{j_2 j_1 0}} \frac{(x_{j_2} - a_{j_1 j_2 1})(x_{j_2} - a_{j_1 j_2 2})(x_{j_2} - a_{j_1 j_2 3})}{(x_{j_1} - a_{j_2 j_1 1})\left[(x_{j_1} - a_{j_2 j_1})^2 + b_{j_2 j_1}\right]}$$

and

$$\begin{aligned}
&a_{j_2 j_1 0}\{(x_{j_1} - a_{j_2 j_1 1})^3 + 2(a_{j_2 j_1 1} - a_{j_2 j_1})(x_{j_1} - a_{j_2 j_1 1})^2\\
&+\left[(a_{j_2 j_1 1} - a_{j_2 j_1})^2 + b_{j_2 j_1}\right](x_{j_1} - a_{j_2 j_1 1})\}dx_{j_1}\\
&= a_{j_1 j_2 0}\left[(x_{j_2} - a_{j_1 j_2 s_1})^3 + \sum\nolimits_{s_2=1, s_2\neq s_1}^{3}(a_{j_1 j_2 s_1} - a_{j_1 j_2 s_2})(x_{j_2} - a_{j_1 j_2 s_1})^2\right.\\
&\left.+\prod\nolimits_{s_2=1, s_2\neq s_1}^{3}(a_{j_1 j_2 s_1} - a_{j_1 j_2 s_2})(x_{j_2} - a_{j_1 j_2 s_1})\right]dx_{j_2}.
\end{aligned}$$

With an initial condition $(x_{j_1 0}, x_{j_2 0})$, the first integral manifold is generated, i.e.,

$$
\begin{aligned}
& a_{j_2 j_1 0}\Big\{\frac{1}{4}\left[(x_{j_1}-a_{j_2 j_1 1})^4-(x_{j_1 0}-a_{j_2 j_1 1})^4\right] \\
& +\frac{2}{3}(a_{j_2 j_1 1}-a_{j_2 j_1})\left[(x_{j_1}-a_{j_2 j_1 1})^2-(x_{j_1 0}-a_{j_2 j_1 1})^2\right] \\
& +\frac{1}{2}\left[(a_{j_2 j_1 1}-a_{j_2 j_1})^2+b_{j_2 j_1}\right]\left[(x_{j_1}-a_{j_2 j_1 1})-(x_{j_1 0}-a_{j_2 j_1 1})\right]\Big\} \\
& =a_{j_1 j_2 0}\Big\{\frac{1}{4}\left[(x_{j_2}-a_{j_1 j_2 s_1})^4-(x_{j_2 0}-a_{j_1 j_2 s_1})^4\right] \\
& +\frac{1}{3}\sum\nolimits_{s_2=1, s_2 \neq s_1}^{3}(a_{j_1 j_2 s_1}-a_{j_1 j_2 s_2})\left[(x_{j_2}-a_{j_1 j_2 s_1})^3-(x_{j_2 0}-a_{j_1 j_2 s_1})^3\right] \\
& +\frac{1}{2}\prod\nolimits_{s_2=1, s_2 \neq s_1}^{3}(a_{j_1 j_2 s_1}-a_{j_1 j_2 s_2})\left[(x_{j_2}-a_{j_1 j_2 s_1})^2-(x_{j_2 0}-a_{j_1 j_2 s_1})^2\right]\Big\}.
\end{aligned}
$$

Consider two cases (I) $x_{j_2}^*=a_{j_1 j_2 s_1}$ and (II)$x_{j_1}^*=a_{j_2 j_1 1}$.

(I) In phase space, at $x_{j_2}^*=a_{j_1 j_2 s_1}(s_1,\, s_2 \in \{1,2,3\},\, s_1 \neq s_2)$ with $\bar{x}_{j_1} \neq a_{j_2 j_1 1}$, we have

$$
\left.\frac{dx_{j_1}}{dx_{j_2}}\right|_{x_{j_2}^*=a_{j_1 j_2 s_1}}=\frac{a_{j_1 j_2 0}}{a_{j_2 j_1 0}}\left.\frac{(x_{j_2}-a_{j_1 j_2 s_1})\prod_{s_2=1, s_2 \neq s_1}^{3}(x_{j_2}-a_{j_1 j_2 s_2})}{(\bar{x}_{j_1}-a_{j_2 j_1 1})\left[(\bar{x}_{j_1}-a_{j_2 j_1})^2+b_{j_2 j_1}\right]}\right|_{x_{j_2}^*=a_{j_1 j_2 s_1}}=0.
$$

If

$$
\left.\frac{d^2 x_{j_1}}{dx_{j_2}^2}\right|_{x_{j_2}^*=a_{j_1 j_2 s_1}}=\frac{a_{j_1 j_2 0}}{a_{j_2 j_1 0}}\frac{\prod_{s_2=1, s_2 \neq s_1}^{3}(a_{j_1 j_2 s_1}-a_{j_1 j_2 s_2})}{(\bar{x}_{j_1}-a_{j_2 j_1 1})\left[(\bar{x}_{j_1}-a_{j_2 j_1})^2+b_{j_2 j_1}\right]}>0,
$$

there is an up-parabola flow at $x_{j_2}^*=a_{j_1 j_2 s_1}$ in the x_{j_1}-direction. If

$$
\left.\frac{d^2 x_{j_1}}{dx_{j_2}^2}\right|_{x_{j_2}^*=a_{j_1 j_2 s_1}}=\frac{a_{j_1 j_2 0}}{a_{j_2 j_1 0}}\frac{\prod_{s_2=1, s_2 \neq s_1}^{3}(a_{j_1 j_2 s_1}-a_{j_1 j_2 s_2})}{(\bar{x}_{j_1}-a_{j_2 j_1 1})\left[(\bar{x}_{j_1}-a_{j_2 j_1})^2+b_{j_2 j_1}\right]}<0,
$$

there is a down-parabola flow at $x_{j_2}^*=a_{j_1 j_2 s_1}$ in the x_{j_1}-direction. Because of

$$
\dot{x}_{j_2}=a_{j_2 j_1 0}(\bar{x}_{j_1}-a_{j_2 j_1 1})\left[(\bar{x}_{j_1}-a_{j_2 j_1})^2+b_{j_2 j_1}\right],
$$

the parabola flows at $x_{j_2}^*=a_{j_1 j_2 s_1}$ in the x_{j_2}-direction are positive and negative for $a_{j_2 j_1 0}(\bar{x}_{j_1}-a_{j_2 j_1 1})>0$ and $a_{j_2 j_1 0}(\bar{x}_{j_1}-a_{j_2 j_1 1})<0$, respectively.

Thus, the equilibrium of $x_{j_2}^*=a_{j_1 j_2 s_1}(s_1,\, s_2 \in \{1,2,3\},\, s_1 \neq s_2)$ on the x_{j_2}-direction has the following properties.

- For $a_{j_1 j_2 0}\prod_{s_2=1, s_2 \neq s_1}^{3}(a_{j_1 j_2 s_1}-a_{j_1 j_2 s_2})>0$ and $a_{j_2 j_1 0}(\bar{x}_{j_1}-a_{j_2 j_1 1})>0$,

$$
(a_{j_1 j_2 s_1}, \dot{x}_{j_2})=\underbrace{(\mathrm{UP}, \mathrm{pF})}_{\text{up-parabola flow }(+)}.
$$

- For $a_{j_1j_20}\prod_{s_2=1,s_2\neq s_1}^{3}(a_{j_1j_2s_1}-a_{j_1j_2s_2})<0$ and $a_{j_2j_10}(\bar{x}_{j_1}-a_{j_2j_11})>0$,

$$(a_{j_1j_2s_1},\dot{x}_{j_2})=\underbrace{(\mathrm{DP},\mathrm{pF})}_{\text{down-parabola flow }(+)}.$$

- For $a_{j_1j_20}\prod_{s_2=1,s_2\neq s_1}^{3}(a_{j_1j_2s_1}-a_{j_1j_2s_2})>0$ and $a_{j_2j_10}(\bar{x}_{j_1}-a_{j_2j_11})<0$,

$$(a_{j_1j_2s_1},\dot{x}_{j_2})=\underbrace{(\mathrm{DP},\mathrm{nF})}_{\text{down-parabola flow }(-)}.$$

- For $a_{j_1j_20}\prod_{s_2=1,s_2\neq s_1}^{3}(a_{j_1j_2s_1}-a_{j_1j_2s_2})<0$ and $a_{j_2j_10}(\bar{x}_{j_1}-a_{j_2j_11})<0$,

$$(a_{j_1j_2s_1},\dot{x}_{j_2})=\underbrace{(\mathrm{UP},\mathrm{nF})}_{\text{up-parabola flow }(-)}.$$

(II) For $x_{j_1}^*=a_{j_2j_11}$ and $\bar{x}_{j_2}\neq a_{j_1j_2s_1},a_{j_2j_1s_2}$ (s_1, $s_2\in\{1,2,3\}$, $s_1\neq s_2$), we have

$$\frac{dx_{j_2}}{dx_{j_1}}\bigg|_{x_{j_1}^*=a_{j_2j_11}}=\frac{a_{j_2j_10}(x_{j_1}-a_{j_2j_11})\left[(x_{j_1}-a_{j_2j_1})^2+b_{j_2j_1}\right]}{a_{j_1j_20}\prod_{l_1=1}^{3}(\bar{x}_{j_2}-a_{j_1j_2s_1})}\bigg|_{x_{j_1}^*=a_{j_2j_11}}=0.$$

If

$$\frac{d^2x_{j_2}}{dx_{j_1}^2}\bigg|_{x_{j_1}^*=a_{j_2j_11}}=\frac{a_{j_2j_10}}{a_{j_1j_20}}\frac{\left[(a_{j_2j_11}-a_{j_2j_1})^2+b_{j_2j_1}\right]}{\prod_{s_1=1}^{3}(\bar{x}_{j_2}-a_{j_1j_2s_1})}>0,$$

there is an up-parabola flow at $x_{j_1}^*=a_{j_2j_11}$ in the x_{j_2}-direction. If

$$\frac{d^2x_{j_2}}{dx_{j_1}^2}\bigg|_{x_{j_1}^*=a_{j_2j_11}}=\frac{a_{j_2j_10}}{a_{j_1j_20}}\frac{\left[(a_{j_2j_11}-a_{j_2j_1})^2+b_{j_2j_1}\right]}{\prod_{s_1=1}^{3}(\bar{x}_{j_2}-a_{j_1j_2s_1})}<0,$$

there is a down-parabola flow at $x_{j_1}^*=a_{j_2j_11}$ in the x_{j_2}-direction. Because

$$\dot{x}_{j_1}=a_{j_1j_20}\prod_{s_1=1}^{3}(\bar{x}_{j_2}-a_{j_1j_2s_1})$$

and

$$(a_{j_2j_11}-a_{j_2j_1})^2+b_{j_2j_1}>0,$$

the flows at $x_{j_1}^*=a_{j_2j_11}$ are positive and negative in the x_{j_1}-direction for $\dot{x}_{j_1}>0$ and $\dot{x}_{j_1}<0$, respectively.

Thus, the equilibrium of $x_{j_1}^*=a_{j_2j_11}$ on the x_{j_2}-direction has the following properties.

- For $a_{j_1j_20}\prod_{s_1=1}^{3}(\overline{x}_{j_2}-a_{j_1j_2s_1})>0$ and $a_{j_2j_10}>0$,

$$(\dot{x}_{j_1},a_{j_2j_11})=\underbrace{(\text{pF},\text{UP})}_{\text{up-parabola flow }(+)}.$$

- For $a_{j_1j_20}\prod_{s_1=1}^{3}(\overline{x}_{j_2}-a_{j_1j_2s_1})<0$ and $a_{j_2j_10}>0$,

$$(\dot{x}_{j_1},a_{j_2j_11})=\underbrace{(\text{nF},\text{DP})}_{\text{down-parabola flow }(-)}.$$

- For $a_{j_1j_20}\prod_{s_1=1}^{3}(\overline{x}_{j_2}-a_{j_1j_2s_1})>0$ and $a_{j_2j_10}<0$,

$$(\dot{x}_{j_1},a_{j_2j_11})=\underbrace{(\text{pF},\text{DP})}_{\text{down-parabola flow }(+)}.$$

- For $a_{j_1j_20}\prod_{s_1=1}^{3}(\overline{x}_{j_2}-a_{j_1j_2s_1})<0$ and $a_{j_2j_10}<0$,

$$(\dot{x}_{j_1},a_{j_2j_11})=\underbrace{(\text{nF},\text{UP})}_{\text{up-parabola flow }(-)}.$$

Therefore, from cases (I) and (II), the equilibrium of $(x_{j_2}^*,x_{j_1}^*)=(a_{j_1j_2s_1},a_{j_2j_11})$ (s_1, $s_2\in\{1,2,3\}$, $s_1\neq s_2$) on the x_{j_2}-direction has the following properties as in Eqs. (4.125)–(4.128).

- For $a_{j_1j_20}\prod_{s_2=1,s_2\neq s_1}^{3}(a_{j_1j_2s_1}-a_{j_1j_2s_2})>0$ and $a_{j_2j_10}>0$,

$$(a_{j_1j_2s_1},a_{j_2j_11})=\underbrace{(\text{UP}_+,\text{UP}_+)}_{\text{positive saddle}}.$$

  The equilibrium of $(x_{j_2}^*,x_{j_1}^*)=(a_{j_1j_2s_1},a_{j_2j_11})$ is a (UP_+,UP_+)-positive saddle.
- For $a_{j_1j_20}\prod_{s_2=1,s_2\neq s_1}^{3}(a_{j_1j_2s_1}-a_{j_1j_2s_2})<0$ and $a_{j_2j_10}>0$,

$$(a_{j_1j_2s_1},a_{j_2j_11})=\underbrace{(\text{DP}_+,\text{DP}_-)}_{\text{CCW center}}.$$

  The equilibrium of $(x_{j_2}^*,x_{j_1}^*)=(a_{j_1j_2s_1},a_{j_2j_11})$ is a (DP_+,DP_-)-counter-clockwise center.
- For $a_{j_1j_20}\prod_{s_2=1,s_2\neq s_1}^{3}(a_{j_1j_2s_1}-a_{j_1j_2s_2})>0$ and $a_{j_2j_10}<0$,

$$(a_{j_1j_2s_1},a_{j_2j_11})=\underbrace{(\text{DP}_-,\text{DP}_+)}_{\text{CW center}}.$$

  The equilibrium of $(x_{j_2}^*,x_{j_1}^*)=(a_{j_1j_2s_1},a_{j_2j_11})$ is a (DP_-,DP_+)-clockwise center.

- For $a_{j_1j_20}\prod_{s_2=1,s_2\neq s_1}^{3}(a_{j_1j_2s_1}-a_{j_1j_2s_2})<0$ and $a_{j_2j_10}<0$,

$$(a_{j_1j_2s_1},a_{j_2j_11})=\underbrace{(\mathrm{UP}_-,\mathrm{UP}_-)}_{\text{negative saddle}}.$$

The equilibrium of $(x_{j_2}^*,x_{j_1}^*)=(a_{j_1j_2s_1},a_{j_2j_11})$ is a (UP$_-$,UP$_-$)-negative saddle.

(xiv) If

$$\Delta_{j_1j_2}>0 \text{ and } \Delta_{j_2j_1}=0,$$

then

$$\begin{aligned}
&\dot{x}_{j_1}=a_{j_1j_20}(x_{j_2}-a_{j_1j_21})(x_{j_2}-a_{j_1j_22})(x_{j_2}-a_{j_1j_23}),\\
&\dot{x}_{j_2}=a_{j_2j_10}(x_{j_1}-a_{j_2j_1l_1})(x_{j_1}-a_{j_2j_1l_2})^2\\
&j_1,j_2\in\{1,2\},j_2\neq j_1.
\end{aligned}$$

In phase space,

$$\frac{dx_{j_1}}{dx_{j_2}}=\frac{a_{j_1j_20}}{a_{j_2j_10}}\frac{(x_{j_2}-a_{j_1j_21})(x_{j_2}-a_{j_1j_22})(x_{j_2}-a_{j_1j_23})}{(x_{j_1}-a_{j_2j_1l_1})(x_{j_1}-a_{j_2j_1l_2})^2}$$

and for $a_{j_1j_2s_1}\neq a_{j_1j_2s_2}$ (s_1, $s_2\in\{1,2,3\}$, $s_1\neq s_2$),

$$\begin{aligned}
&a_{j_2j_10}\big[(x_{j_1}-a_{j_2j_1l_1})^3+2(a_{j_2j_1l_1}-a_{j_2j_1l_2})(x_{j_1}-a_{j_2j_1l_1})^2\\
&+(a_{j_2j_1l_1}-a_{j_2j_1l_2})^2(x_{j_1}-a_{j_2j_1l_1})\big]dx_{j_1}\\
&=a_{j_1j_20}\big[(x_{j_2}-a_{j_1j_2s_1})^3+\textstyle\sum_{s_2=1,s_2\neq s_1}^{3}(a_{j_1j_2s_1}-a_{j_1j_2s_2})(x_{j_2}-a_{j_1j_2s_1})^2\\
&+\textstyle\prod_{s_2=1,s_2\neq s_1}^{3}(a_{j_1j_2s_1}-a_{j_1j_2s_2})(x_{j_2}-a_{j_1j_2s_1})\big]dx_{j_2}.
\end{aligned}$$

With an initial condition (x_{j_10},x_{j_20}), the first integral manifold is generated, i.e.,

$$\begin{aligned}
&a_{j_2j_10}\Big\{\frac{1}{4}\big[(x_{j_1}-a_{j_2j_1l_1})^4-(x_{j_10}-a_{j_2j_1l_1})^4\big]\\
&+\frac{2}{3}(a_{j_2j_1l_1}-a_{j_2j_1l_2})\big[(x_{j_1}-a_{j_2j_1l_1})^3-(x_{j_10}-a_{j_2j_1l_1})^3\big]\\
&+\frac{1}{2}(a_{j_2j_1l_1}-a_{j_2j_1l_2})^2\big[(x_{j_1}-a_{j_2j_1l_1})^2-(x_{j_10}-a_{j_2j_1l_1})^2\big]\Big\}\\
&=a_{j_1j_20}\Big\{\frac{1}{4}\big[(x_{j_2}-a_{j_1j_2s_1})^4-(x_{j_20}-a_{j_1j_2s_1})^4\big]\\
&+\frac{1}{3}\textstyle\sum_{s_2=1,s_2\neq s_1}^{3}(a_{j_1j_2s_1}-a_{j_1j_2s_2})\big[(x_{j_2}-a_{j_1j_2s_1})^3-(x_{j_20}-a_{j_1j_2s_1})^3\big]\\
&+\frac{1}{2}\textstyle\prod_{s_2=1,s_2\neq s_1}^{3}(a_{j_1j_2s_1}-a_{j_1j_2s_2})\big[(x_{j_2}-a_{j_1j_2s_1})^2-(x_{j_20}-a_{j_1j_2s_1})^2\big]\Big\}.
\end{aligned}$$

(xiv$_1$) Consider two cases (I) $x_{j_2}^* = a_{j_1 j_2 s_1}$ and (II)$x_{j_1}^* = a_{j_2 j_1 l_1}$.

(I) In phase space, at $x_{j_2}^* = a_{j_1 j_2 s_1} (s_1, s_2 \in \{1,2,3\}, s_1 \neq s_2)$ and $\overline{x}_{j_1} \neq a_{j_2 j_1 l_1}, a_{j_2 j_1 l_2}$, we have

$$\frac{dx_{j_1}}{dx_{j_2}}\bigg|_{x_{j_2}^* = a_{j_1 j_2 s_1}} = \frac{a_{j_1 j_2 0}}{a_{j_2 j_1 0}} \frac{(x_{j_2} - a_{j_1 j_2 s_1}) \prod_{s_2 = 1, s_2 \neq s_1}^{3} (x_{j_2} - a_{j_1 j_2 s_2})}{(\overline{x}_{j_1} - a_{j_2 j_1 l_1})(\overline{x}_{j_1} - a_{j_2 j_1 l_2})^2}\bigg|_{x_{j_2}^* = a_{j_1 j_2 s_1}} = 0.$$

If

$$\frac{d^2 x_{j_1}}{dx_{j_2}^2}\bigg|_{x_{j_2}^* = a_{j_1 j_2 s_1}} = \frac{a_{j_1 j_2 0}}{a_{j_2 j_1 0}} \frac{\prod_{s_2 = 1, s_2 \neq s_1}^{3} (a_{j_1 j_2 s_1} - a_{j_1 j_2 s_2})}{(\overline{x}_{j_1} - a_{j_2 j_1 l_1})(\overline{x}_{j_1} - a_{j_2 j_1 l_2})^2} > 0,$$

there is an up-parabola flow at $x_{j_2}^* = a_{j_1 j_2 s_1}$ in the x_{j_1}-direction. If

$$\frac{d^2 x_{j_1}}{dx_{j_2}^2}\bigg|_{x_{j_2}^* = a_{j_1 j_2 s_1}} = \frac{a_{j_1 j_2 0}}{a_{j_2 j_1 0}} \frac{\prod_{s_2 = 1, s_2 \neq s_1}^{3} (a_{j_1 j_2 s_1} - a_{j_1 j_2 s_2})}{(\overline{x}_{j_1} - a_{j_2 j_1 l_1})(\overline{x}_{j_1} - a_{j_2 j_1 l_2})^2} < 0,$$

there is a down-parabola flow at $x_{j_2}^* = a_{j_1 j_2 s_1}$ in the x_{j_1}-direction. Because of

$$\dot{x}_{j_2} = a_{j_2 j_1 0} (\overline{x}_{j_1} - a_{j_2 j_1 l_1})(\overline{x}_{j_1} - a_{j_2 j_1 l_2})^2,$$

the flows at $x_{j_2}^* = a_{j_1 j_2 s_1}$ in the x_{j_2}-direction are positive and negative for $\dot{x}_{j_2} > 0$ and $\dot{x}_{j_2} < 0$, respectively.

Thus, the equilibrium of $x_{j_2}^* = a_{j_1 j_2 s_1} (s_1, s_2 \in \{1,2,3\}, s_1 \neq s_2)$ on the x_{j_2}-direction has the following properties.

- For $a_{j_1 j_2 0} \prod_{s_2 = 1, s_2 \neq s_1}^{3} (a_{j_1 j_2 s_1} - a_{j_1 j_2 s_2}) > 0$ and $a_{j_2 j_1 0} (\overline{x}_{j_1} - a_{j_2 j_1 l_1})(\overline{x}_{j_1} - a_{j_2 j_1 l_2})^2 > 0$,

$$(a_{j_1 j_2 s_1}, \dot{x}_{j_2}) = \underbrace{(\text{UP}, \text{pF})}_{\text{up-parabola flow } (+)}.$$

- For $a_{j_1 j_2 0} \prod_{s_2 = 1, s_2 \neq s_1}^{3} (a_{j_1 j_2 s_1} - a_{j_1 j_2 s_2}) < 0$ and $a_{j_2 j_1 0} (\overline{x}_{j_1} - a_{j_2 j_1 l_1})(\overline{x}_{j_1} - a_{j_2 j_1 l_2})^2 > 0$,

$$(a_{j_1 j_2 s_1}, \dot{x}_{j_2}) = \underbrace{(\text{DP}, \text{pF})}_{\text{down-parabola flow } (+)}.$$

- For $a_{j_1 j_2 0} \prod_{s_2 = 1, s_2 \neq s_1}^{3} (a_{j_1 j_2 s_1} - a_{j_1 j_2 s_2}) > 0$ and $a_{j_2 j_1 0} (\overline{x}_{j_1} - a_{j_2 j_1 l_1})(\overline{x}_{j_1} - a_{j_2 j_1 l_2})^2 < 0$,

$$(a_{j_1 j_2 s_1}, \dot{x}_{j_2}) = \underbrace{(\text{DP}, \text{nF})}_{\text{down-parabola flow } (-)}.$$

- For $a_{j_1j_20}\prod_{s_2=1,s_2\neq s_1}^{3}(a_{j_1j_2s_1}-a_{j_1j_2s_2})<0$ and $a_{j_2j_10}(\bar{x}_{j_1}-a_{j_2j_1l_1})(\bar{x}_{j_1}-a_{j_2j_1l_2})^2<0$,

$$(a_{j_1j_2s_1},\dot{x}_{j_2})=\underbrace{(\mathrm{UP},\mathrm{nF})}_{\text{up-parabola flow }(-)}.$$

(II) At $x_{j_1}^*=a_{j_2j_1l_1}$ and $\bar{x}_{j_2}\neq a_{j_1j_21}$, in phase space,

$$\frac{dx_{j_2}}{dx_{j_1}}\bigg|_{x_{j_1}^*=a_{j_2j_1l_1}}=\frac{a_{j_2j_10}}{a_{j_1j_20}}\frac{(x_{j_1}-a_{j_2j_1l_1})(x_{j_1}-a_{j_2j_1l_2})^2}{\prod_{s_1=1}^{3}(\bar{x}_{j_2}-a_{j_1j_2s_1})}\bigg|_{x_{j_1}^*=a_{j_2j_1l_1}}=0.$$

If

$$\frac{d^2x_{j_2}}{dx_{j_1}^2}\bigg|_{x_{j_1}^*=a_{j_2j_1l_1}}=\frac{a_{j_2j_10}}{a_{j_1j_20}}\frac{(a_{j_2j_1l_1}-a_{j_2j_1l_2})^2}{\prod_{s_1=1}^{3}(\bar{x}_{j_2}-a_{j_1j_2s_1})}>0,$$

there is an up-parabola flow at $x_{j_1}^*=a_{j_2j_1l_1}$ on the x_{j_1}-direction. If

$$\frac{d^2x_{j_2}}{dx_{j_1}^2}\bigg|_{x_{j_1}^*=a_{j_2j_1l_1}}=\frac{a_{j_2j_10}}{a_{j_1j_20}}\frac{(a_{j_2j_1l_1}-a_{j_2j_1l_2})^2}{\prod_{s_1=1}^{3}(\bar{x}_{j_2}-a_{j_1j_2s_1})}<0,$$

there is a down-parabola flow at $x_{j_1}^*=a_{j_2j_1l_1}$ on the x_{j_1}-direction.

Because of

$$\dot{x}_{j_1}=a_{j_1j_20}\prod_{s_1=1}^{3}(\bar{x}_{j_2}-a_{j_1j_2s_1})$$

and

$$(a_{j_2j_1l_1}-a_{j_2j_1l_2})^2>0,$$

the flows at $x_{j_1}^*=a_{j_2j_1l_1}$ are positive and negative in the x_{j_1}-direction for $\dot{x}_{j_1}>0$ and $\dot{x}_{j_1}<0$, respectively.

Thus, the equilibrium of $x_{j_1}^*=a_{j_2j_1l_1}$ on the x_{j_2}-direction has the following properties.

- For $a_{j_1j_20}\prod_{s_1=1}^{3}(\bar{x}_{j_2}-a_{j_1j_2s_1})>0$ and $a_{j_2j_10}(a_{j_2j_1l_1}-a_{j_2j_1l_2})^2>0$,

$$(\dot{x}_{j_1},a_{j_2j_1l_1})=\underbrace{(\mathrm{pF},\mathrm{UP})}_{\text{up-parabola flow }(+)}.$$

- For $a_{j_1j_20}\prod_{s_1=1}^{3}(\overline{x}_{j_2}-a_{j_1j_2s_1})<0$ and $a_{j_2j_10}(a_{j_2j_1l_1}-a_{j_2j_1l_2})^2>0$,

$$(\dot{x}_{j_1}, a_{j_2j_1l_1}) = \underbrace{(\text{nF}, \text{DP})}_{\text{down-parabola flow }(-)}.$$

- For $a_{j_1j_20}\prod_{s_1=1}^{3}(\overline{x}_{j_2}-a_{j_1j_2s_1})>0$ and $a_{j_2j_10}(a_{j_2j_1l_1}-a_{j_2j_1l_2})^2<0$,

$$(\dot{x}_{j_1}, a_{j_2j_1l_1}) = \underbrace{(\text{pF}, \text{DP})}_{\text{down-parabola flow }(+)}.$$

- For $a_{j_1j_20}\prod_{s_1=1}^{3}(\overline{x}_{j_2}-a_{j_1j_2s_1})<0$ and $a_{j_2j_10}(a_{j_2j_1l_1}-a_{j_2j_1l_2})^2<0$,

$$(\dot{x}_{j_1}, a_{j_2j_1l_1}) = \underbrace{(\text{nF}, \text{UP})}_{\text{up-parabola flow }(-)}.$$

Therefore, from cases (I) and (II), the equilibrium of $(x_{j_2}^*, x_{j_1}^*) = (a_{j_1j_2s_1}, a_{j_2j_1l_1})$ $(s_1, s_2 \in \{1,2,3\}, s_1 \neq s_2; l_1, l_2 \in \{1,2\}, l_1 \neq l_2)$ has the following properties as in Eqs. (4.132)–(4.135).

- For $a_{j_1j_20}\prod_{s_2=1,s_2\neq s_1}^{3}(a_{j_1j_2s_1}-a_{j_1j_2s_2})>0$ and $a_{j_2j_10}(a_{j_2j_1l_1}-a_{j_2j_1l_2})^2>0$,

$$(a_{j_1j_2s_1}, a_{j_2j_1l_1}) = \underbrace{(\text{UP}_+, \text{UP}_+)}_{\text{positive saddle}}.$$

  The equilibrium of $(x_{j_2}^*, x_{j_1}^*) = (a_{j_1j_2s_1}, a_{j_2j_1l_1})$ is a $(\text{UP}_+,\text{UP}_+)$-positive saddle.

- For $a_{j_1j_20}\prod_{s_2=1,s_2\neq s_1}^{3}(a_{j_1j_2s_1}-a_{j_1j_2s_2})<0$ and $a_{j_2j_10}(a_{j_2j_1l_1}-a_{j_2j_1l_2})^2>0$,

$$(a_{j_1j_2s_1}, a_{j_2j_1l_1}) = \underbrace{(\text{DP}_+, \text{DP}_-)}_{\text{CCW center}}.$$

  The equilibrium of $(x_{j_2}^*, x_{j_1}^*) = (a_{j_1j_2s_1}, a_{j_2j_1l_1})$ is a $(\text{DP}_+,\text{DP}_-)$-counter-clockwise center.

- For $a_{j_1j_20}\prod_{s_2=1,s_2\neq s_1}^{3}(a_{j_1j_2s_1}-a_{j_1j_2s_2})>0$ and $a_{j_2j_10}(a_{j_2j_1l_1}-a_{j_2j_1l_2})^2<0$,

$$(a_{j_1j_2s_1}, a_{j_2j_1l_1}) = \underbrace{(\text{DP}_-, \text{DP}_+)}_{\text{CW center}}.$$

  The equilibrium of $(x_{j_2}^*, x_{j_1}^*) = (a_{j_1j_2s_1}, a_{j_2j_1l_1})$ is a $(\text{DP}_-,\text{DP}_+)$-clockwise center.

- For $a_{j_1j_20}\prod_{s_2=1,s_2\neq s_1}^{3}(a_{j_1j_2s_1}-a_{j_1j_2s_2})<0$ and $a_{j_2j_10}(a_{j_2j_1l_1}-a_{j_2j_1l_2})^2<0$,

$$(a_{j_1j_2s_1},a_{j_2j_1l_1})=\underbrace{(\mathrm{UP}_-,\mathrm{UP}_-)}_{\text{positive saddle}}.$$

The equilibrium of $(x_{j_2}^*,x_{j_1}^*)=(a_{j_1j_2s_1},a_{j_2j_1l_1})$ is a (UP$_-$,UP$_-$)-negative saddle.

(xiv$_2$) In phase space, at $x_{j_1}^*=a_{j_2j_1l_2}$ and $\overline{x}_{j_2}\neq a_{j_1j_2s_1}$ ($s_1=1,2,3$), we have

$$\frac{dx_{j_2}}{dx_{j_1}}\bigg|_{x_{j_1}^*=a_{j_2j_1l_2}}=\frac{a_{j_2j_10}}{a_{j_1j_20}}\frac{(x_{j_1}-a_{j_2j_1l_1})(x_{j_1}-a_{j_2j_1l_2})^2}{\prod_{s_1=1}^{3}(x_{j_2}-a_{j_1j_2s_1})}\bigg|_{x_{j_1}^*=a_{j_2j_1l_2}}=0,$$

$$\frac{d^2x_{j_2}}{dx_{j_1}^2}\bigg|_{x_{j_1}^*=a_{j_2j_1l_2}}=0.$$

If

$$\frac{d^3x_{j_2}}{dx_{j_1}^3}\bigg|_{x_{j_1}^*=a_{j_2j_1l_2}}=\frac{a_{j_1j_20}}{a_{j_2j_10}}\frac{2(a_{j_2j_1l_2}-a_{j_2j_1l_1})}{\prod_{s_1=1}^{3}(\overline{x}_{j_2}-a_{j_1j_2s_1})}>0,$$

there is an increasing-inflection flow at $x_{j_1}^*=a_{j_2j_1l_2}$ in the x_{j_2}-direction. If

$$\frac{d^3x_{j_2}}{dx_{j_1}^3}\bigg|_{x_{j_1}^*=a_{j_2j_11}}=\frac{a_{j_1j_20}}{a_{j_2j_10}}\frac{2(a_{j_2j_1l_2}-a_{j_2j_1l_1})}{\prod_{s_1=1}^{3}(\overline{x}_{j_2}-a_{j_1j_2s_1})}<0,$$

there is a decreasing-inflection flow at $x_{j_1}^*=a_{j_2j_1l_2}$ in the x_{j_1}-direction. Because of

$$\dot{x}_{j_1}=a_{j_1j_20}\prod_{s_1=1}^{3}(\overline{x}_{j_2}-a_{j_1j_2s_1}),$$

the flows at $x_{j_1}^*=a_{j_2j_1l_2}$ in the x_{j_1}-direction are positive and negative for $\dot{x}_{j_1}>0$ and $\dot{x}_{j_1}<0$, respectively.

Thus, the equilibrium of $x_{j_1}^*=a_{j_2j_1l_2}$ and $\overline{x}_{j_2}\neq a_{j_1j_2s_1}$ ($s_1=1,2,3$) on the x_{j_2}-direction has the following properties.

- For $a_{j_1j_20}\prod_{s_1=1}^{3}(\overline{x}_{j_2}-a_{j_1j_2s_1})>0$ and $a_{j_2j_10}(a_{j_2j_1l_2}-a_{j_2j_1l_1})>0$,

$$(\dot{x}_{j_1},a_{j_2j_1l_2})=\underbrace{(\mathrm{pF},\mathrm{II})}_{\text{increasing-inflection flow }(+)}.$$

- For $a_{j_1j_20}\prod_{s_1=1}^{3}(\overline{x}_{j_2}-a_{j_1j_2s_1})<0$ and $a_{j_2j_10}(a_{j_2j_1l_2}-a_{j_2j_1l_1})>0$,

$$(\dot{x}_{j_1}, a_{j_2j_1l_2}) = \underbrace{(\text{nF}, \text{DI})}_{\text{decreasing-inflection flow }(-)}.$$

- For $a_{j_1j_20}\prod_{s_1=1}^{3}(\overline{x}_{j_2}-a_{j_1j_2s_1})>0$ and $a_{j_2j_10}(a_{j_2j_1l_2}-a_{j_2j_1l_1})<0$,

$$(\dot{x}_{j_1}, a_{j_2j_1l_2}) = \underbrace{(\text{pF}, \text{DI})}_{\text{decreasing-inflection flow }(+)}.$$

- For $a_{j_1j_20}\prod_{s_1=1}^{3}(\overline{x}_{j_2}-a_{j_1j_2s_1})<0$ and $a_{j_2j_10}(a_{j_2j_1l_2}-a_{j_2j_1l_1})<0$,

$$(\dot{x}_{j_1}, a_{j_2j_1l_2}) = \underbrace{(\text{nF}, \text{II})}_{\text{increasing-inflection flow }(-)}.$$

Therefore, from the proof of case (xiv$_1$) and the above analysis, the equilibrium of $(x_{j_2}^*, x_{j_1}^*)=(a_{j_1j_2s_1}, a_{j_2j_1l_2})$ (s_1, $s_2 \in \{1,2,3\}$, $s_1 \neq s_2$; l_1, $l_2 \in \{1,2\}$, $l_1 \neq l_2$;) has the following properties as in Eqs. (4.136)–(4.139).

- For $a_{j_1j_20}\prod_{s_2=1,s_2\neq s_1}^{3}(a_{j_1j_2s_1}-a_{j_1j_2s_2})>0$ and $a_{j_2j_10}(a_{j_2j_1l_2}-a_{j_2j_1l_1})>0$,

$$(a_{j_1j_2s_1}, a_{j_2j_1l_2}) = \underbrace{(\text{UP}, \text{US})}_{\text{up-parabola upper-saddle}}.$$

  The equilibrium of $(x_{j_2}^*, x_{j_1}^*)=(a_{j_1j_2s_1}, a_{j_2j_1l_2})$ is a (UP,US)-up-parabola upper-saddle.
- For $a_{j_1j_20}\prod_{s_2=1,s_2\neq s_1}^{3}(a_{j_1j_2s_1}-a_{j_1j_2s_2})<0$ and $a_{j_2j_10}(a_{j_2j_1l_2}-a_{j_2j_1l_1})>0$,

$$(a_{j_1j_2s_1}, a_{j_2j_1l_2}) = \underbrace{(\text{DP}, \text{US})}_{\text{down-parabola upper-saddle}}.$$

  The equilibrium of $(x_{j_2}^*, x_{j_1}^*)=(a_{j_1j_2s_1}, a_{j_2j_1l_2})$ is a (DP,US)-down-parabola upper-saddle.
- For $a_{j_1j_20}\prod_{s_2=1,s_2\neq s_1}^{3}(a_{j_1j_2s_1}-a_{j_1j_2s_2})>0$ and $a_{j_2j_10}(a_{j_2j_1l_2}-a_{j_2j_1l_1})<0$,

$$(a_{j_1j_2s_1}, a_{j_2j_1l_2}) = \underbrace{(\text{DP}, \text{LS})}_{\text{down-parabola lower-saddle}}.$$

  The equilibrium of $(x_{j_2}^*, x_{j_1}^*)=(a_{j_1j_2s_1}, a_{j_2j_1l_2})$ is a (DP,LS)-down-parabola lower-saddle.

- For $a_{j_1j_20}\prod_{s_2=1,s_2\neq s_1}^{3}(a_{j_1j_2s_1}-a_{j_1j_2s_2})<0$ and $a_{j_2j_10}(a_{j_2j_1l_2}-a_{j_2j_1l_1})<0$,

$$(a_{j_1j_2s_1},a_{j_2j_1l_2})=\underbrace{(\mathrm{UP},\mathrm{LS})}_{\text{up-parabola lower-saddle}}.$$

The equilibrium of $(x^*_{j_2},x^*_{j_1})=(a_{j_1j_2s_1},a_{j_2j_1l_2})$ is a (UP,LS)-up-parabola lower-saddle.

(xv) If

$$\Delta_{j_1j_2}>0 \text{ and } \Delta_{j_2j_1}=0, a_{j_2j_1l_1}=a_{j_2j_1l_2}=a_{j_2j_11}$$

then

$$\begin{aligned}
&\dot{x}_{j_1}=a_{j_1j_20}(x_{j_2}-a_{j_1j_21})(x_{j_2}-a_{j_1j_22})(x_{j_2}-a_{j_1j_23}),\\
&\dot{x}_{j_2}=a_{j_2j_10}(x_{j_1}-a_{j_2j_11})^3\\
&j_1,j_2\in\{1,2\},j_2\neq j_1.
\end{aligned}$$

In phase space,

$$\frac{dx_{j_1}}{dx_{j_2}}=\frac{a_{j_1j_20}}{a_{j_2j_10}}\frac{(x_{j_2}-a_{j_1j_21})(x_{j_2}-a_{j_1j_22})(x_{j_2}-a_{j_1j_23})}{(x_{j_1}-a_{j_2j_11})^3}$$

and for $a_{j_1j_2s_1}\neq a_{j_1j_2s_2}$ (s_1, $s_2\in\{1,2,3\}$, $s_1\neq s_2$),

$$\begin{aligned}
&a_{j_2j_10}(x_{j_1}-a_{j_2j_11})^3dx_{j_1}\\
&=a_{j_1j_20}\Big[(x_{j_2}-a_{j_1j_2s_1})^3+\sum\nolimits_{s_2=1,s_2\neq s_1}^{3}(a_{j_1j_2s_1}-a_{j_1j_2s_2})(x_{j_2}-a_{j_1j_2s_1})^2\\
&+\prod\nolimits_{s_2=1,s_2\neq s_1}^{3}(a_{j_1j_2s_1}-a_{j_1j_2s_2})(x_{j_2}-a_{j_1j_2s_1})\Big]dx_{j_2}.
\end{aligned}$$

The first integral manifold is

$$\begin{aligned}
&\frac{1}{4}a_{j_2j_10}\left[(x_{j_1}-a_{j_2j_11})^4-(x_{j_10}-a_{j_2j_11})^4\right]\\
&=a_{j_1j_20}\Big\{\frac{1}{4}\left[(x_{j_2}-a_{j_1j_2s_1})^4-(x_{j_20}-a_{j_1j_2s_1})^4\right]\\
&+\frac{1}{3}\sum\nolimits_{s_2=1,s_2\neq s_1}^{3}(a_{j_1j_2s_1}-a_{j_1j_2s_2})\left[(x_{j_2}-a_{j_1j_2s_1})^3-(x_{j_20}-a_{j_1j_2s_1})^3\right]\\
&+\frac{1}{2}\prod\nolimits_{s_2=1,s_2\neq s_1}^{3}(a_{j_1j_2s_1}-a_{j_1j_2s_2})\left[(x_{j_2}-a_{j_1j_2s_1})^2-(x_{j_20}-a_{j_1j_2s_1})^2\right]\Big\}.
\end{aligned}$$

Consider two cases (I) $x^*_{j_2}=a_{j_1j_2s_1}$ and (II)$x^*_{j_1}=a_{j_2j_11}$.

(I) In phase space, at $x^*_{j_2} = a_{j_1j_2s_1}(s_1,\ s_2 \in \{1,2,3\},\ s_1 \neq s_2)$ with $\bar{x}_{j_1} \neq a_{j_2j_1 1}$, we have

$$\frac{dx_{j_1}}{dx_{j_2}}\bigg|_{x^*_{j_2}=a_{j_1j_2s_1}} = \frac{a_{j_1j_20}}{a_{j_2j_10}} \frac{(x_{j_2}-a_{j_1j_2s_1})\prod_{s_2=1,s_2\neq s_1}^{3}(x_{j_2}-a_{j_1j_2s_2})}{(\bar{x}_{j_1}-a_{j_2j_1 1})^3}\bigg|_{x^*_{j_2}=a_{j_1j_2s_1}} = 0.$$

If

$$\frac{d^2x_{j_1}}{dx^2_{j_2}}\bigg|_{x^*_{j_2}=a_{j_1j_2s_1}} = \frac{a_{j_1j_20}}{a_{j_2j_10}} \frac{\prod_{s_2=1,s_2\neq s_1}^{3}(a_{j_1j_2s_1}-a_{j_1j_2s_2})}{(\bar{x}_{j_1}-a_{j_2j_1 1})^3} > 0,$$

there is an up-parabola flow at $x^*_{j_2} = a_{j_1j_2s_1}$ in the x_{j_1}-direction. If

$$\frac{d^2x_{j_1}}{dx^2_{j_2}}\bigg|_{x^*_{j_2}=a_{j_1j_2s_1}} = \frac{a_{j_1j_20}}{a_{j_2j_10}} \frac{\prod_{s_2=1,s_2\neq s_1}^{3}(a_{j_1j_2s_1}-a_{j_1j_2s_2})}{(\bar{x}_{j_1}-a_{j_2j_1 1})^3} < 0,$$

there is a down-parabola flow at $x^*_{j_2} = a_{j_1j_2s_1}$ in the x_{j_1}-direction. Because of

$$\dot{x}_{j_2} = a_{j_2j_10}(\bar{x}_{j_1} - a_{j_2j_1 1})^3,$$

the flows at $x^*_{j_2} = a_{j_1j_2s_1}$ in the x_{j_2}-direction are positive and negative for $\dot{x}_{j_2} > 0$ and $\dot{x}_{j_2} < 0$, respectively.

Thus, the equilibrium of $x^*_{j_2} = a_{j_1j_2s_1}(s_1,\ s_2 \in \{1,2,3\},\ s_1 \neq s_2)$ on the x_{j_2}-direction has the following properties.

- For $a_{j_1j_20}\prod_{s_2=1,s_2\neq s_1}^{3}(a_{j_1j_2s_1}-a_{j_1j_2s_2}) > 0$ and $a_{j_2j_10}(\bar{x}_{j_1}-a_{j_2j_1 1})^3 > 0$,

$$(a_{j_1j_2s_1}, \dot{x}_{j_2}) = \underbrace{(\text{UP}, \text{pF})}_{\text{up-parabola flow } (+)}.$$

- For $a_{j_1j_20}\prod_{s_2=1,s_2\neq s_1}^{3}(a_{j_1j_2s_1}-a_{j_1j_2s_2}) < 0$ and $a_{j_2j_10}(\bar{x}_{j_1}-a_{j_2j_1 1})^3 > 0$,

$$(a_{j_1j_2s_1}, \dot{x}_{j_2}) = \underbrace{(\text{DP}, \text{pF})}_{\text{down-parabola flow } (+)}.$$

- For $a_{j_1j_20}\prod_{s_2=1,s_2\neq s_1}^{3}(a_{j_1j_2s_1}-a_{j_1j_2s_2}) > 0$ and $a_{j_2j_10}(\bar{x}_{j_1}-a_{j_2j_1 1})^3 < 0$,

$$(a_{j_1j_2s_1}, \dot{x}_{j_2}) = \underbrace{(\text{DP}, \text{nF})}_{\text{down-parabola flow } (-)}.$$

- For $a_{j_1j_20}\prod_{s_2=1,s_2\neq s_1}^{3}(a_{j_1j_2s_1}-a_{j_1j_2s_2})<0$ and $a_{j_2j_10}(\overline{x}_{j_1}-a_{j_2j_11})^3<0$,

$$(a_{j_1j_2s_1},\dot{x}_{j_2})=\underbrace{(\text{UP},\text{nF})}_{\text{up-parabola flow }(-)}.$$

(II) In phase space, at $x_{j_1}^*=a_{j_2j_11}$ and $\overline{x}_{j_2}\neq a_{j_1j_2s_1}$ (s_1 = 1, 2, 3), we have

$$\begin{aligned}
\frac{dx_{j_2}}{dx_{j_1}}\bigg|_{x_{j_1}^*=a_{j_2j_11}}&=\frac{a_{j_2j_10}}{a_{j_1j_20}}\frac{(x_{j_1}-a_{j_2j_11})^3}{\prod_{s_1=1}^{3}(x_{j_2}-a_{j_1j_2s_1})}\bigg|_{x_{j_1}^*=a_{j_2j_11}}=0,\\
\frac{d^2x_{j_2}}{dx_{j_1}^2}\bigg|_{x_{j_1}^*=a_{j_2j_11}}&=\frac{d^3x_{j_2}}{dx_{j_1}^3}\bigg|_{x_{j_1}^*=a_{j_2j_11}}=0.
\end{aligned}$$

If

$$\frac{d^4x_{j_2}}{dx_{j_1}^4}\bigg|_{x_{j_1}^*=a_{j_2j_11}}=\frac{a_{j_1j_20}}{a_{j_2j_10}}\frac{6}{\prod_{s_1=1}^{3}(\overline{x}_{j_2}-a_{j_1j_2s_1})}>0,$$

there is a third-order up-parabola flow at $x_{j_1}^*=a_{j_2j_11}$ in the x_{j_2}-direction. If

$$\frac{d^4x_{j_2}}{dx_{j_1}^4}\bigg|_{x_{j_1}^*=a_{j_2j_11}}=\frac{a_{j_1j_20}}{a_{j_2j_10}}\frac{6}{\prod_{s_1=1}^{3}(\overline{x}_{j_2}-a_{j_1j_2s_1})}<0,$$

there is a third-order down-parabola flow at $x_{j_1}^*=a_{j_2j_11}$ in the x_{j_2}-direction. Because of

$$\dot{x}_{j_1}=a_{j_1j_20}\prod_{s_1=1}^{3}(\overline{x}_{j_2}-a_{j_1j_2s_1}),$$

the flows at $x_{j_1}^*=a_{j_2j_11}$ in the x_{j_1}-direction are positive and negative for $\dot{x}_{j_1}>0$ and $\dot{x}_{j_1}<0$, respectively.

Thus, the equilibrium of $x_{j_1}^*=a_{j_2j_11}$ and $\overline{x}_{j_2}\neq a_{j_1j_2s_1}$ (s_1 = 1, 2, 3) on the x_{j_2}-direction has the following properties.

- For $a_{j_1j_20}\prod_{s_1=1}^{3}(\overline{x}_{j_2}-a_{j_1j_2s_1})>0$ and $a_{j_2j_10}>0$,

$$(\dot{x}_{j_1},a_{j_2j_11})=\underbrace{(\text{pF},3^{\text{rd}}\text{UP})}_{\text{third-order up-parabola flow }(+)}.$$

- For $a_{j_1j_20}\prod_{s_1=1}^{3}(\overline{x}_{j_2}-a_{j_1j_2s_1})<0$ and $a_{j_2j_10}>0$,

$$(\dot{x}_{j_1},a_{j_2j_11})=\underbrace{(\text{nF},3^{\text{rd}}\text{DP})}_{\text{third-order down-parabola flow }(-)}$$

- For $a_{j_1j_20}\prod_{s_1=1}^{3}(\overline{x}_{j_2}-a_{j_1j_2s_1})>0$ and $a_{j_2j_10}<0$,

$$(\dot{x}_{j_1},a_{j_2j_11})=\underbrace{(\text{pF},3^{\text{rd}}\text{DP})}_{\text{third-order down-parabola flow }(+)}$$

- For $a_{j_1j_20}\prod_{s_1=1}^{3}(\overline{x}_{j_2}-a_{j_1j_2s_1})<0$ and $a_{j_2j_10}<0$,

$$(\dot{x}_{j_1},a_{j_2j_11})=\underbrace{(\text{nF},3^{\text{rd}}\text{UP})}_{\text{third-order up-parabola flow }(-)}$$

Therefore, from the proof of case (xiv$_1$) and the above analysis, the equilibrium of $(x_{j_2}^*,x_{j_1}^*)=(a_{j_1j_2s_1},a_{j_2j_11})$ (s_1, $s_2\in\{1,2,3\}$, $s_1\neq s_2$) has the following properties as in Eqs. (4.143)–(4.146).

- For $a_{j_1j_20}\prod_{s_2=1,s_2\neq s_1}^{3}(a_{j_1j_2s_1}-a_{j_1j_2s_2})>0$ and $a_{j_2j_10}>0$,

$$(a_{j_1j_2s_1},a_{j_2j_11})=\underbrace{(\text{UP}_+,3^{\text{rd}}\text{UP}_+)}_{\text{third-order positive saddle}}.$$

  The equilibrium of $(x_{j_2}^*,x_{j_1}^*)=(a_{j_1j_2s_1},a_{j_2j_11})$ is a (UP$_+$,3rdUP$_+$)-third-order positive saddle.

- For $a_{j_1j_20}\prod_{s_2=1,s_2\neq s_1}^{3}(a_{j_1j_2s_1}-a_{j_1j_2s_2})<0$ and $a_{j_2j_10}>0$,

$$(a_{j_1j_2s_1},a_{j_2j_11})=\underbrace{(\text{DP}_+,3^{\text{rd}}\text{DP}_-)}_{\text{third-order CCW center}}.$$

  The equilibrium of $(x_{j_2}^*,x_{j_1}^*)=(a_{j_1j_2s_1},a_{j_2j_11})$ is a (DP$_+$,3rdDP$_-$)-third-order counter-clockwise center.

- For $a_{j_1j_20}\prod_{s_2=1,s_2\neq s_1}^{3}(a_{j_1j_2s_1}-a_{j_1j_2s_2})>0$ and $a_{j_2j_10}<0$,

$$(a_{j_1j_2s_1},a_{j_2j_11})=\underbrace{(\text{DP}_-,3^{\text{rd}}\text{DP}_+)}_{\text{third-order CW center}}.$$

  The equilibrium of $(x_{j_2}^*,x_{j_1}^*)=(a_{j_1j_2s_1},a_{j_2j_11})$ is a (DP$_-$,3rdDP$_+$)-third-order clockwise center.

- For $a_{j_1j_20}\prod_{s_2=1,s_2\neq s_1}^{3}(a_{j_1j_2s_1}-a_{j_1j_2s_2})<0$ and $a_{j_2j_10}<0$,

$$(a_{j_1j_2s_1},a_{j_2j_11})=\underbrace{(\text{UP}_-,3^{\text{rd}}\text{UP}_-)}_{\text{third-order negative saddle}}.$$

  The equilibrium of $(x_{j_2}^*,x_{j_1}^*)=(a_{j_1j_2s_1},a_{j_2j_11})$ is a (UP$_-$,3rdUP$_-$)-third-order negative saddle.

(xvi) If

$$\Delta_{j_1j_2} > 0 \text{ and } \Delta_{j_2j_1} > 0,$$

then

$$\begin{aligned}
&\dot{x}_{j_1} = a_{j_1j_20}(x_{j_2} - a_{j_1j_21})(x_{j_2} - a_{j_1j_22})(x_{j_2} - a_{j_1j_23}),\\
&\dot{x}_{j_2} = a_{j_2j_10}(x_{j_1} - a_{j_2j_11})(x_{j_1} - a_{j_2j_12})(x_{j_1} - a_{j_2j_13}),\\
&j_1, j_2 \in \{1,2\}; j_1 \neq j_2.
\end{aligned}$$

In phase space, for $a_{j_1j_2s_1} \neq a_{j_1j_2s_2}$ (s_1, $s_2 \in \{1,2,3\}$, $s_1 \neq s_2$) and $a_{j_2j_1l_1} \neq a_{j_2j_1l_2}$ (l_1, $l_2 \in \{1,2,3\}$, $l_1 \neq l_2$),

$$\frac{dx_{j_1}}{dx_{j_2}} = \frac{a_{j_1j_20}}{a_{j_2j_10}} \frac{(x_{j_2} - a_{j_1j_21})(x_{j_2} - a_{j_1j_22})(x_{j_2} - a_{j_1j_23})}{(x_{j_1} - a_{j_2j_11})(x_{j_1} - a_{j_2j_12})(x_{j_1} - a_{j_2j_13})},$$

and

$$\begin{aligned}
&a_{j_2j_10}\Big[(x_{j_1} - a_{j_2j_1l_1})^3 + \sum\nolimits_{l_2=1,l_2\neq l_1}^{3} (a_{j_2j_1l_1} - a_{j_2j_1l_2})(x_{j_1} - a_{j_2j_1l_1})^2\\
&+\prod\nolimits_{l_2=1,l_2\neq l_1}^{3} (a_{j_2j_1l_1} - a_{j_2j_1l_2})(x_{j_1} - a_{j_2j_1l_1})\Big]dx_{j_1}\\
&= a_{j_1j_20}\Big[(x_{j_2} - a_{j_1j_2s_1})^3 + \sum\nolimits_{s_2=1,s_2\neq s_1}^{3} (a_{j_1j_2s_1} - a_{j_1j_2s_2})(x_{j_2} - a_{j_1j_2s_1})^2\\
&+\prod\nolimits_{s_2=1,s_2\neq s_1}^{3} (a_{j_1j_2s_1} - a_{j_1j_2s_2})(x_{j_2} - a_{j_1j_2s_1})\Big]dx_{j_2}.
\end{aligned}$$

With an initial condition (x_{j_10}, x_{j_20}), the first integral manifold is generated, i.e.,

$$\begin{aligned}
&a_{j_2j_10}\Big\{\frac{1}{4}\big[(x_{j_1} - a_{j_2j_1l_1})^4 - (x_{j_10} - a_{j_2j_1l_1})^4\big]\\
&+\frac{1}{3}\sum\nolimits_{l_2=1,l_2\neq l_1}^{3} (a_{j_2j_1l_1} - a_{j_2j_1l_2})\big[(x_{j_1} - a_{j_2j_1l_1})^3 - (x_{j_10} - a_{j_2j_1l_1})^3\big]\\
&+\frac{1}{2}\prod\nolimits_{l_2=1,l_2\neq l_1}^{3} (a_{j_2j_1l_1} - a_{j_2j_1l_2})\big[(x_{j_1} - a_{j_2j_1l_1})^2 - (x_{j_10} - a_{j_2j_1l_1})\big]\Big\}\\
&= a_{j_1j_20}\Big\{\frac{1}{4}\big[(x_{j_2} - a_{j_1j_2s_1})^4 - (x_{j_20} - a_{j_1j_2s_1})^4\big]\\
&+\frac{1}{3}\sum\nolimits_{s_2=1,s_2\neq s_1}^{3} (a_{j_1j_2s_1} - a_{j_1j_2s_2})\big[(x_{j_2} - a_{j_1j_2s_1})^3 - (x_{j_20} - a_{j_1j_2s_1})^3\big]\\
&+\frac{1}{2}\prod\nolimits_{s_2=1,s_2\neq s_1}^{3} (a_{j_1j_2s_1} - a_{j_1j_2s_2})\big[(x_{j_2} - a_{j_1j_2s_1})^2 - (x_{j_20} - a_{j_1j_2s_1})^2\big]\Big\}.
\end{aligned}$$

Consider two cases (I) $x^*_{j_2} = a_{j_1j_2s_1}$ and (II)$x^*_{j_1} = a_{j_2j_1l_1}$.

(I) In phase space, at $x_{j_2}^* = a_{j_1 j_2 s_1} (s_1,\ s_2 \in \{1,2,3\},\ s_1 \neq s_2)$ and $\overline{x}_{j_1} \neq a_{j_2 j_1 l_1}$ $(l_1 = 1,$ $2,\ 3)$, we have

$$\frac{dx_{j_1}}{dx_{j_2}}\bigg|_{x_{j_2}^* = a_{j_1 j_2 s_1}} = \frac{a_{j_1 j_2 0}}{a_{j_2 j_1 0}} \frac{(x_{j_2} - a_{j_1 j_2 s_1})\prod_{s_2 = 1, s_2 \neq s_1}^{3} (x_{j_2} - a_{j_1 j_2 s_2})}{\prod_{l_1 = 1}^{3} (\overline{x}_{j_1} - a_{j_2 j_1 l_1})}\bigg|_{x_{j_2}^* = a_{j_1 j_2 s_1}} = 0.$$

If

$$\frac{d^2 x_{j_1}}{dx_{j_2}^2}\bigg|_{x_{j_2}^* = a_{j_1 j_2 s_1}} = \frac{a_{j_1 j_2 0}}{a_{j_2 j_1 0}} \frac{\prod_{s_2 = 1, s_2 \neq s_1}^{3} (a_{j_1 j_2 s_1} - a_{j_1 j_2 s_2})}{\prod_{l_1 = 1}^{3} (\overline{x}_{j_1} - a_{j_2 j_1 l_1})} > 0,$$

there is an up-parabola flow at $x_{j_2}^* = a_{j_1 j_2 s_1}$ in the x_{j_1}-direction. If

$$\frac{d^2 x_{j_1}}{dx_{j_2}^2}\bigg|_{x_{j_2}^* = a_{j_1 j_2 s_1}} = \frac{a_{j_1 j_2 0}}{a_{j_2 j_1 0}} \frac{\prod_{s_2 = 1, s_2 \neq s_1}^{3} (a_{j_1 j_2 s_1} - a_{j_1 j_2 s_2})}{\prod_{l_1 = 1}^{3} (\overline{x}_{j_1} - a_{j_2 j_1 l_1})} < 0,$$

there is a down-parabola flow at $x_{j_2}^* = a_{j_1 j_2 s_1}$ in the x_{j_1}-direction. Because of

$$\dot{x}_{j_2} = a_{j_2 j_1 0} \prod_{l_1 = 1}^{3} (\overline{x}_{j_1} - a_{j_2 j_1 l_1}),$$

the flows at $x_{j_2}^* = a_{j_1 j_2 s_1}$ in the x_{j_2}-direction are positive and negative for $\dot{x}_{j_2} > 0$ and $\dot{x}_{j_2} < 0$, respectively.

Thus, the equilibrium of $x_{j_2}^* = a_{j_1 j_2 s_1} (s_1,\ s_2 \in \{1,2,3\},\ s_1 \neq s_2)$ on the x_{j_2}-direction has the following properties.

- For $a_{j_1 j_2 0} \prod_{s_2 = 1, s_2 \neq s_1}^{3} (a_{j_1 j_2 s_1} - a_{j_1 j_2 s_2}) > 0$ and $a_{j_2 j_1 0} \prod_{l_1 = 1}^{3} (\overline{x}_{j_1} - a_{j_2 j_1 l_1}) > 0,$

$$(a_{j_1 j_2 s_1}, \dot{x}_{j_2}) = \underbrace{(\text{UP}, \text{pF})}_{\text{up-parabola flow } (+)}.$$

- For $a_{j_1 j_2 0} \prod_{s_2 = 1, s_2 \neq s_1}^{3} (a_{j_1 j_2 s_1} - a_{j_1 j_2 s_2}) < 0$ and $a_{j_2 j_1 0} \prod_{l_1 = 1}^{3} (\overline{x}_{j_1} - a_{j_2 j_1 l_1}) > 0,$

$$(a_{j_1 j_2 s_1}, \dot{x}_{j_2}) = \underbrace{(\text{DP}, \text{pF})}_{\text{down-parabola flow } (+)}.$$

- For $a_{j_1 j_2 0} \prod_{s_2 = 1, s_2 \neq s_1}^{3} (a_{j_1 j_2 s_1} - a_{j_1 j_2 s_2}) > 0$ and $a_{j_2 j_1 0} \prod_{l_1 = 1}^{3} (\overline{x}_{j_1} - a_{j_2 j_1 l_1}) < 0,$

$$(a_{j_1 j_2 s_1}, \dot{x}_{j_2}) = \underbrace{(\text{DP}, \text{nF})}_{\text{down-parabola flow } (-)}.$$

- For $a_{j_1j_20}\prod_{s_2=1,s_2\neq s_1}^{3}(a_{j_1j_2s_1}-a_{j_1j_2s_2})<0$ and $a_{j_2j_10}\prod_{l_1=1}^{3}(\bar{x}_{j_1}-a_{j_2j_1l_1})<0$,

$$(a_{j_1j_2s_1},\dot{x}_{j_2})=\underbrace{(\mathrm{UP},\mathrm{nF})}_{\text{up-parabola flow }(-)}.$$

(II) In phase space, at $x_{j_1}^{*}=a_{j_2j_1l_1}(l_1,\ l_2\in\{1,2,3\},\ l_1\neq l_2)$ and $\bar{x}_{j_2}\neq a_{j_1j_2s_1}$ ($s_1=1$, 2, 3), we have

$$\frac{dx_{j_2}}{dx_{j_1}}\bigg|_{x_{j_1}^{*}=a_{j_2j_1l_1}}=\frac{a_{j_2j_10}}{a_{j_1j_20}}\frac{(x_{j_1}-a_{j_2j_1l_1})\prod_{l_2=1,l_2\neq l_1}^{3}(x_{j_1}-a_{j_2j_1l_2})}{\prod_{s_1=1}^{3}(\bar{x}_{j_2}-a_{j_1j_2s_1})}\bigg|_{x_{j_1}^{*}=a_{j_2j_1l_1}}=0.$$

If

$$\frac{d^2x_{j_2}}{dx_{j_1}^2}\bigg|_{x_{j_1}^{*}=a_{j_2j_1l_1}}=\frac{a_{j_2j_10}}{a_{j_1j_20}}\frac{\prod_{l_2=1,l_2\neq l_1}^{3}(a_{j_2j_1l_1}-a_{j_2j_1l_2})}{\prod_{s_1=1}^{3}(\bar{x}_{j_2}-a_{j_1j_2s_1})}>0,$$

there is an up-parabola flow at $x_{j_1}^{*}=a_{j_2j_1l_1}$ in the x_{j_2}-direction. If

$$\frac{d^2x_{j_2}}{dx_{j_1}^2}\bigg|_{x_{j_1}^{*}=a_{j_2j_1l_1}}=\frac{a_{j_2j_10}}{a_{j_1j_20}}\frac{\prod_{l_2=1,l_2\neq l_1}^{3}(a_{j_2j_1l_1}-a_{j_2j_1l_2})}{\prod_{s_1=1}^{3}(\bar{x}_{j_2}-a_{j_1j_2s_1})}<0,$$

there is a down-parabola flow at $x_{j_1}^{*}=a_{j_2j_1l_1}$ in the x_{j_2}-direction. Because of

$$\dot{x}_{j_1}=a_{j_1j_20}\prod_{s_1=1}^{3}(\bar{x}_{j_2}-a_{j_1j_2s_1}),$$

the flows at $x_{j_1}^{*}=a_{j_2j_1l_1}$ in the x_{j_1}-direction are positive and negative for $\dot{x}_{j_1}>0$ and $\dot{x}_{j_1}<0$, respectively.

Thus, the equilibrium of $x_{j_1}^{*}=a_{j_2j_1l_1}\,(l_1,\ l_2\in\{1,2,3\},\ l_1\neq l_2)$ and $\bar{x}_{j_2}\neq a_{j_1j_2s_1}$ ($s_1=1$, 2, 3) on the x_{j_2}-direction has the following properties.

- For $a_{j_1j_20}\prod_{s_1=1}^{3}(\bar{x}_{j_2}-a_{j_1j_2s_1})>0$ and $a_{j_2j_10}\prod_{l_2=1,l_2\neq l_1}^{3}(a_{j_2j_1l_1}-a_{j_2j_1l_2})>0$,

$$(\dot{x}_{j_1},a_{j_2j_1l_1})=\underbrace{(\mathrm{pF},\mathrm{UP})}_{\text{up-parabola flow }(+)}.$$

- For $a_{j_1j_20}\prod_{s_1=1}^{3}(\bar{x}_{j_2}-a_{j_1j_2s_1})<0$ and $a_{j_2j_10}\prod_{l_2=1,l_2\neq l_1}^{3}(a_{j_2j_1l_1}-a_{j_2j_1l_2})>0$,

$$(\dot{x}_{j_1},a_{j_2j_1l_1})=\underbrace{(\mathrm{nF},\mathrm{DP})}_{\text{down-parabola flow }(-)}.$$

- For $a_{j_1j_20}\prod_{s_1=1}^{3}(\overline{x}_{j_2}-a_{j_1j_2s_1})>0$ and $a_{j_2j_10}\prod_{l_2=1,l_2\neq l_1}^{3}(a_{j_2j_1l_1}-a_{j_2j_1l_2})<0$,

$$(\dot{x}_{j_1},a_{j_2j_1l_1})=\underbrace{(\text{pF},\text{DP})}_{\text{down-parabola flow }(+)}.$$

- For $a_{j_1j_20}\prod_{s_1=1}^{3}(\overline{x}_{j_2}-a_{j_1j_2s_1})<0$ and $a_{j_2j_10}\prod_{l_2=1,l_2\neq l_1}^{3}(a_{j_2j_1l_1}-a_{j_2j_1l_2})<0$,

$$(\dot{x}_{j_1},a_{j_2j_1l_1})=\underbrace{(\text{nF},\text{UP})}_{\text{up-parabola flow }(-)}.$$

Therefore, from cases (I) and (II), the equilibrium of $(x_{j_2}^{*},x_{j_1}^{*})=(a_{j_1j_2s_1},a_{j_2j_1l_1})$ (s_1, s_2, l_1, $l_2\in\{1,2,3\}$, $s_1\neq s_2$, $l_1\neq l_2$) has the following properties as in Eqs. (4.150)–(4.153).

- For $a_{j_1j_20}\prod_{s_2=1,s_2\neq s_1}^{3}(a_{j_1j_2s_1}-a_{j_1j_2s_2})>0$ and $a_{j_2j_10}\prod_{l_2=1,l_2\neq l_1}^{3}(a_{j_2j_1l_1}-a_{j_2j_1l_2})>0$,

$$(a_{j_1j_2s_1},a_{j_2j_1l_1})=\underbrace{(\text{UP}_{+},\text{UP}_{+})}_{\text{positive saddle}}.$$

  The equilibrium of $(x_{j_2}^{*},x_{j_1}^{*})=(a_{j_1j_2s_1},a_{j_2j_1l_1})$ is a $(\text{UP}_{+},\text{UP}_{+})$-positive saddle.
- For $a_{j_1j_20}\prod_{s_2=1,s_2\neq s_1}^{3}(a_{j_1j_2s_1}-a_{j_1j_2s_2})<0$ and $a_{j_2j_10}\prod_{l_2=1,l_2\neq l_1}^{3}(a_{j_2j_1l_1}-a_{j_2j_1l_2})>0$,

$$(a_{j_1j_2s_1},a_{j_2j_1l_1})=\underbrace{(\text{DP}_{+},\text{DP}_{-})}_{\text{CCW center}}.$$

  The equilibrium of $(x_{j_2}^{*},x_{j_1}^{*})=(a_{j_1j_2s_1},a_{j_2j_1l_1})$ is a $(\text{DP}_{+},\text{DP}_{-})$-counter-clockwise center.
- For $a_{j_1j_20}\prod_{s_2=1,s_2\neq s_1}^{3}(a_{j_1j_2s_1}-a_{j_1j_2s_2})>0$ and $a_{j_2j_10}\prod_{l_2=1,l_2\neq l_1}^{3}(a_{j_2j_1l_1}-a_{j_2j_1l_2})<0$,

$$(a_{j_1j_2s_1},a_{j_2j_1l_1})=\underbrace{(\text{DP}_{-},\text{DP}_{+})}_{\text{CW center}}.$$

  The equilibrium of $(x_{j_2}^{*},x_{j_1}^{*})=(a_{j_1j_2s_1},a_{j_2j_1l_1})$ is a $(\text{DP}_{-},\text{DP}_{+})$-clockwise center.
- For $a_{j_1j_20}\prod_{s_2=1,s_2\neq s_1}^{3}(a_{j_1j_2s_1}-a_{j_1j_2s_2})<0$ and $a_{j_2j_10}\prod_{l_2=1,l_2\neq l_1}^{3}(a_{j_2j_1l_1}-a_{j_2j_1l_2})<0$,

$$(a_{j_1j_2s_1},a_{j_2j_1l_1})=\underbrace{(\text{UP}_{-},\text{UP}_{-})}_{\text{negative saddle}}.$$

  The equilibrium of $(x_{j_2}^{*},x_{j_1}^{*})=(a_{j_1j_2s_1},a_{j_2j_1l_1})$ is a $(\text{UP}_{-},\text{UP}_{-})$-negative saddle.

(xvi_2) From case (xvi_1), for $(x_{j_2}^{*},x_{j_1}^{*})=(a_{j_1j_2s_1},a_{j_2j_1l_1})$ (s_1, l_1 = 1, 2, 3), the corresponding equilibrium networks have the following properties as in Eqs. (4.154)–(4.157).

- For $a_{j_1j_20}>0$ and $a_{j_2j_10}>0$,

$$\left\{\begin{array}{ccc}(a_{j_1j_2}3,a_{j_2j_1}1) & (a_{j_1j_2}3,a_{j_2j_1}2) & (a_{j_1j_2}3,a_{j_2j_1}3)\\(a_{j_1j_2}2,a_{j_2j_1}1) & (a_{j_1j_2}2,a_{j_2j_1}2) & (a_{j_1j_2}2,a_{j_2j_1}3)\\(a_{j_1j_2}1,a_{j_2j_1}1) & (a_{j_1j_2}1,a_{j_2j_1}2) & (a_{j_1j_2}1,a_{j_2j_1}3)\end{array}\right\}$$
$$=\left\{\begin{array}{ccc}\underbrace{(\mathrm{UP}_+,\mathrm{UP}_+)}_{\text{positive saddle}} & \underbrace{(\mathrm{DP}_-,\mathrm{DP}_+)}_{\text{CW center}} & \underbrace{(\mathrm{UP}_+,\mathrm{UP}_+)}_{\text{positive saddle}}\\ \underbrace{(\mathrm{DP}_+,\mathrm{DP}_-)}_{\text{CCW center}} & \underbrace{(\mathrm{UP}_-,\mathrm{UP}_-)}_{\text{negative saddle}} & \underbrace{(\mathrm{DP}_+,\mathrm{DP}_-)}_{\text{CCW center}}\\ \underbrace{(\mathrm{UP}_+,\mathrm{UP}_+)}_{\text{positive saddle}} & \underbrace{(\mathrm{DP}_-,\mathrm{DP}_+)}_{\text{CW center}} & \underbrace{(\mathrm{UP}_+,\mathrm{UP}_+)}_{\text{positive saddle}}\end{array}\right\}.$$

- For $a_{j_1j_20}<0$ and $a_{j_2j_10}>0$,

$$\left\{\begin{array}{ccc}(a_{j_1j_2}3,a_{j_2j_1}1) & (a_{j_1j_2}3,a_{j_2j_1}2) & (a_{j_1j_2}3,a_{j_2j_1}3)\\(a_{j_1j_2}2,a_{j_2j_1}1) & (a_{j_1j_2}2,a_{j_2j_1}2) & (a_{j_1j_2}2,a_{j_2j_1}3)\\(a_{j_1j_2}1,a_{j_2j_1}1) & (a_{j_1j_2}1,a_{j_2j_1}2) & (a_{j_1j_2}1,a_{j_2j_1}3)\end{array}\right\}$$
$$=\left\{\begin{array}{ccc}\underbrace{(\mathrm{DP}_+,\mathrm{DP}_-)}_{\text{CCW center}} & \underbrace{(\mathrm{UP}_-,\mathrm{UP}_-)}_{\text{negative saddle}} & \underbrace{(\mathrm{DP}_+,\mathrm{DP}_-)}_{\text{CCW center}}\\ \underbrace{(\mathrm{UP}_+,\mathrm{UP}_+)}_{\text{positive saddle}} & \underbrace{(\mathrm{DP}_-,\mathrm{DP}_+)}_{\text{CW center}} & \underbrace{(\mathrm{UP}_+,\mathrm{UP}_+)}_{\text{positive saddle}}\\ \underbrace{(\mathrm{DP}_+,\mathrm{DP}_-)}_{\text{CCW center}} & \underbrace{(\mathrm{UP}_-,\mathrm{UP}_-)}_{\text{negative saddle}} & \underbrace{(\mathrm{DP}_+,\mathrm{DP}_-)}_{\text{CCW center}}\end{array}\right\}.$$

- For $a_{j_1j_20}>0$ and $a_{j_2j_10}<0$,

$$\left\{\begin{array}{ccc}(a_{j_1j_2}3,a_{j_2j_1}1) & (a_{j_1j_2}3,a_{j_2j_1}2) & (a_{j_1j_2}3,a_{j_2j_1}3)\\(a_{j_1j_2}2,a_{j_2j_1}1) & (a_{j_1j_2}2,a_{j_2j_1}2) & (a_{j_1j_2}2,a_{j_2j_1}3)\\(a_{j_1j_2}1,a_{j_2j_1}1) & (a_{j_1j_2}1,a_{j_2j_1}2) & (a_{j_1j_2}1,a_{j_2j_1}3)\end{array}\right\}$$
$$=\left\{\begin{array}{ccc}\underbrace{(\mathrm{DP}_-,\mathrm{DP}_+)}_{\text{CW center}} & \underbrace{(\mathrm{UP}_+,\mathrm{UP}_+)}_{\text{positive saddle}} & \underbrace{(\mathrm{DP}_-,\mathrm{DP}_+)}_{\text{CW center}}\\ \underbrace{(\mathrm{UP}_-,\mathrm{UP}_-)}_{\text{negative saddle}} & \underbrace{(\mathrm{DP}_+,\mathrm{DP}_-)}_{\text{CCW center}} & \underbrace{(\mathrm{UP}_-,\mathrm{UP}_-)}_{\text{negative saddle}}\\ \underbrace{(\mathrm{DP}_-,\mathrm{DP}_+)}_{\text{CW center}} & \underbrace{(\mathrm{UP}_+,\mathrm{UP}_+)}_{\text{positive saddle}} & \underbrace{(\mathrm{DP}_-,\mathrm{DP}_+)}_{\text{CW center}}\end{array}\right\}.$$

- For $a_{j_1j_20} < 0$ and $a_{j_2j_10} < 0$,

$$\left\{\begin{array}{ccc} (a_{j_1j_2 3}, a_{j_2j_1 1}) & (a_{j_1j_2 3}, a_{j_2j_1 2}) & (a_{j_1j_2 3}, a_{j_2j_1 3}) \\ (a_{j_1j_2 2}, a_{j_2j_1 1}) & (a_{j_1j_2 2}, a_{j_2j_1 2}) & (a_{j_1j_2 2}, a_{j_2j_1 3}) \\ (a_{j_1j_2 1}, a_{j_2j_1 1}) & (a_{j_1j_2 1}, a_{j_2j_1 2}) & (a_{j_1j_2 1}, a_{j_2j_1 3}) \end{array}\right\}$$
$$= \left\{\begin{array}{ccc} \underbrace{(\mathrm{UP}_-, \mathrm{UP}_-)}_{\text{negative saddle}} & \underbrace{(\mathrm{DP}_+, \mathrm{DP}_-)}_{\text{CCW center}} & \underbrace{(\mathrm{UP}_-, \mathrm{UP}_-)}_{\text{negative saddle}} \\ \underbrace{(\mathrm{DP}_-, \mathrm{DP}_+)}_{\text{CW center}} & \underbrace{(\mathrm{UP}_+, \mathrm{UP}_+)}_{\text{positive saddle}} & \underbrace{(\mathrm{DP}_-, \mathrm{DP}_+)}_{\text{CW center}} \\ \underbrace{(\mathrm{UP}_-, \mathrm{UP}_-)}_{\text{negative saddle}} & \underbrace{(\mathrm{DP}_+, \mathrm{DP}_-)}_{\text{CCW center}} & \underbrace{(\mathrm{UP}_-, \mathrm{UP}_-)}_{\text{negative saddle}} \end{array}\right\}.$$

(xvii) Under

$$\Delta_{j_1j_2} < 0 \text{ and } \Delta_{j_2j_1} = 0,$$

the standard form is

$$\begin{aligned} \dot{x}_{j_2} &= a_{j_2j_10}(x_{j_1} - a_{j_2j_1l_1})(x_{j_1} - a_{j_2j_1l_2})^2, \\ \dot{x}_{j_1} &= a_{j_1j_20}(x_{j_2} - a_{j_1j_21})\left[(x_{j_2} - a_{j_1j_2})^2 + b_{j_1j_2}\right], \\ & j_1, j_2 \in \{1,2\}; j_1 \neq j_2. \end{aligned}$$

From case (ii$_2$) and case (iv), the equilibrium of $(x^*_{j_2}, x^*_{j_1}) = (a_{j_1j_21}, a_{j_2j_1l_2})$ possesses the following appearing bifurcations as in Eqs. (4.160)–(4.163).

- For $a_{j_1j_20} > 0$ and $a_{j_2j_10}(a_{j_2j_1l_2} - a_{j_2j_1l_1}) > 0$,

$$\underbrace{(a_{j_1j_21}, a_{j_2j_1l_2})}_{(\mathrm{UP}, \mathrm{US})\text{-appearing bifurcation}} \rightleftharpoons \left\{\begin{array}{c} \underbrace{(a_{j_1j_21}, a_{j_2j_1(l_2)_2})}_{(\mathrm{UP}_+, \mathrm{UP}_+)\text{-positive saddle}} \\ \underbrace{(a_{j_1j_21}, a_{j_2j_1(l_2)_1})}_{(\mathrm{DP}_-, \mathrm{DP}_+)\text{-CW center}} \end{array}\right\}.$$

The (UP,US)-up-parabola upper-saddle is a (UP,US)-appearing bifurcation for saddle and center.

- For $a_{j_1j_20} < 0$ and $a_{j_2j_10}(a_{j_2j_1l_2} - a_{j_2j_1l_1}) > 0$,

$$\underbrace{(a_{j_1j_21}, a_{j_2j_1l_2})}_{(\mathrm{DP}, \mathrm{US})\text{-appearing bifurcation}} \rightleftharpoons \left\{\begin{array}{c} \underbrace{(a_{j_1j_21}, a_{j_2j_1(l_2)_2})}_{(\mathrm{DP}_+, \mathrm{DP}_-)\text{-CCW center}} \\ \underbrace{(a_{j_1j_22}, a_{j_2j_1(l_2)_1})}_{(\mathrm{UP}_-, \mathrm{UP}_-)\text{-negative saddle}} \end{array}\right\}.$$

The (DP,US)-down-parabola upper-saddle is a (DP,US)-appearing bifurcation for center and saddle.

- For $a_{j_1j_20} > 0$ and $a_{j_2j_10}(a_{j_2j_1l_2} - a_{j_2j_1l_1}) < 0$,

$$\underbrace{(a_{j_1j_21}, a_{j_2j_1l_2})}_{(\mathrm{DP},\mathrm{LS})\text{-appearing bifurcation}} \rightleftharpoons \left\{\begin{array}{c} \underbrace{(a_{j_1j_21}, a_{j_2j_1(l_2)_2})}_{(\mathrm{DP}_-,\mathrm{DP}_+)\text{-CW center}} \\ \underbrace{(a_{j_1j_21}, a_{j_2j_1(l_2)_1})}_{(\mathrm{UP}_+,\mathrm{UP}_+)\text{-positive saddle}} \end{array}\right\}.$$

The (DP,LS)-down-parabola lower-saddle is a (DP,LS)-appearing bifurcation for center and saddle.

- For $a_{j_1j_20} < 0$ and $a_{j_2j_10}(a_{j_2j_1l_2} - a_{j_2j_1l_1}) < 0$,

$$\underbrace{(a_{j_1j_21}, a_{j_2j_1l_2})}_{(\mathrm{UP},\mathrm{LS})\text{-appearing bifurcation}} \rightleftharpoons \left\{\begin{array}{c} \underbrace{(a_{j_1j_21}, a_{j_2j_1(l_2)_2})}_{(\mathrm{UP}_-,\mathrm{UP}_-)\text{-negative saddle}} \\ \underbrace{(a_{j_1j_21}, a_{j_2j_1(l_2)_1})}_{(\mathrm{DP}_+,\mathrm{DP}_-)\text{-CCW center}} \end{array}\right\}.$$

The (UP,LS)-up-parabola lower-saddle is a (UP,LS)-appearing bifurcation for saddle and center.

– Such parabola upper-saddle and lower-saddle bifurcations are also the switching bifurcations for saddle with center.

(xix) Under

$$\Delta_{j_1j_2} = 0 \text{ and } \Delta_{j_2j_1} < 0, a_{j_1j_2s_1} \neq a_{j_1j_2s_2},$$

the standard form is

$$\begin{aligned} \dot{x}_{j_1} &= a_{j_1j_20}(x_{j_2} - a_{j_1j_2s_1})(x_{j_2} - a_{j_1j_2s_2})^2, \\ \dot{x}_{j_2} &= a_{j_2j_10}(x_{j_1} - a_{j_2j_11})\left[(x_{j_1} - a_{j_2j_1})^2 + b_{j_2j_1}\right] \\ j_1, j_2 &\in \{1,2\}, j_2 \neq j_1. \end{aligned}$$

From case (v_2) and (xiii), the equilibrium state of $x_{j_2}^* = a_{j_1j_2s_2}$ (s_1, $s_2 \in \{1,2\}$, $s_1 \neq s_2$) on the x_{j_2}-direction is for the following appearing bifurcations as in Eqs. (4.166)–(4.169).

- For $a_{j_1j_20}(a_{j_1j_2s_2} - a_{j_1j_2s_1}) > 0$ and $a_{j_2j_10} > 0$,

$$\underbrace{(a_{j_1j_2s_2}, a_{j_2j_11})}_{(\mathrm{US},\mathrm{UP})\text{-appearing bifurcation}} \rightleftharpoons \left\{\begin{array}{c} \underbrace{(a_{j_1j_2(s_2)_2}, a_{j_2j_11})}_{(\mathrm{UP}_+,\mathrm{UP}_+)\text{-positive saddle}} \\ \underbrace{(a_{j_1j_2(s_2)_1}, a_{j_2j_11})}_{(\mathrm{DP}_+,\mathrm{DP}_-)\text{-CCW center}} \end{array}\right\}.$$

The (US,UP)-up-parabola upper-saddle is a (US,UP)-appearing bifurcations for saddle and center.

- For $a_{j_1j_20}(a_{j_1j_2s_2} - a_{j_1j_2s_1}) < 0$ and $a_{j_2j_10} > 0$,

$$\underbrace{(a_{j_1j_2s_2}, a_{j_2j_1}1)}_{(\text{LS},\text{DP})\text{-appearing bifurcation}} \rightleftharpoons \left\{ \begin{array}{c} \underbrace{(a_{j_1j_2(s_2)_2}, a_{j_2j_1}1)}_{(\text{DP}_+,\text{DP}_-)\text{-CCW center}} \\ \underbrace{(a_{j_1j_2(s_2)_1}, a_{j_2j_1}1)}_{(\text{UP}_+,\text{UP}_+)\text{-positive saddle}} \end{array} \right\}.$$

The (LS,DP)-down-parabola lower-saddle is an (LS,DP)-appearing bifurcations for saddle and center.

- For $a_{j_1j_2 0}(a_{j_1j_2s_2} - a_{j_1j_2s_1}) > 0$ and $a_{j_2j_1 0} < 0$,

$$\underbrace{(a_{j_1j_2s_2}, a_{j_2j_1}1)}_{(\text{US},\text{DP})\text{-appearing bifurcation}} \rightleftharpoons \left\{ \begin{array}{c} \underbrace{(a_{j_1j_2(s_2)_2}, a_{j_2j_1}1)}_{(\text{DP}_-,\text{DP}_+)\text{-CW center}} \\ \underbrace{(a_{j_1j_2(s_2)_1}, a_{j_2j_1}1)}_{(\text{UP}_-,\text{UP}_-)\text{-negative saddle}} \end{array} \right\}.$$

The (US,DP)-down-parabola upper-saddle is a (US,DP)-appearing bifurcations for saddle and center.

- For $a_{j_1j_2 0}(a_{j_1j_2s_2} - a_{j_1j_2s_1}) < 0$ and $a_{j_2j_1 0} < 0$,

$$\underbrace{(a_{j_1j_2s_2}, a_{j_2j_1}1)}_{(\text{LS},\text{UP})\text{-appearing bifurcation}} \rightleftharpoons \left\{ \begin{array}{c} \underbrace{(a_{j_1j_2(s_2)_2}, a_{j_2j_1}1)}_{(\text{UP}_-,\text{UP}_-)\text{-negative saddle}} \\ \underbrace{(a_{j_1j_2(s_2)_1}, a_{j_2j_1}1)}_{(\text{DP}_-,\text{DP}_+)\text{-CW center}} \end{array} \right\}.$$

The (LS,UP)-up-parabola lower-saddle is an (LS,UP)-appearing bifurcations for saddle and center.

– Such parabola-saddle bifurcations are also the switching bifurcations for saddle and center.

(xx) Under

$$\Delta_{j_1j_2} = 0 \text{ and } \Delta_{j_2j_1} < 0, a_{j_1j_2s_1} = a_{j_1j_2s_2} = a_{j_1j_2}1$$

the standard form is

$$\begin{aligned} &\dot{x}_{j_1} = a_{j_1j_2 0}(x_{j_2} - a_{j_1j_2}1)^3, \\ &\dot{x}_{j_2} = a_{j_2j_1 0}(x_{j_1} - a_{j_2j_1}1)\left[(x_{j_1} - a_{j_2j_1})^2 + b_{j_2j_1}\right] \\ &j_1, j_2 \in \{1,2\}, j_2 \neq j_1. \end{aligned}$$

From case (vi) and case (xiii), the equilibrium of $(x^*_{j_2}, x^*_{j_1}) = (a_{j_1j_2}1, a_{j_2j_1}1)$ has the following appearing bifurcations as in Eqs. (4.172)–(4.175).

- For $a_{j_1j_2 0} > 0$ and $a_{j_2j_1 0} > 0$,

$$\underbrace{(a_{j_1j_2}1, a_{j_2j_1}1)}_{(\mathrm{UP}_+,\mathrm{UP}_+)\text{-positive saddle}} \rightleftharpoons \underbrace{(a_{j_1j_2}1, a_{j_2j_1}1)}_{(3^{\mathrm{rd}}\mathrm{UP}_+,\mathrm{UP}_+)\text{-appearing bifucation}} \rightleftharpoons \left\{\begin{array}{c} \underbrace{(a_{j_1j_2}3, a_{j_2j_1}1)}_{(\mathrm{UP}_+,\mathrm{UP}_+)\text{-positive saddle}} \\ \underbrace{(a_{j_1j_2}2, a_{j_2j_1}1)}_{(\mathrm{DP}_+,\mathrm{DP}_-)\text{-CCW center}} \\ \underbrace{(a_{j_1j_2}1, a_{j_2j_1}1)}_{(\mathrm{UP}_+,\mathrm{UP}_+)\text{-positive saddle}} \end{array}\right\}.$$

The ($3^{\mathrm{rd}}\mathrm{UP}_+$,$\mathrm{UP}_+$)-third-order positive saddle is a ($3^{\mathrm{rd}}\mathrm{UP}_+$,$\mathrm{UP}_+$)-appearing bifurcation for a saddle to saddle, center, and saddle.

- For $a_{j_1j_2}0 < 0$ and $a_{j_2j_1}0 > 0$,

$$\underbrace{(a_{j_1j_2}1, a_{j_2j_1}1)}_{(\mathrm{DP}_+,\mathrm{DP}_-)\text{-CCW center}} \rightleftharpoons \underbrace{(a_{j_1j_2}1, a_{j_2j_1}1)}_{(3^{\mathrm{rd}}\mathrm{DP}_+,\mathrm{DP}_-)\text{-appearing bifucation}} \rightleftharpoons \left\{\begin{array}{c} \underbrace{(a_{j_1j_2}3, a_{j_2j_1}1)}_{(\mathrm{DP}_+,\mathrm{DP}_-)\text{-CCW center}} \\ \underbrace{(a_{j_1j_2}2, a_{j_2j_1}1)}_{(\mathrm{UP}_+,\mathrm{UP}_+)\text{-positive saddle}} \\ \underbrace{(a_{j_1j_2}1, a_{j_2j_1}1)}_{(\mathrm{DP}_+,\mathrm{DP}_-)\text{-CCW center}} \end{array}\right\}.$$

The ($3^{\mathrm{rd}}\mathrm{DP}_+$,$\mathrm{DP}_-$)-third-order counter-clockwise center is a ($3^{\mathrm{rd}}\mathrm{DP}_+$,$\mathrm{DP}_-$)-appearing bifurcation for a center to center, saddle, and center.

- For $a_{j_1j_2}0 > 0$ and $a_{j_2j_1}0 < 0$,

$$\underbrace{(a_{j_1j_2}1, a_{j_2j_1}1)}_{(\mathrm{DP}_-,\mathrm{DP}_+)\text{-CW center}} \rightleftharpoons \underbrace{(a_{j_1j_2}1, a_{j_2j_1}1)}_{(3^{\mathrm{rd}}\mathrm{DP}_-,\mathrm{DP}_+)\text{-appearing bifucation}} \rightleftharpoons \left\{\begin{array}{c} \underbrace{(a_{j_1j_2}3, a_{j_2j_1}1)}_{(\mathrm{DP}_-,\mathrm{DP}_+)\text{-CW center}} \\ \underbrace{(a_{j_1j_2}2, a_{j_2j_1}1)}_{(\mathrm{UP}_-,\mathrm{UP}_-)\text{-negative saddle}} \\ \underbrace{(a_{j_1j_2}1, a_{j_2j_1}1)}_{(\mathrm{DP}_-,\mathrm{DP}_+)\text{-CW center}} \end{array}\right\}.$$

The ($3^{\mathrm{rd}}\mathrm{DP}_-$,$\mathrm{DP}_+$)-third-order clockwise center is a ($3^{\mathrm{rd}}\mathrm{DP}_-$,$\mathrm{DP}_+$)-appearing bifurcation for a center to center, saddle, and center.

- For $a_{j_1j_2}0 < 0$ and $a_{j_2j_1}0 < 0$,

$$\underbrace{(a_{j_1j_2}1, a_{j_2j_1}1)}_{(\text{UP}_-,\text{UP}_-)\text{-negative saddle}} \rightleftharpoons \underbrace{(a_{j_1j_2}1, a_{j_2j_1}1)}_{(3^{\text{rd}}\text{UP}_-,\text{UP}_-)\text{-appearing bifucation}} \rightleftharpoons \left\{ \begin{array}{c} \underbrace{(a_{j_1j_2}3, a_{j_2j_1}1)}_{(\text{UP}_-,\text{UP}_-)\text{-negative saddle}} \\ \underbrace{(a_{j_1j_2}2, a_{j_2j_1}1)}_{(\text{DP}_-,\text{DP}_+)\text{-CW center}} \\ \underbrace{(a_{j_1j_2}1, a_{j_2j_1}1)}_{(\text{UP}_-,\text{UP}_-)\text{-negative saddle}} \end{array} \right\}.$$

The $(3^{\text{rd}}\text{UP}_-,\text{UP}_-)$-third-order negative saddle is a $(3^{\text{rd}}\text{UP}_-,\text{UP}_-)$-appearing bifurcation for a saddle to the saddle, center, and saddle.

- Such third-order saddle bifurcations are also the switching bifurcations for saddle, center, and saddle.
- Such third-order center bifurcations are also the switching bifurcations for center, saddle, and center.

(xxi) Under

$$\Delta_{j_1j_2} = 0 \text{ and } \Delta_{j_2j_1} = 0, a_{j_2j_1s_1} \neq a_{j_2j_1s_2}, a_{j_1j_2l_1} \neq a_{j_1j_2l_2},$$

the standard form is

$$\begin{aligned} \dot{x}_{j_1} &= a_{j_1j_2}0(x_{j_2} - a_{j_1j_2s_1})(x_{j_2} - a_{j_1j_2s_2})^2, \\ \dot{x}_{j_2} &= a_{j_2j_1}0(x_{j_1} - a_{j_2j_1l_1})(x_{j_1} - a_{j_2j_1l_2})^2, \\ j_1, j_2 &\in \{1,2\}; j_1 \neq j_2. \end{aligned}$$

(xxi_1) From case (vii_2) and (xiv_1), the equilibrium of $(x^*_{j_2}, x^*_{j_1}) = (a_{j_1j_2s_2}, a_{j_2j_1l_1})$ has the following appearing bifurcations as in Eqs. (4.178)–(4.181).

- For $a_{j_1j_2}0(a_{j_1j_2s_2} - a_{j_1j_2s_1}) > 0$ and $a_{j_2j_1}0(a_{j_2j_1l_1} - a_{j_2j_1l_2})^2 > 0$,

$$\underbrace{(a_{j_1j_2s_2}, a_{j_2j_1l_1})}_{(\text{US},\text{UP})\text{-appearing bifurcation}} \rightleftharpoons \left\{ \begin{array}{c} \underbrace{(a_{j_1j_2(s_2)_1}, a_{j_2j_1l_1})}_{(\text{UP}_+,\text{UP}_+)\text{-positive saddle}} \\ \underbrace{(a_{j_1j_2(s_2)_1}, a_{j_2j_1l_1})}_{(\text{DP}_+,\text{DP}_-)\text{-CCW center}} \end{array} \right\}. \quad (4.178)$$

The (US,UP)-up-parabola upper-saddle is a (US,UP)-appearing bifurcation for center and saddle.

- For $a_{j_1j_2}0(a_{j_1j_2s_2} - a_{j_1j_2s_1}) < 0$ and $a_{j_2j_1}0(a_{j_2j_1l_1} - a_{j_2j_1l_2})^2 > 0$,

$$\underbrace{(a_{j_1j_2s_2}, a_{j_2j_1l_1})}_{(\text{LS},\text{DP})\text{-appearing bifurcation}} \rightleftharpoons \left\{\begin{array}{c} \underbrace{(a_{j_1j_2(s_2)_2}, a_{j_2j_1l_1})}_{(\text{DP}_+,\text{DP}_-)\text{-CCW center}} \\ \underbrace{(a_{j_1j_2(s_2)_1}, a_{j_2j_1l_1})}_{(\text{UP}_+,\text{UP}_+)\text{-positive saddle}} \end{array}\right\}.$$

The (LS,DP)-down-parabola lower-saddle is an (LS,DP)-appearing bifurcation for center and saddle.

- For $a_{j_1j_20}(a_{j_1j_2s_2} - a_{j_1j_2s_1}) > 0$ and $a_{j_2j_10}(a_{j_2j_1l_1} - a_{j_2j_1l_2})^2 < 0$,

$$\underbrace{(a_{j_1j_2s_2}, a_{j_2j_1l_1})}_{(\text{US},\text{DP})\text{-appearing bifurcation}} \rightleftharpoons \left\{\begin{array}{c} \underbrace{(a_{j_1j_2(s_2)_2}, a_{j_2j_1l_1})}_{(\text{DP}_-,\text{DP}_+)\text{-CW center}} \\ \underbrace{(a_{j_1j_2(s_2)_1}, a_{j_2j_1l_1})}_{(\text{UP}_-,\text{UP}_-)\text{-negative saddle}} \end{array}\right\}.$$

The (US,DP)-down-parabola upper-saddle is a (US,DP)-appearing bifurcation for center and saddle.

- For $a_{j_1j_20}(a_{j_1j_2s_2} - a_{j_1j_2s_1}) < 0$ and $a_{j_2j_10}(a_{j_2j_1l_1} - a_{j_2j_1l_2})^2 < 0$,

$$\underbrace{(a_{j_1j_2s_2}, a_{j_2j_1l_1})}_{(\text{LS},\text{UP})\text{-appearing bifurcation}} \rightleftharpoons \left\{\begin{array}{c} \underbrace{(a_{j_1j_2(s_2)_2}, a_{j_2j_1l_2})}_{(\text{UP}_-,\text{UP}_-)\text{-negative saddle}} \\ \underbrace{(a_{j_1j_2(s_2)_1}, a_{j_2j_1l_1})}_{(\text{DP}_-,\text{DP}_+)\text{-CW center}} \end{array}\right\}.$$

The (LS,UP)-up-parabola lower-saddle is an (LS,UP)-appearing bifurcation for center and saddle.

– Such parabola-saddle bifurcations are also the switching bifurcations for saddle and center.

(xxi_2) From case (vii_3) and case (xi_1), the equilibrium of $(x_{j_2}^*, x_{j_1}^*) = (a_{j_1j_2s_1}, a_{j_2j_1l_2})$ has the following appearing bifurcations as in Eqs. (4.182)–(4.185).

- For $a_{j_1j_20}(a_{j_1j_2s_1} - a_{j_1j_2s_2})^2 > 0$ and $a_{j_2j_10}(a_{j_2j_1l_2} - a_{j_2j_1l_1}) > 0$,

$$\underbrace{(a_{j_1j_2s_1}, a_{j_2j_1l_2})}_{(\text{UP},\text{US})\text{-appearing bifurcation}} \rightleftharpoons \left\{\begin{array}{c} \underbrace{(a_{j_1j_2s_1}, a_{j_2j_1(l_2)_2})}_{(\text{UP}_+,\text{UP}_+)\text{-positive saddle}} \\ \underbrace{(a_{j_1j_2s_1}, a_{j_2j_1(l_2)_1})}_{(\text{DP}_-,\text{DP}_+)\text{-CW center}} \end{array}\right\}.$$

The (UP,US)-up-parabola upper-saddle is a (UP,US)-appearing bifurcation for center and saddle.

- For $a_{j_1j_20}(a_{j_1j_2s_1} - a_{j_1j_2s_2})^2 < 0$ and $a_{j_2j_10}(a_{j_2j_1l_2} - a_{j_2j_1l_1}) > 0$,

$$\underbrace{(a_{j_1j_2s_1}, a_{j_2j_1l_2})}_{(\mathrm{DP},\mathrm{US})\text{-appearing bifurcation}} \rightleftharpoons \left\{ \begin{array}{c} \underbrace{(a_{j_1j_2s_1}, a_{j_2j_1(l_1)_2})}_{(\mathrm{DP}_+,\mathrm{DP}_-)\text{-CCW center}} \\ \underbrace{(a_{j_1j_2s_1}, a_{j_2j_1(l_2)_1})}_{(\mathrm{UP}_-,\mathrm{UP}_-)\text{-negative saddle}} \end{array} \right\}.$$

The (DP,US)-down-parabola upper-saddle is a (DP,US)-appearing bifurcation for center and saddle.

- For $a_{j_1j_20}(a_{j_1j_2s_1} - a_{j_1j_2s_2})^2 > 0$ and $a_{j_2j_10}(a_{j_2j_1l_2} - a_{j_2j_1l_1}) < 0$,

$$\underbrace{(a_{j_1j_2s_1}, a_{j_2j_1l_2})}_{(\mathrm{DP},\mathrm{LS})\text{-appearing bifurcation}} \rightleftharpoons \left\{ \begin{array}{c} \underbrace{(a_{j_1j_2s_1}, a_{j_2j_1(l_2)_2})}_{(\mathrm{DP}_-,\mathrm{DP}_+)\text{-CW center}} \\ \underbrace{(a_{j_1j_2s_1}, a_{j_2j_1(l_2)_1})}_{(\mathrm{UP}_+,\mathrm{UP}_+)\text{-positive saddle}} \end{array} \right\}.$$

The (DP,LS)-down-parabola lower-saddle is a (DP,LS)-appearing bifurcation for center and saddle.

- For $a_{j_1j_20}(a_{j_1j_2s_1} - a_{j_1j_2s_2})^2 < 0$ and $a_{j_2j_10}(a_{j_2j_1l_2} - a_{j_2j_1l_1}) < 0$,

$$\underbrace{(a_{j_1j_2s_1}, a_{j_2j_1l_2})}_{(\mathrm{UP},\mathrm{LS})\text{-appearing bifurcation}} \rightleftharpoons \left\{ \begin{array}{c} \underbrace{(a_{j_1j_2s_1}, a_{j_2j_1(l_2)_2})}_{(\mathrm{UP}_-,\mathrm{UP}_-)\text{-negative saddle}} \\ \underbrace{(a_{j_1j_2s_1}, a_{j_2j_1(l_2)_1})}_{(\mathrm{DP}_+,\mathrm{DP}_-)\text{-CCW center}} \end{array} \right\}.$$

The (UP,LS)-up-parabola lower-saddle is a (UP,LS)-appearing bifurcation for center and saddle.

– Such parabola upper-saddle and lower-saddle bifurcations are also the switching bifurcations for saddle and center.

(xxi_3) From case (vii_2) and (xvi_1), the equilibrium of $(x^*_{j_2}, x^*_{j_1}) = (a_{j_1j_2s_2}, a_{j_2j_1l_2})$ has the following bifurcations as in Eqs. (4.186)–(4.189).

- For $a_{j_1j_20}(a_{j_1j_2s_2} - a_{j_1j_2s_1}) > 0$ and $a_{j_2j_10}(a_{j_2j_1l_2} - a_{j_2j_1l_1}) > 0$,

$$\underbrace{(a_{j_1j_2s_2}, a_{j_2j_1l_2})}_{(\mathrm{II}_+,\mathrm{II}_+)\text{-appearing bifurcation}} \rightleftharpoons \left\{ \begin{array}{cc} \underbrace{(a_{j_1j_2(s_2)_2}, a_{j_2j_1(l_2)_1})}_{(\mathrm{DP}_-,\mathrm{DP}_+)\text{-CW center}} & \underbrace{(a_{j_1j_2(s_2)_2}, a_{j_2j_1(l_2)_2})}_{(\mathrm{UP}_+,\mathrm{UP}_+)\text{-positive saddle}} \\ \underbrace{(a_{j_1j_2(s_2)_1}, a_{j_2j_1(l_2)_1})}_{(\mathrm{UP}_-,\mathrm{UP}_-)\text{-negative saddle}} & \underbrace{(a_{j_1j_2(s_2)_1}, a_{j_2j_1(l_2)_2})}_{(\mathrm{DP}_+,\mathrm{DP}_-)\text{-CCW center}} \end{array} \right\}.$$

The (II_+,II_+)-double increasing-inflection saddle is an (II_+,II_+)-appearing bifurcation for a network of saddles and centers.

- For $a_{j_1j_20}(a_{j_1j_2s_2} - a_{j_1j_2s_1}) < 0$ and $a_{j_2j_10}(a_{j_2j_1l_2} - a_{j_2j_1l_1}) > 0$,

$$\underbrace{(a_{j_1j_2s_2}, a_{j_2j_1l_2})}_{(\mathrm{DI}_+,\mathrm{DI}_-)\text{-appearing bifurcation}} \rightleftharpoons \left\{\begin{array}{cc} \underbrace{(a_{j_1j_2(s_2)_2}, a_{j_2j_1(l_2)_1})}_{(\mathrm{UP}_-,\mathrm{UP}_-)\text{-negative saddle}} & \underbrace{(a_{j_1j_2(s_2)_2}, a_{j_2j_1(l_2)_2})}_{(\mathrm{DP}_+,\mathrm{DP}_-)\text{-CCW center}} \\ \underbrace{(a_{j_1j_2(s_2)_1}, a_{j_2j_1(l_2)_1})}_{(\mathrm{DP}_-,\mathrm{DP}_+)\text{-CW center}} & \underbrace{(a_{j_1j_2(s_2)_1}, a_{j_2j_1(l_2)_2})}_{(\mathrm{UP}_+,\mathrm{UP}_+)\text{-positive saddle}} \end{array}\right\}.$$

The (DI_+,DI_-)-double decreasing-inflection saddle is a (DI_+,DI_-)-appearing bifurcation for a network of saddles and centers.

- For $a_{j_1j_20}(a_{j_1j_2s_2} - a_{j_1j_2s_1}) > 0$ and $a_{j_2j_10}(a_{j_2j_1l_2} - a_{j_2j_1l_1}) < 0$,

$$\underbrace{(a_{j_1j_2s_2}, a_{j_2j_1l_2})}_{(\mathrm{DI}_-,\mathrm{DI}_+)\text{-appearing bifurcation}} \rightleftharpoons \left\{\begin{array}{cc} \underbrace{(a_{j_1j_2(s_2)_2}, a_{j_2j_1(l_2)_1})}_{(\mathrm{UP}_+,\mathrm{UP}_+)\text{-positive saddle}} & \underbrace{(a_{j_1j_2(s_2)_2}, a_{j_2j_1(l_2)_2})}_{(\mathrm{DP}_-,\mathrm{DP}_+)\text{-CW center}} \\ \underbrace{(a_{j_1j_2(s_2)_1}, a_{j_2j_1(l_2)_1})}_{(\mathrm{DP}_+,\mathrm{DP}_-)\text{-CCW center}} & \underbrace{(a_{j_1j_2(s_2)_1}, a_{j_2j_1(l_2)_2})}_{(\mathrm{UP}_-,\mathrm{UP}_-)\text{-negative saddle}} \end{array}\right\}.$$

The (DI_-,DI_+)-double decreasing-inflection saddle is a (DI_-,DI_+)-appearing bifurcation for a network of saddles and centers.

- For $a_{j_1j_20}(a_{j_1j_2s_2} - a_{j_1j_2s_1}) < 0$ and $a_{j_2j_10}(a_{j_2j_1l_2} - a_{j_2j_1l_1}) < 0$,

$$\underbrace{(a_{j_1j_2s_2}, a_{j_2j_1l_2})}_{(\mathrm{II}_-,\mathrm{II}_-)\text{-appearing bifurcation}} \rightleftharpoons \left\{\begin{array}{cc} \underbrace{(a_{j_1j_2(s_2)_2}, a_{j_2j_1(l_2)_1})}_{(\mathrm{DP}_+,\mathrm{DP}_-)\text{-CCW center}} & \underbrace{(a_{j_1j_2(s_2)_2}, a_{j_2j_1(l_2)_2})}_{(\mathrm{UP}_-,\mathrm{UP}_-)\text{-negative saddle}} \\ \underbrace{(a_{j_1j_2(s_2)_1}, a_{j_2j_1(l_2)_1})}_{(\mathrm{UP}_+,\mathrm{UP}_+)\text{-positive saddle}} & \underbrace{(a_{j_1j_2(s_2)_1}, a_{j_2j_1(l_2)_2})}_{(\mathrm{DP}_-,\mathrm{DP}_+)\text{-CW center}} \end{array}\right\}.$$

The (II_-,II_-)-double increasing-inflection saddle is an (II_-,II_-)-appearing bifurcation for a network of saddles and centers.

– Such double inflection saddles are also the switching bifurcations for two networks of saddles and centers.

(xxii) Under

$$\Delta_{j_1j_2} = 0 \text{ and } \Delta_{j_2j_1} = 0, a_{j_2j_1s_1} = a_{j_2j_1s_2} = a_{j_2j_11}, a_{j_1j_2l_1} \neq a_{j_1j_2l_2}$$

the standard form is

$$\begin{aligned} &\dot{x}_{j_1} = a_{j_1j_20}(x_{j_2} - a_{j_1j_21})^3, \\ &\dot{x}_{j_2} = a_{j_2j_10}(x_{j_1} - a_{j_2j_1l_1})(x_{j_1} - a_{j_2j_1l_2})^2, \\ &j_1, j_2 \in \{1,2\}; j_1 \neq j_2. \end{aligned}$$

(xxii_1) From cases (viii_1) and (xiv_1), the equilibrium of $(x_{j_2}^*, x_{j_1}^*) = (a_{j_1j_21}, a_{j_2j_1l_1})$ has the following bifurcations as in Eqs. (4.192)–(4.195).

- For $a_{j_1j_20} > 0$ and $a_{j_2j_10}(a_{j_2j_1l_1} - a_{j_2j_1l_2})^2 > 0$,

$$\underbrace{(a_{j_1j_2}1, a_{j_2j_1l_1})}_{(\mathrm{UP}_+,\,\mathrm{UP}_+)\text{-positive saddle}} \rightleftharpoons \underbrace{(a_{j_1j_2}1, a_{j_2j_1l_1})}_{(3^{\mathrm{rd}}\mathrm{UP}_+,\,\mathrm{UP}_+)\text{-appearing bifucation}} \rightleftharpoons \left\{ \begin{array}{c} \underbrace{(a_{j_1j_2}3, a_{j_2j_1l_1})}_{(\mathrm{UP}_+,\,\mathrm{UP}_+)\text{-positive saddle}} \\ \underbrace{(a_{j_1j_2}2, a_{j_2j_1l_1})}_{(\mathrm{DP}_+,\,\mathrm{DP}_-)\text{-CCW center}} \\ \underbrace{(a_{j_1j_2}1, a_{j_2j_1l_1})}_{(\mathrm{UP}_+,\,\mathrm{UP}_+)\text{-positive saddle}} \end{array} \right\}.$$

The (3rdUP₊,UP₊)-third-order positive saddle is a (3rdUP₊,UP₊)-appearing bifurcation from a saddle to saddle, center, and saddle.

- For $a_{j_1j_2}0 < 0$ and $a_{j_2j_1}0(a_{j_2j_1l_1} - a_{j_2j_1l_2})^2 > 0$,

$$\underbrace{(a_{j_1j_2}1, a_{j_2j_1l_1})}_{(\mathrm{DP}_+,\,\mathrm{DP}_-)\text{-CCW center}} \rightleftharpoons \underbrace{(a_{j_1j_2}1, a_{j_2j_1l_1})}_{(3^{\mathrm{rd}}\mathrm{DP}_+,\,\mathrm{DP}_-)\text{-appearing bifucation}} \rightleftharpoons \left\{ \begin{array}{c} \underbrace{(a_{j_1j_2}3, a_{j_2j_1l_1})}_{(\mathrm{DP}_+,\,\mathrm{DP}_-)\text{-CCW center}} \\ \underbrace{(a_{j_1j_2}2, a_{j_2j_1l_1})}_{(\mathrm{UP}_+,\,\mathrm{UP}_+)\text{-positive saddle}} \\ \underbrace{(a_{j_1j_2}1, a_{j_2j_1l_1})}_{(\mathrm{DP}_+,\,\mathrm{DP}_-)\text{-CCW center}} \end{array} \right\}.$$

The $(3^{\mathrm{rd}}\mathrm{DP}_+,\mathrm{DP}_-)$-third-order counter-clockwise center is a$(3^{\mathrm{rd}}\mathrm{DP}_+,\mathrm{DP}_-)$-appearing bifurcation from a center to center, saddle, clockwise center.

- For $a_{j_1j_2}0 > 0$ and $a_{j_2j_1}0(a_{j_2j_1l_1} - a_{j_2j_1l_2})^2 < 0$,

$$\underbrace{(a_{j_1j_2}1, a_{j_2j_1l_1})}_{(\mathrm{DP}_-,\,\mathrm{DP}_+)\text{-CW center}} \rightleftharpoons \underbrace{(a_{j_1j_2}1, a_{j_2j_1l_1})}_{(3^{\mathrm{rd}}\mathrm{DP}_-,\,\mathrm{DP}_+)\text{-appearing bifucation}} \rightleftharpoons \left\{ \begin{array}{c} \underbrace{(a_{j_1j_2}3, a_{j_2j_1l_1})}_{(\mathrm{DP}_-,\,\mathrm{DP}_+)\text{-CW center}} \\ \underbrace{(a_{j_1j_2}2, a_{j_2j_1l_1})}_{(\mathrm{UP}_-,\,\mathrm{UP}_-)\text{-negative saddle}} \\ \underbrace{(a_{j_1j_2}1, a_{j_2j_1l_1})}_{(\mathrm{DP}_-,\,\mathrm{DP}_+)\text{-CW center}} \end{array} \right\}.$$

The $(3^{\mathrm{rd}}\mathrm{DP}_-,\mathrm{DP}_+)$-third-order clockwise center is a $(3^{\mathrm{rd}}\mathrm{DP}_-,\mathrm{DP}_+)$-appearing bifurcation from a center to center, saddle, and center.

- For $a_{j_1j_2}0 < 0$ and $a_{j_2j_1}0(a_{j_2j_1l_1} - a_{j_2j_1l_2})^2 < 0$,

$$\underbrace{(a_{j_1j_2}1, a_{j_2j_1}l_1)}_{(\text{UP}_-,\text{UP}_-)\text{-negative saddle}} \rightleftharpoons \underbrace{(a_{j_1j_2}1, a_{j_2j_1}l_1)}_{(3^{\text{rd}}\text{UP}_-,\text{UP}_-)\text{-appearing bifucation}} \rightleftharpoons \left\{ \begin{array}{c} \underbrace{(a_{j_1j_2}3, a_{j_2j_1}l_1)}_{(\text{UP}_-,\text{UP}_-)\text{-negative saddle}} \\ \underbrace{(a_{j_1j_2}2, a_{j_2j_1}l_1)}_{(\text{DP}_-,\text{DP}_+)\text{-CW center}} \\ \underbrace{(a_{j_1j_2}1, a_{j_2j_1}l_1)}_{(\text{UP}_-,\text{UP}_-)\text{-negative saddle}} \end{array} \right\}.$$

The (3^{rd}UP$_-$,UP$_-$)-third-order negative saddle is a (3^{rd}UP$_-$,UP$_-$)-appearing bifurcation from a saddle to saddle, center, and saddle.

($xxii_2$) From cases ($viii_2$) and (xiv_2), the equilibrium of $(x_{j_2}^*, x_{j_1}^*) = (a_{j_1j_2}1, a_{j_2j_1}l_2)$ has the following appearing bifurcations as in Eqs. (4.196)–(4.199).

- For $a_{j_1j_2}0 > 0$ and $a_{j_2j_1}0(a_{j_2j_1}l_2 - a_{j_2j_1}l_1) > 0$,

$$\underbrace{(a_{j_1j_2}1, a_{j_2j_1}l_2)}_{(3^{\text{rd}}\text{UP},\text{US})\text{-appearing bifurcation}} \rightleftharpoons \left\{ \begin{array}{cc} \underbrace{(a_{j_1j_2}3, a_{j_2j_1}(l_2)_1)}_{(\text{DP}_-,\text{DP}_+)\text{-CW center}} & \underbrace{(a_{j_1j_2}3, a_{j_2j_1}(l_2)_2)}_{(\text{UP}_+,\text{UP}_+)\text{-positive saddle}} \\ \underbrace{(a_{j_1j_2}2, a_{j_2j_1}(l_2)_1)}_{(\text{UP}_-,\text{UP}_-)\text{-negative saddle}} & \underbrace{(a_{j_1j_2}2, a_{j_2j_1}(l_2)_2)}_{(\text{DP}_+,\text{DP}_-)\text{-CCW center}} \\ \underbrace{(a_{j_1j_2}1, a_{j_2j_1}(l_2)_1)}_{(\text{DP}_-,\text{DP}_+)\text{-CW center}} & \underbrace{(a_{j_1j_2}1, a_{j_2j_1}(l_2)_2)}_{(\text{UP}_+,\text{UP}_+)\text{-positive saddle}} \end{array} \right\}.$$

  The (3^{rd}UP,US)-third-order-up-parabola upper-saddle is a (3^{rd}UP,US)-appearing bifurcation for a 3×2 network of saddles and centers.
- For $a_{j_1j_2}0 < 0$ and $a_{j_2j_1}0(a_{j_2j_1}l_2 - a_{j_2j_1}l_1) > 0$,

$$\underbrace{(a_{j_1j_2}1, a_{j_2j_1}l_2)}_{(3^{\text{rd}}\text{DP},\text{US})\text{-appearing bifucation}} \rightleftharpoons \left\{ \begin{array}{cc} \underbrace{(a_{j_1j_2}3, a_{j_2j_1}(l_2)_1)}_{(\text{UP}_-,\text{UP}_-)\text{-negative saddle}} & \underbrace{(a_{j_1j_2}3, a_{j_2j_1}(l_2)_2)}_{(\text{DP}_+,\text{DP}_-)\text{-CCW center}} \\ \underbrace{(a_{j_1j_2}2, a_{j_2j_1}(l_2)_1)}_{(\text{DP}_-,\text{DP}_+)\text{-CW center}} & \underbrace{(a_{j_1j_2}2, a_{j_2j_1}(l_2)_2)}_{(\text{UP}_+,\text{UP}_+)\text{-positive saddle}} \\ \underbrace{(a_{j_1j_2}1, a_{j_2j_1}(l_2)_1)}_{(\text{UP}_-,\text{UP}_-)\text{-negative saddle}} & \underbrace{(a_{j_1j_2}1, a_{j_2j_1}(l_2)_2)}_{(\text{DP}_+,\text{DP}_-)\text{-CCW center}} \end{array} \right\}.$$

  The (3^{rd}DP,US)-third-order-down-parabola upper-saddle is a (3^{rd}DP,US)-appearing bifurcation for a saddle and center network.
- For $a_{j_1j_2}0 > 0$ and $a_{j_2j_1}0(a_{j_2j_1}l_2 - a_{j_2j_1}l_1) < 0$,

$$\underbrace{(a_{j_1j_2}1, a_{j_2j_1l_2})}_{(3^{\text{rd}}\text{DP, LS})\text{-appearing bifucation}} \rightleftharpoons \left\{\begin{array}{cc} \underbrace{(a_{j_1j_2}3, a_{j_2j_1(l_2)_1})}_{(\text{UP}_+, \text{UP}_+)\text{-positive saddle}} & \underbrace{(a_{j_1j_2}3, a_{j_2j_1(l_2)_2})}_{(\text{DP}_-, \text{DP}_+)\text{-CW center}} \\ \underbrace{(a_{j_1j_2}2, a_{j_2j_1(l_2)_1})}_{(\text{DP}_+, \text{DP}_-)\text{-CCW center}} & \underbrace{(a_{j_1j_2}2, a_{j_2j_1(l_2)_2})}_{(\text{UP}_-, \text{UP}_-)\text{-negative saddle}} \\ \underbrace{(a_{j_1j_2}1, a_{j_2j_1(l_2)_1})}_{(\text{UP}_+, \text{UP}_+)\text{-positive saddle}} & \underbrace{(a_{j_1j_2}1, a_{j_2j_1(l_2)_2})}_{(\text{DP}_-, \text{DP}_+)\text{-CW center}} \end{array}\right\}.$$

The (3^{rd}DP,LS)-third-order down-parabola lower-saddle is a (3^{rd}DP,LS)-appearing bifurcation for a 3×2 network of saddles and centers.

- For $a_{j_1j_2}0 < 0$ and $a_{j_2j_1}0(a_{j_2j_1l_2} - a_{j_2j_1l_1}) < 0$,

$$\underbrace{(a_{j_1j_2}1, a_{j_2j_1l_2})}_{(3^{\text{rd}}\text{UP, LS})\text{-appearing bifurcation}} \rightleftharpoons \left\{\begin{array}{cc} \underbrace{(a_{j_1j_2}3, a_{j_2j_1(l_2)_1})}_{(\text{DP}_+, \text{DP}_-)\text{-CCW center}} & \underbrace{(a_{j_1j_2}3, a_{j_2j_1(l_2)_2})}_{(\text{UP}_-, \text{UP}_-)\text{-negative saddle}} \\ \underbrace{(a_{j_1j_2}2, a_{j_2j_1(l_2)_1})}_{(\text{UP}_+, \text{UP}_+)\text{-positive saddle}} & \underbrace{(a_{j_1j_2}2, a_{j_2j_1(l_2)_2})}_{(\text{DP}_-, \text{DP}_+)\text{-CW center}} \\ \underbrace{(a_{j_1j_2}1, a_{j_2j_1(l_2)_1})}_{(\text{DP}_+, \text{DP}_-)\text{-CCW center}} & \underbrace{(a_{j_1j_2}1, a_{j_2j_1(l_2)_2})}_{(\text{UP}_-, \text{UP}_-)\text{-negative saddle}} \end{array}\right\}.$$

The (3^{rd}UP,LS)-third-order-up-parabola lower-saddle is a (3^{rd}UP,LS)-appearing bifurcation for a 3×2 network of saddles and centers.

(xxiii) If

$$\Delta_{j_1j_2} = 0 \text{ and } \Delta_{j_2j_1} = 0; a_{j_2j_1s_1} = a_{j_2j_1s_2} = a_{j_2j_1}1, a_{j_1j_2l_1} = a_{j_1j_2l_2} = a_{j_1j_2}1,$$

the standard form is

$$\begin{aligned} \dot{x}_{j_1} &= a_{j_1j_2}0(x_{j_2} - a_{j_1j_2}1)^3, \\ \dot{x}_{j_2} &= a_{j_2j_1}0(x_{j_1} - a_{j_2j_1}1)^3, \\ j_1, j_2 &\in \{1, 2\}; j_1 \neq j_2. \end{aligned}$$

From case (x) and case (xvi), the equilibrium of $(x^*_{j_2}, x^*_{j_1}) = (a_{j_1j_2}1, a_{j_2j_1}1)$ has the following appearing bifurcations as Eqs. (4.202)–(4.205).

- For $a_{j_1j_2}0 > 0$ and $a_{j_2j_1}0 > 0$,

$$\underbrace{(a_{j_1j_2}1,a_{j_2j_1}1)}_{(\mathrm{UP}_+,\mathrm{UP}_+)\text{-positive saddle}} \rightleftharpoons \underbrace{(a_{j_1j_2}1,a_{j_2j_1}1)}_{(3^{\mathrm{rd}}\mathrm{UP}_+,3^{\mathrm{rd}}\mathrm{UP}_+)\text{-appearing bifurcation}}$$

$$\rightleftharpoons \left\{\begin{array}{ccc}
\underbrace{(a_{j_1j_2}3,a_{j_2j_1}1)}_{(\mathrm{UP}_+,\mathrm{UP}_+)\text{-positive saddle}} & \underbrace{(a_{j_1j_2}3,a_{j_2j_1}2)}_{(\mathrm{DP}_-,\mathrm{DP}_+)\text{-CW center}} & \underbrace{(a_{j_1j_2}3,a_{j_2j_1}3)}_{(\mathrm{UP}_+,\mathrm{UP}_+)\text{-positive saddle}} \\
\underbrace{(a_{j_1j_2}2,a_{j_2j_1}1)}_{(\mathrm{DP}_+,\mathrm{DP}_-)\text{-CCW center}} & \underbrace{(a_{j_1j_2}2,a_{j_2j_1}2)}_{(\mathrm{UP}_-,\mathrm{UP}_-)\text{-negative saddle}} & \underbrace{(a_{j_1j_2}2,a_{j_2j_1}3)}_{(\mathrm{DP}_+,\mathrm{DP}_-)\text{-CCW center}} \\
\underbrace{(a_{j_1j_2}1,a_{j_2j_1}1)}_{(\mathrm{UP}_+,\mathrm{UP}_+)\text{-positive saddle}} & \underbrace{(a_{j_1j_2}1,a_{j_2j_1}2)}_{(\mathrm{DP}_-,\mathrm{DP}_+)\text{-CW center}} & \underbrace{(a_{j_1j_2}1,a_{j_2j_1}3)}_{(\mathrm{UP}_+,\mathrm{UP}_+)\text{-positive saddle}}
\end{array}\right\}.$$

The $(3^{\mathrm{rd}}\mathrm{UP}_+,3^{\mathrm{rd}}\mathrm{UP}_+)$-double third-order positive saddle is a $(3^{\mathrm{rd}}\mathrm{UP}_+,3^{\mathrm{rd}}\ \mathrm{UP}_+)$-appearing bifurcation from a saddle to a 3×3 network of saddles and centers.

- For $a_{j_1j_2}0<0$ and $a_{j_2j_1}0>0$,

$$\underbrace{(a_{j_1j_2}1,a_{j_2j_1}1)}_{(\mathrm{DP}_+,\mathrm{DP}_-)\text{-CCW center}} \rightleftharpoons \underbrace{(a_{j_1j_2}1,a_{j_2j_1}1)}_{(3^{\mathrm{rd}}\mathrm{DP}_+,3^{\mathrm{rd}}\mathrm{DP}_-)\text{-appearingbifucation}}$$

$$\rightleftharpoons \left\{\begin{array}{ccc}
\underbrace{(a_{j_1j_2}3,a_{j_2j_1}1)}_{(\mathrm{DP}_+,\mathrm{DP}_-)\text{-CCW center}} & \underbrace{(a_{j_1j_2}3,a_{j_2j_1}2)}_{(\mathrm{UP}_-,\mathrm{UP}_-)\text{-negative saddle}} & \underbrace{(a_{j_1j_2}3,a_{j_2j_1}3)}_{(\mathrm{DP}_+,\mathrm{DP}_-)\text{-CCW center}} \\
\underbrace{(a_{j_1j_2}2,a_{j_2j_1}1)}_{(\mathrm{UP}_+,\mathrm{UP}_+)\text{-positive saddle}} & \underbrace{(a_{j_1j_2}2,a_{j_2j_1}2)}_{(\mathrm{DP}_-,\mathrm{DP}_+)\text{-CW center}} & \underbrace{(a_{j_1j_2}2,a_{j_2j_1}3)}_{(\mathrm{UP}_+,\mathrm{UP}_+)\text{-positive saddle}} \\
\underbrace{(a_{j_1j_2}1,a_{j_2j_1}1)}_{(\mathrm{DP}_+,\mathrm{DP}_-)\text{-CCW center}} & \underbrace{(a_{j_1j_2}1,a_{j_2j_1}2)}_{(\mathrm{UP}_-,\mathrm{UP}_-)\text{-negative saddle}} & \underbrace{(a_{j_1j_2}1,a_{j_2j_1}3)}_{(\mathrm{DP}_+,\mathrm{UP}_-)\text{-CCW center}}
\end{array}\right\}.$$

The $(3^{\mathrm{rd}}\mathrm{DP}_+,3^{\mathrm{rd}}\mathrm{DP}_-)$-double third-order counter-clockwise center is a $(3^{\mathrm{rd}}\mathrm{DP}_+,3^{\mathrm{rd}}\mathrm{DP}_-)$-appearing bifurcation for a center to a 3×3 network of saddles and centers.

- For $a_{j_1j_2}0>0$ and $a_{j_2j_1}0<0$,

$$\underbrace{(a_{j_1j_2}1,a_{j_2j_1}1)}_{(\mathrm{DP}_-,\mathrm{DP}_+)\text{-CW center}} \rightleftharpoons \underbrace{(a_{j_1j_2}1,a_{j_2j_1}1)}_{(3^{\mathrm{rd}}\mathrm{DP}_+,3^{\mathrm{rd}}\mathrm{DP}_-)\text{-appearing bifucation}}$$

$$\rightleftharpoons \left\{\begin{array}{ccc}
\underbrace{(a_{j_1j_2}3,a_{j_2j_1}1)}_{(\mathrm{DP}_-,\mathrm{DP}_+)\text{-CW center}} & \underbrace{(a_{j_1j_2}3,a_{j_2j_1}2)}_{(\mathrm{UP}_+,\mathrm{UP}_+)\text{-positive saddle}} & \underbrace{(a_{j_1j_2}3,a_{j_2j_1}3)}_{(\mathrm{DP}_-,\mathrm{DP}_+)\text{-CW center}} \\
\underbrace{(a_{j_1j_2}2,a_{j_2j_1}1)}_{(\mathrm{UP}_-,\mathrm{UP}_-)\text{-negative saddle}} & \underbrace{(a_{j_1j_2}2,a_{j_2j_1}2)}_{(\mathrm{DP}_+,\mathrm{DP}_-)\text{-CCW center}} & \underbrace{(a_{j_1j_2}2,a_{j_2j_1}3)}_{(\mathrm{UP}_-,\mathrm{UP}_-)\text{-negative saddle}} \\
\underbrace{(a_{j_1j_2}1,a_{j_2j_1}1)}_{(\mathrm{DP}_-,\mathrm{DP}_+)\text{-CW center}} & \underbrace{(a_{j_1j_2}1,a_{j_2j_1}2)}_{(\mathrm{UP}_+,\mathrm{UP}_+)\text{-positive saddle}} & \underbrace{(a_{j_1j_2}1,a_{j_2j_1}3)}_{(\mathrm{DP}_-,\mathrm{DP}_+)\text{-CW center}}
\end{array}\right\}.$$

The $(3^{\mathrm{rd}}\mathrm{DP}_+,3^{\mathrm{rd}}\mathrm{DP}_-)$-double third-order clockwise center is a $(3^{\mathrm{rd}}\mathrm{DP},\ 3^{\mathrm{rd}}\mathrm{DP}_-)$ -appearing bifurcation from a center to a 3×3 network of saddles and centers.

- For $a_{j_1j_20} < 0$ and $a_{j_2j_10} < 0$,

$$\underbrace{(a_{j_1j_21}, a_{j_2j_11})}_{(\text{UP}_-,\text{UP}_-)\text{-negative saddle}} \rightleftharpoons \underbrace{(a_{j_1j_21}, a_{j_2j_11})}_{(3^{\text{rd}}\text{UP}_-,3^{\text{rd}}\text{UP}_-)\text{-appearing bifurcation}}$$
$$\rightleftharpoons \left\{ \begin{array}{ccc} \underbrace{(a_{j_1j_23}, a_{j_2j_11})}_{(\text{UP}_-,\text{UP}_-)\text{-negative saddle}} & \underbrace{(a_{j_1j_23}, a_{j_2j_12})}_{(\text{DP}_+,\text{DP}_-)\text{-CCW center}} & \underbrace{(a_{j_1j_23}, a_{j_2j_13})}_{(\text{UP}_-,\text{UP}_-)\text{-negative saddle}} \\ \underbrace{(a_{j_1j_22}, a_{j_2j_11})}_{(\text{DP}_-,\text{DP}_+)\text{-CW center}} & \underbrace{(a_{j_1j_22}, a_{j_2j_12})}_{(\text{UP}_+,\text{UP}_+)\text{-positive saddle}} & \underbrace{(a_{j_1j_22}, a_{j_2j_13})}_{(\text{DP}_-,\text{DP}_+)\text{-CW center}} \\ \underbrace{(a_{j_1j_21}, a_{j_2j_11})}_{(\text{UP}_-,\text{UP}_-)\text{-negative saddle}} & \underbrace{(a_{j_1j_21}, a_{j_2j_12})}_{(\text{DP}_+,\text{DP}_-)\text{-CCW center}} & \underbrace{(a_{j_1j_21}, a_{j_2j_13})}_{(\text{UP}_-,\text{UP}_-)\text{-negative saddle}} \end{array} \right\}.$$

The (3^{rd}UP_-,3^{rd}UP_-)-double third-order negative saddle is a (3^{rd}UP_-,3^{rd}UP_-)- appearing bifurcation for a saddle to a 3×3 network of saddles and centers.

(xxiv) Under

$$\Delta_{j_1j_2} = 0 \text{ and } \Delta_{j_2j_1} > 0, a_{j_1j_2s_1} \neq a_{j_1j_2s_2},$$

the standard form is

$$\begin{aligned} &\dot{x}_{j_2} = a_{j_2j_10}(x_{j_1} - a_{j_2j_11})(x_{j_1} - a_{j_2j_12})(x_{j_1} - a_{j_2j_13}), \\ &\dot{x}_{j_1} = a_{j_1j_20}(x_{j_2} - a_{j_1j_2s_1})(x_{j_2} - a_{j_1j_2s_2})^2, \\ &j_1,j_2 \in \{1,2\}; j_1 \neq j_2. \end{aligned}$$

From cases (xi$_2$) and (xvi), the equilibrium of $(x_{j_2}^*, x_{j_1}^*) = (a_{j_1j_2s_2}, a_{j_2j_1l_1})$ (s_1, $s_2 \in \{1,2\}$, $s_1 \neq s_2$; l_1, $l_2 \in \{1,2,3\}$, $l_1 \neq l_2$) has the following appearing bifurcation properties as in Eqs. (4.208)–(4.211).

- For $a_{j_1j_20}(a_{j_1j_2s_2} - a_{j_1j_2s_1}) > 0$ and $a_{j_2j_10}\prod_{l_2=1,l_2\neq l_1}^{3}(a_{j_2j_1l_1} - a_{j_2j_1l_2}) > 0$,

$$\underbrace{(a_{j_1j_2s_2}, a_{j_2j_1l_1})}_{(\text{US},\text{UP})\text{-appearing bifurcation}} \rightleftharpoons \left\{ \begin{array}{c} \underbrace{(a_{j_1j_2(s_2)_2}, a_{j_2j_1l_1})}_{(\text{UP}_+,\text{UP}_+)\text{-positive saddle}} \\ \underbrace{(a_{j_1j_2(s_2)_1}, a_{j_2j_1l_1})}_{(\text{DP}_-,\text{DP}_+)\text{-CCW center}} \end{array} \right\}.$$

The (US,UP)-up-parabola upper-saddle is a (US,UP)-appearing bifurcation for saddle and center.

- For $a_{j_1j_20}(a_{j_1j_2s_2} - a_{j_1j_2s_1}) < 0$ and $a_{j_2j_10}\prod_{l_2=1,l_2\neq l_1}^{3}(a_{j_2j_1l_1} - a_{j_2j_1l_2}) > 0$,

$$\underbrace{(a_{j_1j_2s_2}, a_{j_2j_1l_1})}_{(\mathrm{LS},\mathrm{DP})\text{-appearing bifurcation}} \rightleftharpoons \left\{ \begin{array}{c} \underbrace{(a_{j_1j_2(s_2)_2}, a_{j_2j_1l_1})}_{(\mathrm{DP}_+,\mathrm{DP}_-)\text{-CCW center}} \\ \underbrace{(a_{j_1j_2(s_2)_1}, a_{j_2j_1l_1})}_{(\mathrm{UP}_+,\mathrm{UP}_+)\text{-positive saddle}} \end{array} \right\}.$$

The (LS,DP)-down-parabola lower-saddle is an (LS,DP)-appearing bifurcation for saddle and center.

- For $a_{j_1j_20}(a_{j_1j_2s_2} - a_{j_1j_2s_1}) > 0$ and $a_{j_2j_10}\prod_{l_2=1,l_2\neq l_1}^{3}(a_{j_2j_1l_1} - a_{j_2j_1l_2}) < 0$,

$$\underbrace{(a_{j_1j_2s_2}, a_{j_2j_1l_1})}_{(\mathrm{US},\mathrm{DP})\text{-appearing bifurcation}} \rightleftharpoons \left\{ \begin{array}{c} \underbrace{(a_{j_1j_2(s_2)_2}, a_{j_2j_1l_1})}_{(\mathrm{DP}_-,\mathrm{DP}_+)\text{-CW center}} \\ \underbrace{(a_{j_1j_2(s_2)_1}, a_{j_2j_1l_1})}_{(\mathrm{UP}_-,\mathrm{UP}_-)\text{-negative saddle}} \end{array} \right\}.$$

The (US,DP)-down-parabola upper-saddle is a (US,DP)-appearing bifurcation for saddle and center.

- For $a_{j_1j_20}(a_{j_1j_2s_2} - a_{j_1j_2s_1}) < 0$ and $a_{j_2j_10}\prod_{l_2=1,l_2\neq l_1}^{3}(a_{j_2j_1l_1} - a_{j_2j_1l_2}) < 0$,

$$\underbrace{(a_{j_1j_2s_2}, a_{j_2j_1l_1})}_{(\mathrm{LS},\mathrm{UP})\text{-appearing bifurcation}} \rightleftharpoons \left\{ \begin{array}{c} \underbrace{(a_{j_1j_2(s_2)_2}, a_{j_2j_1l_1})}_{(\mathrm{UP}_-,\mathrm{UP}_-)\text{-negative saddle}} \\ \underbrace{(a_{j_1j_2(s_2)_1}, a_{j_2j_1l_1})}_{(\mathrm{DP}_+,\mathrm{DP}_-)\text{-CW center}} \end{array} \right\}.$$

The (LS,UP)-up-parabola lower-saddle is an (LS,UP)-appearing bifurcation for saddle and center.

(xxv) Under

$$\Delta_{j_1j_2} = 0 \text{ and } \Delta_{j_2j_1} > 0, a_{j_1j_2s_1} = a_{j_1j_2s_2} = a_{j_1j_21},$$

the standard form is

$$\begin{aligned} &\dot{x}_{j_2} = a_{j_2j_10}(x_{j_1} - a_{j_2j_11})(x_{j_1} - a_{j_2j_12})(x_{j_1} - a_{j_2j_13}), \\ &\dot{x}_{j_1} = a_{j_1j_20}(x_{j_2} - a_{j_1j_21})^3, \\ &j_1, j_2 \in \{1,2\}; j_1 \neq j_2. \end{aligned}$$

From cases (xiii) and (xvi$_1$), the equilibrium of $(x^*_{j_2}, x^*_{j_1}) = (a_{j_1j_21}, a_{j_2j_1l_1})$ (l_1, $l_2 \in \{1,2,3\}$, $l_1 \neq l_2$) has the following bifurcations as in Eqs. (4.214)–(4.217).

- For $a_{j_1j_20} > 0$ and $a_{j_2j_10}\prod_{l_2=1,l_2\neq l_1}^{3}(a_{j_2j_il_1} - a_{j_2j_1l_2}) > 0$,

$$\underbrace{(a_{j_1j_2}1,a_{j_2j_1l_1})}_{(\mathrm{UP}_+,\mathrm{UP}_+)\text{-positive saddle}} \rightleftharpoons \underbrace{(a_{j_1j_2}1,a_{j_2j_1l_1})}_{(3^{\mathrm{rd}}\mathrm{UP}_+,\mathrm{UP}_+)\text{-appearing bifucation}} \rightleftharpoons \left\{\begin{array}{c} \underbrace{(a_{j_1j_2}3,a_{j_2j_1l_1})}_{(\mathrm{UP}_+,\mathrm{UP}_+)\text{-positive saddle}} \\ \underbrace{(a_{j_1j_2}2,a_{j_2j_1l_1})}_{(\mathrm{DP}_+,\mathrm{DP}_-)\text{-CCW center}} \\ \underbrace{(a_{j_1j_2}1,a_{j_2j_1l_1})}_{(\mathrm{UP}_+,\mathrm{UP}_+)\text{-positive saddle}} \end{array}\right\}.$$

The ($3^{\mathrm{rd}}\mathrm{UP}_+,\mathrm{UP}_+$)-third-order positive saddle is a ($3^{\mathrm{rd}}\mathrm{UP}_+,\mathrm{UP}_+$)-appearing bifurcation from a saddle to saddle, center, and saddle.

- For $a_{j_1j_2 0}<0$ and $a_{j_2j_1 0}\prod_{l_2=1,l_2\neq l_1}^{3}(a_{j_2j_1l_1}-a_{j_2j_1l_2})>0$,

$$\underbrace{(a_{j_1j_2}1,a_{j_2j_1l_1})}_{(\mathrm{DP}_+,\mathrm{DP}_-)\text{-CCW center}} \rightleftharpoons \underbrace{(a_{j_1j_2}1,a_{j_2j_1l_1})}_{(3^{\mathrm{rd}}\mathrm{DP}_+,\mathrm{DP}_-)\text{-appearing bifucation}} \rightleftharpoons \left\{\begin{array}{c} \underbrace{(a_{j_1j_2}3,a_{j_2j_1l_1})}_{(\mathrm{DP}_+,\mathrm{DP}_-)\text{-CCW center}} \\ \underbrace{(a_{j_1j_2}2,a_{j_2j_1l_1})}_{(\mathrm{UP}_+,\mathrm{UP}_+)\text{-positive saddle}} \\ \underbrace{(a_{j_1j_2}1,a_{j_2j_1l_1})}_{(\mathrm{DP}_+,\mathrm{DP}_-)\text{-CCW center}} \end{array}\right\}.$$

The ($3^{\mathrm{rd}}\mathrm{DP}_+,\mathrm{DP}_-$)-third-order counter-clockwise center is a ($3^{\mathrm{rd}}\mathrm{DP}_+,\mathrm{DP}_-$)-appearing bifurcation from a center to center, saddle, and center.

- For $a_{j_1j_2 0}>0$ and $a_{j_2j_1 0}\prod_{l_2=1,l_2\neq l_1}^{3}(a_{j_2j_1l_1}-a_{j_2j_ll_2})<0$,

$$\underbrace{(a_{j_1j_2}1,a_{j_2j_1l_1})}_{(\mathrm{DP}_-,\mathrm{DP}_+)\text{-CW center}} \rightleftharpoons \underbrace{(a_{j_1j_2}1,a_{j_2j_1l_1})}_{(3^{\mathrm{rd}}\mathrm{DP}_-,\mathrm{DP}_+)\text{-appearing bifucation}} \rightleftharpoons \left\{\begin{array}{c} \underbrace{(a_{j_1j_2}3,a_{j_2j_1l_1})}_{(\mathrm{DP}_-,\mathrm{DP}_+)\text{-CW center}} \\ \underbrace{(a_{j_1j_2}2,a_{j_2j_1l_1})}_{(\mathrm{UP}_-,\mathrm{UP}_-)\text{-negative saddle}} \\ \underbrace{(a_{j_1j_2}1,a_{j_2j_1l_1})}_{(\mathrm{DP}_-,\mathrm{DP}_+)\text{-CW center}} \end{array}\right\}.$$

The ($3^{\mathrm{rd}}\mathrm{DP}_-,\mathrm{DP}_+$)-third-order clockwise center is a ($3^{\mathrm{rd}}\mathrm{DP}_-,\mathrm{DP}_+$)-appearing bifurcation from a center to center, saddle, and center.

- For $a_{j_1j_2 0}<0$ and $a_{j_2j_1 0}\prod_{l_2=1,l_2\neq l_1}^{3}(a_{j_2j_1l_1}-a_{j_2j_1l_2})<0$,

$$\underbrace{(a_{j_1j_21},a_{j_2j_1l_1})}_{(\text{UP}_-,\text{UP}_-)\text{-negative saddle}} \rightleftharpoons \underbrace{(a_{j_1j_21},a_{j_2j_1l_1})}_{(3^{\text{rd}}\text{UP}_-,\text{UP}_-)\text{-appearing bifucation}} \rightleftharpoons \left\{ \begin{array}{c} \underbrace{(a_{j_1j_23},a_{j_2j_1l_1})}_{(\text{UP}_-,\text{UP}_-)\text{-negative saddle}} \\ \underbrace{(a_{j_1j_22},a_{j_2j_1l_1})}_{(\text{DP}_-,\text{DP}_+)\text{-CW center}} \\ \underbrace{(a_{j_1j_21},a_{j_2j_1l_1})}_{(\text{UP}_-,\text{UP}_-)\text{-negative saddle}} \end{array} \right\}.$$

The $(3^{\text{rd}}\text{UP}_-,\text{UP}_-)$-third-order negative saddle is a $(3^{\text{rd}}\text{UP}_-,\text{UP}_-)$-appearing bifurcation from a saddle to saddle, center, and saddle.

In the end, from the above local analysis and the first integral of such a dynamical system, this theorem is proved. ■

4.3 Singular Equilibriums

From the previous theorem, the bifurcations and global dynamics of a cubic dynamical system possessing two variable-crossing, univariate cubic vector fields. The singularity and bifurcations for saddle and centers of equilibriums will be presented.

4.3.1 Double Third-Order Saddles and Centers

Consider a dynamical system as

$$\begin{aligned} \dot{x}_1 &= a_{120}(x_2 - a_{121})^3, \\ \dot{x}_2 &= a_{210}(x_1 - a_{211})^3, \end{aligned} \tag{4.218}$$

and the corresponding first integral manifold is determined by

$$\begin{aligned} &\frac{1}{4}a_{120}\left[(x_2 - a_{121})^4 - (x_{20} - a_{121})^4\right] \\ &= \frac{1}{4}a_{210}\left[(x_1 - a_{211})^4 - (x_{10} - a_{211})^4\right]. \end{aligned} \tag{4.219}$$

The corresponding phase portraits near the third-order parabola-upper-saddle and parabola-lower-saddle bifurcations are presented in Fig. 4.1. From the first integral manifold, the phase portraits for two-dimensional system for $x_1^* = a_{211}$ on the x_1-direction for the x_2-direction and $x_2^* = a_{121}$ on the x_2-direction for the x_1-direction are sketched. The equilibriums of $(x_1^*, x_2^*) = (a_{211}, a_{121})$ has the following properties. Figure 4.1a–d presented a $(3^{\text{rd}}\text{UP}_+, 3^{\text{rd}}\text{UP}_+)$-double third-order positive saddle $(a_{120} > 0, a_{210} > 0)$, a $(3^{\text{rd}}\text{DP}_+, 3^{\text{rd}}\text{DP}_-)$-double third-order counter-clockwise center $(a_{120} < 0, a_{210} > 0)$, a $(3^{\text{rd}}\text{DP}_-, 3^{\text{rd}}\text{DP}_+)$-double third-order clockwise center $(a_{120} > 0, a_{210} < 0)$, and a $(3^{\text{rd}}\text{UP}_-, 3^{\text{rd}}\text{UP}_-)$-double-third-order negative saddle $(a_{120} < 0, a_{210} < 0)$, respectively. The double third-order positive and negative saddles and clockwise and counter-clockwise centers are as follows.

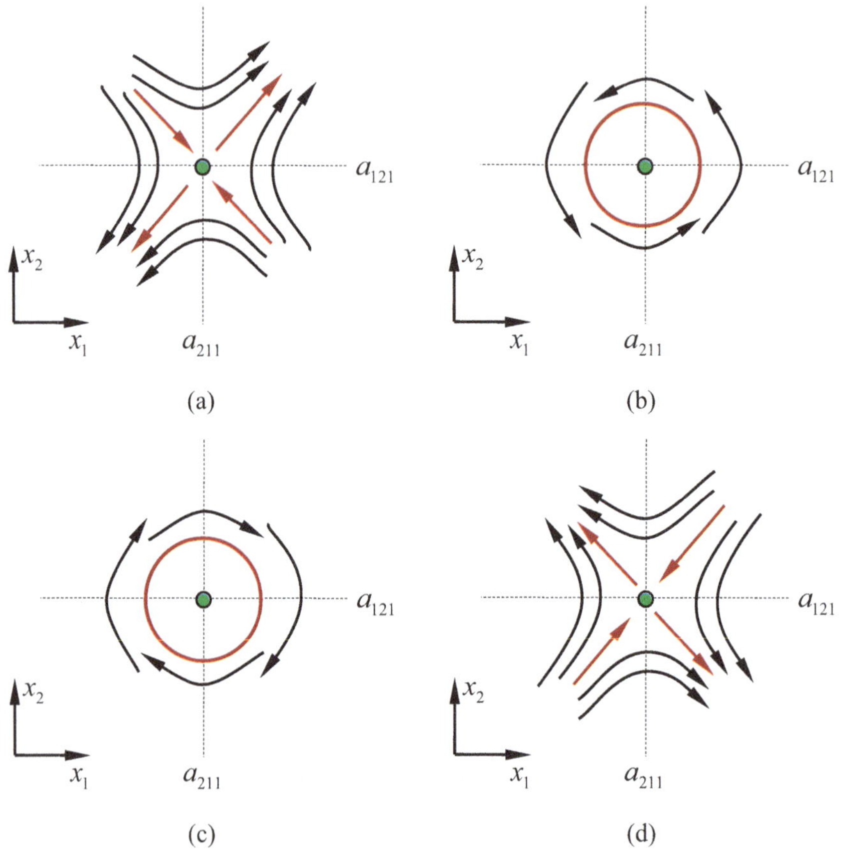

Fig. 4.1 Phase portraits for two-dimensional systems with $x_1^* = a_{211}$ on the x_1-direction for the x_2-direction and $x_2^* = a_{121}$ on the x_2-direction for the x_1-direction. The equilibriums of $(x_1^*, x_2^*) = (a_{211}, a_{121})$: (**a**) a $(3^{\mathrm{rd}}\mathrm{UP}_+, 3^{\mathrm{rd}}\mathrm{UP}_+)$-double third-order positive saddle ($a_{120} > 0$, $a_{210} > 0$), (**b**) a $(3^{\mathrm{rd}}\mathrm{DP}_+, 3^{\mathrm{rd}}\mathrm{DP}_-)$-double third-order counter-clockwise center ($a_{120} < 0$, $a_{210} > 0$), (**c**) a $(3^{\mathrm{rd}}\mathrm{DP}_-, 3^{\mathrm{rd}}\mathrm{DP}_+)$-double third-order clockwise center ($a_{120} > 0, a_{210} < 0$), (**d**) a $(3^{\mathrm{rd}}\mathrm{UP}_-, 3^{\mathrm{rd}}\mathrm{UP}_-)$,-double third-order negative saddle ($a_{120} < 0, a_{210} < 0$)

$$(a_{121}, a_{211}) = \begin{cases} \underbrace{(3^{\mathrm{rd}}\mathrm{UP}_+, 3^{\mathrm{rd}}\mathrm{UP}_+)}_{\text{double third-order positive saddle}} & \text{for } a_{120} > 0, a_{210} > 0; \\ \underbrace{(3^{\mathrm{rd}}\mathrm{DP}_+, 3^{\mathrm{rd}}\mathrm{DP}_-)}_{\text{double third-order CCW center}} & \text{for } a_{120} < 0, a_{210} > 0; \\ \underbrace{(3^{\mathrm{rd}}\mathrm{DP}_-, 3^{\mathrm{rd}}\mathrm{DP}_+)}_{\text{double third-order CW center}} & \text{for } a_{120} > 0, a_{210} < 0; \\ \underbrace{(3^{\mathrm{rd}}\mathrm{UP}_-, 3^{\mathrm{rd}}\mathrm{UP}_-)}_{\text{double third-order negative saddle}} & \text{for } a_{120} < 0, a_{210} < 0. \end{cases} \tag{4.220}$$

4.3.2 Third-Order Saddles, Centers, and Parabola-Saddles

(i) Consider a dynamical system as

$$\begin{aligned}\dot{x}_1 &= a_{120}(x_2 - a_{121})^3,\\ \dot{x}_2 &= a_{210}(x_1 - a_{211})(x_1 - a_{212})^2,\end{aligned} \tag{4.221}$$

and the corresponding first integral manifold is determined by

$$\begin{aligned}&\frac{1}{4}a_{120}\left[(x_2 - a_{121})^4 - (x_{20} - a_{121})^4\right]\\ &= a_{210}\Big\{\frac{1}{4}\left[(x_1 - a_{211})^4 - (x_{10} - a_{211})^4\right]\\ &+ \frac{2}{3}(a_{211} - a_{212})\left[(x_1 - a_{211})^3 - (x_{10} - a_{211})^3\right]\\ &+ \frac{1}{2}(a_{211} - a_{212})^2\left[(x_1 - a_{211})^2 - (x_{10} - a_{211})^2\right]\Big\}.\end{aligned} \tag{4.222}$$

From the first integral manifold, the phase portraits for the two-dimensional cubic system for $x_1^* = a_{211}, a_{212}$ on the x_1-direction for the x_2-direction and $x_2^* = a_{121}$ on the x_2-direction for the x_1-direction are sketched in Fig. 4.2. In Fig. 4.2a–d, the equilibriums are a (3^{rd}UP$_+$, UP$_+$)-third-order positive saddle and (3^{rd}UP,US)-third-order up-parabola upper-saddle ($a_{120} > 0, a_{210} > 0$); a (3^{rd}DP$_+$, DP$_-$)-third-order counter-clockwise center and (3^{rd}DP,US)-third-order down-parabola upper-saddle ($a_{120} < 0, a_{210} > 0$); a (3^{rd}DP$_-$, DP$_+$)-third-order clockwise center and (3^{rd}DP,LS)-third-order down-parabola lower-saddle ($a_{120} > 0, a_{210} < 0$); a (3^{rd}UP$_-$, UP$_-$)-third-order negative saddle and (3^{rd}UP,LS)-third-order up-parabola lower-saddle ($a_{120} < 0, a_{210} < 0$). The following properties of equilibriums of $(x_2^*, x_1^*) = (a_{121}, a_{211}), (a_{121}, a_{212})$ are summarized as follows.

$$\left\{\begin{array}{l}(a_{121}, a_{211})\\ (a_{121}, a_{212})\end{array}\right\} = \left\{\begin{array}{c}\underbrace{(3^{\text{rd}}\text{UP}_+, \text{UP}_+)}_{\text{thrid-order positive saddle}}\\ \underbrace{(3^{\text{rd}}\text{UP}, \text{US})}_{\text{third-order up-parabola upper-saddle}}\end{array}\right\} \text{for } a_{120} > 0, a_{210} > 0, \tag{4.223}$$

$$\left\{\begin{array}{l}(a_{121}, a_{211})\\ (a_{121}, a_{212})\end{array}\right\} = \left\{\begin{array}{c}\underbrace{(3^{\text{rd}}\text{DP}_+, \text{DP}_-)}_{\text{thrid-order CCWcenter}}\\ \underbrace{(3^{\text{rd}}\text{DP}, \text{US})}_{\text{third-order down-parabola upper-saddle}}\end{array}\right\} \text{for } a_{120} < 0, a_{210} > 0, \tag{4.224}$$

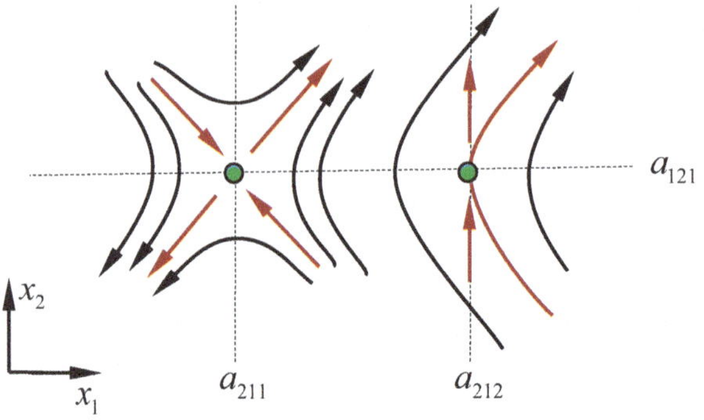

(a)

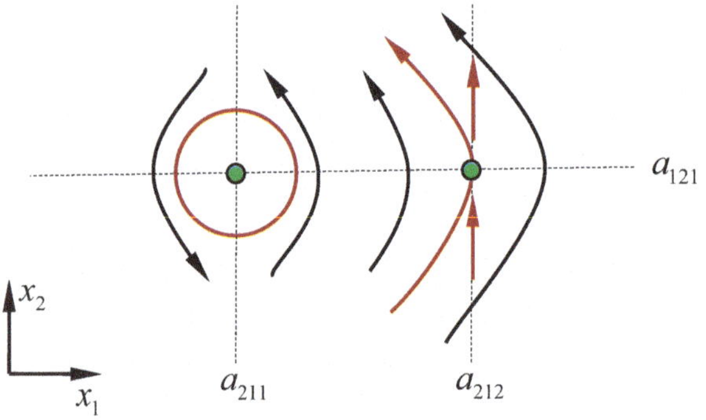

(b)

Fig. 4.2 Phase portraits for two-dimensional systems with $x_1^* = a_{211}$ on the x_1-direction for the x_2-direction and $x_2^* = a_{121}$ on the x_2-direction for the x_1-direction. The equilibriums of $(x_2^*, x_1^*) = (a_{121}, a_{211}), (a_{121}, a_{212})$: **(a)** a (3rdUP$_+$, UP$_+$)-third-order positive saddle and a (3rdUP, US)-third-order up-parabola upper-saddle ($a_{120} > 0, a_{210} > 0$), **(b)** a (3rdDP$_-$, DP$_+$)-third-order counter-clockwise center and a (3rdDP,US)-third-order down-parabola upper-saddle ($a_{120} < 0, a_{210} > 0$), **(c)** a (3rdDP$_-$, DP$_+$)-third-order clockwise center and a (3rdDP,LS)-third-order down-parabola lower-saddle ($a_{120} > 0, a_{210} < 0$), **(d)** a (3rdUP$_-$, UP$_+$)-third-order negative saddle and a (3rdUP, LS)-third-order up-parabola lower-saddle ($a_{120} < 0, a_{210} < 0$)

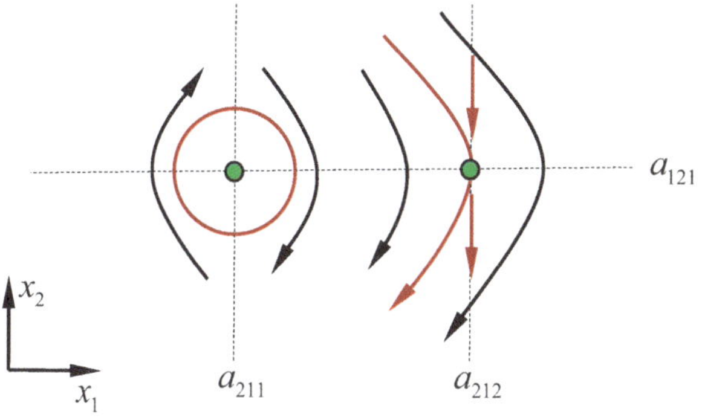

(c)

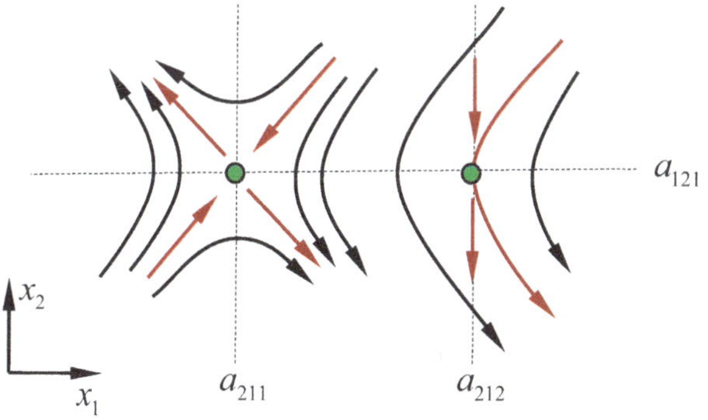

(d)

Fig. 4.2 (continued)

$$\left\{\begin{array}{l}(a_{121}, a_{211})\\(a_{121}, a_{212})\end{array}\right\} = \left\{\begin{array}{c}\underbrace{(3^{\mathrm{rd}}\mathrm{DP}_{-}, \mathrm{DP}_{+})}_{\text{thrid-order CW center}}\\ \underbrace{(3^{\mathrm{rd}}\mathrm{DP}, \mathrm{LS})}_{\text{third-order down-parabola lower-saddle}}\end{array}\right\} \text{ for } a_{120} > 0, a_{210} < 0, \quad (4.225)$$

$$\left\{\begin{array}{l}(a_{121}, a_{211})\\(a_{121}, a_{212})\end{array}\right\} = \left\{\begin{array}{c}\underbrace{(3^{\mathrm{rd}}\mathrm{UP}_{-}, \mathrm{UP}_{-})}_{\text{third-order negative saddle}}\\ \underbrace{(3^{\mathrm{rd}}\mathrm{UP}, \mathrm{LS})}_{\text{third-order up-parabola lower-saddle}}\end{array}\right\} \text{ for } a_{120} < 0, a_{210} < 0. \quad (4.226)$$

(ii) Consider a dynamical system as

$$\begin{aligned}\dot{x}_1 &= a_{120}(x_2 - a_{121})^3,\\ \dot{x}_2 &= a_{210}(x_1 - a_{211})^2(x_1 - a_{212}),\end{aligned} \tag{4.227}$$

and the corresponding first integral manifold is determined by

$$\begin{aligned}&\frac{1}{4}a_{120}\left[(x_2 - a_{121})^4 - (x_{20} - a_{121})^4\right]\\ &= a_{210}\Big\{\frac{1}{4}\left[(x_1 - a_{212})^4 - (x_{10} - a_{212})^4\right]\\ &+ \frac{2}{3}(a_{212} - a_{211})\left[(x_1 - a_{212})^3 - (x_{10} - a_{212})^3\right]\\ &+ \frac{1}{2}(a_{212} - a_{211})^2\left[(x_1 - a_{212})^2 - (x_{10} - a_{212})^2\right]\Big\}.\end{aligned} \tag{4.228}$$

The following properties of equilibriums of $(x_2^*, x_1^*) = (a_{121}, a_{211}), (a_{121}, a_{212})$ are summarized as follow.

$$\begin{Bmatrix}(a_{121}, a_{211})\\ (a_{121}, a_{212})\end{Bmatrix} = \begin{Bmatrix}\underbrace{(3^{\text{rd}}\text{DP}, \text{LS})}_{\text{third-order down-parabola lower-saddle}}\\ \underbrace{(3^{\text{rd}}\text{UP}_+, \text{UP}_+)}_{\text{third-order positive saddle}}\end{Bmatrix} \text{ for } a_{120} > 0, a_{210} > 0, \tag{4.229}$$

$$\begin{Bmatrix}(a_{121}, a_{211})\\ (a_{121}, a_{212})\end{Bmatrix} = \begin{Bmatrix}\underbrace{(3^{\text{rd}}\text{UP}, \text{LS})}_{\text{third-order up-parabola lower-saddle}}\\ \underbrace{(3^{\text{rd}}\text{DP}_+, \text{DP}_-)}_{\text{thrid-order CCWcenter}}\end{Bmatrix} \text{ for } a_{120} < 0, a_{210} > 0, \tag{4.230}$$

$$\begin{Bmatrix}(a_{121}, a_{211})\\ (a_{121}, a_{212})\end{Bmatrix} = \begin{Bmatrix}\underbrace{(3^{\text{rd}}\text{UP}, \text{US})}_{\text{third-order up-parabola upper-saddle}}\\ \underbrace{(3^{\text{rd}}\text{DP}_-, \text{DP}_+)}_{\text{third-order CW center}}\end{Bmatrix} \text{ for } a_{120} > 0, a_{210} < 0, \tag{4.231}$$

$$\begin{Bmatrix}(a_{121}, a_{211})\\ (a_{121}, a_{212})\end{Bmatrix} = \begin{Bmatrix}\underbrace{(3^{\text{rd}}\text{DP}, \text{US})}_{\text{third-order down-parabola upper-saddle}}\\ \underbrace{(3^{\text{rd}}\text{UP}_-, \text{UP}_-)}_{\text{third-order negative saddlle}}\end{Bmatrix} \text{ for } a_{120} < 0, a_{210} < 0. \tag{4.232}$$

From the first integral manifold, the phase portraits for the two-dimensional cubic system for $x_1^* = a_{211}, a_{212}$ on the x_1-direction for the x_2-direction and $x_2^* = a_{121}$ on the x_2-direction for the x_1-direction are sketched in Fig. 4.3. In Fig. 4.3a–d, the equilibriums are a (3^{rd}DP,LS)-third-order down-parabola lower-saddle (3^{rd}UP$_+$, UP$_+$)-

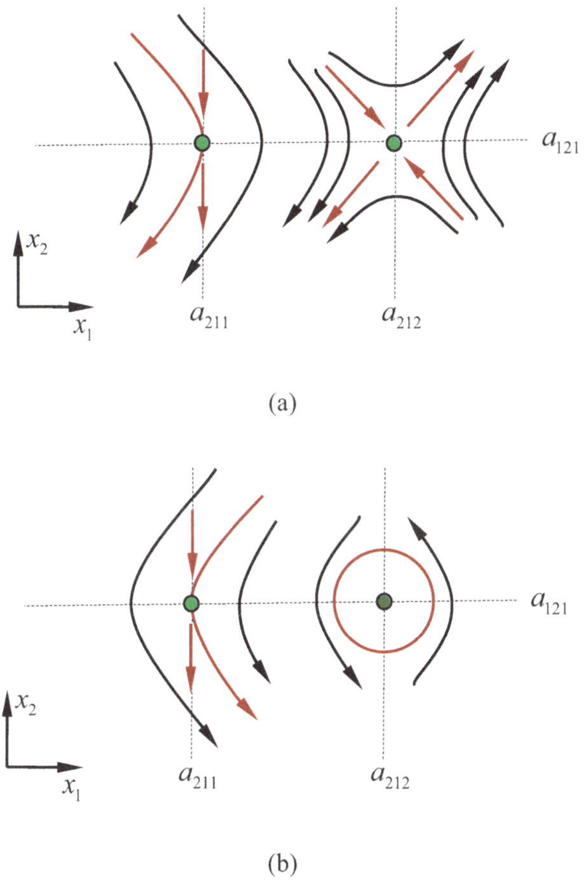

Fig. 4.3 Phase portraits for two-dimensional systems with $x_1^* = a_{211}$ on the x_1-direction for the x_2-direction and $x_2^* = a_{121}$ on the x_2-direction for the x_1-direction. The equilibriums of $(x_1^*, x_2^*) = (a_{211}, a_{121}), (a_{211}, a_{122})$: (**a**) a (3rdDP, LS)-third-order down-parabola lower-saddle and a (3rdUP$_+$, UP$_+$)-third-order positive saddle ($a_{120} > 0, a_{210} > 0$), (**b**) a (3rdUP,LS)-third-order up-parabola lower-saddle and a (3rdDP$_+$, DP$_-$)-third-order counter-clockwise center and ($a_{120} < 0, a_{210} > 0$), (**c**) a (3rdUP,US)-third-order up-parabola upper-saddle and a (3rdDP$_-$, DP$_+$)-third-order clockwise center ($a_{120} > 0, a_{210} < 0$), (**d**) a (3rdDP,US)-third-order down-parabola upper-saddle and a (3rdUP$_-$, UP$_-$)-third-order negative saddle ($a_{120} < 0, a_{210} < 0$)

third-order positive saddle ($a_{120} > 0, a_{210} > 0$); a (3rdUP,LS)-third-order up-parabola lower-saddle and a (3rdDP$_+$, DP$_-$)-third-order counter-clockwise center ($a_{120} < 0$, $a_{210} > 0$); a (3rdUP,US)-third-order up-parabola upper-saddle and a (3rdDP$_-$, DP$_+$)-third-order clockwise center ($a_{120} > 0, a_{210} < 0$); a (3rdDP,US)-third-order down-parabola upper-saddle and a (3rdUP$_-$, UP$_-$) -third-order negative saddle ($a_{120} < 0$, $a_{210} < 0$).

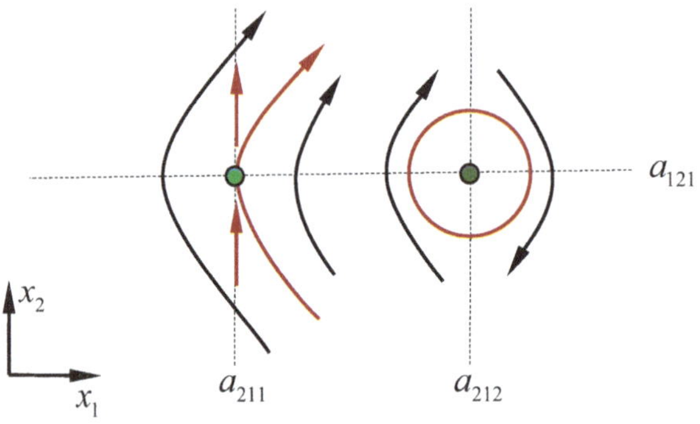

(c)

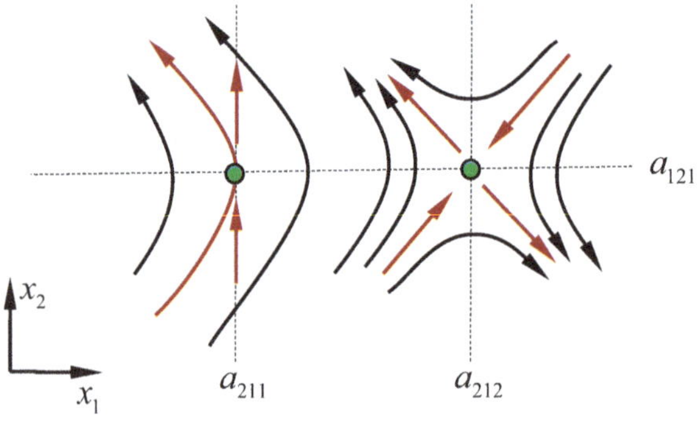

(d)

Fig. 4.3 (continued)

The corresponding switching bifurcations for the two cubic systems are

$$\left\{\begin{array}{c}\underbrace{(3^{\rm rd}{\rm UP}_+,{\rm UP}_+)}_{\text{third-order positive saddle}}\\ \underbrace{(3^{\rm rd}{\rm UP},{\rm US})}_{\text{third-order up-parabola upper-saddle}}\end{array}\right\}\rightleftharpoons\underbrace{(3^{\rm rd}{\rm UP}_+,3^{\rm rd}{\rm UP}_+)}_{\text{double third-order positive saddle}}$$
$$\rightleftharpoons\left\{\begin{array}{c}\underbrace{(3^{\rm rd}{\rm DP},{\rm LS})}_{\text{third-order down-parabola lower-saddle}}\\ \underbrace{(3^{\rm rd}{\rm UP}_+,{\rm UP}_+)}_{\text{third-order positive saddle}}\end{array}\right\}\text{for } a_{120}>0, a_{210}>0, \tag{4.233}$$

$$\left\{\begin{array}{c}\underbrace{(3^{\rm rd}{\rm DP}_{+},{\rm DP}_{-})}_{\text{third-order CCW center}}\\ \underbrace{(3^{\rm rd}{\rm DP},{\rm US})}_{\text{third-order down-parabola upper-saddle}}\end{array}\right\}\rightleftharpoons\underbrace{(3^{\rm rd}{\rm DP}_{+},3^{\rm rd}{\rm DP}_{-})}_{\text{double third-order CCW center}}$$
$$\rightleftharpoons\left\{\begin{array}{c}\underbrace{(3^{\rm rd}{\rm UP},{\rm LS})}_{\text{third-order up-parabola lower-saddle}}\\ \underbrace{(3^{\rm rd}{\rm DP}_{+},{\rm DP}_{-})}_{\text{third-order CCW center}}\end{array}\right\}\text{for } a_{120}<0, a_{210}>0, \tag{4.234}$$

$$\left\{\begin{array}{c}\underbrace{(3^{\rm rd}{\rm DP}_{-},{\rm DP}_{+})}_{\text{third-order CW center}}\\ \underbrace{(3^{\rm rd}{\rm DP},{\rm LS})}_{\text{third-order down-parabola lower-saddle}}\end{array}\right\}\rightleftharpoons\underbrace{(3^{\rm rd}{\rm DP}_{-},3^{\rm rd}{\rm DP}_{+})}_{\text{double third-order CW center}}$$
$$\rightleftharpoons\left\{\begin{array}{c}\underbrace{(3^{\rm rd}{\rm UP},{\rm US})}_{\text{third-order up-parabola upper-saddle}}\\ \underbrace{(3^{\rm rd}{\rm DP}_{-},{\rm DP}_{+})}_{\text{thrid-order CW center}}\end{array}\right\}\text{for } a_{120}>0, a_{210}<0, \tag{4.235}$$

$$\left\{\begin{array}{c}\underbrace{(3^{\rm rd}{\rm UP}_{-},{\rm UP}_{-})}_{\text{third-order negative saddle}}\\ \underbrace{(3^{\rm rd}{\rm DP},{\rm LS})}_{\text{third-order down-parabola lower-saddle}}\end{array}\right\}\rightleftharpoons\underbrace{(3^{\rm rd}{\rm UP}_{-},3^{\rm rd}{\rm UP}_{-})}_{\text{double third-order negative saddle}}$$
$$\rightleftharpoons\left\{\begin{array}{c}\underbrace{(3^{\rm rd}{\rm DP},{\rm US})}_{\text{third-order down-parabola upper-saddle}}\\ \underbrace{(3^{\rm rd}{\rm UP}_{-},{\rm UP}_{-})}_{\text{third-order negative saddle}}\end{array}\right\}\text{for } a_{120}<0, a_{210}<0. \tag{4.236}$$

(iii) Consider a dynamical system as

$$\begin{aligned}\dot{x}_1&=a_{120}(x_2-a_{121})(x_2-a_{122})^2,\\ \dot{x}_2&=a_{210}(x_1-a_{211})^3,\end{aligned} \tag{4.237}$$

and the corresponding first integral manifold is determined by

$$\begin{aligned}&a_{120}\Big\{\frac{1}{4}\left[(x_2-a_{121})^4-(x_{20}-a_{121})^4\right]\\ &+\frac{2}{3}(a_{121}-a_{122})\left[(x_2-a_{121})^3-(x_{20}-a_{121})^3\right]\\ &+\frac{1}{2}(a_{121}-a_{122})^2\left[(x_2-a_{121})^2-(x_{20}-a_{121})^2\right]\Big\}\\ &=\frac{1}{4}a_{210}\left[(x_1-a_{211})^4-(x_{10}-a_{211})^4\right].\end{aligned} \tag{4.238}$$

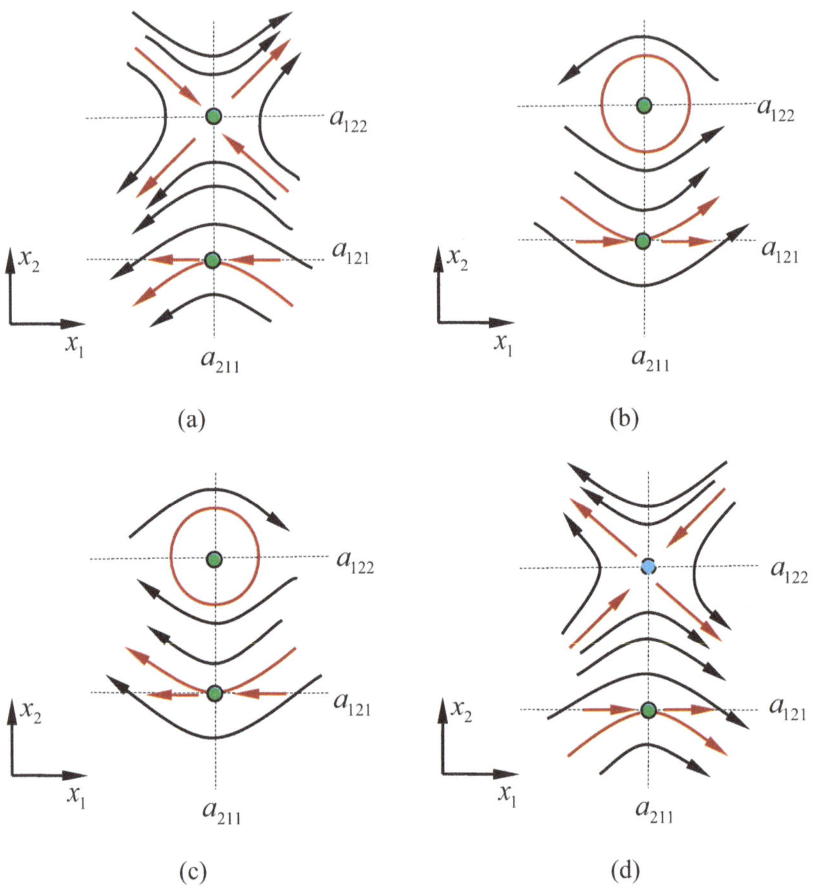

Fig. 4.4 Phase portraits for two-dimensional systems on the x_1-direction with $x_1^* = a_{211}$ and on the x_2-direction with $x_2^* = a_{121}, a_{122}$. The equilibriums of $(x_2^*, x_1^*) = (a_{121}, a_{211}), (a_{122}, a_{211})$, : **(a)** (LS,3rdDP)-third-order down-parabola lower-saddle and (UP$_+$, 3rdUP$_+$)-third-order positive saddle $(a_{120} > 0, a_{210} > 0)$, **(b)** (US,3rdUP)-third-order up-parabola upper-saddle and (DP$_+$, 3rdDP$_-$)-third-order CCW center $(a_{120} < 0, a_{210} > 0)$, **(c)** (US,3rdUP)-third-order up-parabola upper-saddle and (DP$_+$, 3rdDP$_-$)-third-order CW center $(a_{120} > 0, a_{210} < 0)$, **(d)** (US,3rdDP)-third-order down-parabola upper-saddle and (UP$_-$, 3rdUP$_-$)-third-order negative saddle $(a_{120} < 0, a_{210} < 0)$

From the first integral manifold, the phase portraits for the two-dimensional cubic system for $x_1^* = a_{211}$ on the x_1-direction for the x_2-direction and $x_2^* = a_{121}, a_{122}$ on the x_2-direction for the x_1-direction are sketched in Fig. 4.4. In Fig. 4.4a–d, the equilibriums are an (LS, 3rdDP)-third-order down-parabola lower-saddle and (UP$_+$, 3rdUP$_+$)-third-order positive saddle $(a_{120} > 0, a_{210} > 0)$; a (US,3rdUP)-third-order up-parabola upper-saddle and (DP$_-$, 3rdDP$_+$)-third-order counter-clockwise center and $(a_{120} < 0, a_{210} > 0)$; an (LS,3rdUP)-third-order up-parabola lower-saddle and (DP$_-$, 3rdDP$_+$)-third-order clockwise center $(a_{120} > 0, a_{210} < 0)$; a (US, 3rdDP)-

third-order down-parabola upper-saddle and $(\mathrm{UP}_-, 3^{\mathrm{rd}}\mathrm{UP}_-)$-third-order negative saddle $(a_{120} < 0, a_{210} < 0)$. The following properties of equilibriums of $(x_2^*, x_1^*) = (a_{121}, a_{211}), (a_{122}, a_{211})$ are summarized as follows.

$$\left\{\begin{array}{c}(a_{122}, a_{211})\\(a_{121}, a_{211})\end{array}\right\} = \left\{\begin{array}{c}\underbrace{(\mathrm{UP}_+, 3^{\mathrm{rd}}\mathrm{UP}_+)}_{\text{third-order positive saddle}}\\ \underbrace{(\mathrm{LS}, 3^{\mathrm{rd}}\mathrm{DP})}_{\text{third-order down-parabola lower-saddle}}\end{array}\right\} \text{for } a_{120} > 0, a_{210} > 0, \quad (4.239)$$

$$\left\{\begin{array}{c}(a_{121}, a_{211})\\(a_{121}, a_{212})\end{array}\right\} = \left\{\begin{array}{c}\underbrace{(\mathrm{DP}_+, 3^{\mathrm{rd}}\mathrm{DP}_-)}_{\text{thrid-order CCW center}}\\ \underbrace{(\mathrm{US}, 3^{\mathrm{rd}}\mathrm{UP})}_{\text{third-order up-parabola upper-saddle}}\end{array}\right\} \text{for } a_{120} < 0, a_{210} > 0, \quad (4.240)$$

$$\left\{\begin{array}{c}(a_{121}, a_{211})\\(a_{121}, a_{212})\end{array}\right\} = \left\{\begin{array}{c}\underbrace{(\mathrm{DP}_-, 3^{\mathrm{rd}}\mathrm{DP}_+)}_{\text{thrid-order CW center}}\\ \underbrace{(\mathrm{LS}, 3^{\mathrm{rd}}\mathrm{UP})}_{\text{third-order up-parabola lower-saddle}}\end{array}\right\} \text{for } a_{120} > 0, a_{210} < 0, \quad (4.241)$$

$$\left\{\begin{array}{c}(a_{121}, a_{211})\\(a_{121}, a_{212})\end{array}\right\} = \left\{\begin{array}{c}\underbrace{(\mathrm{UP}_-, 3^{\mathrm{rd}}\mathrm{UP}_-)}_{\text{third-order negative saddle}}\\ \underbrace{(\mathrm{US}, 3^{\mathrm{rd}}\mathrm{DP})}_{\text{third-order down-parabola upper-saddle}}\end{array}\right\} \text{for } a_{120} < 0, a_{210} < 0. \quad (4.244)$$

(iv) Consider a dynamical system as

$$\begin{aligned}\dot{x}_1 &= a_{120}(x_2 - a_{121})^2(x_2 - a_{122}),\\ \dot{x}_2 &= a_{210}(x_1 - a_{211})^3,\end{aligned} \quad (4.245)$$

and the corresponding first integral manifold is determined by

$$\begin{aligned}&a_{120}\Big\{\frac{1}{4}\left[(x_2 - a_{122})^4 - (x_{20} - a_{122})^4\right]\\ &+\frac{2}{3}(a_{122} - a_{122})\left[(x_1 - a_{122})^3 - (x_{10} - a_{122})^3\right]\\ &+\frac{1}{2}(a_{122} - a_{121})^2\left[(x_1 - a_{122})^2 - (x_{10} - a_{122})^2\right\}\\ &= \frac{1}{4}a_{210}\left[(x_1 - a_{211})^4 - (x_{10} - a_{211})^4\right].\end{aligned} \quad (4.246)$$

The properties of equilibriums of $(x_2^*, x_1^*) = (a_{121}, a_{211}), (a_{121}, a_{212})$ are given as

$$\left\{\begin{array}{l}(a_{122}, a_{211})\\(a_{121}, a_{211})\end{array}\right\} = \left\{\begin{array}{c}\underbrace{(\mathrm{US}, 3^{\mathrm{rd}}\mathrm{UP})}_{\text{third-order down-parabola lower-saddle}}\\ \underbrace{(\mathrm{UP}_+, 3^{\mathrm{rd}}\mathrm{UP}_+)}_{\text{third-order positive saddle}}\end{array}\right\} \text{ for } a_{120} > 0, a_{210} > 0, \quad (4.247)$$

$$\left\{\begin{array}{l}(a_{122}, a_{211})\\(a_{121}, a_{211})\end{array}\right\} = \left\{\begin{array}{c}\underbrace{(\mathrm{LS}, 3^{\mathrm{rd}}\mathrm{DP})}_{\text{third-order down-parabola lower-saddle}}\\ \underbrace{(\mathrm{DP}_+, 3^{\mathrm{rd}}\mathrm{DP}_-)}_{\text{thrid-order CCW center}}\end{array}\right\} \text{ for } a_{120} < 0, a_{210} > 0, \quad (4.248)$$

$$\left\{\begin{array}{l}(a_{122}, a_{211})\\(a_{121}, a_{211})\end{array}\right\} = \left\{\begin{array}{c}\underbrace{(\mathrm{LS}, 3^{\mathrm{rd}}\mathrm{DP})}_{\text{third-order down-parabola lower-saddle}}\\ \underbrace{(\mathrm{DP}_+, 3^{\mathrm{rd}}\mathrm{DP}_-)}_{\text{thrid-order CCW center}}\end{array}\right\} \text{ for } a_{120} > 0, a_{210} < 0, \quad (4.249)$$

$$\left\{\begin{array}{l}(a_{122}, a_{211})\\(a_{121}, a_{211})\end{array}\right\} = \left\{\begin{array}{c}\underbrace{(\mathrm{LS}, 3^{\mathrm{rd}}\mathrm{UP})}_{\text{third-order up-parabola lower-saddle}}\\ \underbrace{(\mathrm{UP}_-, 3^{\mathrm{rd}}\mathrm{UP}_-)}_{\text{third-order negative saddle}}\end{array}\right\} \text{ for } a_{120} < 0, a_{210} < 0. \quad (4.250)$$

From the first integral manifold, the phase portraits for the two-dimensional cubic system for $x_1^* = a_{211}$ on the x_1-direction for the x_2-direction and $x_2^* = a_{121}, a_{122}$ on the x_2-direction for the x_1-direction are sketched in Fig. 4.5. In Fig. 4.5a–d, the equilibriums are a (UP$_+$, 3rdUP$_+$)-third-order positive saddle and (US,3rdUP)-third-order up-parabola upper-saddle ($a_{120} > 0, a_{210} > 0$); a (DP$_+$, 3rdDP$_-$)-third-order counter-clockwise center and an (LS, 3rdDP)-third-order down-parabola lower-saddle ($a_{120} < 0, a_{210} > 0$); a (3rdDP$_-$, DP$_+$)-third-order clockwise center and a (3rdUP, US)-third-order up-parabola upper-saddle ($a_{120} > 0, a_{210} < 0$); a (3rdUP$_-$, UP$_-$)-third-order negative saddle and a (3rdDP,US)-third-order down-parabola upper-saddle ($a_{120} < 0, a_{210} < 0$).

The corresponding switching bifurcations for the two cubic systems are

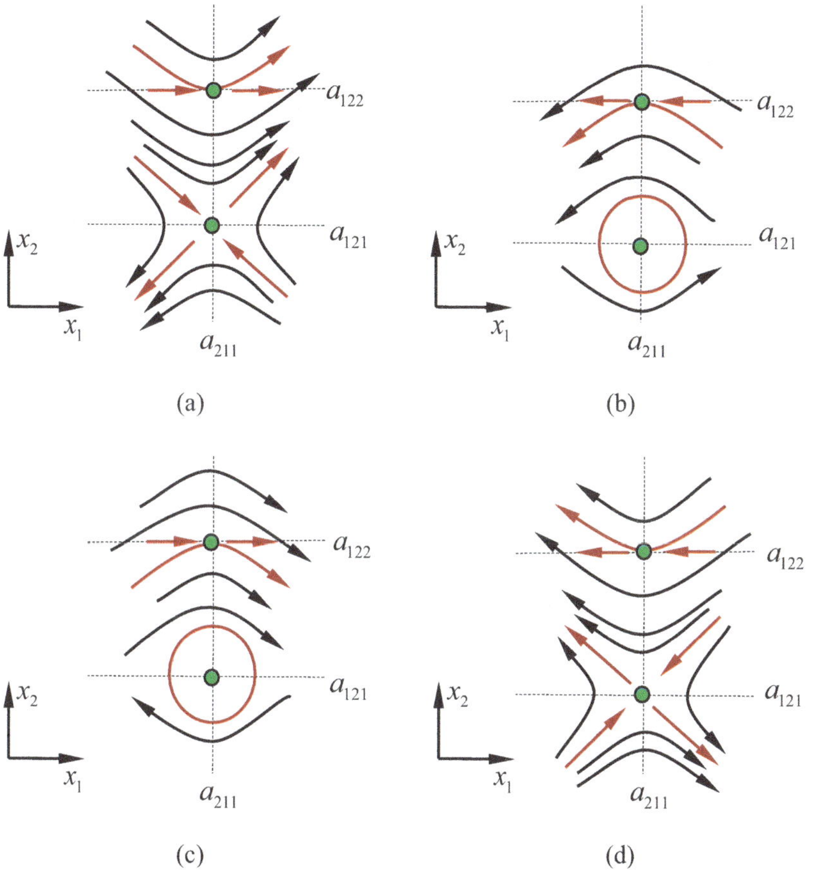

Fig. 4.5 Phase portraits for two-dimensional systems on the x_1-direction with $x_1^* = a_{211}$ and on the x_2-direction with $x_2^* = a_{121}, a_{122}$. The equilibriums of $(x_2^*, x_1^*) = (a_{121}, a_{211}), (a_{122}, a_{211})$: (**a**) (UP$_+$, 3rdUP$_+$)-third-order positive saddle and (US,3rdUP)-third-order up-parabola upper-saddle ($a_{120} > 0, a_{210} > 0$), (**b**) (DP$_+$, 3rdDP$_-$)-third-order CCW center and (LS,3rdDP)-third-order down-parabola lower-saddle ($a_{120} < 0, a_{210} > 0$), (**c**) (DP$_+$, 3rdDP$_-$)-third-order CW center and (US,3rdDP)-third-order down-parabola upper-saddle ($a_{120} > 0, a_{210} < 0$), (**d**) (UP$_-$, 3rdUP$_-$)-third-order negative saddle and (LS,3rdUP)-third-order up-parabola lower-saddle ($a_{120} < 0, a_{210} < 0$)

$$\left\{\begin{array}{c}\underbrace{(\mathrm{UP}_{+},3^{\mathrm{rd}}\mathrm{UP}_{+})}_{\text{third-order positive saddle}}\\ \underbrace{(\mathrm{LS},3^{\mathrm{rd}}\mathrm{DP})}_{\text{third-order down-parabola lower-saddle}}\end{array}\right\}\rightleftharpoons\underbrace{(3^{\mathrm{rd}}\mathrm{UP}_{+},3^{\mathrm{rd}}\mathrm{UP}_{+})}_{\text{double third-order positive saddle}}$$
$$\rightleftharpoons\left\{\begin{array}{c}\underbrace{(\mathrm{US},3^{\mathrm{rd}}\mathrm{UP})}_{\text{third-order up-parabola upper-saddle}}\\ \underbrace{(\mathrm{UP}_{+},3^{\mathrm{rd}}\mathrm{UP}_{+})}_{\text{third-order positive saddle}}\end{array}\right\}\text{for } a_{120}>0, a_{210}>0, \tag{4.251}$$

$$\left\{\begin{array}{c}\underbrace{(\mathrm{DP}_{+},3^{\mathrm{rd}}\mathrm{DP}_{-})}_{\text{thrid-order CCW center}}\\ \underbrace{(\mathrm{US},3^{\mathrm{rd}}\mathrm{UP})}_{\text{third-order up-parabola upper-saddle}}\end{array}\right\}\rightleftharpoons\underbrace{(3^{\mathrm{rd}}\mathrm{DP}_{+},3^{\mathrm{rd}}\mathrm{DP}_{-})}_{\text{double third-order CCW center}}$$
$$\rightleftharpoons\left\{\begin{array}{c}\underbrace{(\mathrm{LS},3^{\mathrm{rd}}\mathrm{DP})}_{\text{third-order down-parabola lower-saddle}}\\ \underbrace{(\mathrm{DP}_{+},3^{\mathrm{rd}}\mathrm{DP}_{-})}_{\text{thrid-order CCW center}}\end{array}\right\}\text{for } a_{120}<0, a_{210}>0, \tag{4.252}$$

$$\left\{\begin{array}{c}\underbrace{(\mathrm{DP}_{-},3^{\mathrm{rd}}\mathrm{DP}_{+})}_{\text{third-order CW center}}\\ \underbrace{(\mathrm{LS},3^{\mathrm{rd}}\mathrm{UP})}_{\text{third-order up-parabola lower-saddle}}\end{array}\right\}\rightleftharpoons\underbrace{(3^{\mathrm{rd}}\mathrm{DP}_{-},3^{\mathrm{rd}}\mathrm{DP}_{+})}_{\text{double third-order CW center}}$$
$$\rightleftharpoons\left\{\begin{array}{c}\underbrace{(\mathrm{US},3^{\mathrm{rd}}\mathrm{DP})}_{\text{third-order down-parabola upper-saddle}}\\ \underbrace{(\mathrm{DP}_{-},3^{\mathrm{rd}}\mathrm{DP}_{+})}_{\text{thrid-order CW center}}\end{array}\right\}\text{for } a_{120}>0, a_{210}<0, \tag{4.253}$$

$$\left\{\begin{array}{c}\underbrace{(\mathrm{UP}_{-},3^{\mathrm{rd}}\mathrm{UP}_{-})}_{\text{third-order negative saddle}}\\ \underbrace{(\mathrm{US},3^{\mathrm{rd}}\mathrm{DP})}_{\text{third-order down-parabola upper-saddle}}\end{array}\right\}\rightleftharpoons\underbrace{(3^{\mathrm{rd}}\mathrm{UP}_{-},3^{\mathrm{rd}}\mathrm{UP}_{-})}_{\text{double third-order negative saddle}}$$
$$\rightleftharpoons\left\{\begin{array}{c}\underbrace{(\mathrm{LS},3^{\mathrm{rd}}\mathrm{UP})}_{\text{third-order up-parabola lower-saddle}}\\ \underbrace{(\mathrm{UP}_{-},3^{\mathrm{rd}}\mathrm{UP}_{-})}_{\text{third-order negative saddle}}\end{array}\right\}\text{for } a_{120}<0, a_{210}<0. \tag{4.254}$$

4.3.3 A Series of Third-Order Saddle and Centers

Consider a dynamical system as

$$\begin{aligned}\dot{x}_2 &= a_{210}(x_1 - a_{211})(x_1 - a_{212})(x_1 - a_{213}),\\ \dot{x}_1 &= a_{120}(x_2 - a_{121})^3.\end{aligned} \tag{4.255}$$

and the first integral manifold is given by

$$\begin{aligned}&\frac{1}{4}a_{120}\left[(x_2 - a_{121})^4 - (x_{20} - a_{121})^4\right]\\ &= a_{210}\Big\{\frac{1}{4}\left[(x_1 - a_{21l_1})^4 - (x_1 - a_{21l_1})^4\right]\\ &+\frac{1}{3}\sum\nolimits_{l_2=1,l_2\neq l_1}^{3}(a_{21l_1} - a_{21l_2})\left[(x_1 - a_{211})^3 - (x_{10} - a_{211})^3\right]\\ &+\frac{1}{2}\prod\nolimits_{2=1,l_2\neq l_1}^{3}(a_{21l_1} - a_{21l_2})\left[(x_1 - a_{211})^2 - (x_{10} - a_{211})^2\right]\Big\}.\end{aligned} \tag{4.256}$$

The phase portraits of third-order center and saddle are presented in Fig. 4.6. Phase portraits for two-dimensional cubic systems near the third-order center and third-order saddle bifurcations of $(x_2^*, x_1^*) = (a_{121}, a_{211}), (a_{121}, a_{212}), (a_{121}, a_{213})$ are presented in Fig. 4.6a–d for four cases. The equilibriums of $(x_2^*, x_1^*) = (a_{121}, a_{211}), (a_{121}, a_{212}), (a_{121}, a_{213})$ are: (a) $(3^{\text{rd}}\text{UP}_+, \text{UP}_+)$-third-order positive saddle, $(3^{\text{rd}}\text{DP}_-, \text{DP}_+)$-third-order clockwise center and $(3^{\text{rd}}\text{UP}_+, \text{UP}_+)$-third-order positive saddle ($a_{120} > 0$, $a_{210} > 0$), (b) $(3^{\text{rd}}\text{DP}_+, \text{DP}_-)$-third-order clockwise center, $(3^{\text{rd}}\text{UP}_-, \text{UP}_-)$-third-order negative saddle and $(3^{\text{rd}}\text{DP}_+, \text{DP}_-)$-third-order clockwise center ($a_{120} < 0$, $a_{210} > 0$), (c) $(3^{\text{rd}}\text{DP}_+, \text{DP}_-)$-third-order counter-clockwise center, $(3^{\text{rd}}\text{UP}_+, \text{UP}_+)$-parabola-saddle, and $(3^{\text{rd}}\text{DP}_+, \text{DP}_-)$-third-order counter-clockwise center ($a_{120} > 0$, $a_{210} < 0$), (d) $(3^{\text{rd}}\text{UP}_-, \text{UP}_-)$-third-order negative saddle, $(3^{\text{rd}}\text{DP}_+, \text{DP}_-)$-third-order counter-clockwise center and $(3^{\text{rd}}\text{UP}_-, \text{UP}_-)$-third-order negative saddle ($a_{120} < 0$, $a_{210} < 0$).

(i) For $a_{120} > 0$ and $a_{210} > 0$,

$$\left\{\begin{matrix}(a_{121}, a_{211})\\ (a_{121}, a_{212})\\ (a_{121}, a_{213})\end{matrix}\right\} = \left\{\begin{matrix}\underbrace{(3^{\text{rd}}\text{UP}_+, \text{UP}_+)}_{\text{third-order positive-saddle}}\\ \underbrace{(3^{\text{rd}}\text{DP}_-, \text{DP}_+)}_{\text{third-order CW-center}}\\ \underbrace{(3^{\text{rd}}\text{UP}_+, \text{UP}_+)}_{\text{third-order positive-saddle}}\end{matrix}\right\}. \tag{4.257}$$

(ii) For $a_{120} < 0$ and $a_{210} > 0$,

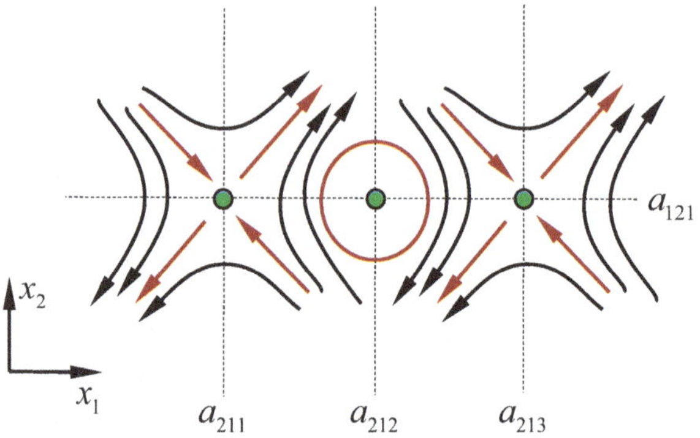

(a)

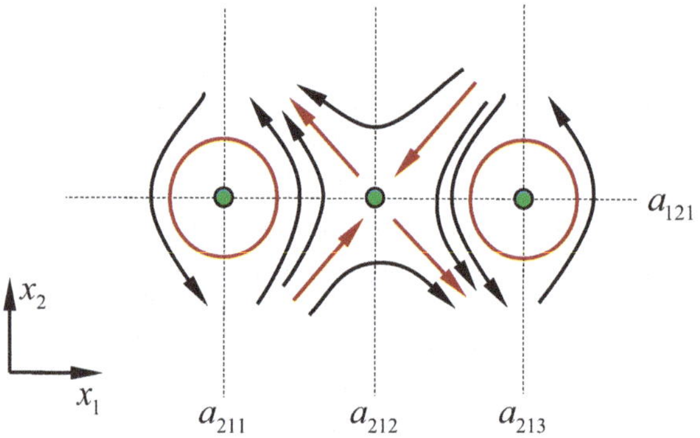

(b)

Fig. 4.6 Phase portraits for two-dimensional systems on the x_1-direction with $x_1^* = a_{211}, a_{212}$ for the x_2-direction and on the x_2-direction with $x_2^* = a_{121}$ for the x_1-direction. The equilibriums of $(x_2^*, x_1^*) = (a_{121}, a_{211}), (a_{121}, a_{212}), (a_{121}, a_{213})$: (**a**) ($3^{\text{rd}}\text{UP}_+, \text{UP}_+$)-third-order positive saddle, ($3^{\text{rd}}\text{DP}_-, \text{DP}_+$)-third-order CW center and ($3^{\text{rd}}\text{UP}_+, \text{UP}_+$)-third-order positive saddle ($a_{120} > 0$, $a_{210} > 0$), (**b**) ($3^{\text{rd}}\text{DP}_+, \text{DP}_-$)-third-order CCW center, ($3^{\text{rd}}\text{UP}_-, \text{UP}_-$)-third-order negative saddle and ($3^{\text{rd}}\text{DP}_+, \text{DP}_-$)-third-order CCW center ($a_{210} > 0$, $a_{120} < 0$), (**c**) ($3^{\text{rd}}\text{DP}_+, \text{DP}_-$)-third-order CW center, ($3^{\text{rd}}\text{UP}_+, \text{UP}_+$)-parabola-saddle and ($3^{\text{rd}}\text{DP}_+, \text{DP}_-$)-third-order CW center ($a_{120} > 0$, $a_{210} < 0$), (**d**) ($3^{\text{rd}}\text{UP}_-, \text{UP}_-$) -third-order negative saddle, ($3^{\text{rd}}\text{DP}_+, \text{DP}_-$)-third-order CCW center and ($3^{\text{rd}}\text{UP}_-, \text{UP}_-$)-third-order negative saddle ($a_{210} < 0$, $a_{120} < 0$)

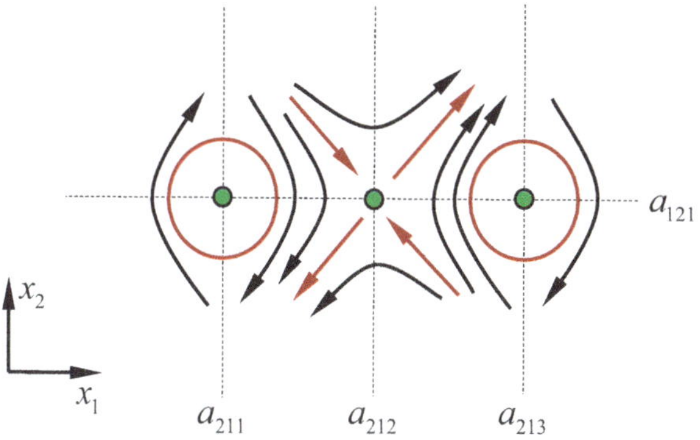

(c)

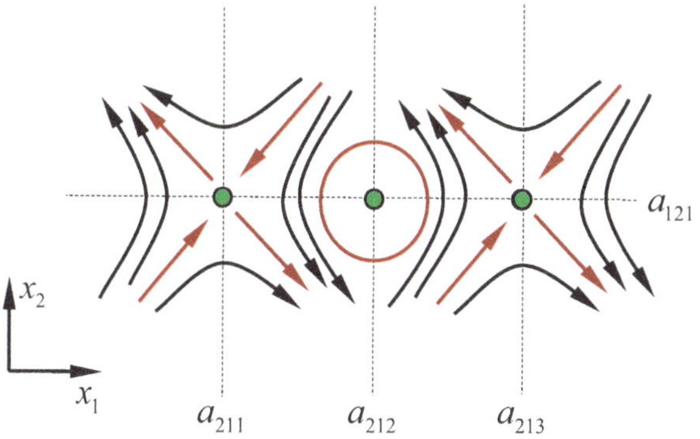

(d)

Fig. 4.6 (continued)

$$\begin{Bmatrix} (a_{121}, a_{211}) \\ (a_{121}, a_{212}) \\ (a_{121}, a_{213}) \end{Bmatrix} = \begin{Bmatrix} \underbrace{(3^{\text{rd}}\text{DP}_{+}, \text{DP}_{-})}_{\text{third-order CCW center}} \\ \underbrace{(3^{\text{rd}}\text{UP}_{-}, \text{UP}_{-})}_{\text{third-order negative-saddle}} \\ \underbrace{(3^{\text{rd}}\text{DP}_{+}, \text{DP}_{-})}_{\text{third-order CCW center}} \end{Bmatrix}. \tag{4.258}$$

(iii) For $a_{120} > 0$ and $a_{210} < 0$,

$$\left\{\begin{array}{c}(a_{121},a_{211})\\(a_{121},a_{212})\\(a_{121},a_{213})\end{array}\right\}=\left\{\begin{array}{c}\underbrace{(3^{\text{rd}}\text{DP}_{-},\text{DP}_{+})}_{\text{third-order CW center}}\\\underbrace{(3^{\text{rd}}\text{UP}_{+},\text{UP}_{+})}_{\text{third-order positive-saddle}}\\\underbrace{(3^{\text{rd}}\text{DP}_{-},\text{DP}_{+})}_{\text{third-order CW center}}\end{array}\right\}. \tag{4.259}$$

(iv) For $a_{120}<0$ and $a_{210}<0$,

$$\left\{\begin{array}{c}(a_{121},a_{211})\\(a_{121},a_{212})\\(a_{121},a_{213})\end{array}\right\}=\left\{\begin{array}{c}\underbrace{(3^{\text{rd}}\text{UP}_{-},\text{UP}_{-})}_{\text{third-order negative-saddle}}\\\underbrace{(3^{\text{rd}}\text{DP}_{+},\text{DP}_{-})}_{\text{third-order CCW-center}}\\\underbrace{(3^{\text{rd}}\text{UP}_{-},\text{UP}_{-})}_{\text{third-order negative-saddle}}\end{array}\right\}. \tag{4.260}$$

The global flows for the up-parabola saddles and down-parabola saddles are also determined by the double third-order saddles and centers, as shown in Fig. 4.1.

Similarly, consider a quadratic dynamical system as

$$\begin{aligned}\dot{x}_2&=a_{210}(x_1-a_{211})^3,\\\dot{x}_1&=a_{120}(x_2-a_{121})(x_2-a_{122})(x_2-a_{123})\end{aligned} \tag{4.261}$$

and the first integral manifold is given by

$$\begin{aligned}&\frac{1}{4}a_{210}\left[(x_1-a_{211})^4-(x_{10}-a_{211})^4\right]\\&=a_{120}\Big\{\frac{1}{4}\left[(x_2-a_{12s_1})^4-(x_{20}-a_{12s_1})^4\right]\\&+\frac{1}{3}\sum\nolimits_{s_2=1,s_2\neq s_1}^{3}(a_{12s_1}-a_{12s_2})\left[(x_2-a_{12s_1})^3-(x_{20}-a_{12s_1})^3\right]\\&+\frac{1}{2}\prod\nolimits_{s_2=1,s_2\neq s_1}^{3}(a_{12s_1}-a_{12s_2})\left[(x_2-a_{12s_1})^2-(x_{20}-a_{12s_1})^2\right]\Big\}.\end{aligned} \tag{4.262}$$

The phase portraits of third-order canters and saddles are presented in Fig. 4.7. Phase portraits for two-dimensional cubic systems near the e bifurcation with two equilibriums of $(x_1^*,x_2^*)=(a_{211},a_{121}),(a_{211},a_{122})$ are illustrated. In Fig. 4.7a–d, the four cases are the $(\text{DP}_{+},3^{\text{rd}}\text{DP}_{-})$-third-order counter-clockwise center and $(\text{UP}_{+},3^{\text{rd}}\text{UP}_{+})$-third-order positive saddle ($a_{120}>0$,$a_{210}>0$); the $(\text{UP}_{+},3^{\text{rd}}\text{UP}_{+})$-third-order positive saddle and $(\text{DP}_{+},3^{\text{rd}}\text{DP}_{-})$-third-order counter-clockwise center ($a_{120}<0$,$a_{210}>0$); the $(3^{\text{rd}}\text{UP}_{-},\text{UP}_{-})$ -third-order negative saddle and $(3^{\text{rd}}\text{DP}_{-}$,

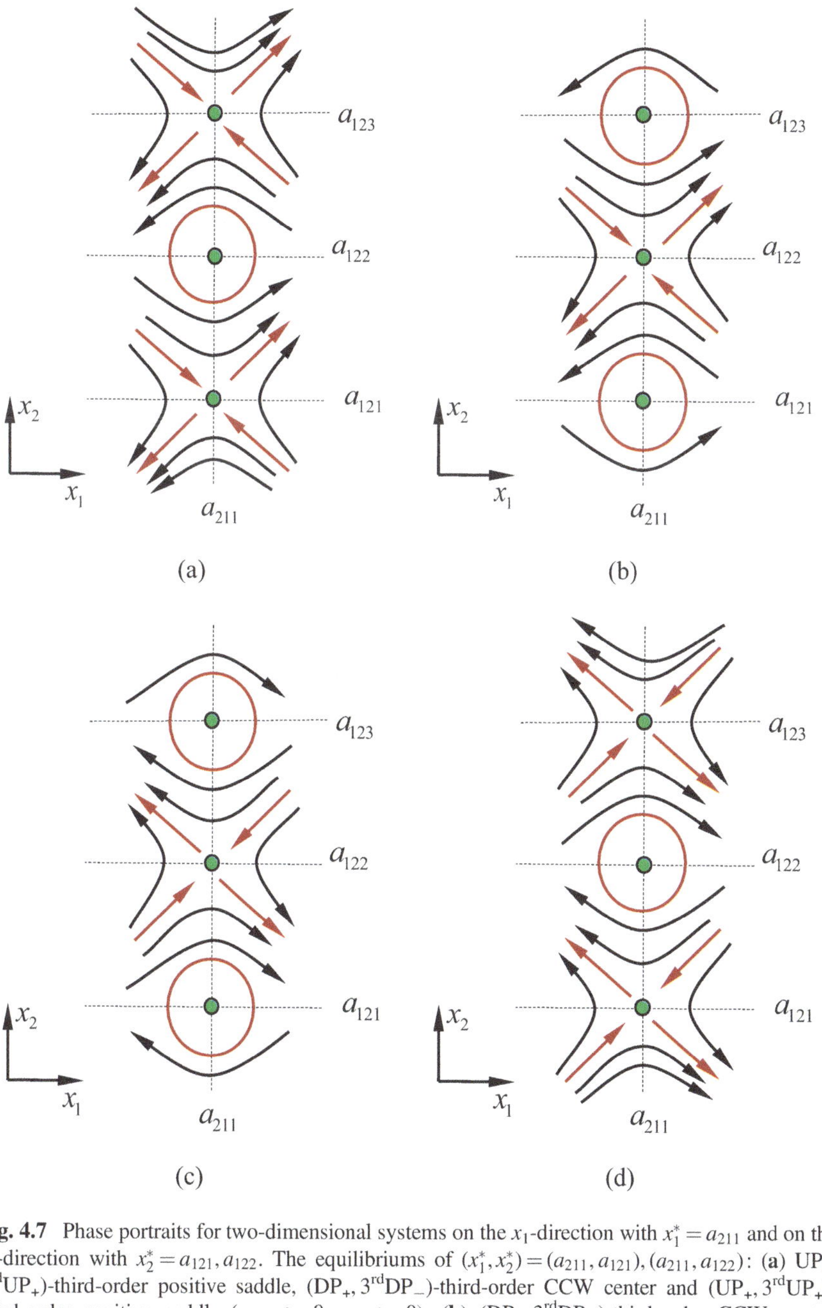

Fig. 4.7 Phase portraits for two-dimensional systems on the x_1-direction with $x_1^* = a_{211}$ and on the x_2-direction with $x_2^* = a_{121}, a_{122}$. The equilibriums of $(x_1^*, x_2^*) = (a_{211}, a_{121}), (a_{211}, a_{122})$: (**a**) UP_+, $3^{\mathrm{rd}}\mathrm{UP}_+$)-third-order positive saddle, (DP_+, $3^{\mathrm{rd}}\mathrm{DP}_-$)-third-order CCW center and (UP_+, $3^{\mathrm{rd}}\mathrm{UP}_+$)-third-order positive saddle ($a_{120} > 0, a_{210} > 0$), (**b**) (DP_+, $3^{\mathrm{rd}}\mathrm{DP}_-$)-third-order CCW center, (UP_+, $3^{\mathrm{rd}}\mathrm{UP}_+$)-third-order positive saddle and (DP_+, $3^{\mathrm{rd}}\mathrm{DP}_-$)-third-order CCW center ($a_{120} < 0$, $a_{210} > 0$), (**c**) ($3^{\mathrm{rd}}\mathrm{DP}_-$, DP_+)-third-order CW center, ($3^{\mathrm{rd}}\mathrm{UP}_-$, UP_-)-third-order negative saddle and ($3^{\mathrm{rd}}\mathrm{DP}_-$, DP_+)-third-order CW center ($a_{120} > 0$, $a_{210} < 0$), (**d**) (UP_-, $3^{\mathrm{rd}}\mathrm{UP}_-$)-third-order negative saddle, (DP_-, $3^{\mathrm{rd}}\mathrm{DP}_+$)-third-order CW center and (UP_-, $3^{\mathrm{rd}}\mathrm{UP}_-$)-third-order negative saddle ($a_{120} < 0$, $a_{210} < 0$)

DP_+)-third-order clockwise center $a_{120} > 0$, $a_{210} < 0$); the (DP_-, $3^{\mathrm{rd}}\mathrm{DP}_+$)-third-order clockwise. center and (UP_-, $3^{\mathrm{rd}}\mathrm{UP}_-$)-third-order negative saddle ($a_{120} < 0$, $a_{210} < 0$). The cases are summarized as follows.

(i) For $a_{120} > 0$ and $a_{210} > 0$,

$$\begin{Bmatrix} (a_{123}, a_{211}) \\ (a_{122}, a_{211}) \\ (a_{121}, a_{211}) \end{Bmatrix} = \begin{Bmatrix} \underbrace{(\mathrm{UP}_+, 3^{\mathrm{rd}}\mathrm{UP}_+)}_{\text{positive-saddle}} \\ \underbrace{(\mathrm{DP}_+, 3^{\mathrm{rd}}\mathrm{DP}_-)}_{\text{CCW-center}} \\ \underbrace{(\mathrm{UP}_+, 3^{\mathrm{rd}}\mathrm{UP}_+)}_{\text{positive-saddle}} \end{Bmatrix}. \tag{4.263}$$

(ii) For $a_{120} < 0$ and $a_{210} > 0$,

$$\begin{Bmatrix} (a_{123}, a_{211}) \\ (a_{122}, a_{211}) \\ (a_{121}, a_{211}) \end{Bmatrix} = \begin{Bmatrix} \underbrace{(\mathrm{DP}_+, 3^{\mathrm{rd}}\mathrm{DP}_-)}_{\text{CCW center}} \\ \underbrace{(\mathrm{UP}_+, 3^{\mathrm{rd}}\mathrm{UP}_+)}_{\text{positive-saddle}} \\ \underbrace{(\mathrm{DP}_+, 3^{\mathrm{rd}}\mathrm{DP}_-)}_{\text{CCW center}} \end{Bmatrix}. \tag{4.264}$$

(iii) For $a_{120} > 0$ and $a_{210} < 0$,

$$\begin{Bmatrix} (a_{123}, a_{211}) \\ (a_{122}, a_{211}) \\ (a_{121}, a_{211}) \end{Bmatrix} = \begin{Bmatrix} \underbrace{(\mathrm{DP}_-, 3^{\mathrm{rd}}\mathrm{DP}_+)}_{\text{CW center}} \\ \underbrace{(\mathrm{UP}_-, 3^{\mathrm{rd}}\mathrm{UP}_-)}_{\text{negative saddle}} \\ \underbrace{(\mathrm{DP}_-, 3^{\mathrm{rd}}\mathrm{DP}_+)}_{\text{CW center}} \end{Bmatrix}. \tag{4.265}$$

(iv) For $a_{120} < 0$ and $a_{210} < 0$,

$$\left\{\begin{array}{l}(a_{123}, a_{211})\\(a_{122}, a_{211})\\(a_{121}, a_{211})\end{array}\right\} = \left\{\begin{array}{c}\underbrace{(\mathrm{UP}_{-}, 3^{\mathrm{rd}}\mathrm{UP}_{-})}_{\text{negative-sadde}}\\\underbrace{(\mathrm{DP}_{-}, 3^{\mathrm{rd}}\mathrm{DP}_{+})}_{\text{CW-center}}\\\underbrace{(\mathrm{UP}_{-}, 3^{\mathrm{rd}}\mathrm{UP}_{-})}_{\text{negative-saddle}}\end{array}\right\}. \tag{4.266}$$

The global flows for the up-parabola flows and down-parabola flows are also determined by the double third-order saddles and centers, as shown in Fig. 4.1.

- The double third-order center is the appearing (switching) bifurcation from a third-order center to the third-order center, saddle, and center.
- The double third-order saddle is the appearing (switching) bifurcation from a third-order saddle to the third-order saddle, center, and saddle.

4.3.4 *Inflection-Saddles, Parabola-Saddles, Saddle (Center)*

Consider a dynamical system as

$$\begin{aligned}\dot{x}_1 &= a_{120}(x_2 - a_{12s_1})(x_2 - a_{12s_2})^2,\\\dot{x}_2 &= a_{210}(x_1 - a_{21l_1})(x_1 - a_{21l_2})^2,\end{aligned} \tag{4.266}$$

and the corresponding first integral manifold is determined by

$$\begin{aligned}&a_{120}\Big\{\frac{1}{4}\big[(x_2 - a_{12s_1})^4 - (x_{20} - a_{12s_1})^4\big]\\&+\frac{2}{3}(a_{12s_1} - a_{12s_2})\big[(x_2 - a_{12s_1})^3 - (x_{20} - a_{12s_1})^3\big]\\&+\frac{1}{2}(a_{12s_1} - a_{12s_2})^2\big[(x_2 - a_{12s_1})^2 - (x_{20} - a_{12s_1})^2\big]\Big\}\\&= a_{210}\Big\{\frac{1}{4}\big[(x_1 - a_{21l_1})^4 - (x_{10} - a_{21l_1})^4\big]\\&+\frac{2}{3}(a_{21l_1} - a_{21l_2})\big[(x_1 - a_{21l_1})^3 - (x_{10} - a_{2l_1})^3\big]\\&+\frac{1}{2}(a_{21l_1} - a_{21l_2})^2\big[(x_1 - a_{21l_1})^2 - (x_{10} - a_{21l_1})^2\big]\Big\}.\end{aligned} \tag{4.267}$$

(i) For $a_{12s_1} = a_{121}, a_{12s_2} = a_{122}, a_{21l_1} = a_{211}, a_{21l_2} = a_{212}$, the following properties of equilibriums of $(x_2^*, x_1^*) = (a_{121}, a_{211}), (a_{121}, a_{212}), (a_{122}, a_{211}), (a_{122}, a_{212})$ are summarized as follows.

$$\left\{\begin{array}{cc}(a_{122},a_{211}) & (a_{122},a_{212})\\(a_{121},a_{211}) & (a_{121},a_{212})\end{array}\right\}=\left\{\begin{array}{cc}\underbrace{(\mathrm{US},\mathrm{UP})}_{\text{up-parabola upper-saddle}} & \underbrace{(\mathrm{II}_{+},\mathrm{II}_{+})}_{\text{double-inflection saddle}}\\\underbrace{(\mathrm{UP}_{+},\mathrm{UP}_{+})}_{\text{positive saddle}} & \underbrace{(\mathrm{UP},\mathrm{US})}_{\text{up-parabola upper-saddle}}\end{array}\right\}$$
$$\text{for } a_{120}>0, a_{210}>0, \tag{4.268}$$

$$\left\{\begin{array}{cc}(a_{122},a_{211}) & (a_{122},a_{212})\\(a_{121},a_{211}) & (a_{121},a_{212})\end{array}\right\}=\left\{\begin{array}{cc}\underbrace{(\mathrm{LS},\mathrm{DP})}_{\text{down-parabola lower-saddle}} & \underbrace{(\mathrm{DI}_{+},\mathrm{DI}_{-})}_{\text{double-inflection saddle}}\\\underbrace{(\mathrm{DP}_{+},\mathrm{DP}_{-})}_{\text{CCW center}} & \underbrace{(\mathrm{DP},\mathrm{US})}_{\text{down-parabola upper-saddle}}\end{array}\right\}$$
$$\text{for } a_{120}<0, a_{210}>0, \tag{4.269}$$

$$\left\{\begin{array}{cc}(a_{122},a_{211}) & (a_{122},a_{212})\\(a_{121},a_{211}) & (a_{121},a_{212})\end{array}\right\}=\left\{\begin{array}{cc}\underbrace{(\mathrm{US},\mathrm{DP})}_{\text{down-parabola upper-saddle}} & \underbrace{(\mathrm{DI}_{-},\mathrm{DI}_{+})}_{\text{double-inflection saddle}}\\\underbrace{(\mathrm{DP}_{-},\mathrm{DP}_{+})}_{\text{CW center}} & \underbrace{(\mathrm{DP},\mathrm{LS})}_{\text{down-parabola lower-saddle}}\end{array}\right\}$$
$$\text{for } a_{120}>0, a_{210}<0, \tag{4.270}$$

$$\left\{\begin{array}{cc}(a_{122},a_{211}) & (a_{122},a_{212})\\(a_{121},a_{211}) & (a_{121},a_{212})\end{array}\right\}=\left\{\begin{array}{cc}\underbrace{(\mathrm{LS},\mathrm{UP})}_{\text{up-parabola lower-saddle}} & \underbrace{(\mathrm{II}_{-},\mathrm{II}_{-})}_{\text{double-inflection saddle}}\\\underbrace{(\mathrm{UP}_{-},\mathrm{UP}_{-})}_{\text{negative saddle}} & \underbrace{(\mathrm{UP},\mathrm{LS})}_{\text{up-parabola lower-saddle}}\end{array}\right\}$$
$$\text{for } a_{120}<0, a_{210}<0. \tag{4.271}$$

From the first integral manifold, the phase portraits for the two-dimensional cubic system for $x_1^* = a_{211}, a_{212}$ on the x_1-direction for the x_2-direction and $x_2^* = a_{121}, a_{122}$ on the x_2-direction for the x_1-direction are sketched in Fig. 4.8a–d for ($a_{120} > 0$, $a_{210} > 0$), ($a_{120} < 0$, $a_{210} > 0$), ($a_{120} > 0$, $a_{210} < 0$), ($a_{120} < 0$, $a_{210} < 0$).

(ii) For $a_{12s_1} = a_{122}, a_{12s_2} = a_{121}, a_{21l_1} = a_{211}, a_{21l_2} = a_{212}$, the following properties of equilibriums of $(x_2^*, x_1^*) = (a_{121}, a_{211}), (a_{121}, a_{212}), (a_{122}, a_{211}), (a_{122}, a_{212})$ are summarized as follows.

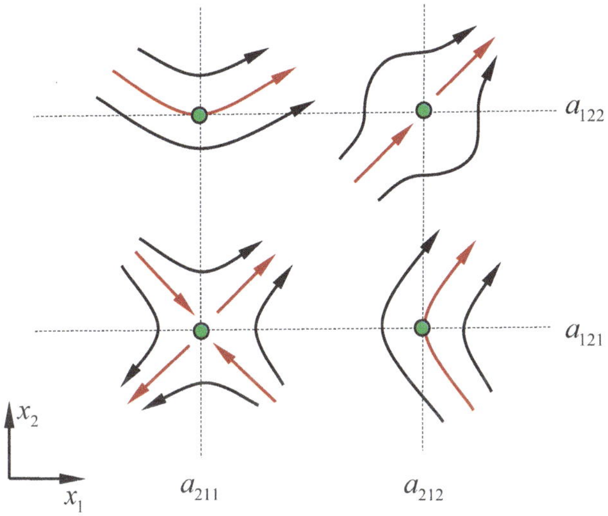

(a)

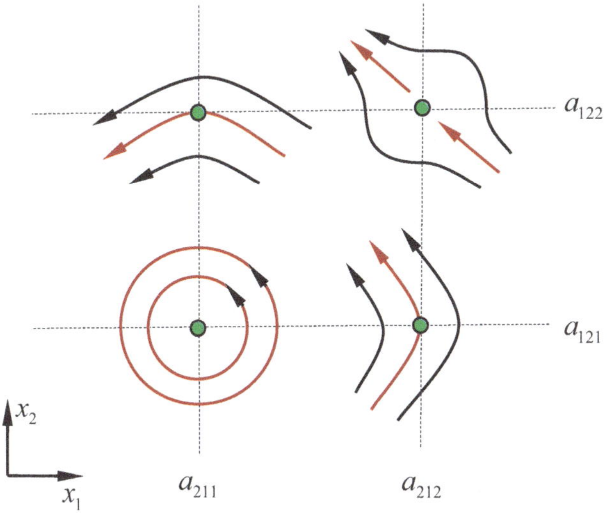

(b)

Fig. 4.8 Phase portraits for two-dimensional systems with $x_1^* = a_{211}, a_{212}$ on the x_1-direction for the x_2-direction and $x_2^* = a_{121}, a_{122}$ on the x_2-direction for the x_1-direction. The equilibriums of $(x_2^*, x_1^*) = (a_{121}, a_{211}), (a_{121}, a_{212}); (a_{122}, a_{211}), (a_{122}, a_{212})$: **(a)** $(a_{120} > 0, a_{210} > 0)$, **(b)** $(a_{120} < 0, a_{210} > 0)$, **(c)** $(a_{120} > 0, a_{210} < 0)$, **(d)** $(a_{120} < 0, a_{210} < 0)$

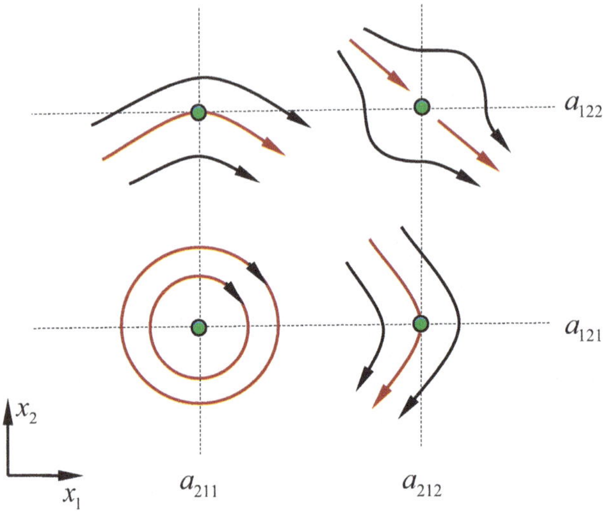

(c)

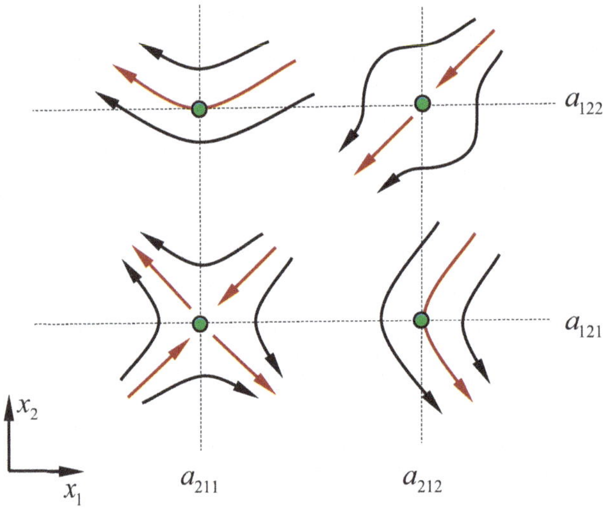

(d)

Fig. 4.8 (continued)

$$\left\{\begin{matrix} (a_{122}, a_{211}) & (a_{122}, a_{212}) \\ (a_{121}, a_{211}) & (a_{121}, a_{212}) \end{matrix}\right\} = \left\{\begin{matrix} \underbrace{(\mathrm{UP}_+, \mathrm{UP}_+)}_{\text{positive saddle}} & \underbrace{(\mathrm{UP}, \mathrm{US})}_{\text{up-parabola upper-saddle}} \\ \underbrace{(\mathrm{LS}, \mathrm{DP})}_{\text{down-parabola lower-saddle}} & \underbrace{(\mathrm{DI}_+, \mathrm{DI}_-)}_{\text{double-inflection saddle}} \end{matrix}\right\}$$

for $a_{120} > 0, a_{210} > 0$,

(4.272)

$$\left\{\begin{matrix} (a_{122}, a_{211}) & (a_{122}, a_{212}) \\ (a_{121}, a_{211}) & (a_{121}, a_{212}) \end{matrix}\right\} = \left\{\begin{matrix} \underbrace{(\mathrm{DP}_+, \mathrm{DP}_-)}_{\text{CCW center}} & \underbrace{(\mathrm{DP}, \mathrm{US})}_{\text{down-parabola upper-saddle}} \\ \underbrace{(\mathrm{US}, \mathrm{UP})}_{\text{up-parabola upper-saddle}} & \underbrace{(\mathrm{II}_+, \mathrm{II}_+)}_{\text{double-inflection saddle}} \end{matrix}\right\}$$

for $a_{120} < 0, a_{210} > 0$,

(4.273)

$$\left\{\begin{matrix} (a_{122}, a_{211}) & (a_{122}, a_{212}) \\ (a_{121}, a_{211}) & (a_{121}, a_{212}) \end{matrix}\right\} = \left\{\begin{matrix} \underbrace{(\mathrm{DP}_-, \mathrm{DP}_+)}_{\text{CW center}} & \underbrace{(\mathrm{DP}, \mathrm{LS})}_{\text{down-parabola lower-saddle}} \\ \underbrace{(\mathrm{LS}, \mathrm{UP})}_{\text{up-parabola lower-saddle}} & \underbrace{(\mathrm{II}_-, \mathrm{II}_-)}_{\text{double-inflection saddle}} \end{matrix}\right\}$$

for $a_{120} > 0, a_{210} < 0$,

(4.274)

$$\left\{\begin{matrix} (a_{122}, a_{211}) & (a_{122}, a_{212}) \\ (a_{121}, a_{211}) & (a_{121}, a_{212}) \end{matrix}\right\} = \left\{\begin{matrix} \underbrace{(\mathrm{UP}_-, \mathrm{UP}_-)}_{\text{negative saddle}} & \underbrace{(\mathrm{UP}, \mathrm{LS})}_{\text{up-parabola lower-saddle}} \\ \underbrace{(\mathrm{US}, \mathrm{DP})}_{\text{down-parabola upper-saddle}} & \underbrace{(\mathrm{DI}_-, \mathrm{DI}_+)}_{\text{double-inflection saddle}} \end{matrix}\right\}$$

for $a_{120} < 0, a_{210} < 0$.

(4.275)

From the first integral manifold, the phase portraits for the two-dimensional cubic system for $x_1^* = a_{211}, a_{212}$ on the x_1-direction for the x_2-direction and $x_2^* = a_{121}, a_{122}$ on the x_2-direction for the x_1-direction are sketched in Fig. 4.9a–d for ($a_{120} > 0$, $a_{210} > 0$), ($a_{120} < 0, a_{210} > 0$), ($a_{120} > 0, a_{210} < 0$), ($a_{120} < 0, a_{210} < 0$).

(iii) For $a_{12s_1} = a_{121}, a_{12s_2} = a_{122}, a_{21l_1} = a_{212}, a_{21l_2} = a_{211}$, the following properties of equilibriums of $(x_2^*, x_1^*) = (a_{121}, a_{211}), (a_{121}, a_{212}), (a_{122}, a_{211}), (a_{122}, a_{212})$ are summarized as follows.

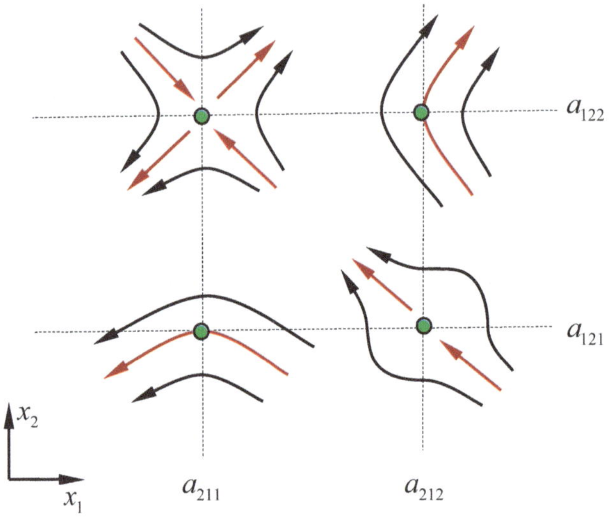

(a)

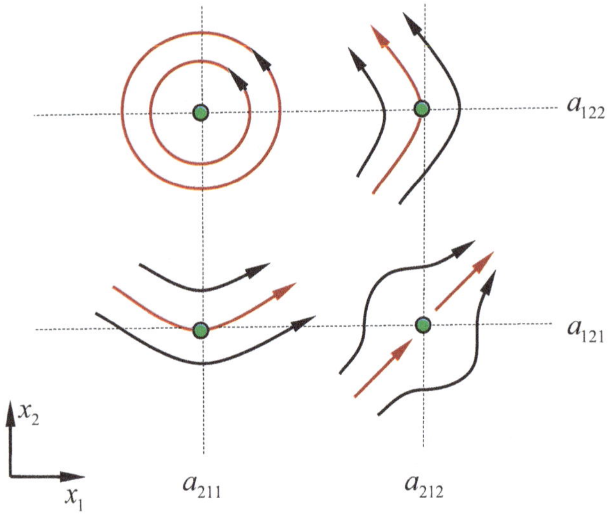

(b)

Fig. 4.9 Phase portraits for two-dimensional systems with $x_1^* = a_{211}, a_{212}$ on the x_1-direction for the x_2-direction and $x_2^* = a_{121}, a_{122}$ on the x_2-direction for the x_1-direction. The equilibriums of $(x_2^*, x_1^*) = (a_{121}, a_{211}), (a_{121}, a_{212}); (a_{122}, a_{211}), (a_{122}, a_{212})$: **(a)** $(a_{120} > 0, a_{210} > 0)$, **(b)** $(a_{120} < 0, a_{210} > 0)$, **(c)** $(a_{120} > 0, a_{210} < 0)$, **(d)** $(a_{120} < 0, a_{210} < 0)$

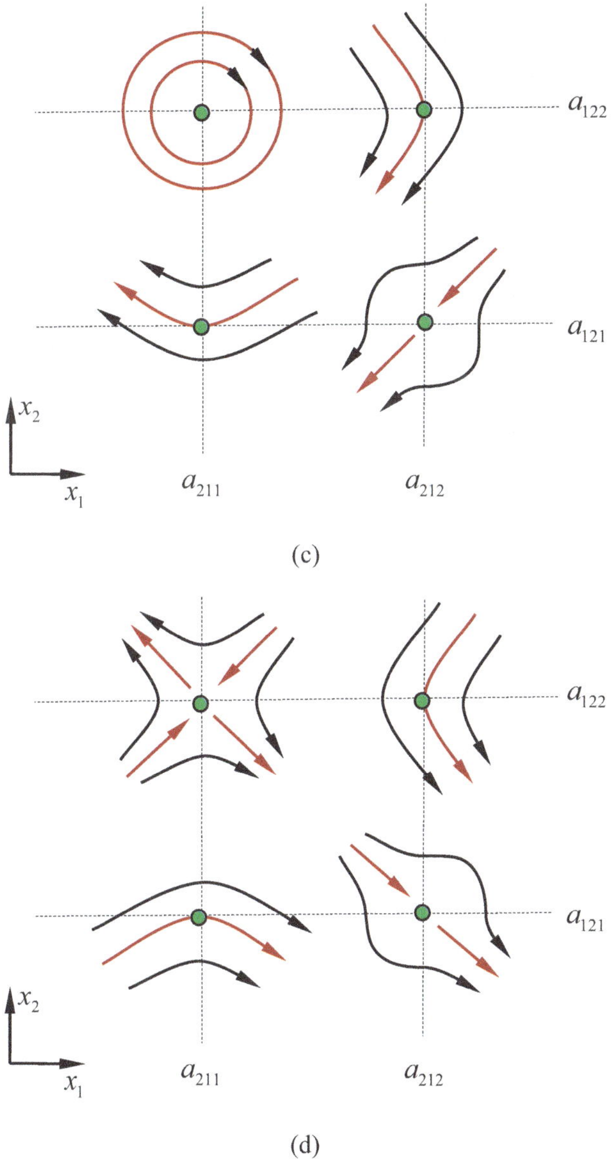

Fig. 4.9 (continued)

$$\left\{\begin{array}{cc}(a_{122},a_{211}) & (a_{122},a_{212})\\ (a_{121},a_{211}) & (a_{121},a_{212})\end{array}\right\}=\left\{\begin{array}{cc}\underbrace{(\mathrm{DI}_-,\mathrm{DI}_+)}_{\text{double-inflection saddle}} & \underbrace{(\mathrm{US},\mathrm{UP})}_{\text{up-parabola upper-saddle}}\\ \underbrace{(\mathrm{DP},\mathrm{LS})}_{\text{down-parabola lower-saddle}} & \underbrace{(\mathrm{UP}_+,\mathrm{UP}_+)}_{\text{positive saddle}}\end{array}\right\}$$

for $a_{120}>0, a_{210}>0,$

(4.276)

$$\left\{\begin{array}{cc}(a_{122},a_{211}) & (a_{122},a_{212})\\ (a_{121},a_{211}) & (a_{121},a_{212})\end{array}\right\}=\left\{\begin{array}{cc}\underbrace{(\mathrm{II}_-\mathrm{II}_-)}_{\text{double-inflection saddle}} & \underbrace{(\mathrm{LS},\mathrm{DP})}_{\text{down-parabola lower-saddle}}\\ \underbrace{(\mathrm{UP},\mathrm{LS})}_{\text{up-parabola lower-ssadde}} & \underbrace{(\mathrm{DP}_+,\mathrm{DP}_-)}_{\text{CCW center}}\end{array}\right\}$$

for $a_{120}<0, a_{210}>0,$

(4.277)

$$\left\{\begin{array}{cc}(a_{122},a_{211}) & (a_{122},a_{212})\\ (a_{121},a_{211}) & (a_{121},a_{212})\end{array}\right\}=\left\{\begin{array}{cc}\underbrace{(\mathrm{II}_+,\mathrm{II}_+)}_{\text{double-inflection saddle}} & \underbrace{(\mathrm{US},\mathrm{DP})}_{\text{down-parabola upper-saddle}}\\ \underbrace{(\mathrm{UP},\mathrm{US})}_{\text{up-parabola upper-saddle}} & \underbrace{(\mathrm{DP}_-,\mathrm{DP}_+)}_{\text{CW center}}\end{array}\right\}$$

for $a_{120}>0, a_{210}<0,$

(4.278)

$$\left\{\begin{array}{cc}(a_{122},a_{211}) & (a_{122},a_{212})\\ (a_{121},a_{211}) & (a_{121},a_{212})\end{array}\right\}=\left\{\begin{array}{cc}\underbrace{(\mathrm{DI}_-,\mathrm{DI}_+)}_{\text{double-infection saddle}} & \underbrace{(\mathrm{LS},\mathrm{UP})}_{\text{up-parabola lower-saddle}}\\ \underbrace{(\mathrm{DP},\mathrm{US})}_{\text{down-parabola upper-saddle}} & \underbrace{(\mathrm{UP}_-,\mathrm{UP}_-)}_{\text{negative saddle}}\end{array}\right\}$$

for $a_{120}<0, a_{210}<0.$

(4.279)

From the first integral manifold, the phase portraits for the two-dimensional cubic system for $x_1^*=a_{211},a_{212}$ on the x_1-direction for the x_2-direction and $x_2^*=a_{121},a_{122}$ on the x_2-direction for the x_1-direction are sketched in Fig. 4.10a–d for $(a_{120}>0,$ $a_{210}>0)$, $(a_{120}<0,a_{210}>0)$, $(a_{120}>0,a_{210}<0)$, $(a_{120}<0,a_{210}<0)$.

(iv) For $a_{12s_1}=a_{122},a_{12s_2}=a_{121},a_{21l_1}=a_{212},a_{21l_2}=a_{211}$, the following properties of equilibriums of $(x_2^*,x_1^*)=(a_{121},a_{211}),(a_{121},a_{212}),(a_{122},a_{211}),(a_{122},a_{212})$ are summarized as follows.

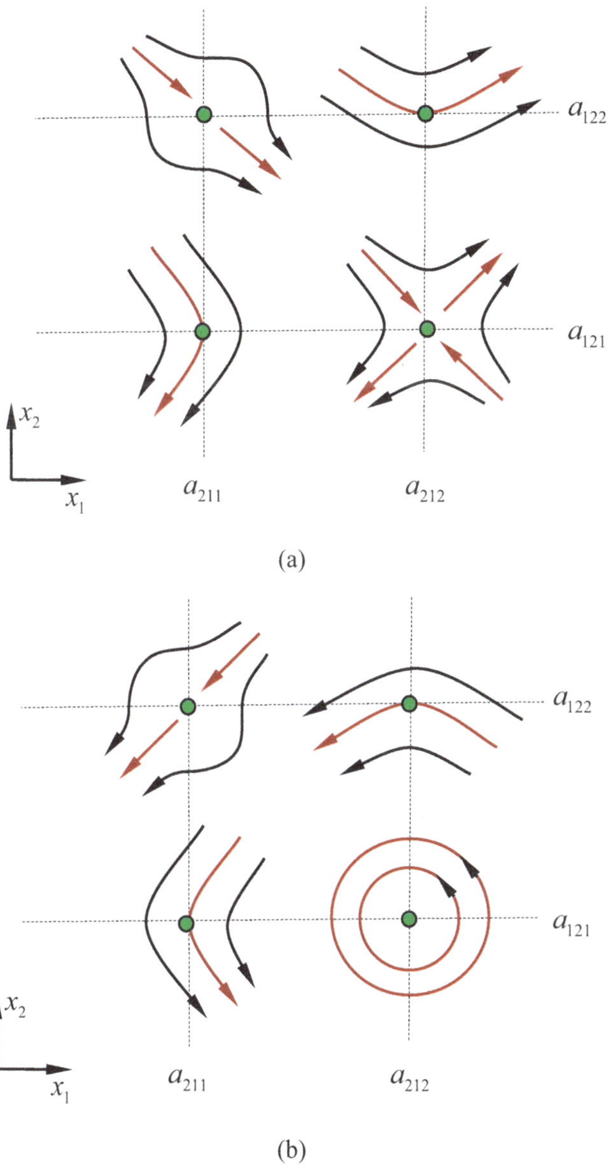

Fig. 4.10 Phase portraits for two-dimensional systems with $x_1^* = a_{211}, a_{212}$ on the x_1-direction for the x_2-direction and $x_2^* = a_{121}, a_{122}$ on the x_2-direction for the x_1-direction. The equilibriums of $(x_2^*, x_1^*) = (a_{121}, a_{211}), (a_{121}, a_{212}); (a_{122}, a_{211}), (a_{122}, a_{212})$: **(a)** $(a_{120} > 0, a_{210} > 0)$, **(b)** $(a_{120} < 0, a_{210} > 0)$, **(c)** $(a_{120} > 0, a_{210} < 0)$, **(d)** $(a_{120} < 0, a_{210} < 0)$

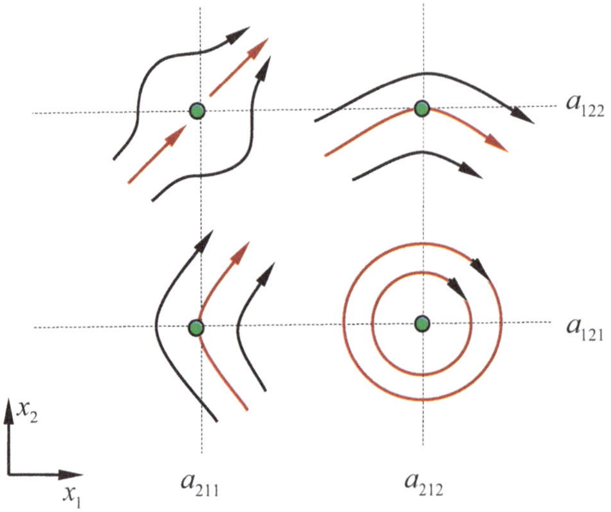

(c)

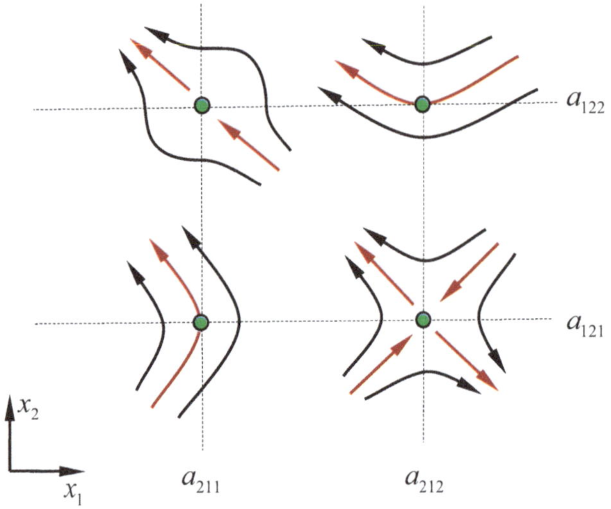

(d)

Fig. 4.10 (continued)

$$\left\{\begin{array}{ll}(a_{122},a_{211}) & (a_{122},a_{212})\\ (a_{121},a_{211}) & (a_{121},a_{212})\end{array}\right\}=\left\{\begin{array}{cc}\underbrace{(\mathrm{DP},\mathrm{LS})}_{\text{down-parabola lower-saddle}} & \underbrace{(\mathrm{UP}_+,\mathrm{UP}_+)}_{\text{positive saddle}}\\ \underbrace{(\mathrm{II}_-,\mathrm{II}_-)}_{\text{double-inflection saddle}} & \underbrace{(\mathrm{LS},\mathrm{DP})}_{\text{down-parabola lower-saddle}}\end{array}\right\}$$

for $a_{120}>0, a_{210}>0,$

(4.280)

$$\left\{\begin{array}{ll}(a_{122},a_{211}) & (a_{122},a_{212})\\ (a_{121},a_{211}) & (a_{121},a_{212})\end{array}\right\}=\left\{\begin{array}{cc}\underbrace{(\mathrm{UP},\mathrm{LS})}_{\text{up-parabola lower-saddle}} & \underbrace{(\mathrm{DP}_+,\mathrm{DP}_-)}_{\text{CCW center}}\\ \underbrace{(\mathrm{DI}_-,\mathrm{DI}_+)}_{\text{double-inflection saddle}} & \underbrace{(\mathrm{US},\mathrm{UP})}_{\text{up-parabola upper-saddle}}\end{array}\right\}$$

for $a_{120}<0, a_{210}>0,$

(4.281)

$$\left\{\begin{array}{ll}(a_{122},a_{211}) & (a_{122},a_{212})\\ (a_{121},a_{211}) & (a_{121},a_{212})\end{array}\right\}=\left\{\begin{array}{cc}\underbrace{(\mathrm{UP},\mathrm{US})}_{\text{up-parabola upper-saddle}} & \underbrace{(\mathrm{DP}_-,\mathrm{DP}_+)}_{\text{CW center}}\\ \underbrace{(\mathrm{DI}_+,\mathrm{DI}_-)}_{\text{double-inflection saddle}} & \underbrace{(\mathrm{US},\mathrm{UP})}_{\text{up-parabola lower-saddle}}\end{array}\right\}$$

for $a_{120}>0, a_{210}<0,$

(4.282)

$$\left\{\begin{array}{ll}(a_{122},a_{211}) & (a_{122},a_{212})\\ (a_{121},a_{211}) & (a_{121},a_{212})\end{array}\right\}=\left\{\begin{array}{cc}\underbrace{(\mathrm{DP},\mathrm{LS})}_{\text{down-parabola lower-saddle}} & \underbrace{(\mathrm{UP}_-,\mathrm{UP}_-)}_{\text{negative saddle}}\\ \underbrace{(\mathrm{II}_+,\mathrm{II}_+)}_{\text{double-inflection saddle}} & \underbrace{(\mathrm{US},\mathrm{DP})}_{\text{down-parabola upper-saddle}}\end{array}\right\}$$

for $a_{120}<0, a_{210}<0.$

(4.283)

From the first integral manifold, the phase portraits for the two-dimensional cubic system for $x_1^*=a_{211},a_{212}$ on the x_1-direction for the x_2-direction and $x_2^*=a_{121}$ on the x_2-direction for the x_1-direction are sketched in Fig. 4.11a–d for $(a_{120}>0,a_{210}>0)$, $(a_{120}<0,a_{210}>0)$, $(a_{120}>0,a_{210}<0)$, $(a_{120}<0,a_{210}<0)$.

4.4 Limit Cycles and Homoclinic Networks

Consider a quadratic dynamical system as

$$\begin{aligned}\dot{x}_1&=a_{120}(x_2-a_{121})(x_2-a_{122})(x_2-a_{123}),\\ \dot{x}_2&=a_{210}(x_1-a_{211})(x_1-a_{212})(x_1-a_{213}),\end{aligned} \tag{4.284}$$

and the first integral manifold is determined by

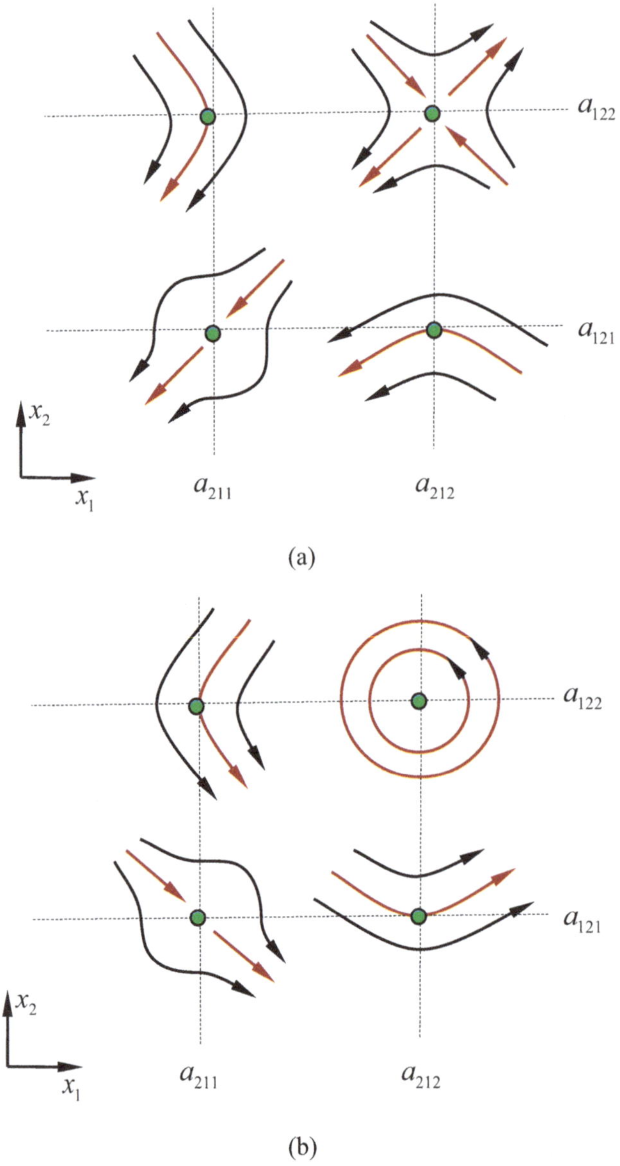

Fig. 4.11 Phase portraits for two-dimensional systems with $x_1^* = a_{211}, a_{212}$ on the x_1-direction for the x_2-direction and $x_2^* = a_{121}, a_{122}$ on the x_2-direction for the x_1-direction. The equilibriums of $(x_2^*, x_1^*) = (a_{121}, a_{211}), (a_{121}, a_{212}); (a_{122}, a_{211}), (a_{122}, a_{212})$: **(a)** $(a_{120} > 0, a_{210} > 0)$, **(b)** $(a_{120} < 0, a_{210} > 0)$, **(c)** $(a_{120} > 0, a_{210} < 0)$, **(d)** $(a_{120} < 0, a_{210} < 0)$

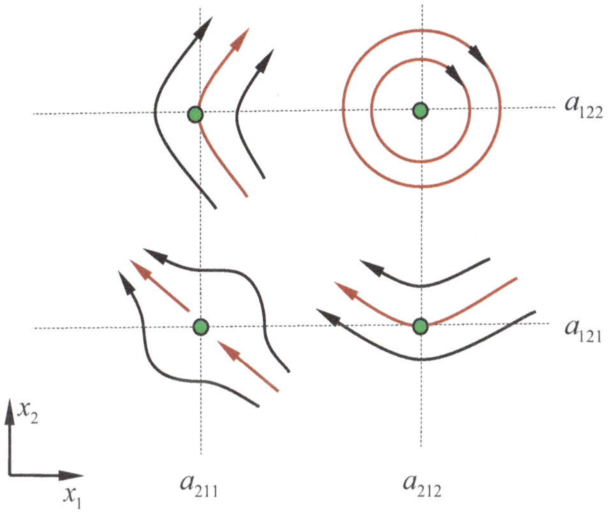

(c)

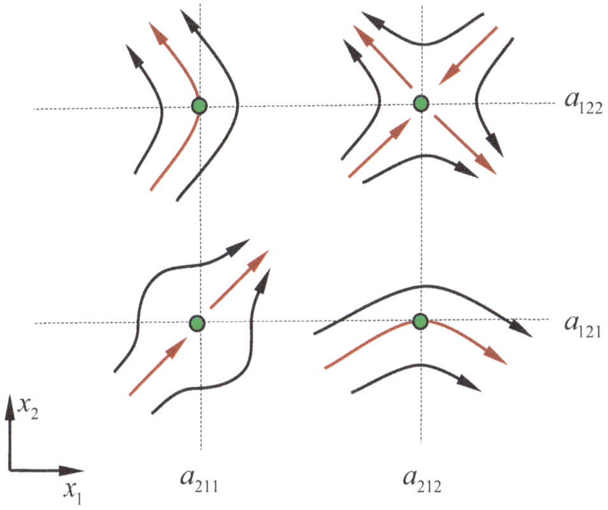

(d)

Fig. 4.11 (continued)

Table 4.1 Local saddles and centers for two-dimensional cubic nonlinear systems

Fig. 4.12	(a)	(b)	(c)	(d)
(a_{120}, a_{210})	(+,+)	(−,+)	(+,−)	(−,−)
(a_{121}, a_{211})	positive saddle	CCW center	CW center	negative saddle
(a_{121}, a_{212})	CW center	negative saddle	positive saddle	CCW center
(a_{121}, a_{213})	positive saddle	CCW center	CW center	negative saddle
(a_{122}, a_{211})	CCW center	positive saddle	negative saddle	CW center
(a_{122}, a_{212})	negative saddle	CW center	CCW center	positive saddle
(a_{122}, a_{213})	CCW center	positive saddle	negative saddle	CW center
(a_{123}, a_{211})	positive saddle	CCW center	CW center	negative saddle
(a_{123}, a_{212})	CW center	negative saddle	positive saddle	CCW center
(a_{123}, a_{213})	positive saddle	CCW center	CW center	negative saddle

$$
\begin{aligned}
&a_{120}\Big\{\frac{1}{4}\left[(x_2-a_{121})^4-(x_{20}-a_{121})^4\right]\\
&+\frac{1}{3}(2a_{121}-a_{122}-a_{123})\left[(x_2-a_{121})^3-(x_{20}-a_{121})^3\right]\\
&+\frac{1}{2}(a_{121}-a_{122})(a_{121}-a_{123})\left[(x_2-a_{121})^2-(x_{20}-a_{121})^2\right]\Big\}\\
&=a_{210}\Big\{\frac{1}{4}\left[(x_1-a_{211})^4-(x_{10}-a_{211})^4\right]\\
&+\frac{1}{3}(2a_{211}-a_{212}-a_{213})\left[(x_1-a_{211})^3-(x_{10}-a_{211})^3\right]\\
&+\frac{1}{2}(a_{211}-a_{212})(a_{211}-a_{213})\left[(x_1-a_{211})^2-(x_{10}-a_{211})^2\right]\Big\}.
\end{aligned}
\tag{4.285}
$$

The equilibrium properties in the homoclinic networks are listed in Table 4.1, which can be presented as follows.

(i) For $a_{120} > 0$ and $a_{210} > 0$,

$$
\begin{aligned}
&\begin{Bmatrix}
(a_{123},a_{211}) & (a_{123},a_{212}) & (a_{123},a_{213})\\
(a_{122},a_{211}) & (a_{122},a_{212}) & (a_{122},a_{213})\\
(a_{121},a_{211}) & (a_{121},a_{212}) & (a_{121},a_{213})
\end{Bmatrix}\\
&=\begin{Bmatrix}
\underbrace{(\mathrm{UP}_+,\mathrm{UP}_+)}_{\text{positive saddle}} & \underbrace{(\mathrm{DP}_-,\mathrm{DP}_+)}_{\text{CW center}} & \underbrace{(\mathrm{UP}_+,\mathrm{UP}_+)}_{\text{positive saddle}}\\
\underbrace{(\mathrm{DP}_+,\mathrm{DP}_-)}_{\text{CCW center}} & \underbrace{(\mathrm{UP}_-,\mathrm{UP}_-)}_{\text{negative saddle}} & \underbrace{(\mathrm{DP}_+,\mathrm{DP}_-)}_{\text{CCW center}}\\
\underbrace{(\mathrm{UP}_+,\mathrm{UP}_+)}_{\text{positive saddle}} & \underbrace{(\mathrm{DP}_-,\mathrm{DP}_+)}_{\text{CW center}} & \underbrace{(\mathrm{UP}_+,\mathrm{UP}_+)}_{\text{negative saddle}}
\end{Bmatrix}.
\end{aligned}
\tag{4.286}
$$

(ii) For $a_{120} < 0$ and $a_{210} > 0$,

$$\begin{Bmatrix} (a_{123}, a_{211}) & (a_{123}, a_{212}) & (a_{123}, a_{213}) \\ (a_{122}, a_{211}) & (a_{122}, a_{212}) & (a_{122}, a_{213}) \\ (a_{121}, a_{211}) & (a_{121}, a_{212}) & (a_{121}, a_{213}) \end{Bmatrix}$$
$$= \left\{ \begin{array}{ccc} \underbrace{(\mathrm{DP}_+, \mathrm{DP}_-)}_{\text{CCW center}} & \underbrace{(\mathrm{UP}_-, \mathrm{UP}_-)}_{\text{negative saddle}} & \underbrace{(\mathrm{DP}_+, \mathrm{DP}_-)}_{\text{CCW center}} \\ \underbrace{(\mathrm{UP}_+, \mathrm{UP}_+)}_{\text{positive saddle}} & \underbrace{(\mathrm{DP}_-, \mathrm{DP}_+)}_{\text{CW center}} & \underbrace{(\mathrm{UP}_+, \mathrm{UP}_+)}_{\text{positive saddle}} \\ \underbrace{(\mathrm{DP}_+, \mathrm{DP}_-)}_{\text{CCW center}} & \underbrace{(\mathrm{UP}_-, \mathrm{UP}_-)}_{\text{negative saddle}} & \underbrace{(\mathrm{DP}_+, \mathrm{DP}_-)}_{\text{CCW center}} \end{array} \right\}. \tag{4.287}$$

(iii) For $a_{120} > 0$ and $a_{210} < 0$,

$$\begin{Bmatrix} (a_{123}, a_{211}) & (a_{123}, a_{212}) & (a_{123}, a_{213}) \\ (a_{122}, a_{211}) & (a_{122}, a_{212}) & (a_{122}, a_{213}) \\ (a_{121}, a_{211}) & (a_{121}, a_{212}) & (a_{121}, a_{213}) \end{Bmatrix}$$
$$= \left\{ \begin{array}{ccc} \underbrace{(\mathrm{DP}_-, \mathrm{DP}_+)}_{\text{CW center}} & \underbrace{(\mathrm{UP}_+, \mathrm{UP}_+)}_{\text{positive sadde}} & \underbrace{(\mathrm{DP}_-, \mathrm{DP}_+)}_{\text{CW center}} \\ \underbrace{(\mathrm{UP}_-, \mathrm{UP}_-)}_{\text{negative saddle}} & \underbrace{(\mathrm{DP}_+, \mathrm{DP}_-)}_{\text{CCW center}} & \underbrace{(\mathrm{UP}_-, \mathrm{UP}_-)}_{\text{negative saddle}} \\ \underbrace{(\mathrm{DP}_-, \mathrm{DP}_+)}_{\text{CW center}} & \underbrace{(\mathrm{UP}_+, \mathrm{UP}_+)}_{\text{positive saddle}} & \underbrace{(\mathrm{DP}_-, \mathrm{DP}_+)}_{\text{CW center}} \end{array} \right\}. \tag{4.288}$$

(iv) For $a_{120} < 0$ and $a_{210} < 0$,

$$\begin{Bmatrix} (a_{123},a_{211}) & (a_{123},a_{212}) & (a_{123},a_{213}) \\ (a_{122},a_{211}) & (a_{122},a_{212}) & (a_{122},a_{213}) \\ (a_{121},a_{211}) & (a_{121},a_{212}) & (a_{121},a_{213}) \end{Bmatrix} = \begin{Bmatrix} \underbrace{(\mathrm{UP}_-,\mathrm{UP}_-)}_{\text{negative saddle}} & \underbrace{(\mathrm{DP}_+,\mathrm{DP}_-)}_{\text{CW center}} & \underbrace{(\mathrm{UP}_-,\mathrm{UP}_-)}_{\text{negative saddle}} \\ \underbrace{(\mathrm{DP}_-,\mathrm{DP}_+)}_{\text{CW center}} & \underbrace{(\mathrm{UP}_+,\mathrm{UP}_+)}_{\text{positive saddle}} & \underbrace{(\mathrm{DP}_-,\mathrm{DP}_+)}_{\text{CW center}} \\ \underbrace{(\mathrm{UP}_-,\mathrm{UP}_-)}_{\text{negative saddle}} & \underbrace{(\mathrm{DP}_+,\mathrm{DP}_-)}_{\text{CCW center}} & \underbrace{(\mathrm{UP}_-,\mathrm{UP}_-)}_{\text{negative saddle}} \end{Bmatrix}. \tag{4.289}$$

For the above four cases, without the local centers and saddles, the global characteristics of such a dynamical system with the four equilibriums are given by the double-inflections flows, as shown in Fig. 4.12. The global flows are given by the $(3^{\mathrm{rd}}\mathrm{UP}_+, 3^{\mathrm{rd}}\mathrm{UP}_+)$-double third-order positive saddle $(a_{120} > 0, a_{210} > 0)$, the $(3^{\mathrm{rd}}\mathrm{DP}_+, 3^{\mathrm{rd}}\mathrm{DP}_-)$-double third-order counter-clockwise center $(a_{120} < 0, a_{210} > 0)$, the $(3^{\mathrm{rd}}\mathrm{DP}_-, 3^{\mathrm{rd}}\mathrm{DP}_+)$-third-order clockwise center $(a_{120} > 0, a_{210} < 0)$, the $(3^{\mathrm{rd}}\mathrm{UP}_-, 3^{\mathrm{rd}}\mathrm{UP}_-)$-double third-order negative saddle $(a_{120} < 0, a_{210} < 0)$. The four local equilibriums inside of the homoclinic orbits are treated as a local point. The flows near the six equilibriums are called a local network of the center and saddle with limit cycle and homoclinic orbits. The local network is formed by the homoclinic orbits, and the saddles are nodes for such a local network.

4.5 Bifurcation Routes

The appearing bifurcations of the third-order positive and negative saddles and third-order CCW and CW centers occur at the x_2-direction, which are listed as follows.

(i) For $a_{j_1j_20} > 0$ and $a_{j_2j_10} > 0$,

$$\underbrace{(\mathrm{UP}_+,\mathrm{UP}_+)}_{\text{positive saddle}} \rightleftharpoons \underbrace{(\mathrm{UP}_+,3^{\mathrm{rd}}\mathrm{UP}_+)}_{\text{third-order positive saddle}} \rightleftharpoons \begin{Bmatrix} \underbrace{(\mathrm{UP}_+,\mathrm{UP}_+)}_{\text{positive saddle}} \\ \underbrace{(\mathrm{DP}_-,\mathrm{DP}_+)}_{\text{CW center}} \\ \underbrace{(\mathrm{UP}_+,\mathrm{UP}_+)}_{\text{positive saddle}} \end{Bmatrix}. \tag{4.290}$$

(ii) For $a_{j_1j_20} < 0$ and $a_{j_2j_10} > 0$,

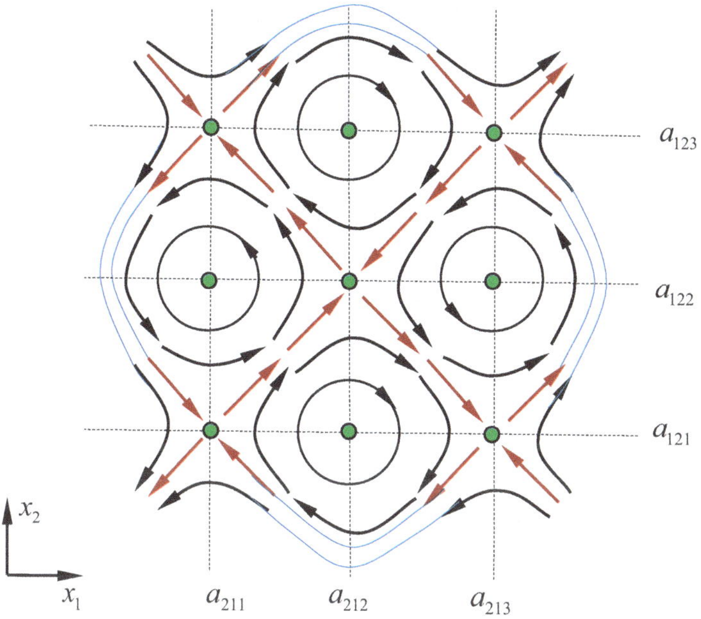

(a)

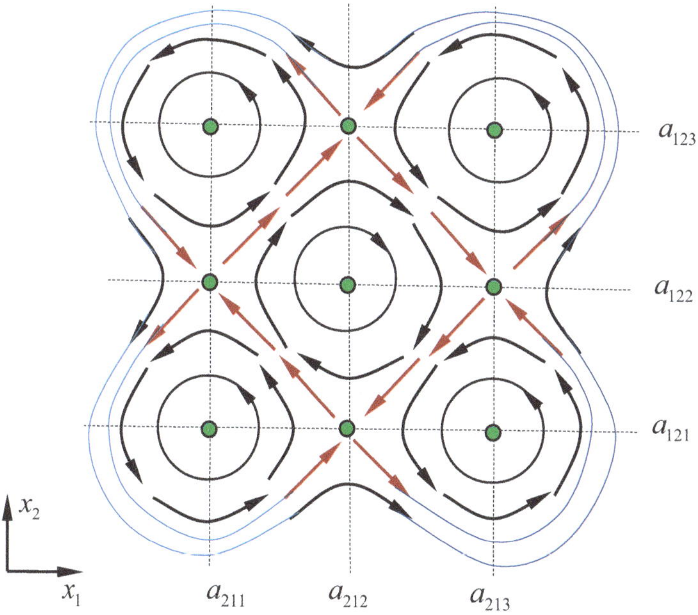

(b)

Fig. 4.12 Phase portraits for two-dimensional systems on the x_1-direction with $x_1^* = a_{211}, a_{212}, a_{213}$ and on the x_2-direction with $x_1^* = a_{121}, a_{122}, a_{123}$. The four sets of nine simple saddle and center: (**a**) $(a_{120} > 0, a_{210} > 0)$, (**b**) $(a_{120} < 0, a_{210} > 0)$, (**c**) $(a_{120} > 0, a_{210} < 0)$, (**d**) $(a_{120} < 0, a_{210} < 0)$

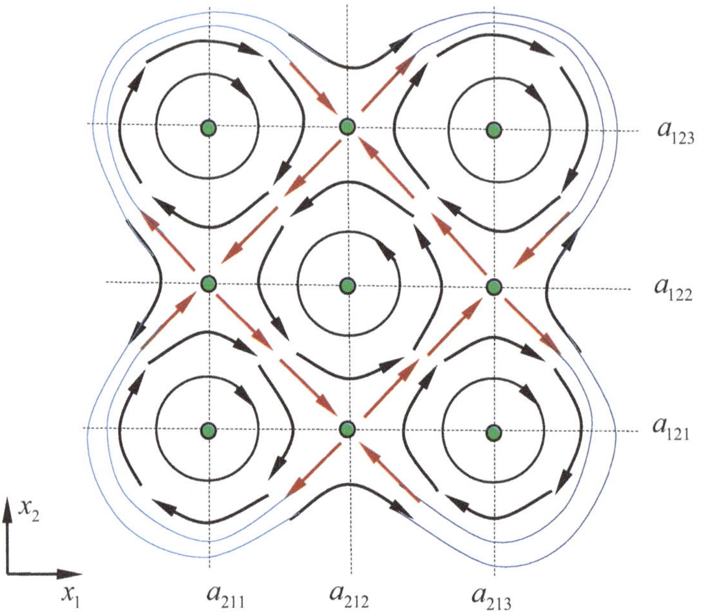

(c)

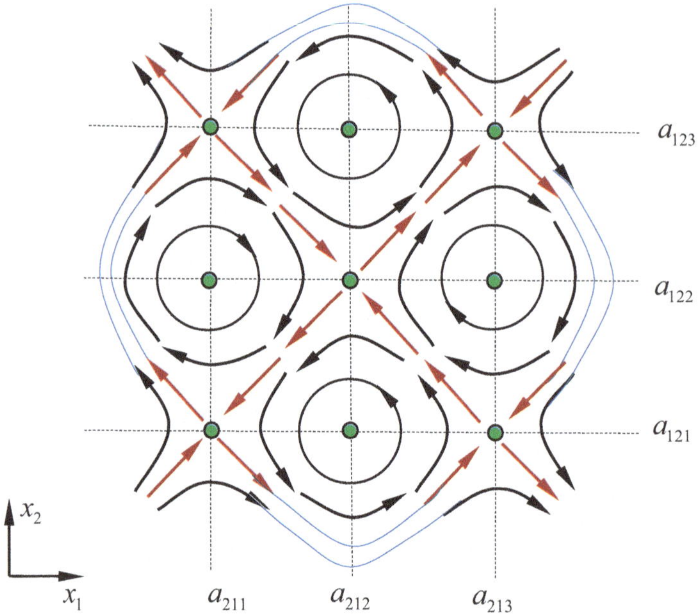

(d)

Fig. 4.12 (continued)

$$\underbrace{(\mathrm{DP}_+,\mathrm{DP}_-)}_{\text{CCW center}} \rightleftharpoons \underbrace{(\mathrm{DP}_+,3^{\mathrm{rd}}\mathrm{DP}_-)}_{\text{third-order CCW center}} \rightleftharpoons \left\{ \begin{array}{c} \underbrace{(\mathrm{DP}_+,\mathrm{DP}_-)}_{\text{CCW center}} \\ \underbrace{(\mathrm{UP}_-,\mathrm{UP}_-)}_{\text{negative saddle}} \\ \underbrace{(\mathrm{DP}_+,\mathrm{DP}_-)}_{\text{CCW center}} \end{array} \right\}. \tag{4.291}$$

(iii) For $a_{j_1j_20} > 0$ and $a_{j_2j_10} < 0$,

$$\underbrace{(\mathrm{DP}_-,\mathrm{DP}_+)}_{\text{CW center}} \rightleftharpoons \underbrace{(\mathrm{DP}_-,3^{\mathrm{rd}}\mathrm{DP}_+)}_{\text{third-order CW center}} \rightleftharpoons \left\{ \begin{array}{c} \underbrace{(\mathrm{DP}_-,\mathrm{DP}_+)}_{\text{CW center}} \\ \underbrace{(\mathrm{UP}_+,\mathrm{UP}_+)}_{\text{positive saddle}} \\ \underbrace{(\mathrm{DP}_-,\mathrm{DP}_+)}_{\text{CW center}} \end{array} \right\}. \tag{4.292}$$

(iv) For $a_{j_1j_20} < 0$ and $a_{j_2j_10} < 0$,

$$\underbrace{(\mathrm{UP}_-,\mathrm{UP}_-)}_{\text{negative saddle}} \rightleftharpoons \underbrace{(\mathrm{UP}_-,3^{\mathrm{rd}}\mathrm{UP}_-)}_{\text{third-order negative saddle}} \rightleftharpoons \left\{ \begin{array}{c} \underbrace{(\mathrm{UP}_-,\mathrm{UP}_-)}_{\text{negative saddle}} \\ \underbrace{(\mathrm{DP}_+,\mathrm{DP}_-)}_{\text{CCW center}} \\ \underbrace{(\mathrm{UP}_-,\mathrm{UP}_-)}_{\text{negative saddle}} \end{array} \right\}. \tag{4.293}$$

- The $(\mathrm{UP}_+, 3^{\mathrm{rd}}\mathrm{UP}_+)$-third-order saddle bifurcation is from a positive saddle to positive saddle, CW center, and positive saddle.
- The $(\mathrm{DP}_+, 3^{\mathrm{rd}}\mathrm{DP}_-)$-third-order counter-clockwise center bifurcation is from a CCW center to CCW center, negative saddle, and CCW center.
- The $(\mathrm{DP}_-, 3^{\mathrm{rd}}\mathrm{DP}_+)$-third-order clockwise center bifurcation is from a CW center to CW center, positive saddle, and CW center.
- The $(\mathrm{UP}_-, 3^{\mathrm{rd}}\mathrm{UP}_-)$-third-order negative saddle bifurcation is from a negative saddle to negative saddle, CCW center, and negative saddle.

The appearing bifurcations of the third-order positive and negative saddles and third-order CCW and CW centers occur at the x_1-direction, which are listed as follows.

(i) For $a_{j_1 j_2 0} > 0$ and $a_{j_2 j_1 0} > 0$,

$$\underbrace{(\mathrm{UP}_+, \mathrm{UP}_+)}_{\text{positive saddle}} \rightleftharpoons \underbrace{(3^{\text{rd}}\mathrm{UP}_+, \mathrm{UP}_+)}_{\text{third-order positive saddle}} \rightleftharpoons \left\{ \begin{array}{c} \underbrace{(\mathrm{UP}_+, \mathrm{UP}_+)}_{\text{positive saddle}} \\ \underbrace{(\mathrm{DP}_+, \mathrm{DP}_-)}_{\text{CCW center}} \\ \underbrace{(\mathrm{UP}_+, \mathrm{UP}_+)}_{\text{positive saddle}} \end{array} \right\}. \tag{4.294}$$

(ii) For $a_{j_1 j_2 0} < 0$ and $a_{j_2 j_1 0} > 0$,

$$\underbrace{(\mathrm{DP}_+, \mathrm{DP}_-)}_{\text{CCW center}} \rightleftharpoons \underbrace{(3^{\text{rd}}\mathrm{DP}_+, \mathrm{DP}_-)}_{\text{third-order CCW center}} \rightleftharpoons \left\{ \begin{array}{c} \underbrace{(\mathrm{DP}_+, \mathrm{DP}_-)}_{\text{CCW center}} \\ \underbrace{(\mathrm{UP}_+, \mathrm{UP}_+)}_{\text{positive saddle}} \\ \underbrace{(\mathrm{DP}_+, \mathrm{DP}_-)}_{\text{CCW center}} \end{array} \right\}. \tag{4.295}$$

(iii) For $a_{j_1 j_2 0} > 0$ and $a_{j_2 j_1 0} < 0$,

$$\underbrace{(\mathrm{DP}_-, \mathrm{DP}_+)}_{\text{CW center}} \rightleftharpoons \underbrace{(3^{\text{rd}}\mathrm{DP}_-, \mathrm{DP}_+)}_{\text{third-order CW center}} \rightleftharpoons \left\{ \begin{array}{c} \underbrace{(\mathrm{DP}_-, \mathrm{DP}_+)}_{\text{CW center}} \\ \underbrace{(\mathrm{UP}_-, \mathrm{UP}_-)}_{\text{negative saddle}} \\ \underbrace{(\mathrm{DP}_-, \mathrm{DP}_+)}_{\text{CW center}} \end{array} \right\}. \tag{4.296}$$

(iv) For $a_{j_1 j_2 0} < 0$ and $a_{j_2 j_1 0} < 0$,

$$\underbrace{(\mathrm{UP}_-, \mathrm{UP}_-)}_{\text{negative saddle}} \rightleftharpoons \underbrace{(3^{\text{rd}}\mathrm{UP}_-, \mathrm{UP}_-)}_{\text{third-order negative saddle}} \rightleftharpoons \left\{ \begin{array}{c} \underbrace{(\mathrm{UP}_-, \mathrm{UP}_-)}_{\text{negative saddle}} \\ \underbrace{(\mathrm{DP}_-, \mathrm{DP}_+)}_{\text{CW center}} \\ \underbrace{(\mathrm{UP}_-, \mathrm{UP}_-)}_{\text{negative saddle}} \end{array} \right\} \tag{4.297}$$

- The $(3^{rd}UP_{+}, UP_{+})$-third-order saddle bifurcation is from a positive saddle to positive saddle, CCW center, and positive saddle.
- The $(3^{rd}DP_{+}, DP_{-})$-third-order counter-clockwise center bifurcation is from a CCW center to CCW center, positive saddle, and CCW center.
- The $(3^{rd}DP_{-}, DP_{+})$-third-order clockwise center bifurcation is from a CW center to CW center, negative saddle, and CW center.
- The $(3^{rd}UP_{-}, UP_{-})$-third-order negative saddle bifurcation is from a negative saddle to negative saddle, CW center, and negative saddle.

Through the two times switching of the third-order saddles and centers, the homoclinic networks of saddles and centers are generated. In Fig. 4.13, the appearing bifurcations of the double-third-order positive saddle, double-third-order CCW center, and double-third-order CW center, the double-third-order negative saddle are illustrated.

In Fig. 4.13(a), the $(3^{rd}UP_{+}, 3^{rd}UP_{+})$-double-third-order positive saddle appearing bifurcation through two times are presented ($a_{j_1j_20} > 0$, $a_{j_2j_10} > 0$). The first $(3^{rd}UP_{+}, UP_{+})$ third-order positive saddle appearing bifurcation from the (UP_{+}, UP_{+})-positive saddle generates the (UP_{+}, UP_{+})-positive saddle, (DP_{+}, DP_{-})-CCW center and (UP_{+}, UP_{+})-positive saddle. Further, from the (UP_{+}, UP_{+})-positive saddle, the $(UP_{+}, 3^{rd}UP_{+})$-third-order positive saddle appearing bifurcation generates (UP_{+}, UP_{+})-positive saddle, (DP_{+}, DP_{-})-CW center and (UP_{+}, UP_{+})-positive saddle. The $(DP_{-}, 3^{rd}DP_{+})$-third-order CW center appearing bifurcation from the (DP_{-}, DP_{+})-CW center generates (DP_{-}, DP_{+})-CW center, (UP_{+}, UP_{+})-positive sink and (DP_{-}, DP_{+})-CW center.

The $(3^{rd}DP_{+}, 3^{rd}DP_{-})$-double-third-order CCW center appearing bifurcation for ($a_{j_1j_20} < 0$, $a_{j_2j_10} > 0$) are presented in Fig. 4.13(b). The first $(3^{rd}DP_{+}, DP_{-})$-third-order CCW center appearing bifurcation from the (DP_{+}, DP_{-})-CCW center generates (DP_{+}, DP_{-})-CCW center, (UP_{+}, UP_{+})-positive saddle and (DP_{+}, DP_{-})-CCW center. The $(DP_{+}, 3^{rd}DP_{-})$-third-order CCW center appearing bifurcation from the (DP_{+}, DP_{-})-CCW center produces (DP_{+}, DP_{-})-CCW center, (UP_{-}, UP_{-})-negative sink and (DP_{+}, DP_{-})-CCW center. From the (UP_{-}, UP_{-})-negative saddle, the $(UP_{-}, 3^{rd}UP_{-})$-third-order negative saddle appearing bifurcation generates the (UP_{-}, UP_{-})-negative saddle, (DP_{-}, DP_{+})-CCW center and (UP_{-}, UP_{-})-negative saddle.

The $(3^{rd}DP_{-}, 3^{rd}DP_{+})$-double-third-order CW center switching bifurcation for ($a_{j_1j_20} > 0$, $a_{j_2j_10} < 0$) are presented in Fig. 4.13(c). The first $(3^{rd}DP_{-}, DP_{+})$-third-order CW center switching bifurcation from the (DP_{-}, DP_{+})-CW center generates the (DP_{-}, DP_{+})-CW center, (UP_{-}, UP_{-})-negative saddle, and (DP_{-}, DP_{+})-CW center. The $(DP_{-}, 3^{rd}DP_{+})$-third-order CW center appearing bifurcation from the (DP_{-}, DP_{+})-CW center generates the (DP_{-}, DP_{+})-CW center, (UP_{+}, UP_{+})-positive sink and (DP_{-}, DP_{+})-CW center. From the (UP_{+}, UP_{+})-positive saddle, the $(UP_{+}, 3^{rd}UP_{+})$-third-order positive saddle appearing bifurcation generates the (UP_{+}, UP_{+})-positive saddle, (DP_{+}, DP_{-})-CW center and (UP_{+}, UP_{+})-positive saddle.

In Fig. 4.13(d), the $(3^{rd}UP_{-}, 3^{rd}UP_{-})$-double-third-order negative saddle switching bifurcation through two times are presented ($a_{j_1j_20} < 0$, $a_{j_2j_10} < 0$). The

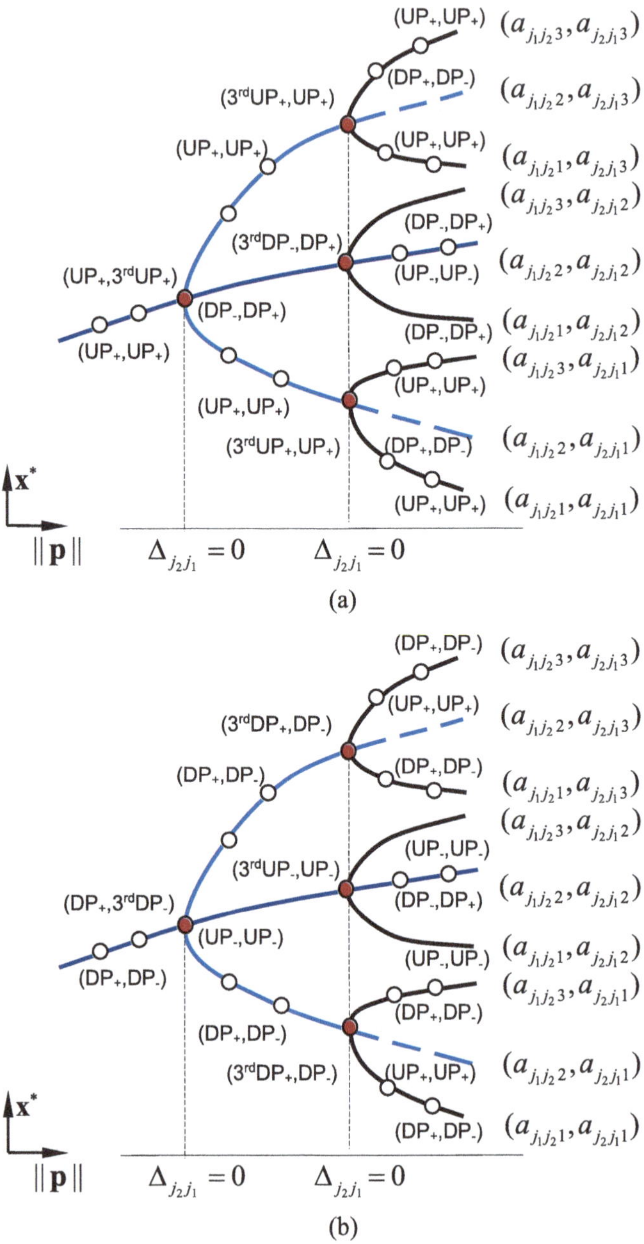

Fig. 4.13 The switching bifurcations for $x_1^* = a_{j_2j_1 1}, a_{j_2j_1 2}, a_{j_2j_1 3}$ and $x_{j_2}^* = a_{j_1j_2 1}, a_{j_1j_2 2}, a_{j_1j_2 2}$: (**a**) ($3^{\mathrm{rd}}\mathrm{UP}_+, 3^{\mathrm{rd}}\mathrm{UP}_+$)-positive saddle ($a_{j_1j_2 0} > 0$, $a_{j_2j_1 0} > 0$), (**b**) ($3^{\mathrm{rd}}\mathrm{DP}_+, 3^{\mathrm{rd}}\mathrm{DP}_-$)-CCW center ($a_{j_1j_2 0} < 0$, $a_{j_2j_1 0} > 0$), (**c**) ($3^{\mathrm{rd}}\mathrm{DP}_-, 3^{\mathrm{rd}}\mathrm{DP}_+$)-CW center ($a_{j_1j_2 0} > 0$, $a_{j_2j_1 0} < 0$), (**d**) ($3^{\mathrm{rd}}\mathrm{UP}_-, 3^{\mathrm{rd}}\mathrm{UP}_-$) -negative saddle ($a_{j_1j_2 0} < 0$, $a_{j_2j_1 0} < 0$). UP: up-parabola, DP: down-parabola

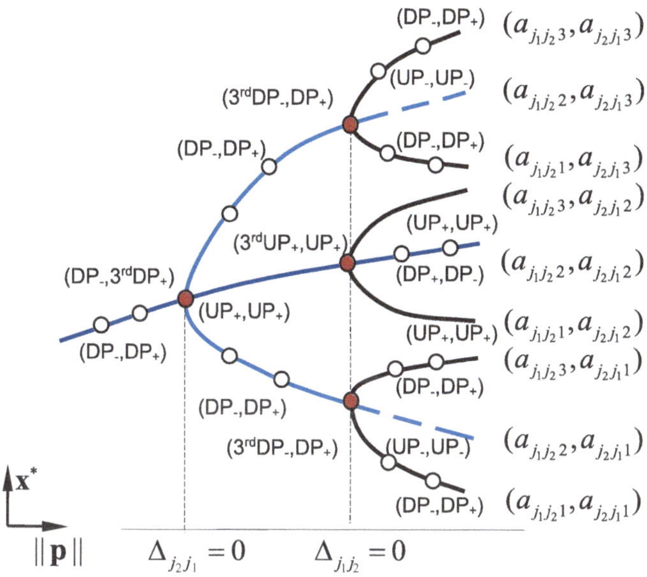

(c)

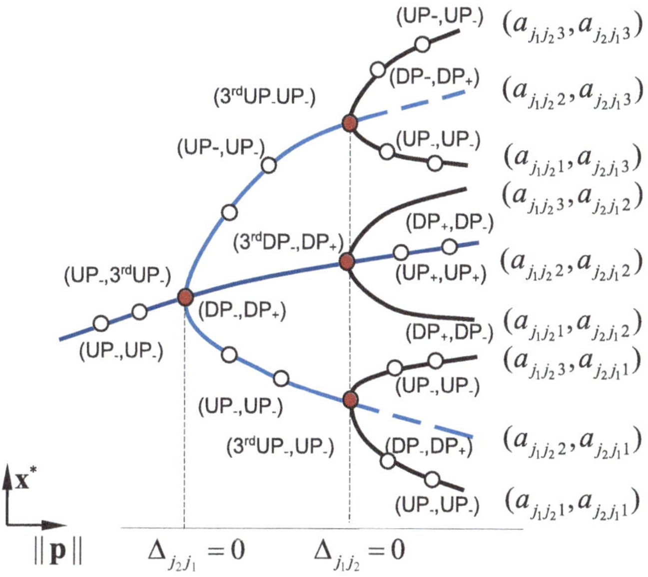

(d)

Fig. 4.13 (continued)

first $(3^{rd}UP_-, 3^{rd}UP_-)$-third-order negative saddle appearing bifurcation from the (UP_-, UP_-)-negative saddle generates the (UP_-, UP_-)-negative saddle, (DP_-, DP_+)-CW center and (UP_-, UP_-)-negative saddle. Further, from the (UP_-, UP_-)-negative saddle, the $(UP_-, 3^{rd}UP_-)$-third-order negative saddle appearing bifurcation generates the (UP_-, UP_-)-negative saddle, (DP_-, DP_+)-CCW center and (UP_-, UP_-)-negative saddle. The $(DP_+, 3^{rd}DP_-)$-third-order CCW center appearing bifurcation from the (DP_+, DP_-)-CCW center generates the (DP_+, DP_-)-CCW center, (UP_-, UP_-)-negative sink and (DP_+, DP_-)-CCW center.

The 3×3 homoclinic networks of saddles and centers are generated through the double third-order saddles and centers.

(i) For $a_{j_1j_20} > 0$ and $a_{j_2j_10} > 0$, the $(3^{rd}UP_+, 3^{rd}UP_+)$-positive saddle appearing bifurcation of equilibriums is from a (UP_+, UP_+)-positive saddle to a homoclinic network of the CCW and CW centers with positive and negative saddles, i.e.,

$$
\begin{aligned}
\underbrace{(UP_+, UP_+)}_{\text{positive saddle}} &\rightleftharpoons \underbrace{(3^{rd}UP_+, 3^{rd}UP_+)}_{\text{double third-order positive saddle}} \\
&\rightleftharpoons \left\{
\begin{array}{ccc}
\underbrace{(UP_+, UP_+)}_{\text{positive saddle}} & \underbrace{(DP_-, DP_+)}_{\text{CW center}} & \underbrace{(UP_+, UP_+)}_{\text{positive saddle}} \\
\underbrace{(DP_+, DP_-)}_{\text{CCW center}} & \underbrace{(UP_-, UP_-)}_{\text{negative saddle}} & \underbrace{(DP_+, DP_-)}_{\text{CCW center}} \\
\underbrace{(UP_+, UP_+)}_{\text{positive saddle}} & \underbrace{(DP_-, DP_+)}_{\text{CW center}} & \underbrace{(UP_+, UP_+)}_{\text{positive saddle}}
\end{array}
\right\}.
\end{aligned}
\tag{4.298}
$$

(ii) For $a_{j_1j_10} < 0$ and $a_{j_2j_20} > 0$, the $(3^{rd}DP_+, 3^{rd}DP_-)$-CCW center appearing bifurcation of equilibriums is from (DP_+, DP_-)-CCW center to a homoclinic network of the CCW and CW centers with positive and negative saddles, i.e.,

$$
\begin{aligned}
\underbrace{(DP_+, DP_-)}_{\text{CCW center}} &\rightleftharpoons \underbrace{(3^{rd}DP_+, 3^{rd}DP_-)}_{\text{double third-order CCW center}} \\
&\rightleftharpoons \left\{
\begin{array}{ccc}
\underbrace{(DP_+, DP_-)}_{\text{CCW center}} & \underbrace{(UP_-, UP_-)}_{\text{negative saddle}} & \underbrace{(DP_+, DP_-)}_{\text{CCW center}} \\
\underbrace{(UP_+, UP_+)}_{\text{positive saddle}} & \underbrace{(DP_-, DP_+)}_{\text{CW center}} & \underbrace{(UP_+, UP_+)}_{\text{positive saddle}} \\
\underbrace{(DP_+, DP_-)}_{\text{CCW center}} & \underbrace{(UP_-, UP_-)}_{\text{neative saddle}} & \underbrace{(DP_+, DP_-)}_{\text{CCW center}}
\end{array}
\right\}.
\end{aligned}
\tag{4.299}
$$

(iii) For $a_{j_1j_10}>0$ and $a_{j_2j_20}<0$, the ($3^{\rm rd}$DP$_-$, $3^{\rm rd}$DP$_+$)-CW center appearing bifurcation of equilibriums is from (DP$_-$, DP$_+$)-CW center to a homoclinic network of the CCW and CW centers with positive and negative saddles, i.e.,

$$\begin{aligned}\underbrace{(\mathrm{DP}_-,\mathrm{DP}_+)}_{\text{CW center}} &\rightleftharpoons \underbrace{(3^{\rm rd}\mathrm{DP}_-,3^{\rm rd}\mathrm{DP}_+)}_{\text{double third-order CW center}}\\ &\rightleftharpoons \left\{\begin{matrix}\underbrace{(\mathrm{DP}_-,\mathrm{DP}_+)}_{\text{CW center}} & \underbrace{(\mathrm{UP}_+,\mathrm{UP}_+)}_{\text{positive saddle}} & \underbrace{(\mathrm{DP}_-,\mathrm{DP}_+)}_{\text{CW center}}\\ \underbrace{(\mathrm{UP}_-,\mathrm{UP}_-)}_{\text{negative saddle}} & \underbrace{(\mathrm{DP}_+,\mathrm{DP}_-)}_{\text{CCW center}} & \underbrace{(\mathrm{UP}_-,\mathrm{UP}_-)}_{\text{negative saddle}}\\ \underbrace{(\mathrm{DP}_-,\mathrm{DP}_+)}_{\text{CW center}} & \underbrace{(\mathrm{UP}_+,\mathrm{UP}_+)}_{\text{positive saddle}} & \underbrace{(\mathrm{DP}_-,\mathrm{DP}_+)}_{\text{CW center}}\end{matrix}\right\}.\end{aligned} \tag{4.300}$$

(iv) For $a_{j_1j_10}<0$ and $a_{j_2j_20}<0$, the ($3^{\rm rd}$UP$_-$, $3^{\rm rd}$UP$_-$)-negative saddle appearing bifurcation of equilibriums is from a (UP$_-$, UP$_-$)-negative saddle to a homoclinic network of the CCW and CW centers with positive and negative saddles, i.e.,

$$\begin{aligned}\underbrace{(\mathrm{UP}_-,\mathrm{UP}_-)}_{\text{negative saddle}} &\rightleftharpoons \underbrace{(3^{\rm rd}\mathrm{UP}_-,3^{\rm rd}\mathrm{UP}_-)}_{\text{double third-order negative saddle}}\\ &\rightleftharpoons \left\{\begin{matrix}\underbrace{(\mathrm{UP}_-,\mathrm{UP}_-)}_{\text{negative saddle}} & \underbrace{(\mathrm{DP}_+,\mathrm{DP}_-)}_{\text{CCW center}} & \underbrace{(\mathrm{UP}_-,\mathrm{UP}_-)}_{\text{negative saddle}}\\ \underbrace{(\mathrm{DP}_-,\mathrm{DP}_+)}_{\text{CW center}} & \underbrace{(\mathrm{UP}_+,\mathrm{UP}_+)}_{\text{positive saddle}} & \underbrace{(\mathrm{DP}_-,\mathrm{DP}_+)}_{\text{CW center}}\\ \underbrace{(\mathrm{UP}_-,\mathrm{UP}_-)}_{\text{negative saddle}} & \underbrace{(\mathrm{DP}_+,\mathrm{DP}_-)}_{\text{CCW center}} & \underbrace{(\mathrm{UP}_-,\mathrm{UP}_-)}_{\text{negative saddle}}\end{matrix}\right\}.\end{aligned} \tag{4.301}$$

In Fig. 4.14, the double third-order centers and saddle appearing bifurcations are sketched for homoclinic networks of CCW and CW centers with positive and negative saddles.

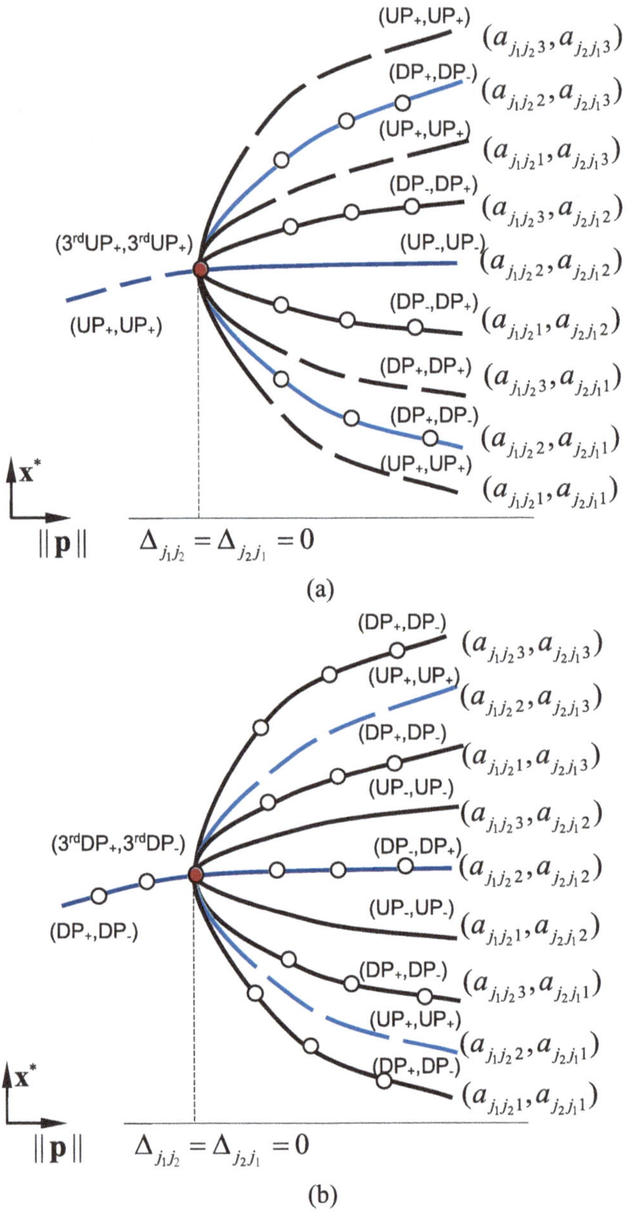

Fig. 4.14 The switching bifurcation diagram for $x_{j_1}^*=a_{j_2j_1}1, a_{j_2j_1}2, a_{j_2j_1}3$ and $x_{j_2}^*=a_{j_1j_2}1, a_{j_1j_2}2, a_{j_1j_2}3$: (**a**) ($3^{\rm rd}UP_+$, $3^{\rm rd}UP_+$)-positive-saddle ($a_{j_1j_2 0}>0$, $a_{j_2j_1 0}>0$), (**b**) ($3^{\rm rd}DP_+$, $3^{\rm rd}DP_-$)-CCW center ($a_{j_1j_2 0}<0$, $a_{j_2j_1 0}>0$), (**c**) ($3^{\rm rd}DP_-$, $3^{\rm rd}DP_+$)-CW center ($a_{j_1j_2 0}>0$, $a_{j_2j_1 0}<0$), (**d**) ($3^{\rm rd}SI$,$3^{\rm rd}SI$)-sink ($a_{j_1j_2 0}<0$, $a_{j_2j_1 0}<0$)

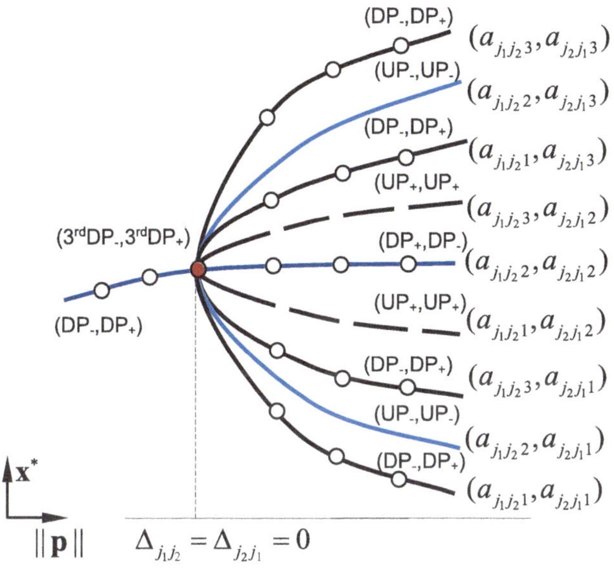

(c)

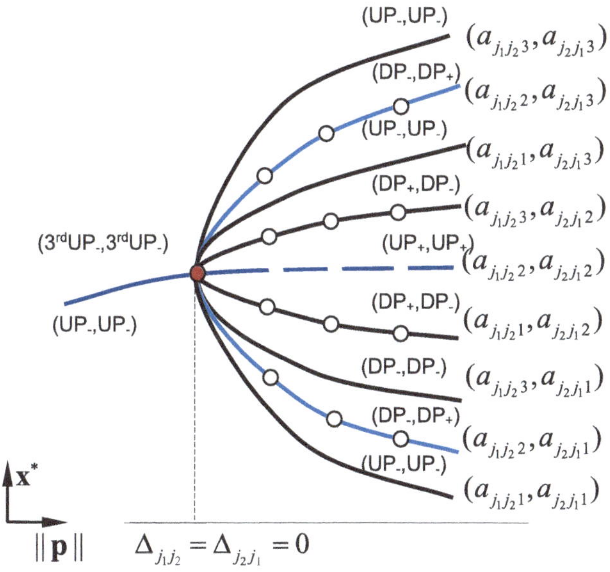

(d)

Fig. 4.14 (continued)

Index

A. C. J. Luo, *Two-dimensional Crossing-Variable Cubic Nonlinear Systems*,
https://doi.org/10.1007/978-3-031-62810-8

The manufacturer's authorised representative in the EU is Springer Nature Customer Service Centre GmbH, Europaplatz 3, 69115 Heidelberg, Germany. If you have any concerns regarding our products, please contact ProductSafety@springernature.com

Printed and bound by CPI Group (UK) Ltd, Croydon, CR0 4YY
01/10/2026
02225960-0016